Federated Learning

Heiko Ludwig • Nathalie Baracaldo

Editors

Federated Learning

A Comprehensive Overview of Methods
and Applications

 Springer

Editors
Heiko Ludwig
IBM Research – Almaden
San Jose, CA, USA

Nathalie Baracaldo
IBM Research – Almaden
San Jose, CA, USA

ISBN 978-3-030-96898-4   ISBN 978-3-030-96896-0   (eBook)
https://doi.org/10.1007/978-3-030-96896-0

This Springer imprint is published by the registered company Springer Nature Switzerland AG
The registered company address is: Gewerbestrasse 11, 6330 Cham, Switzerland

Preface

Machine learning has made great strides over the past two decades and has been adopted in many application domains. Successful machine learning depends largely on access to quality data, both labeled and unlabeled.

Concerns related to data privacy, security, and sovereignty have caused public and technical discussion on how to use data for machine learning purposes consistent with regulatory and stakeholder interests. These concerns and legislation have led to the realization that collecting training data in large central repositories may be at odds with maintaining privacy for data owners.

While distributed learning or model fusion has been discussed since at least a decade, federated machine learning (FL) as a concept has been popularized by MacMahan and others since 2017. In the subsequent years, much research has been conducted – both, in academia and the industry – and, at the time of writing this book, the first viable commercial frameworks for federated learning are coming to the market.

This book aims to capture the research progress and state of the art that has been made in the past years, from the initial conception of the field to first applications and commercial use. To get this broad and deep overview, we invited leading researchers to address the different perspectives of federated learning: the core machine learning perspective, privacy and security, distributed systems, and specific application domains.

The book's title, *Federated Learning: A Comprehensive Overview of Methods and Applications*, outlines its scope. It presents in depth the most important issues and approaches to federated learning for researchers and practitioners. Some chapters contain a variety of technical content that is relevant to understand the intricacies of the algorithms and paradigms that make it possible to deploy federated learning in multiple enterprise settings. Other chapters focus on providing clarity on how to select privacy and security solutions in a way that can be tailored to specific use cases, while others take into consideration the pragmatics of the systems where the federated learning process will run.

Given the inherent cross-disciplinary nature of the topic, we encounter different notational conventions in different chapters of the book. What might be parties in federated machine learning may be called clients in the distributed systems perspectives. In the introductory chapter of this book, we lay out the primary terminology we use, and each chapter explains how the discipline-specific terminology maps to the common one when it is introduced, if this is the case. With this approach, we make this book understandable to readers from diverse backgrounds while staying true to the conventions of the specific disciplines involved.

Taken as a whole, this book enables the reader to get a broad state-of-the-art summary of the most recent research developments.

Editing this book, and writing some of the chapters, required the help of many, who we want to acknowledge. IBM Research gave us the opportunity to work in this exciting field, not just academically but also to put this technology into practice and make it part of a product. We learned invaluable lessons on the journey, and we have much to thank to our colleagues at IBM. In particular, we want to acknowledge our director, *Sandeep Gopisetty*, for giving us the space to work on this book: Gegi Thomas, who made sure our research contributions make their way into the product: and our team members.

The chapter authors provide the substance of this book and were patient with us with requests for changes to their chapters.

We owe greatest thanks to our families, who patiently put up with us devoting time to the book rather than them over the year of writing and editing this book. Heiko is deeply thankful to his wife, *Beatriz Raggio*, for making these sacrifices and supporting him throughout. Nathalie is profoundly thankful to her husband and sons, *Santiago* and *Matthias Bock*, for their love and support and for cheering for all her projects, including this one. She also thanks her parents, Adriana and Jesus; this and many more achievements would not be possible without their amazing and continuous support.

San Jose, CA, USA   Heiko Ludwig
September 2021   Nathalie Baracaldo

Contents

Chapter 1
Introduction to Federated Learning

Heiko Ludwig and Nathalie Baracaldo

Abstract *Federated learning (FL)* is an approach to machine learning in which the training data is not managed centrally. Data is retained by data parties that participate in the FL process and is not shared with any other entity. This makes FL an increasingly popular solution for machine learning tasks for which bringing data together in a centralized repository is problematic, either for privacy, regulatory, or practical reasons. In this chapter, we introduce the basic concepts of FL, provide an overview of its application use cases, and discuss it from a machine learning, distributed computing, and privacy perspective. We also provide an introduction to dive deep into the matter covered in the subsequent chapters.

1.1 Overview

Machine learning (ML) has become a crucial technique to develop cognitive and analytic functionality difficult and ineffective to develop algorithmically. Applications in computer vision, speech recognition, and natural language understanding progressed by leaps and bounds with the advent of Deep Neural Networks (DNNs) and the computational hardware to train complex networks effectively. Also, classical ML techniques such as decision trees, linear regression, and support vector models (SVMs) found increased use, in particular related to structured data.

The application of ML critically depends on the availability of high-quality training data. Sometimes, though, privacy considerations prevent training data to be brought to a central data repository to be curated and managed for the ML process. *Federated learning (FL)* is an approach first proposed by this name in [28] to train ML models on training data in disparate locations, not requiring the central collection of data.

H. Ludwig (✉) · N. Baracaldo
IBM Research – Almaden, San Jose, CA, USA
e-mail: hludwig@us.ibm.com; baracald@us.ibm.com

© The Author(s), under exclusive license to Springer Nature Switzerland AG 2022
H. Ludwig, N. Baracaldo (eds.), *Federated Learning*,
https://doi.org/10.1007/978-3-030-96896-0_1

An important driver for this reluctance to use a central data repository has been consumer privacy regulation in different jurisdictions. The European Union's General Data Protection Regulation (GDPR) [50], the Health Insurance Portability and Accountability Act (HIPAA) [53], and the California Consumer Privacy Act (CCPA) [48] are example regulatory frameworks for the collection and use of consumer data. In addition, news coverage of data breaches have raised awareness of the liability entailed by storing sensitive consumer data [9, 42, 43, 51]. FL facilitates using data without actually having to store it in a central repository, mitigating this risk. Regulation also restricts the movement of data between jurisdictions such as different countries. This is motivated by considering data protection in other countries potentially insufficient or related to national security, requiring critical data to remain onshore [40]. National and regional regulations pose challenges for international companies with subsidiaries in different markets but want to train a model using all their data. Beyond regulatory requirements, learning from data in disparate locations might also be just practical. Poor communication connections and the sheer amount of data collected by sensors or in telecommunication devices can make central data collection infeasible. FL also enables different companies to work together creating models for mutual benefit without revealing their trade secrets.

How does FL work then? In the FL approach, a set of distinct *parties*, who control their respective training data, collaborate to train a machine learning model. They do this without sharing their training data with other parties—or any other third-party entity. Parties to the collaboration are also called *clients* or *devices* in the literature. Parties can be a variety of things, including consumer devices such as smart phones or cars, but also cloud services of different providers, data centers processing enterprise data in different countries, application silos within a company, or embedded systems such as manufacturing robots in an automotive plant.

While the FL collaboration can be conducted in different ways, its most common form is outlined in Fig. 1.1. In this approach, an *aggregator*, sometimes called a *server* or *coordinator*, facilitates the collaboration. Parties perform a local training process based on their private training data. When their local training is completed, they send their model parameters to the aggregator as model updates. The type of model updates depends on the type of machine learning model to be trained; for a neural network, for example, the model updates might be the weights of the network. Once the aggregator has received the model updates from the parties, they can then be merged into a common model, a process we call *model fusion*. In the neural network example, this can be as simple as averaging the weights, as proposed in the FedAvg algorithm [38]. The resulting merged model is then distributed again to the parties as model update to form the basis of the next *round* of learning. This process can be repeated over multiple rounds until the training process converges. The role of the aggregator is to coordinate the learning process and information exchange between the parties and to perform the fusion algorithm to merge the model parameters from each party into a common model. The result of the FL process is a model that is based on the training data of all parties, while the training data is never shared.

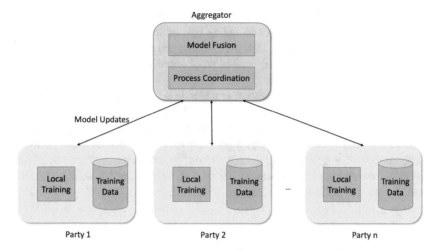

Fig. 1.1 Federated learning overview

The FL approach appears to be related to distributed learning on clusters [15], which is a common approach for large ML tasks. Distributed learning uses a cluster of compute nodes to share the computational effort of ML and thereby accelerate the learning process. Distributed learning typically uses a parameter server to aggregate results from nodes, not unlike in the federated model. However, it is different in some important ways. In FL, the *distribution and quantity of data* are not controlled centrally and might be unknown if all training data is kept private. We cannot make assumptions about the independence and identical distribution (IID) of data among the parties. Likewise, some parties may have more data than others, leading to an imbalance of the datasets among the parties. In distributed learning, data is managed centrally and distributed to different nodes in shards, with a central entity being aware of the stochastic properties of the data. Imbalance and non-IIDness of party data must be accounted for when designing FL training algorithms.

In contrast, in FL, the number of parties may vary, depending on the use case. Training a model on datasets in different data centers of a multi-national company may have fewer than ten parties. This is often called an *enterprise* [35] or *cross-silo use case* [26]. Training on data of a mobile phone application may have hundreds of millions of parties contributing. This is typically referred to as the *cross-device use case* [26]. In the enterprise use case, it is generally important to consider model updates from all or most parties in every round. In the device use case, every FL round will only include a—potentially large—sub-sample of the total set of devices. In the enterprise use case, the FL process considers the identity of the parties involved and can use this in the training and verification process. In the cross-device use case, party identity is generally not important, and a single party might be involved only in one round of training.

In the device use case, more than in the enterprise scenario, communication failure of some devices can be assumed, given the large number of participants. Cell phones can be off or a device might be in an area of poor network coverage. This can be managed by sampling parties and setting up time limits to perform aggregation, or other mitigation techniques. In the enterprise use case, communication failure has to be carefully managed as individual party contributions are relevant given the small number of participants.

In the remainder of this chapter, we will provide an overview of FL. We provide a formal introduction to the main concepts used in the next section. After that, we discuss FL from three important perspectives, each in a separate section: First, we discuss FL from the machine learning perspective; then, we cover the security and privacy perspective by outlining threats and mitigation techniques; and finally, we present an overview of the systems' perspective of federated learning. This will provide a starting point to the remainder of the book.

1.2 Concepts and Terminology

Like any machine learning task, FL trains a model $\mathcal{M}$ representing a predictive function f on training data D. $\mathcal{M}$ can have the structure of a neural network or any other, non-neural model. In contrast to centralized machine learning, D is partitioned between n parties $P = \{P_1, P_2, \ldots, P_n\}$, where each party $P_k \in P$ owns a private training dataset D_k. An FL process involves an *aggregator* A and a set of parties P. It is important to note that D_k can only be accessed by party P_k. In other words, no party has knowledge of any other dataset than its own, and A has no knowledge of any dataset.

How the FL process is conducted at this abstract level is shown in Fig. 1.2. To train a global machine learning model $\mathcal{M}$, the aggregator and the parties perform a *Federated Learning algorithm* that is executed in a distributed way at the aggregator

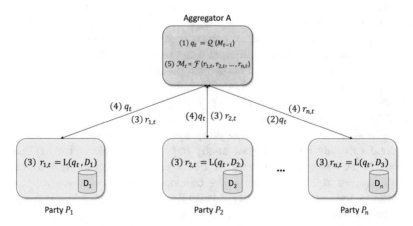

Fig. 1.2 Federated learning concepts

and the parties. The main algorithmic components are each party's local training function $\mathcal{L}$, which performs the local training on dataset D_k, and the aggregator's fusion function $\mathcal{F}$, which combines the results of each party's $\mathcal{L}$ into a new joint model. There can be a set of iterations of local training and fusion, which we call *rounds*, using the index t. The execution of the algorithm is coordinated by sending messages between parties and aggregator. The overall process runs as follows:

1. The process starts at the aggregator. To train the model, the aggregator uses a function Q that takes as input the model of the previous round of the training M_{t-1} at round t, and generates a query q_t for the current round. When the process starts, M_0 may be empty or just randomly seeded. Also, some FL algorithms may include additional inputs for Q and may tailor queries to each party, but for simplicity of discussion and without loss of generality, we use this simpler approach.
2. The query q_t is sent to the parties and requests information about their respective local model or aggregated information about each party's dataset. Example queries include requests for gradients or model weights of a neural network, or counts for decision trees.
3. When receiving q_t, the local training process performs the local training function $\mathcal{L}$ that takes as input query q_t and the local dataset D_k and outputs a *model update* $r_{k,t}$. Usually, the query q_t contains information that the party can use to initialize the local training process. This includes, for example, model weights of the new, common model M_t to initialize local training, or other information for different model types.
4. When $\mathcal{L}$ is completed, $r_{k,t}$ is sent back from party p_k to the aggregator A, which collects all the $r_{k,t}$ from all parties.
5. When all expected parties' model updates $R_t = (r_{1,t}, r_{2,t}, \ldots, r_{n,t})$ are received by the aggregator, they are processed by applying fusion function $\mathcal{F}$ that takes as input R_t and returns M_t.

This process can be executed over multiple rounds and continues until a termination criterion is met, e.g., the maximum number of training rounds t_{max} has elapsed, resulting in a final global model $M = M_{tmax}$. The number of rounds required can vary highly, from a single model merge of a Naive Bayes approach to many rounds of training for typical gradient-based machine learning algorithms.

The local training function $\mathcal{L}$, the fusion function $\mathcal{F}$, and the query generation function Q are typically a complimentary set that is designed to work together. $\mathcal{L}$ interacts with the actual dataset and performs the local training, generating the model update $r_{k,t}$. The content of R_t is the input to $\mathcal{F}$ and, thus, must be interpreted by $\mathcal{F}$, which creates the next model M_t of this input. If another round is required, Q then creates another query.

In further sections, we will describe in detail how this process takes place in the case of training neural networks, decision trees, and gradient-boosted trees.

We can introduce different variants to this basic approach to FL: In the case of cross-device FL, the number of parties is often large, in the millions. Not all parties participate in every round. In this case, Q determines not only the query but also which parties $P_s \subset P$ to include in the next round of querying. The party selection can be random, based on party characteristics, or on the merits of prior contributions.

Also, queries to each party might be different, with $\mathcal{F}$ needing to integrate the results of different queries in the creation of a new model M_t.

While an approach with a single aggregator is most commonly used and practical for most scenarios, other alternative FL architectures have been proposed. For example, every party P_k might have its own, associated aggregator A_k, querying the other parties; the set of parties might be partitioned between aggregators, and a hierarchical aggregation process may take place. In the remainder of the introduction, we focus on the common single aggregator configuration.

1.3 Machine Learning Perspective

In this section, we look at federated learning from its machine learning perspective. The choices of the approach to a federated learning system—such as what information to send in queries—influence the machine learning behavior. We discuss this in the following subsections for different machine learning paradigms.

1.3.1 Deep Neural Networks

DNNs have become very popular and lend themselves to FL in a relatively straightforward way in its basic approach by training locally at each party and fusing local training results at the aggregator. Local training $\mathcal{L}$ often corresponds to the regular—centralized—training of a neural network at party P_k and its parameters w_k in each round t. We optimize at each party P_k

$$w_k^* = arg \ min_{w_k} \frac{1}{|D_k|} \sum_{(x_i, y_i) \in D_k} l(w_k, x_i, y_i), \tag{1.1}$$

minimizing the loss function l for parameters w_k, the weight vector of the neural network, on the party's training dataset D_k. If a gradient descent algorithm is used, in each epoch τ of a given round t, w_k is updated as follows:

$$w_k^{t,\tau} := w_k^{t,\tau-1} - \eta_k \nabla l(w_k^{t,\tau-1}, X_k, Y_k). \tag{1.2}$$

The loss function l is computed based on the local data D_k, comprised of samples X_k and labels Y_k, and can be any suitable function such as the commonly used Mean-Squared Error (MSE). The parameters for this round $w_k^{t,\tau}$ are updated using the party-specific learning rate η_k. Each local round is seeded with a new model update from the aggregator, $w_k^{t,0}$, which provides the new starting point for local training for each round.

We can make choices regarding the party-local hyperparameters when setting up a federated learning system or a specific FL project, for example:

- Which *batch size* should we choose for the party-local gradient descent algorithm: one, i.e., the original Stochastic Gradient Descent (SGD); the whole set; or a suitable mini-batch size?
- How many *local epochs* are run before sending model updates $r_{k,t}$ to the aggregator? Should all parties use the same number of epochs in each round? Training just a single epoch in each party prevents local models w_k from diverging much from each other but causes much more network traffic and frequent aggregation activity. Running multiple epochs, or even a different number of epochs in different parties, can cause larger differences but can be used to adapt to differences in parties' computational capabilities and sizes in training datasets.
- Which *learning rate* η_k do we choose for each party? Differences in data distributions between parties can make different learning rates beneficial.
- Other optimization algorithms might use different local hyperparameters such as momentum or decay rates [27].

Let us consider the case of a simple *Federated SGD*, as described in [38], in which—like in centralized SGD—each new sample leads to a model upgrade. The aggregator would choose a party P_k and send a query $q_{t,k}<w_t>$ to the chosen party. P_k picks the next training sample $(x_i, y_i) \in D_k$ and performs its local training $\mathcal{L}$, computing its loss gradient $\nabla l(w_t, x_i, y_i)$ for this sample. We will refer to gradients of a party P_k in a particular round t as

$$g_{k,t} := \frac{1}{|D_k|} \sum_{(x_i,y_i)\in D_k} \nabla l(w_k, x_i, y_i), \qquad (1.3)$$

the average gradient of the training data samples in D_k. Party P_k returns it as reply $r_{k,t}<g_{t,k}>$ to the aggregator. The aggregator then computes the new query content with model weights based on the reply from P_k and the aggregator's learning rate:

$$w_{t+1} := w_t - \eta_a g_{k,t}. \qquad (1.4)$$

Then the next round begins with the aggregator choosing another party to contribute. In this simplistic way, it is quite ineffective because it introduces communication overhead and does not take advantage of concurrent training. To make Federated SGD more effective, we can train with mini-batches at each party, increasing computation at each party per round. We can also train concurrently at all or a subset of parties $P_s \subset P$, averaging the replied gradients of the parties when computing new model weights:

$$w_{t+1} := w_t - \eta_a \frac{1}{|K|} \sum_K g_{k,t}. \tag{1.5}$$

While this is more effective than the naive approach, it still involves much communication with the aggregator and potential coordination delays at least once in every epoch when the batch size is the full D_k or multiple times when we use mini-batches.

FedAvg, as proposed in [28], is more effective by taking advantage of independent processing at each party. Each party runs multiple epochs before replying. Rather than replying with gradients, parties can compute a new set of weights $w_{t,k}$ directly at each party P_k, using a common learning rate η, and reply with $r_{k,t}<w_{k,t}, n_k>$, their model, and the number of samples. The aggregator's fusion algorithm $\mathcal{F}$ averages the parameters of each party, weighted by the number of samples at each party, for the next round:

$$w_{t+1} := \sum_{k \in K} \frac{n_k}{n} w_{k,t}. \tag{1.6}$$

Experiments show that this approach performs well for different model types [38]. FedAvg uses most of the variables laid out in equation (1.2), but we can imagine introducing other parameters such as local or variable learning rates for gradient descent algorithms.

Further approaches can expand on these basic FL fusion and local training algorithms to adapt to different properties of data distributions, client selection, and privacy requirements. The paper in [32] proposes a momentum-based FL approach to accelerate convergence, inspired by centralized ML optimization such as [27]. Stateful optimization algorithms such as ADMM generally are only applicable if all parties in a collaboration participate every time, retaining state at the party [7]. Different approaches, including [18] and [17], adapt ADMM to a practical FL setting. FedProx [31] introduces a proximal regularization term to address data heterogeneity among parties for non-IID use cases. Other approaches such as [36] go beyond gradient descent methods for optimization.

For each FL approach that addresses specific aspects of data heterogeneity, model structure, and parties, we need to define an algorithm that comprises of $\mathcal{L}, \mathcal{F}$, and the protocol for interaction between parties and the aggregator, i.e., the format of q_k and r_k. In the remainder of the book, we find different state-of-the-art approaches to deal with data and model heterogeneity aspects.

1.3.2 Classical Machine Learning Models

Classical machine learning techniques can also be applied to federated learning scenarios. Some of these techniques can be approached very similarly to DNNs. Others have to be entirely rethought for decentralized training.

Linear models, including regression and classification, can be trained in FL by adapting the training process similarly to the way the neural networks training process was adjusted. Training data with a feature vector $x_i = (x_i^1, x_i^2, \ldots, x_i^m)$ can be used to train a predictor for a linear regression, for example, of the shape

$$y_i = w_1 x_i^1 + w_2 x_i^2 + \ldots + w_m x_i^m + b. \tag{1.7}$$

It predicts y_i for m linear variables x_i^j and bias b and entails minimizing a loss function for weight vector w and b. w is typically much smaller than the weight vector of a DNN. With data D partitioned among parties as D_k, we can follow the approaches outlined in the previous section. We train at each party, minimizing locally the loss function $l(w_k, b_k, x_i, y_i)$ for the local training data. Like with DNNs, we have a choice how to fuse local models into a global model. Using, for example, FedAvg as the fusion function $\mathcal{F}$, we can then compute locally the new local model weights as

$$w_{k,t+1} := w_t - \eta_k \nabla l(w_{k,t+1}, X_k, Y_k), \tag{1.8}$$

applying a party-specific learning rate η_k to the gradients of the weights. All parties send their model weights to the aggregator where the weights are averaged, and the new model $\mathcal{M}$, defined by (w_t, b_t), is redistributed to the parties. We can apply other fusion methods as well such as Federated SGD or any of the advanced methods discussed in the previous subsection. This often converges faster than in the case of DNNs due to the smaller w. Other classic linear models such as logistic regression or linear *Support Vector Machine (SVM)* [20] can be transformed into a federated learning approach in a similar way.

Decision trees and more advanced tree-based models require a different approach to Federated Learning than model types with a static parameter structure such as the ones we discussed to this point. Decision trees are an established classification model type and is commonly used for classification problems [46]. It is particularly relevant in domains where the explainability of the decision is societally important such as in healthcare, finance, and other areas where regulation requires demonstrating on which criteria decisions are based on. While DNNs and linear models can be trained locally and local parameters can be merged at the aggregator, no good fusion algorithms have been proposed to merge independently trained tree models into a single decision tree.

The white paper [35] describes a federated approach for the ID3 algorithm [46] in which the tree formation takes place at the aggregator and the role of parties is to respond with counts to the proposed class splits based on their local training data. It works for numeric and categorical data. In its centralized, original version, an ID3 decision tree computes the information gain for each feature splitting the training dataset into the given classes. It chooses the feature with the most information gain and computes the values for this feature that splits D best. One attribute typically does not split D sufficiently. For each branch of the tree just created, we apply the same approach recursively. We ask which next feature will split the data subset in each sub-tree best by computing the information gain of each sub-tree dataset with respect to the remaining features. The algorithm continues refining the classification recursively until it stops when all members of a tree node have the same class label or a maximum depth is met.

In a federated version, the fusion function $\mathcal{F}$ at the aggregator computes the information gain and chooses the next feature to grow the tree. To obtain the input to compute the information gain, the aggregator queries all parties with proposed features and split values. The parties count the members of each proposed sub-tree and their labels as their local training function $\mathcal{F}$ and return these counts as reply to the aggregator. The aggregator adds up the counts for each proposed feature from all parties and then proceeds to compute the information gain on these aggregated counts. As in the centralized version, the next best feature is chosen and the sub-tree is split again, and so forth.

In this approach, the aggregator takes a prominent role and performs much of the computation, while the parties mainly provide counts related to features and split values. Like in other federated learning approaches, training data never leaves any party. Depending on the quantity of the training dataset and the number of class members, this might require further privacy-preserving measures to ensure not too much information is disclosed in this simple approach. Nevertheless, this is a good example of how federated learning can be approached differently than in case of DNNs and linear models.

Decision tree ensemble methods often provide better model performance than individual decision trees. *Random forests* [8] and, in particular, *gradient-boosted trees* such as the popular XGBoost [13] are used successfully in different applications and also in Kaggle competitions, providing better predictive accuracy. Federated random forest algorithms can pursue a similar approach than decision trees, growing individual trees in the aggregator and then using data collection from parties. A subset of features is selected randomly for each addition, creating the next tree of the ensemble, which is then again queried from the parties. More complex algorithms are proposed for scenarios in which not all parties have the same set of features for each data record in question, e.g., [34] and [20]. This scenario is referred to as vertical federated learning (for more, see the next subsection) and requires cryptographic techniques to match the record of each party to the same entity.

Gradient-boosted trees add to the ensemble in areas of the decision space where predictions were made poorly, as opposed to randomly in the random forest. To determine where to start with the next tree of the ensemble, the loss function has to be computed for all training data samples in D_k, which are located at the parties. Like in the other tree-based algorithms, tree growth and decision-making on ensembles take place at the aggregator. However, additionally, the parties need to include gradients and Hessians in their reply to the aggregator to make the choice of the next tree of the ensemble. The aggregator's fusion function also needs a quantile approximation, e.g., a histogram of training data samples in potential classes. Federated gradient-boosted trees, like their centrally trained counterparts, often have good accuracy and may overfit less than other tree-based learning algorithms. The approach proposed by Ong et al. [45] uses party-adaptive quantile sketches to reduce information disclosure. Other approaches to federated XGBoost use cryptographic methods and a secure multi-party computation approach for interaction and loss computation [14, 33]. This entails a higher training time at comparable model performance and is suitable for enterprise scenarios requiring very stringent privacy. The overview presented in [63] provides an interesting discussion on privacy trade-offs in federated gradient boosting that can also be applied to the simpler tree-based models.

Chapter 2 covers in more detail multiple algorithms to train tree-based models, including gradient-boosted trees.

With this brief tour, we provided an overview of the most popular classic and neural network approaches. We see that federated versions of common machine learning algorithms can be created by carefully considering which computation to take place at the aggregator, which at the parties, and what interaction between parties and aggregator is needed.

1.3.3 Horizontal, Vertical Federated Learning and Split Learning

So far, when discussing the distribution of data among parties, we generally assumed that all parties' training data comprises the same features for each sample and parties have data pertaining to different samples. For example, hospital A has health records and images on some patients; a second hospital B on other patients, as illustrated in Fig. 1.3.

In case of neural networks, we assumed that each party has samples of equivalent size and content.

However, in some cases, parties may have different features referring to the same entity. Using again the health care example, primary care physicians might have electronic health records relating to patients' visits over time, while a radiologist has images relating to a patient's disease. An orthopedic surgeon might have surgery records on patients. When looking for predictors of health outcomes for orthopedic surgery, it might be beneficial to base the prediction on data of all three parties, primary care, radiologist, and orthopedic surgeon. In this case, only one party, the orthopedic surgeon, might have the actual label: the outcome of the surgery. We call this dataset *vertically partitioned*.

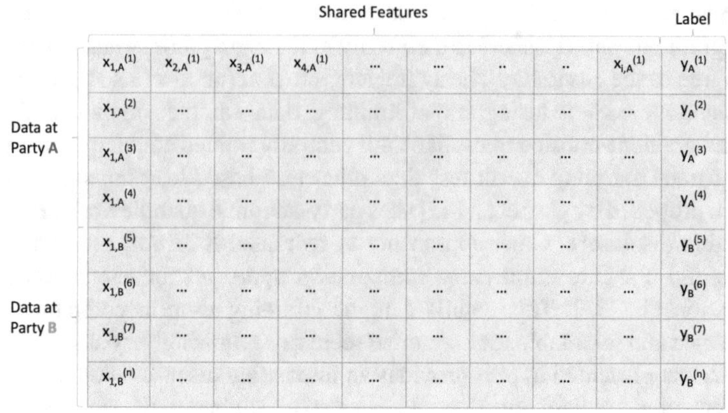

	Shared Features								Label
	$x_{1,A}^{(1)}$	$x_{2,A}^{(1)}$	$x_{3,A}^{(1)}$	$x_{4,A}^{(1)}$	...	...	..	..	$x_{i,A}^{(1)}$	$y_A^{(1)}$

Data at Party A

$x_{1,A}^{(1)}$	$x_{2,A}^{(1)}$	$x_{3,A}^{(1)}$	$x_{4,A}^{(1)}$	...	...	..	..	$x_{i,A}^{(1)}$	$y_A^{(1)}$
$x_{1,A}^{(2)}$	...	...	...	...	...	...	...	...	$y_A^{(2)}$
$x_{1,A}^{(3)}$	...	...	...	...	...	...	...	...	$y_A^{(3)}$
$x_{1,A}^{(4)}$	...	...	...	...	...	...	...	...	$y_A^{(4)}$
$x_{1,B}^{(5)}$	...	...	...	...	...	...	...	...	$y_B^{(5)}$
$x_{1,B}^{(6)}$	...	...	...	...	...	...	...	...	$y_B^{(6)}$
$x_{1,B}^{(7)}$	...	...	...	...	...	...	...	...	$y_B^{(7)}$
$x_{1,B}^{(n)}$	...	...	...	...	...	...	...	...	$y_B^{(n)}$

Data at Party B (rows 5–8)

Fig. 1.3 Horizontally partitioned data

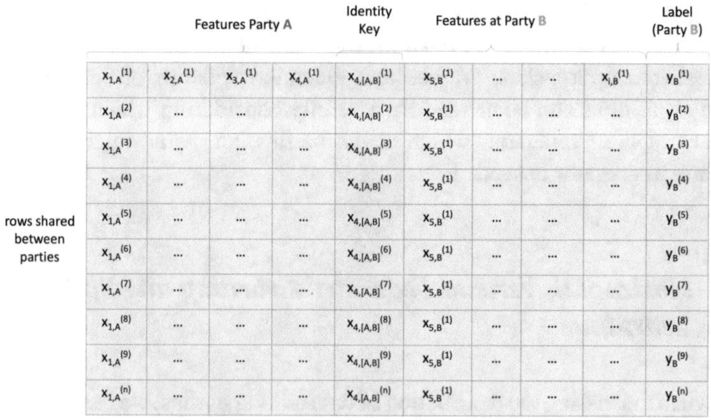

rows shared between parties

Features Party A				Identity Key	Features at Party B				Label (Party B)
$x_{1,A}^{(1)}$	$x_{2,A}^{(1)}$	$x_{3,A}^{(1)}$	$x_{4,A}^{(1)}$	$x_{4,[A,B]}^{(1)}$	$x_{5,B}^{(1)}$	...	..	$x_{i,B}^{(1)}$	$y_B^{(1)}$
$x_{1,A}^{(2)}$	...	...	...	$x_{4,[A,B]}^{(2)}$	$x_{5,B}^{(1)}$	...	...	...	$y_B^{(2)}$
$x_{1,A}^{(3)}$	...	...	...	$x_{4,[A,B]}^{(3)}$	$x_{5,B}^{(1)}$	...	...	...	$y_B^{(3)}$
$x_{1,A}^{(4)}$	...	...	...	$x_{4,[A,B]}^{(4)}$	$x_{5,B}^{(1)}$	...	...	...	$y_B^{(4)}$
$x_{1,A}^{(5)}$	...	...	...	$x_{4,[A,B]}^{(5)}$	$x_{5,B}^{(1)}$	...	...	...	$y_B^{(5)}$
$x_{1,A}^{(6)}$	...	...	...	$x_{4,[A,B]}^{(6)}$	$x_{5,B}^{(1)}$	...	...	...	$y_B^{(6)}$
$x_{1,A}^{(7)}$	...	...	...	$x_{4,[A,B]}^{(7)}$	$x_{5,B}^{(1)}$	...	...	...	$y_B^{(7)}$
$x_{1,A}^{(8)}$	...	...	...	$x_{4,[A,B]}^{(8)}$	$x_{5,B}^{(1)}$	...	...	...	$y_B^{(8)}$
$x_{1,A}^{(9)}$	...	...	...	$x_{4,[A,B]}^{(9)}$	$x_{5,B}^{(1)}$	...	...	...	$y_B^{(9)}$
$x_{1,A}^{(n)}$	...	...	...	$x_{4,[A,B]}^{(n)}$	$x_{5,B}^{(1)}$	...	...	...	$y_B^{(n)}$

Fig. 1.4 Vertically partitioned data

Figure 1.4 illustrates vertical partitioning, with features overlapping in an *identity key* to match the records of both parties, e.g., a government identifier. Since not all relevant features are present in any party, learning cannot take place independently at each party. Furthermore, identity keys have to be matched to understand how the features at each party complement each other. To preserve data privacy at each party, we need a cryptographic approach to match data and perform the learning process. Hardy et al. proposed a seminal early approach based on partial homomorphic encryption [24] with others such as Xu et al. [62] proposing much more effective variants, reducing communication, and computing requirements to an extent that it becomes viable in actual enterprise practice. Vertical FL is covered in detail in Chap. 18. Later in this chapter, we will discuss the security and privacy of federated learning in more depth.

Somewhat related to vertical federated learning, **Split Learning** has been proposed by Vepakomma et al. [55], among others [49]. In split learning, a DNN is partitioned between a client and a server in a way that the client maintains the "upper" part of the DNN down to a *split layer* and the server has the split layer and those below. In its basic form, the client has the input data and the server the labels. When using SGD as the training algorithm, the forward pass begins at the client with the input and is propagated to the server at the split layer. Back propagation takes place from the server to the client via the split layer. With this approach, one party's data can also be kept private, while another party has a part of the model structure. Split learning can be varied to the client also having the label, with the last, fully connected layer being on the client side via a second split layer, or multiple clients having vertically partitioned data and communicating with the server using partitions of the split layer. The latter case can be regarded as a generalization of vertical federated learning. Chapter 19 discusses split learning in more depth.

1.3.4   Model Personalization

Model personalization refers to the adaption of a (federally trained) global model to the data distribution of the specific parties participating in the FL process. While the participation in an FL process enables all parties to benefit from a large pool of training data, it is sometimes beneficial to personalize the final model to ensure it reflects the data owned by a specific party. This is relevant in particular if parties correspond to individual users or organizations. In a naive case, individual parties can run additional local training epochs on local data to the end of an FL process. Wang et al. propose an approach to evaluate the benefit of personalization for each party [56].

Mansour et al. [37] analyze three different approaches to personalization: user clustering, training on interpolated data (between global and local), and model interpolation. The first approach requires a relaxation of privacy requirements or advanced privacy techniques to cluster users based on training data. Data interpolation is based on the creation of a global dataset. While all approaches work, model interpolation has the widest applicability from a privacy perspective. Grimberg et al. propose a method to optimize averaging a global model and a local model for personalization purposes by determining optimized weights, expanding on the approaches discussed before [22].

While the approaches to personalization are still evolving, this is an important complement to the FL process. Chapters 4 and 5 discuss model personalization in depth.

1.4   Security and Privacy

By leaving data where it is, FL provides an inherent level of privacy at the outset. However, there is still potential for infringing on privacy of data. It is important to understand different threat models that may arise during the application of FL to ensure that relevant risks are mitigated appropriately with the right defenses. In this section, we provide an overview of vulnerabilities of FL and sketch the corresponding mitigation techniques.

Figure 1.5 presents a characterization of potential threats of FL as well as the potential adversaries.

Let us analyze the risk by first understanding the potential attack surfaces that an adversary may exploit. A well-set-up FL system makes use of secure-and-authenticated channels to ensure all messages exchanged between the parties and aggregator cannot be intercepted by other entities while preventing impersonation. Therefore, we can assume that the aggregator and parties are the only two entities that can access the messages exchanged between them and the artifacts produced during the training process. With this in mind, we can categorize potential adversaries as *insiders* or *outsiders*. Insider adversaries are entities involved in the training process who have access to the artifacts produced during training and messages directed at them. All other potential adversaries are considered outsiders. In this classification, entities that receive the final model produced by the FL training process are considered outsiders.

We can categorize threats to FL into *manipulation* and *inference* threats, where *manipulation threats* are those where an insider's objective is to influence the model to their advantage by manipulating any of the artifacts she can access during the training process, while *inference threats* are those where an insider or outsider tries to extract private information about the training data. In the following, we explain some of these attacks in more detail.

1.4.1   Manipulation Attacks

There are multiple types of manipulation attacks where the main goal of an insider adversary is to manipulate the model produced during the FL training to her advan-

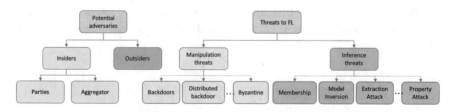

Fig. 1.5 Potential adversaries and threats

tage. In some cases, an adversary may want to cause targeted misclassifications, while in other cases, she may want to decrease the model performance to make it unusable. Backdoor [2, 23] and Byzantine attacks [29] are two examples of targeted and untargeted attacks, respectively. Backdoor attacks create targeted misclassifications, while byzantine attacks cause poor model performance. Byzantine attacks may be carried out by a single party or by multiple colluding parties and may be as simple as injecting random noise [57] or as elaborate as running optimizations to circumvent existing defenses [4, 60]. Label flipping attacks, where one or more malicious parties flip some of the labels, are another popular way to decrease the model performance [19].

In the FL literature, an insider carrying out a manipulation attack is often assumed to be a malicious party [57, 59]. However, a malicious aggregator may also carry out this type of attack. This would require the aggregator to train for a few epochs the aggregated model with poisoned samples and then sending the new manipulated model to the parties. There are also attacks where multiple colluding parties agree to manipulate the model updates so the final model causes targeted classifications [65] or poor behavior.

Unfortunately, manipulation attacks in FL are not easy to detect. First, not all data is available for a potential defender to run defenses that are frequently applied in centralized settings.[1] Second, data heterogeneity has been shown to affect the robustness of FL [65] making it difficult to distinguish between malicious model updates that should not be included and benign model updates that inclusion would benefit the final model. Third, a successful attack does not require long rounds of manipulation; it is possible to have high attack success rates by timing the attack correctly [65]. Finally, as defenses evolve, so do the attacks; adaptive attacks have been designed to circumvent some of the proposed defenses [4, 60], creating a familiar competition between attackers and defenders.

Most defense approaches assume the aggregator is the defender and may filter out malicious model updates. Model updates received by the aggregator are inspected to determine updates that are *too different*. Defenses in this category use multiple distance metrics and some assume a certain number of parties are always malicious [5, 12, 64]. However, a substantially different model updates may not always be an attack; it may be organically generated by a party whose data exhibits substantial non-IIDness with respect to the other parties. The difficulty in knowing when *unusual* updates are benign or malicious is clearly exacerbated by the fact that the aggregator does not have access to the training data. To overcome this difficulty, some approaches have been developed with the assumption that the aggregator can obtain a dataset with similar distribution as the one held by the parties [58]. This, however, may be difficult to obtain for certain use cases. Another approach [54] does not discard unusual updates altogether while training neural networks, but rather

---

[1] Backdoor attacks and Byzantine attacks exist both in centralized and distributed learning. Existing defenses for traditional cases often make use of the entire training data and, to date, are still unable to fully achieve a 100% detection rate.

adapts some layers of the neural network to prevent overfitting. A substantially different approach was presented in [3], where accountability is used to deter attacks. This approach stores a non-repudiable record of the complete training process. Transparency is provided by ensuring all parties are accountable for their model updates as well as for their training process, while the aggregator is also accountable for the way it fusions them.

An overview of manipulation attacks and defenses will be presented in Chap. 16, and Chap. 17 focuses on understanding Byzantine attacks and defenses when training neural networks.

1.4.2   Inference Attacks

Training without sharing data is one of the drivers and the most important advantages of applying FL. Recall that model updates are the only data shared with the aggregator, and private training data is never revealed. This design ensures that simple exposure of private information is not an issue in FL systems. Hence, privacy leakage can only take place through inference.

Inference attacks take advantage of artifacts produced during or after the FL process to try to *deduce* private information. Inference threats are not new to machine learning. In fact, a large body of work has documented ways in which an adversary, who only has access to an ML model, can infer private information about its training data. Attacks in this black-box setting include:

- *Membership inference attacks*, where an adversary can determine if a particular sample was used to train a model. This is a privacy violation when, for example, the model used data coming from certain social group, e.g., political or sexual affiliation or disease.
- *Model inversion attacks*, where the adversary wants to find the representative of each class. In a face recognition system, this, for example, may reveal the face of a person.
- *Extraction attacks*, where the adversary's goal is to obtain all the samples used during the training process.
- *Property inference attacks*, where properties independent of the training tasks may be revealed.

In the FL setting, outsiders get access to the final ML model, and insiders may get access to intermediate models; hence, both insiders and outsiders may be able to carry out the attacks listed above.

In addition and interestingly, the exchanged model updates that at first glance may seem to be innocuous can also be used by insiders to infer private information. Attacks that use model updates include [21, 25, 39, 41, 67, 68] and in some cases exhibit faster and higher success rates than the ones carried out using the model. Model-update-based attacks can be carried out by curious parties or a malicious aggregator. These attacks can be *passive*, where the adversary uniquely inspects the produced artifacts, or *active* when it acts to speed up the inference.

Given these risks of privacy exposure, several techniques to protect the FL process have been proposed. They include using differential privacy (DP) [16], secure multi-party computation techniques [6, 44, 61, 66], a combination of both [52], and the use of trusted execution environments [11, 30], among others.

DP is a rigorous mathematical framework where an algorithm may be described as differentially private if and only if the inclusion of a single instance in the training dataset causes only statistically insignificant changes to the algorithm's output. The methodology adds *noise* through a DP *mechanism* that is tailored to both the dataset and the query that will be answered with the data. DP provides a provable mathematical guarantee; however, it may reduce the model accuracy substantially. Another popular technique to prevent inference attacks is the use of secure multi-party computation, where a curious aggregator cannot get to know the individual model updates received from parties, but can still obtain the final fusion result (in plaintext or ciphertext). These techniques include masking [6], Paillier [44, 66], Threshold Paillier [52], and Functional encryption [61], to name a few. All these techniques have slightly different threat models and, therefore, are suitable for difference scenarios.

Existing defenses offer different protection and target diverse inference attacks. It is important to ensure that the defense is selected according to the use case at hand to ensure the right level of protection is achieved. In fully trusted cases, no additional protection may be needed, for example, when a single company is training a model with data coming from multiple data centers. In a consortium of competitors, however, the risk of inference may be too high leading to the use of one or more protection mechanisms.

Threats and defenses to inference attacks are addressed by multiple chapters in this book. Chapter analyzes the inference risks to FL systems. It presents existing attacks and defenses demonstrating that the level of protection offered by each defense is suitable for slightly different cases. The chapter also presents an analysis to help determine how to match different scenarios and trust assumptions into suitable defenses. Chapter 14 provides a more in-depth review of a defense based on trusted execution environments, and Chap. 15 presents in detail the mechanics of gradient-based data extraction attacks.

1.5   Federated Learning Systems

An FL process is ultimately executed on a distributed system on which parties and the aggregator run. The parts of this system must satisfy the computational, memory, and networking requirements of parties, aggregator, and the communication between them. Since the local model training is performed where the data is located, we must pay close attention to the resources available at the point of training the parties. Aggregators are mostly run in a data center environment, at least in the commonly used single aggregator architecture. Still, they need to have the right resources and ability to scale when dealing with a large number of parties. Finally, network connectivity and bandwidth requirements might differ based on model size, model-update content, frequency and use of cryptographic protocols, as discussed in the previous section. Hence, the systems requirements for federated learning are quite different than those for centralized learning approaches.

Party clients: The most obvious distinction to centralized ML systems lies in the fact that a party might not be located on a system we would typically choose as an ML platform. While this might not be as problematic when parties are data centers in different jurisdictions, it is more problematic in embedded systems, edge computing, and mobile phones. Three different types of functionalities potentially draw resources:

- The *local machine learning process* may require significant compute and memory if the model is large. This is the case in particular for large DNNs, e.g., for large language models. It might require GPU support, which might not be available in embedded systems or even in remote data centers or software-as-a-service-related data stores. However, classic techniques might be viable even on small devices such as Rasberry Pis®, and there are small footprint packages such as Tensorflow Lite® that require less on-party storage and memory.
- The federated learning party client drives the local machine learning model and communicates with the aggregator. In most cases, though, it has a small footprint that can be accommodated even in small edge devices.
- However, if a *cryptographic protocol* is used, e.g., a secure multi-party computation (SMC) implementation based on a threshold Paillier cryptosystem, it might increase computation cost of the party client by several orders of magnitude. Most encryption and decryption techniques can be paralleled and, hence, be supported by a GPU or specialized hardware.

Aggregator servers: An aggregator is typically located in a data center environment that has access to ample resources. However, scaling to a large number of parties entails a number of challenges:

To communicate with a large number of parties, the aggregator has to be able to maintain a large number of connections. Pooling connections is a well-established approach for all kinds of systems that can be used here in a similar way.

Performing a fusion algorithm at the aggregator often incurs moderate computational costs. Simple fusion algorithms such as FedAvg, discussed in Sect. 1.3, perform quite simple averaging operations. Other fusion algorithms might be more complex but will often have lower computational requirements than the local training at a party. However, in the case of large DNNs and a large number of parties, the size of the set of, say, weight vectors received as replies from parties can be very large. One party's weight vector can be up to tens of megabytes. Dealing with many thousands of parties can be too much to perform average computation in memory in one compute node.

Different approaches have been proposed to address scaling the aggregator computationally: The weights can be made persistent, and the fusion algorithm can be performed in a parallel computation approach, e.g., using Hadoop or Spark. Other approaches use the commutative properties of addition and partition parties into groups. These groups are assigned to an aggregator, each computing averages over this group. A primary aggregator then aggregates the results of local aggregation, weighted by the number of parties at each aggregator. So et al. [47] propose one such approach, and there are different variants, including multi-level aggregation. For very large sets of parties, sub-sampling of parties is often used at each round and can complement the other approaches.

Tree-based FL algorithms usually pose more computational demands on the aggregator and less on the parties.

Communication: Quantity and quality of communication between aggregators and parties must be considered in FL design. In data center and Cloud settings, we can often assume bandwidth to be sufficient and connections to be reliable. FL processes can take quite some time. Hence, communication protocols need to be robust to occasional disconnection. An important practical consideration in enterprise contexts is the connection direction. IT departments in organizations have tightly controlled processes to open networking ports. Choosing a networking protocol that does not require parties to open ports but have them initialize the connection to the aggregator will accelerate implementations of FL systems.

Embedded systems, edge devices, and mobile systems pose a bigger challenge. Some party systems might be intermittently connected, e.g., in vehicles, or have poor bandwidth, being low-cost devices. This can be problematic for the FL process. If parties do not respond in time for the next round, we need a strategy to manage these drop-offs. We need to establish a *quorum* that may be specific for a given use case. We also need an approach when parties re-join. While quora are simple means of drop-off management, other approaches such as TIFL propose an active straggler management, grouping parties by response time and querying them less frequently [10]. Systemic differences in response time can even lead to bias in the model [1].

Intermittent or low-bandwidth communication can also be addressed algorithmically, for example, by reducing the number of rounds, compressing models, and fusing more divergent models. Chapters 6 and 7 discuss this in more detail.

The use of secure computation methods such as SMC may increase both message size and quantity and may pose an issue for devices with poor connections. Furthermore, some SMC protocols for vertical federated learning may require peer-to-peer communication between parties, which is problematic in two ways: It requires parties to expose ports to their peers, which is an implementation obstacle in enterprises. If mitigated by routing all traffic through the aggregator or another intermediary, this again doubles network traffic. Hence, while SMC is often a very good approach to preserve privacy, it comes with significant resource requirements.

Design choices and trade-offs: When implementing an FL system, we often need to trade off available resources with a suitable algorithmic approach. If we can make a choice about the hardware available to a party, we can choose one that suits the

ML approach we choose. We can add an embedded system with a strong GPU to a vehicle or manufacturing robot or add GPUs to the data centers we want to have participate in the federation. That is not always possible. In situations in which the compute platform at parties is given, we can use a ML approach that suits our resources. While DNNs are resource intensive at the party side, tree-based model such as federated XGBoost is not as demanding. Also, algorithms can be adapted to system constraints.

1.6 Summary and Conclusion

This chapter provides an introduction to federated learning. We discussed the main motivations to bring training to the data, rather than bringing all the data together, as it is the case in centralized ML. The need for complying with privacy regulation, secrecy of data, and pragmatic considerations such as network quality are the main drivers. We introduced the main concepts of parties and aggregator and then went through the main perspectives on FL we need to consider: the machine learning perspective, the security and privacy perspective, and then the systems perspective. All of those perspectives go hand in hand to design a FL system suitable for its task.

We took a particular perspective on the needs of enterprises implementing FL. This includes the need to support both neural networks and classical approaches, heterogeneity of data and systems at parties, and the need for vertical FL when different categories of data are kept in different systems. This is somewhat different from the application of FL in mobile devices, which are mostly more homogeneous but pose different issues of scale.

The remainder of this book addresses all of those aspects in more depth:

- *Part I* looks at the machine learning perspective of FL and discusses tree-based models, efficiency, personalization, and fairness.
- *Part II* addresses the systems perspective in more depth.
- *Part III* includes five chapters covering privacy and security. Inference and manipulation attacks are described in detail and more information on the defenses, when to apply them, is provided.
- *Part IV* contains chapters covering in detail vertical FL as well as split learning.
- *Part V* showcases work on applications of FL and requirements for important application domains such as healthcare and finance.

This scope of the book provides an overview of the state-of-the-art of FL in the enterprise for researchers and practitioners looking for an in-depth background.

References

1. Abay A, Zhou Y, Baracaldo N, Rajamoni S, Chuba E, Ludwig H (2020) Mitigating bias in federated learning. arXiv preprint arXiv:201202447
2. Bagdasaryan E, Veit A, Hua Y, Estrin D, Shmatikov V (2020) How to backdoor federated learning. In: International conference on artificial intelligence and statistics. PMLR, pp 2938–2948
3. Balta D, Sellami M, Kuhn P, Schöpp U, Buchinger M, Baracaldo N, Anwar A, Sinn M, Purcell M, Altakrouri B (2021) Accountable federated machine learning in government: engineering and management insights
4. Baruch M, Baruch G, Goldberg Y (2019) A little is enough: circumventing defenses for distributed learning. arXiv preprint arXiv:190206156
5. Blanchard P, Mhamdi EME, Guerraoui R, Stainer J (2017) Byzantine-tolerant machine learning. 1703.02757
6. Bonawitz KA, Ivanov V, Kreuter B, Marcedone A, McMahan HB, Patel S, Ramage D, Segal A, Seth K (2016) Practical secure aggregation for federated learning on user-held data. In: NIPS workshop on private multi-party machine learning. https://arxiv.org/abs/1611.04482
7. Boyd S, Parikh N, Chu E (2011) Distributed optimization and statistical learning via the alternating direction method of multipliers. Now Publishers Inc., Hanover
8. Breiman L (2001) Random forests. Mach Learn 45(1):5–32
9. Business Insider (2018) Macy's is warning customers that their information might have been stolen in a data breach. https://www.businessinsider.com/macys-bloomingdales-hack-disclosed-2018-7
10. Chai Z, Ali A, Zawad S, Truex S, Anwar A, Baracaldo N, Zhou Y, Ludwig H, Yan F, Cheng Y (2020) TIFL: a tier-based federated learning system. In: Proceedings of the 29th international symposium on high-performance parallel and distributed computing, pp 125–136
11. Chamani JG, Papadopoulos D (2020) Mitigating leakage in federated learning with trusted hardware. arXiv preprint arXiv:201104948
12. Charikar M, Steinhardt J, Valiant G (2016) Learning from untrusted data. 1611.02315
13. Chen T, He T, Benesty M, Khotilovich V, Tang Y, Cho H et al (2015) XGBoost: extreme gradient boosting. R package version 04-2 1(4)
14. Cheng K, Fan T, Jin Y, Liu Y, Chen T, Yang Q (2019) SecureBoost: a lossless federated learning framework. arXiv preprint arXiv:190108755
15. Dean J, Corrado GS, Monga R, Chen K, Devin M, Le QV, Mao MZ, Ranzato M, Senior A, Tucker P, Yang K, Ng AY (2012) Large scale distributed deep networks. In: NIPS
16. Dwork C (2008) Differential privacy: a survey of results. In: International conference on theory and applications of models of computation. Springer, pp 1–19
17. Elgabli A, Park J, Ahmed S, Bennis M (2020) L-FGADMM: layer-wise federated group ADMM for communication efficient decentralized deep learning. In: 2020 IEEE wireless communications and networking conference (WCNC). IEEE, pp 1–6
18. Elgabli A, Park J, Bedi AS, Bennis M, Aggarwal V (2020) GADMM: fast and communication efficient framework for distributed machine learning. J Mach Learn Res 21(76):1–39
19. Fang M, Cao X, Jia J, Gong N (2020) Local model poisoning attacks to byzantine-robust federated learning. In: 29th {USENIX} security symposium ({USENIX} security 20), pp 1605–1622
20. Ge N, Li G, Zhang L, Liu YLY (2021) Failure prediction in production line based on federated learning: an empirical study. arXiv preprint arXiv:210111715
21. Geiping J, Bauermeister H, Dröge H, Moeller M (2020) Inverting gradients—how easy is it to break privacy in federated learning? In: Part of advances in neural information processing systems (NeurIPS 2020), vol 33
22. Grimberg F, Hartley MA, Karimireddy SP, Jaggi M (2021) Optimal model averaging: towards personalized collaborative learning. In: Proceedings of the international workshop on federated learning for user privacy and data confidentiality. https://fl-icml.github.io/2021/papers

23. Gu T, Dolan-Gavitt B, Garg S (2017) BadNets: identifying vulnerabilities in the machine learning model supply chain. arXiv preprint arXiv:170806733

24. Hardy S, Henecka W, Ivey-Law H, Nock R, Patrini G, Smith G, Thorne B (2017) Private federated learning on vertically partitioned data via entity resolution and additively homomorphic encryption. arXiv preprint arXiv:171110677

25. Jin X, Du R, Chen PY, Chen T (2020) CAFE: catastrophic data leakage in federated learning

26. Kairouz P, McMahan HB, Avent B, Bellet A, Bennis M, Bhagoji AN, Bonawitz K, Charles Z, Cormode G, Cummings R et al (2019) Advances and open problems in federated learning. arXiv preprint arXiv:191204977

27. Kingma DP, Ba J (2017) Adam: a method for stochastic optimization. 1412.6980

28. Konečný J, McMahan HB, Yu FX, Richtárik P, Suresh AT, Bacon D (2016) Federated learning: strategies for improving communication efficiency. arXiv preprint arXiv:161005492

29. Lamport L, Shostak R, Pease M (1982) The byzantine generals problem. ACM Trans Program Lang Syst 4(3):382–401

30. Law A, Leung C, Poddar R, Popa RA, Shi C, Sima O, Yu C, Zhang X, Zheng W (2020) Secure collaborative training and inference for XGBoost. In: Proceedings of the 2020 workshop on privacy-preserving machine learning in practice, pp 21–26

31. Li T, Sahu AK, Zaheer M, Sanjabi M, Talwalkar A, Smith V (2018) Federated optimization in heterogeneous networks. arXiv preprint arXiv:181206127

32. Liu W, Chen L, Chen Y, Zhang W (2020) Accelerating federated learning via momentum gradient descent. IEEE Trans Parallel Distrib Syst 31(8):1754–1766

33. Liu Y, Ma Z, Liu X, Ma S, Nepal S, Deng R (2019) Boosting privately: privacy-preserving federated extreme boosting for mobile crowdsensing. arXiv preprint arXiv:190710218

34. Liu Y, Liu Y, Liu Z, Liang Y, Meng C, Zhang J, Zheng Y (2020) Federated forest. IEEE Trans Big Data, p. 1

35. Ludwig H, Baracaldo N, Thomas G, Zhou Y, Anwar A, Rajamoni S, Ong Y, Radhakrishnan J, Verma A, Sinn M et al (2020) IBM federated learning: an enterprise framework white paper v0. 1. arXiv preprint arXiv:200710987

36. Malinovskiy G, Kovalev D, Gasanov E, Condat L, Richtarik P (2020) From local SGD to local fixed-point methods for federated learning. In: International conference on machine learning. PMLR, pp 6692–6701

37. Mansour Y, Mohri M, Ro J, Suresh AT (2020) Three approaches for personalization with applications to federated learning. arXiv preprint arXiv:200210619

38. McMahan B, Moore E, Ramage D, Hampson S, y Arcas BA (2017) Communication-efficient learning of deep networks from decentralized data. In: Artificial intelligence and statistics. PMLR, pp 1273–1282

39. Melis L, Song C, De Cristofaro E, Shmatikov V (2019) Exploiting unintended feature leakage in collaborative learning. In: 2019 IEEE symposium on security and privacy (SP). IEEE, pp 691–706

40. Meltzer J (2020) The Court of Justice of the European Union in Schrems II: the impact of GDPR on data flows and national security. https://voxeu.org/article/impact-gdpr-data-flows-and-national-security

41. Nasr M, Shokri R, Houmansadr A (2018) Comprehensive privacy analysis of deep learning: stand-alone and federated learning under passive and active white-box inference attacks

42. NBC News (2018) Yahoo to pay $50 million, offer credit monitoring for massive security breach. https://www.nbcnews.com/tech/tech-news/yahoo-pay-50m-offer-credit-monitoring-massive-security-breach-n923531

43. New York Times (2018) Facebook security breach exposes accounts of 50 million users. https://www.nytimes.com/2018/09/28/technology/facebook-hack-data-breach.html

44. Nikolaenko V, Weinsberg U, Ioannidis S, Joye M, Boneh D, Taft N (2013) Privacy-preserving ridge regression on hundreds of millions of records. In: 2013 IEEE symposium on security and privacy. IEEE, pp 334–348

45. Ong YJ, Zhou Y, Baracaldo N, Ludwig H (2020) Adaptive histogram-based gradient boosted trees for federated learning. arXiv preprint arXiv:201206670

46. Quinlan JR (1986) Induction of decision trees. Mach Learn 1(1):81–106
47. So J, Güler B, Avestimehr AS (2021) Turbo-aggregate: breaking the quadratic aggregation barrier in secure federated learning. IEEE J Sel Areas Inf Theory 2(1):479–489
48. State of California (2018) California Consumer Privacy Act of 2018
49. Thapa C, Chamikara MAP, Camtepe S (2020) SplitFed: when federated learning meets split learning. arXiv preprint arXiv:200412088
50. The European Parliament and Council (2016) Regulation (EU) 2016/679 of the European Parliament and of the Council of 27th of April 2016 on the protection of natural persons with regard to the processing of personal data and on the free movement of such data, and repealing directive 95/46
51. The Wall Street Journal (2018) Google exposed user data, feared repercussions of disclosing to public. https://www.wsj.com/articles/google-exposed-user-data-feared-repercussions-of-disclosing-to-public-1539017194
52. Truex S, Baracaldo N, Anwar A, Steinke T, Ludwig H, Zhang R, Zhou Y (2019) A hybrid approach to privacy-preserving federated learning. In: Proceedings of the 12th ACM workshop on artificial intelligence and security, pp 1–11
53. United States (1996) Health Insurance Portability and Accountability Act of 1996. U.S. Government Printing Office, Washington, DC
54. Varma K, Zhou Y, Baracaldo N, Anwar A (2021) LEGATO: a LayerwisE Gradient AggregaTiOn algorithm for mitigating byzantine attacks in federated learning. In: 2021 IEEE 14th international conference on cloud computing (CLOUD)
55. Vepakomma P, Gupta O, Swedish T, Raskar R (2018) Split learning for health: distributed deep learning without sharing raw patient data. arXiv preprint arXiv:181200564
56. Wang K, Mathews R, Kiddon C, Eichner H, Beaufays F, Ramage D (2019) Federated evaluation of on-device personalization. arXiv preprint arXiv:191010252
57. Xie C, Koyejo O, Gupta I (2018) Generalized byzantine-tolerant SGD. 1802.10116
58. Xie C, Koyejo O, Gupta I (2018) Zeno: distributed stochastic gradient descent with suspicion-based fault-tolerance. 1805.10032
59. Xie C, Huang K, Chen PY, Li B (2019) DBA: distributed backdoor attacks against federated learning. In: International conference on learning representations
60. Xie C, Koyejo S, Gupta I (2019) Fall of empires: breaking byzantine-tolerant SGD by inner product manipulation. 1903.03936
61. Xu R, Baracaldo N, Zhou Y, Anwar A, Ludwig H (2019) HybridAlpha: an efficient approach for privacy-preserving federated learning. In: Proceedings of the 12th ACM workshop on artificial intelligence and security, pp 13–23
62. Xu R, Baracaldo N, Zhou Y, Anwar A, Joshi J, Ludwig H (2021) FedV: Privacy-preserving federated learning over vertically partitioned data. arXiv preprint arXiv:210303918
63. Yang M, Song L, Xu J, Li C, Tan G (2019) The tradeoff between privacy and accuracy in anomaly detection using federated XGBoost. arXiv preprint arXiv:190707157
64. Yin D, Chen Y, Ramchandran K, Bartlett P (2018) Byzantine-robust distributed learning: towards optimal statistical rates. 1803.01498
65. Zawad S, Ali A, Chen PY, Anwar A, Zhou Y, Baracaldo N, Tian Y, Yan F (2021) Curse or redemption? How data heterogeneity affects the robustness of federated learning. In: Proceedings of the AAAI conference on artificial intelligence, vol 35, pp 10807–10814
66. Zhang C, Li S, Xia J, Wang W, Yan F, Liu Y (2020) BatchCrypt: efficient homomorphic encryption for cross-silo federated learning. In: 2020 USENIX annual technical conference (USENIX ATC 20), pp 493–506
67. Zhao B, Mopuri KR, Bilen H (2020) iDLG: improved deep leakage from gradients. arXiv preprint arXiv:200102610
68. Zhu L, Han S (2020) Deep leakage from gradients. In: Federated learning. Springer, Cham, pp 17–31

Part I
Federated Learning as a Machine Learning Problem

Part I of this book addresses a number of specific issues of federated learning from a machine learning perspective, going beyond the general overview that was presented in the introduction. The chapters look at specific model types in the context of federated learning, the issue of model personalization for a specific party, how to adapt the federated learning process under conditions of limited communication process, and at bias and fairness.

Chapters 2 and 3 cover two types of models particularly important for applications in the enterprise where tabular and text data are widely available. Chapter 2 takes a deep dive into methodologies to adapt tree-based model training to a federated setting. The chapter covers multiple algorithm designs for both horizontal and vertical federated learning. Chapter 3 focuses on text-based models that require producing embeddings, which are also in common use for graphs.

Personalization has become an extremely important aspect for federated learning systems where engaging in a federation helps improve model generalization, but where at the same time, the final model for each party is different and tailored to the specific party's needs. Chapter 4 provides a thorough overview of the existing penalization techniques and research challenges in this area. Chapter 5 focuses on a methodology that produces personalized and robust models.

The next two chapters address techniques for dealing with situations where communication between the party and the aggregator is limited or expensive. Chapter 6 discusses multiple techniques to reduce the frequency with which model updates are exchanged, while also reducing the amount of data exchanged in each round, e.g., by model compression or pruning. Chapter 7 reviews and analyzes model fusion approaches that can merge models that have been produced independently or with little interaction, overcoming structural and model differences both for classic and neural model types.

Finally, Chap. 8 focuses on the important aspect of social fairness in federated learning. The chapter provides an overview of sources of undesired bias in federated learning. This includes bias as we find it in centralized machine learning but also novel sources of bias being brought about by a federated training process. When it comes to mitigation techniques, in traditional, centralized machine learning, fairness mitigation usually requires analyzing all the training data. This is of course not possible in federated learning settings. Chapter 8 also discusses how to adapt to some of the existing techniques to mitigate unwanted bias.

Chapter 2
Tree-Based Models for Federated Learning Systems

Yuya Jeremy Ong, Nathalie Baracaldo, and Yi Zhou

Abstract Many Federated Learning algorithms have been focused on linear models, kernel-based, and neural-network-based models. However, recent interest in tree-based models such as Random Forest and Gradient Boosted Trees such as XGBoost has started to be explored due to their simplicity, robust performance, and interpretability in various applications. In this chapter, we introduce recent innovations, techniques, and implementations specifically for tree-based algorithms. We highlight how these tree-based methods differ from many of the existing FL methods and some of the key advantages they have compared to other Federated Learning algorithms.

2.1 Introduction

Federated Learning (FL) [29] has emerged as a new paradigm for training machine learning models collaboratively across a federation of multiple distributed parties without revealing the underlying raw data. Its applications have already been demonstrated and implemented within the consumer setting to protect personal data [17], as well as in the B2B enterprise settings [27] where constraints such as government policies and regulations (e.g., GDPR, HIPPA, CCPA) and/or model training through cross-enterprise collaborations are necessary for joint industrial endeavors. In particular, the adoption of Federated Learning systems has become widely popular due to concerns over potential risks and threats of data leakage and privacy violations being posed to individual consumers, organizations, businesses, and government entities.

As a result, this gave rise to various methods and techniques being proposed for common machine learning methods such as linear models [19], kernel-based models [7], and deep neural networks [2]. In particular, the growth of deep neural-network-

Y. J. Ong (✉) · N. Baracaldo · Y. Zhou
IBM Research – Almaden, San Jose, CA, USA
e-mail: yuyajong@ibm.com; baracald@us.ibm.com; yi.zhou@ibm.com

© The Author(s), under exclusive license to Springer Nature Switzerland AG 2022   27
H. Ludwig, N. Baracaldo (eds.), *Federated Learning*,
https://doi.org/10.1007/978-3-030-96896-0_2

based approach has grown to popularity due to the conflated increase in attention within the machine learning community and demonstrated successes in centralized learning scenarios. Hence, the literature within the Federated Learning space today has focused on deep neural networks. However, very few work has addressed alternative model architectures—in particular **Decision tree-based algorithms** such as *ID3 Decision Trees* and *Gradient Boosted Decision Trees (GBDT)*.

Tree-based models have been widely adopted within the machine learning community due to its simple, robust, and highly interpretable structures that are generated. In practice, they have been utilized in various domain areas such as finance [30], medical [24], biological sciences [9], and business transactions [41]— to name a few use cases. These algorithms have also been widely popular among various data science competition such as Kaggle,[1] where many winning solutions have at least adopted some form of decision tree-based algorithm as part of their solution, demonstrating robust performance in various real-world applications.

However, to implement these tree-based models within a Federated Learning setting required researchers to solve a number of different fundamental challenges, which significantly differ from the other model types that are commonly used within Federated Learning. These challenges include building models where the underlying structures are not initially defined, as opposed to neural networks where the model architectures are pre-determined. Another key challenge includes what type of fusion processes is necessary for aggregators to grow the model structures that are derived from various contributing parties within the federation. As a result of these new emerging challenges, this has spawned a new direction interesting discoveries and research directions within the Federated Learning research space. Along with its fair share of challenges, the field has also uncovered a set of key advantages such as direct interpretability of the model's decision, robust support for training over non-IID data distributions across different parties, and lower computational overhead. Many of the inherent advantages of tree-based models have naturally carried over to align with many of the inherent objectives and even complement the benefits of Federated Learning.

In this chapter, we focus on introducing the paradigm of tree-based methods for Federated Learning through showcasing its benefits, providing a survey of the state-of-the-art, and its practical implementation and applications in greater detail.

2.1.1   Tree-Based Models

The tree-based models are a class of machine learning algorithms that utilizes a decision tree structure, depicted in Fig. 2.1, as its model representation, which makes its decisions by partitioning the inputs' feature space based on a recursive series of binary "if-then" decision thresholds. For example, these "if-then" statements can

---

[1] https://www.kaggle.com/

Fig. 2.1 Example of a
decision tree structure

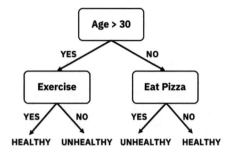

take the form of either input features that are *binary*, such as "does this person exercise" or "does this person eat pizza," versus *continuous* input features such as "is this person's age greater than 30." This model transforms the input data from the root node into the target leaf nodes through recursively traversing a set of branch points within a decision tree to derive the model's prediction, which resides at the leaf nodes, where the outcome here in Fig. 2.1 determines whether the person is "healthy" or "unhealthy."

Machine learning algorithms that can construct these types of decision tree-like structures have been widely adopted in various tasks such as *classification*, *regression*, and *ranking tasks*. The objective of these machine learning methods is to grow the tree structure by searching for the optimal split point to generate decision tree structures that generate accurate predictions from an input data sample. These algorithms produce more than one decision tree structure through generating an *ensemble* of decision tree structures, which are predominantly performed by either *bootstrap aggregation (i.e., bagging)* or *boosting*. *Bagging* builds multiple trees in parallel through repeated resampling with replacement and using an averaged aggregation process to combine the prediction outcomes of multiple decision tree structures [3], while *boosting* incrementally builds a new tree that is recursively built based on the error's of the previous tree's errors [14, 18]. In later sections of this chapter, we will introduce two of the most commonly utilized tree algorithms: ID3 Decision Trees [33] and XGBoost [4], an optimized variant of the Gradient Boosted Decision Tree algorithm [13].

2.1.2   Key Research Challenges of Tree-Based Models in FL

To adopt tree-based machine learning algorithms for Federated Learning, this requires re-evaluating the process of how decision trees are constructed due to fundamental differences in the underlying representation of the resulting model's structure. This paradigm of machine learning therefore introduces a whole set of new challenges, as the implementation of fusion processes for decision tree-based structures is non-trivial in a distributed learning setting, along with the additional challenges of preventing information leakage that can be potentially inferred from the generated tree structures.

First, as opposed to linear models, kernel-based models, and deep neural-network-based approaches, where the inherent structure of the model's parameters is defined as a finite structure, decision trees' model representations are not pre-defined but constructed dynamically from a bottom-up approach. This raises the question as to how fusion algorithms should be designed within a Federated Learning context, since the decisions of an input feature closer to the root node of the tree are optimized based on searching for the feature and split values with the greatest separability as an objective. Since the aggregator does not have full and direct access to the raw distribution of the entire dataset, this makes the process of searching for the optimal split points non-trivial, especially for Gradient Boosted Decision Trees where there is a dependence on the prior tree structures that were generated previously. In cases where the party's data are non-IID (i.e., independent and identically distributed), locally constructing such trees and attempting to fuse them within the aggregator are non-trivial tasks. This leads to different decision paths that are specifically dependent on the prior split choices made from the previous decision nodes, which ultimately can result in drastically different tree structures.

Another challenge in implementing decision tree-based algorithms for Federated Learning involves what type of information that would be exchanged between each of the parties and the aggregator. While Federated Learning approaches for neural networks and linear models exchange model parameters such as model weights, these parameters are somewhat latent in nature making it hard to identify properties of the raw data distribution. However, tree-based parameters that are comprised primarily of a collection of decision split nodes based on the feature's value are much more sensitive in nature to exchange over the network, as they explicitly reveal the immediate representation of the raw data. Hence, an alternative representation of the raw data must be used to prevent unnecessary exposure of the local party's information from being leaked to potential adversaries. Alternatively, potential data sources that can be exchanged can include: surrogate values that approximate close to the raw distribution, gradients, and Hessians, and noisy samples of the raw data distribution, for example. Other measures to prevent data leakage such as encryption methods can also be employed to provide near or complete lossless accurate information exchange while retaining high fidelity of the data. However, regardless of the representation of the data type that has been selected, this brings various costs toward additional computation and network overhead concerns. Thus, selectively considering the different trade-offs between privacy, utility, and application use case requirements is the necessary factor to identify when implementing tree-based models in practice.

2.1.3   Advantages of Tree-Based Models in FL

Despite the major challenges that are required to enable tree-based algorithms to work within a Federated Learning setting, we discuss some of the key advantages of tree-based algorithms and how these methods compliment the major advantages in Federated Learning systems.

First, decision tree algorithms provide the right balance between complexity and interpretability, which allows for both highly robust model performance while retaining clear visibility into the model's decision process. Decision trees can model non-linear mappings between the inputs and outputs, while using a clear rule-based-like decision structure, which allows for decision-makers to clearly identify the factors the models used to make its predictions. This is especially important in highly regulated settings such as finance, medical, and other government use cases for Federated Learning, where these aspects are also evaluated in addition to the data privacy and security concerns which Federated Learning mitigates.

Another favorable aspect of decision trees that complements well with Federated Learning settings is the ability to handle both categorical and numerical features out of the box. For example, algorithms such as Federated Gradient Decision Trees can natively handle missing data, which is a very common form of data that requires preprocessing in various domains. Another additional feature of tree-based algorithm is the in-built feature selection methods that are implicit part of the modeling process. This allows for joint modeling on the dataset for the predictive task while being able to preprocess the data across different parties to identify key features that are relevant across all parties. In short, these features allow for modeling without much data preprocessing, which is a non-trivial task to perform if the data is especially decentralized in a Federated Learning system. To implement such data preprocessing routines in Federated Learning, this would incur additional compute and network communication costs if non-tree-based algorithms were used. Hence, tree-based algorithms can significantly cut down on any additional communication costs, while jointly being able to handle various data types.

Finally, another key advantage of tree-based algorithms is the overall robust model performance demonstrated across various machine learning tasks— *especially in handling Non-IID (independent and identically distributed) data* [31]. This is especially important in cases of Federated Learning where the data is distributed and where the entire view of the dataset cannot be seen globally. This, in turn, can lead to working with datasets where the data distribution is not well balanced between each individual parties, and thus can make it increasingly difficult in cases when training models jointly across all devices in Federated Learning. However, algorithms such as Gradient Boosted Decision Tree models are known to handle non-IID data well and often provide robust performances, as we will demonstrate in later sections of the chapter.

In the remaining sections of this chapter, we focus on evaluating the current state-of-the-art of tree-based models in Federated Learning, the fundamental methodologies, and core building blocks behind how to train these tree-based models in a Federated Learning setting and preview some of the other possible research directions that are in the horizon for tree-based models for Federated Learning. The rest of the chapter is organized as follows:

- **Section 2.1**: Provides an introduction to tree-based models for Federated Learning and highlights some of the key challenges and advantages over alternative Federated Learning algorithms.
- **Section 2.2**: Provides a high-level survey of different state-of-the-art approaches of tree-based algorithms proposed within the Federated Learning space.
- **Section 2.3**: Introduces key notations as well as a brief overview of the ID3 Decision Tree and XGBoost algorithm necessary for understanding the implementation under a Federated Learning setting.
- **Section 2.4**: We introduce the core algorithm behind the ID3 Decision Tree implementation in Federated Learning.
- **Section 2.5**: We introduce the core algorithm behind the Gradient Boosted Trees (XGBoost) algorithm for Federated Learning.
- **Section 2.6**: We identify some key research directions and future extensions for tree-based models for Federated Learning.
- **Section 2.7**: We conclude and summarize the key ideas in this chapter.

2.2 Survey of Tree-Based Methods for FL

In this section, we survey some of the proposed methods that have been published recently in the literature around tree-based methods for Federated Learning. We summarize some of the major trends found across different papers and identify notable differences between the methods of implementations across different works. We organize the list of proposed methods in the state-of-the-art, as of writing this chapter, in Table 2.1, where we highlight some of the key differences in each of the methods. In our evaluation, we find that the main differentiation between the proposed methods is based on: (i) whether they are based on horizontal or vertical Federated Learning approaches, (ii) what type of tree-based learning algorithm is implemented, (iii) what information is exchanged between parties and the aggregator, and (iv) what security measures and protocols are implemented to protect user's private data. Furthermore, we also look at which Federated Learning frameworks and products provide an implementation for tree-based models for Federated Learning.

Table 2.1 Summary of tree-based models for Federated Learning

Paper	Horizontal or vertical FL	Tree algorithm	Exchanged entities	Security measure
Giacomelli et al. (2019) [15]	Horizontal	RF	Tree model	HE
Yang et al. (2019) [43]	Horizontal	XGBoost (GBDT)	G & H	K-Anon
Liu et al. (2019) [26]	Horizontal	XGBoost (GBDT)	G & H	SS
Li et al. (2020) [23]	Horizontal	XGBoost (GBDT)	G & H	LSH
Truex et al. (2018) [38]	Horizontal	RF	Counts	DP & SMC
Sjöberg et al. (2019) [35]	Horizontal	Deep neural decision forest	Model params	FedAvg
Peltari (2020) [32]	Horizontal	RF	Tree model	DP
Liu et al. (2020) [25]	Horizontal	Extra trees	Splits	Local DP
Souza et al. (2020) [8]	Horizontal	RF	Tree model	Blockchain
Yamamoto et al. (2020) [42]	Horizontal	XGBoost (GBDT)	G & H	Encryption
Wang et al. (2020) [39]	Horizontal	XGBoost (GBDT)	G & H	Encryption
Ong et al. (2020) [31]	Horizontal	XGBoost (GBDT)	G & H and Splits	Hist Approx.
Cheng et al. (2019) [6]	Vertical	XGBoost (GBDT)	G & H	HE
Liu et al. (2019) [25]	Vertical	RF	Tree nodes	HE
Feng et al. (2019) [12]	Vertical	LightGBM (GBDT)	Encrypted data	HE
Wu et al. (2020) [40]	Vertical	RF & GBDT	G & H	SMC + HE
Zhang et al. (2020) [44]	Vertical	XGBoost (GBDT)	Splits	Encryption
Fang et al. (2020) [11]	Vertical	XGBoost (GBDT)	G & H and Splits	HE + SS
Leung et al. (2020) [22]	Vertical	XGBoost (GBDT)	Encrypted data	Sec. Enclave + Obliviousness
Xie et al. (2021) [46]	Vertical	XGBoost (GBDT)	G & H and Splits	SS
Chen et al. (2021) [50]	Vertical	RF & GBDT	Tree model	HE
Tian et al. (2020) [37]	Horiz. & Vert.	GBDT	G & H	Sec Agg + DP

2.2.1 Horizontal vs. Vertical FL

One of the most important distinctions in Federated Learning solutions is based on whether they can be used *vertically* or *horizontally*. Federated Learning algorithms can be categorized into two different types, *horizontal* or *vertical* Federated Learning, depending on what dimensions are common across the participants within the training of the model. In horizontal Federated Learning, parties share the same set of features, while in vertical Federated Learning, parties share the same set of data sample identifiers. Depending on the data that is structured across the party, the resulting communication topology of the Federated Learning system as well as the method and type of information exchanged can differ greatly. In Chap. 1, a formal definition of vertical and horizontal is introduced in detail.

In the literature for tree-based Federated Learning systems, the majority of approaches are based on horizontal data partitions. Relatively, very few methods such as SecureBoost [6], S-XGB and HESS-XGB [11], SecureGBM [12], and others consider the vertical partition of the data. In most of the vertical-based Federated Learning systems, feature alignment must be performed in some fashion to perform the learning process. However, given the underlying structure of the tree algorithm is based on finding the optimal feature to split on, the task of finding the most optimal feature to split on for tree-based algorithms in a vertical setting is a crucially important problem, as the representation of the model is dependent on the availability of such feature existing in the local party's dataset.

2.2.2 Tree-Based Algorithm Types in Federated Learning

As shown in Table 2.1, the majority of algorithms implemented in Federated Learning are based on Gradient Boosted Decision Trees (GBDT), using either an optimized variant of the original algorithm such as XGBoost [4] or LightGBM [20]. XGBoost or eXtreeme Gradient Boosted Decision Trees is a variant of GBDT where the algorithm implements various different optimizations including histogram approximations using a weighted quantile sketch method and the use of Taylor-based approximations to compute approximate the loss computations [4]. On the other hand, LightGBM is another implementation of GBDT where boosting is performed at a leaf-wise tree growth and handles categorical features much more efficiently than the implementation of XGBoost [20]. Other models such as Random Forest (RF) are also another commonly implemented architecture, based on the bagging techniques implemented within tree-based models for Federated Learning. In the literature, there exist other alternative learning algorithms for tree-based models such as Deep Neural Decision Forests [35] and Extra Trees [25], which have also been proposed in the state-of-the-art methods for tree-based algorithms for Federated Learning. With constraints placed over how information is distributed among different parties, the trend for alternative tree-based algorithms is being explored as methods for learning a tree-based structure.

2.2.3   Handling Security Requirements for Tree-Based Federated Learning

Applying Federated Learning under different scenarios may require substantially different protection mechanisms. For example, consider a multi-cloud environment where all the parties are owned by the same company and the data used for model training is not sensitive. In this case, there is no need for stringent security measures. We call these types of scenarios *Trusted Federations*. However, for federations where parties are embodied by competitors or where the training data include highly sensitive information, additional protections are needed. We refer to these types of scenarios as *Protected Federations*.

Trusted Federations: Several approaches have been proposed to address the needs of these types of federations. This paradigm involves the use of dimensionality reduction-based approaches that reduces the overall fidelity of the data to prevent other parties from having an actual view of the raw data—instead having some surrogate representation or approximation of the raw data. Examples of approaches that employ this method include: Li et al. [23] that implements a Locality Sensitivity Hashing (LSH) method, Yang et al. [43] proposed a clustering-based k-anonymity scheme, and Ong et al. [31] that utilizes a party-adaptive histogram approximation mechanism, which will be further introduced in later sections of this chapter. This form of data obfuscation is slightly different from the additive form of statistical methods such as differential privacy where noise is added to the data; however, it follows a very similar principle for hiding the raw data distribution from potential adversaries.

Protected Federations: To protect data within a Federated Learning context to address issues concerning data privacy and security, various methods have been proposed to prevent adversaries from obtaining direct unauthorized access to the raw data distribution of those participating in the joint training of a machine learning model or inference of private data. The two predominant paradigms in Federated Learning are based on *statistical methods* and/or *encryption-based methods*. An in-depth evaluation of these different threat models and their mitigation methods is presented in Chap. 14.

Statistical methods of privacy protection employ techniques such as k-anonymity [36] and differential privacy (DP) [10] that are two commonly employed techniques for tree-based Federated Learning methods. Examples of methods that employ these schemes include Yang et al. [43], Truex et al. [38], Peltari [32], Liu et al. [25], and Tian et al. [37]. These methods define some statistical measure on privacy guarantees with an upper-bound error margin for some defined privacy budget.

Although these methods do not require as much computational overhead compared to encryption-based methods, the major disadvantage of these schemes primarily degrades the overall accuracy performance of the machine learning task at hand due to additional additive errors introduced into the data.

On the other hand, cryptographic-based approaches utilize encryption as a mechanism to perform data operations directly on encrypted data sources through methods such as homomorphic encryption (HE), Secret Sharing (SS), and Secure Multi-Party Computation (SMC).[2] Examples of encryption-based security protocols include: SecureBoost [6], FedXGB [26], and HESS-XGB [11], which employ homomorphic encryption. Alternatively, methods such as S-XGB [11] and PrivColl [22] use Secret Sharing. Here, the major trade-off to take into consideration is the additional computational and network communication overhead incurred during the process of encryption and decryption of data, which increases the overall runtime necessary for training a model, in exchange for near lossless data being transmitted, which translates to better model performance.

2.2.4   Implementations of Tree-Based Models in FL

Compared to the state-of-the-art methods in linear models, kernel-based methods, and deep neural-network-based methods, we find that there are a relatively small number of real-world implementations of tree-based methods for Federated Learning that are readily available as open source solutions and/or commercially available in the market today. As of writing, the current Federated Learning frameworks that provide implementations of Gradient Boosted Trees include FATE[3] [6] by WeBank and Secure XGBoost from UC Berkley's MC2[4] framework. However, as far as robust product offering in the marketplace, IBM Federated Learning [27] offers an enterprise-grade solution for implementing and deploying a Federated Learning version of XGBoost [31] in production.

2.3   Preliminaries on Decision Trees and Gradient Boosting

In this section, we provide the readers with the standard notation, especially on the setup of the FL system used throughout this chapter and the sufficient background necessary for the discussion on implementing ID3 Decision Trees and XGBoost algorithms in Federated Learning. We refer the reader to additional references when necessary for further information behind each algorithm.

---

[2] For an in-depth look at the various cryptographic techniques, see Chap. 14

[3] https://fate.fedai.org/

[4] https://github.com/mc2-project/secure-xgboost

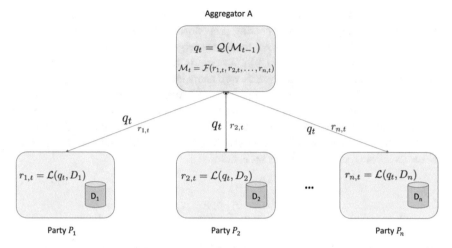

Fig. 2.2 The Federated Learning system architecture

2.3.1 The Federated Learning System

We define our Federated Learning system based on the setup from the Introductory chapter, where we consider a FL system with n parties, $\mathcal{P} = \{P_1, P_2, \ldots, P_n\}$, with a disjoint dataset $\mathcal{D}_1, \mathcal{D}_2, \ldots, \mathcal{D}_n$ each sharing the same number of features (*i.e., horizontal FL*), m, and an aggregator $\mathcal{A}$, which orchestrates the training procedure. Figure 2.2 demonstrates the high-level architecture of our Federated Learning setup, where for each training round, the aggregator issues a query Q to the available parties in the system, and a party $\mathcal{P}_i$ replies to that query with r_i based on a computed value from its local dataset $\mathcal{D}_i$ and replies back to the aggregator. The aggregator then collects the responses from each party and fuses the responses to update the global ML model, defined by $M = \mathcal{F}(r_1, r_2, \ldots, r_n)$, which resides on the aggregator side. This process gets performed for multiple iterations until some termination criteria, such as model convergence or a user-defined heuristic, have been met.

2.3.2 Preliminaries on Centralized ID3 Models

In this section, we briefly introduce the **ID3 (Iterative Dichotomiser)** decision tree algorithm, an algorithm that generates a classification decision tree structure as its output. We refer readers who wish to understand the algorithm further to Quinlan's original paper [33].

The fundamental process behind training a decision tree-based algorithm entails the following steps: (1) determining the best feature to which we partition the training data into, (2) performing the split of the training data into subsets based on the chosen feature, and (3) repeating procedures (1) and (2) repeatedly for each subset of the data until the subsets have converged to a pre-determined level of uniformity based on a target variable. To determine the "best" feature to perform the split on, the ID3 [33] algorithm, as well as its later variants, C4.5 [34] and C5.0, maximizes on *information gain* to determine the optimal feature to split on. For a given feature candidate f, the information gain metric can measure the difference between the *entropy* of the current data against the weighted sum of the entropy values for each data subset if f were used as split feature criteria. Given our dataset, $\mathcal{D}$ (or its subset), entropy for that set is defined as

$$E(\mathcal{D}) = \sum_{i=1}^{|C|} p_i \log_2 (p_i),$$

where C is the set of target class values and p_i is the probability that a given data sample instance from $\mathcal{D}$ is of a given ith class. Thus, to determine the best feature to select requires identifying class-wise probabilities, which can be determined through count values.

Concretely, the computational steps for ID3, to a dataset, $\mathcal{D}$, starts from the root node of the tree. On each iteration of the algorithm, the method computes the given entropy for the data for each feature. It then determines the "best" feature to split the data on using the computed entropy metric and subsequently splits the data to produce a subset of the dataset $\mathcal{D}$. This split generates a node in the tree. This procedure is then recursively performed for every new child leaf node of the tree.

This recursive procedure terminates when: (i) every element in the remaining subset of $\mathcal{D}$ is of the same class attribute; (ii) there are no more attributes to select from and the data samples do not belong in the same class attribute; (iii) there are no samples in the subset of $\mathcal{D}$. This occurs when there are no class attributes within the parent set that match the current selected feature attribute, hence generating a leaf node with the most common class attribute in that given data subset.

2.3.3   Preliminaries on Gradient Boosting

In this section, we now briefly introduce the **eXtreme Gradient Boosted (XGBoost)** algorithm, which is based on a highly optimized and efficient variant of the Gradient Boosted Decision Tree defined by Friedman et al. [13]. We refer the readers to [4] for further details.

Given a dataset $\mathcal{D}$ with n samples and m features, $\mathcal{D} = \{(x_i, y_i)\}_{i=1}^{n}$, where $x_i \in \mathbb{R}^m$ and $y_i \in \mathbb{R}$, the predictions output from the XGBoost model, $\hat{y}_i$, are defined as an additive tree-based ensemble model, $\phi(x_i)$, comprising of K additive functions, f_k, defined as

$$\hat{y}_i = \phi(x_i) = \sum_{k=1}^{K} f_k(x_i), \ f_k \in \mathcal{F},$$

where $\mathcal{F} = \{f(x) = w_{q(x)}\}$ is a collection of Classification and Regression Trees (CART), such that the function $q(x)$ maps each input feature x to one of T leaves in the tree by a weight vector, $w \in \mathbb{R}^T$.

Given the defined model prediction above, the XGBoost algorithm minimizes the following regularized loss function:

$$\tilde{\mathcal{L}} = \sum_i l(y_i, \hat{y}_i) + \sum_k \Omega(f_k),$$

where $l(y_i, \hat{y}_i)$ is the loss function of the ith sample between the prediction $\hat{y}_i$ and the target value y_i, and $\Omega(f_k) = \gamma T + \frac{1}{2}\lambda\|w\|^2$ is the regularization component. This component discourages each kth tree, f_k, from over-fitting through hyperparameters λ, the regularization parameter penalizing the weight vector w, and γ, a term penalizing the tree from growing additional leaves.

To approximate the loss function, a second-order Taylor expansion function is used, as defined by

$$\mathcal{L}^{(t)} \simeq \sum_{i=1}^{n} [l(y_i, \hat{y}_i^{(t-1)}) + g_i f_t(x_i) + \frac{1}{2}h_i f_t^2(x_i)] + \Omega(f_t).$$

As the tree is trained in a recursively additive manner, each iteration index of the training process is denoted as t; hence, $\mathcal{L}^{(t)}$ denotes the tth loss of the training process. Here, we also define the gradient and the second-order gradient, or the Hessian, respectively, as follows:

$$g_i = \partial_{\hat{y}_i^{(t-1)}} l(y_i, \hat{y}_i^{(t-1)})$$

$$h_i = \partial^2_{\hat{y}_i^{(t-1)}} l(y_i, \hat{y}_i^{(t-1)}).$$

Given the derived gradients and Hessians for a given $q(x)$, we can compute the optimal weights of leaf j using

$$w_j^* = -\frac{G_j}{H_j + \lambda},$$

where $G_j = \sum_{i \in I_j} g_i$ and $H_j = \sum_{i \in I_j} h_i$ are the total summation of the gradients and Hessians for each of the specific data sample indices, I_j, respectively. To efficiently compute the optimal weights w_j^*, we can greedily maximize the gain score to search for best partition value for a leaf node at each iteration efficiently. This gain score is defined as follows:

$$Gain = \frac{1}{2}\left[\frac{G_L^2}{H_L + \lambda} + \frac{G_R^2}{H_R + \lambda} - \frac{(G_L + G_R)^2}{(H_L + H_R) + \lambda}\right] - \gamma.$$

Here, L and R correspondingly consider the sum of the gradients and Hessians based on the specific index of the left and right children of the given leaf node, I_L and I_R, respectively.

In the next two sections of this chapter, we introduce two of the tree-based algorithms, ID3 and XGBoost, now implemented within a Federated Learning system. In particular, we demonstrate how each of the different security techniques is implemented for the two different Federated Learning scenarios that are introduced in Sect. 2.2.3.

2.4   Decision Trees for Federated Learning

In this section, we introduce an ID3 Decision Tree implementation for a Federated Learning system, as proposed by Truex et al. [38]. This implementation combines differential privacy and secure Multi-Party Computation (SMC) as a hybrid approach for privacy-preserving tree-based model, providing robust formal privacy guarantees through two combined approaches. The pseudocode implementation is outlined in Algorithm 2.1.

> We also present some of the optional steps, which include additional security measures and different techniques, as indicated in a gray box as such.

As described previously in the Preliminaries Section (Sect. 2.3.2), to obtain a decision tree structure from the data for the ID3 algorithm entails determining the best feature to perform the partition of the data. In the case of a centralized setting, this would be trivial as one can easily obtain the necessary count statistics from a single data distribution source. However, when the data source is fragmented and distributed across different parties with further privacy limitations imposed, the method for obtaining the statistics necessary for growing the decision tree becomes non-trivial. This raises both challenges and opportunities for how we can effectively obtain these counts to find the best partition while respecting privacy limitations of the local parties' data distribution.

To implement this within a Federated-Learning-based setting, we would need to decompose the process of training a tree-based model and assign those respective processes to the aggregator and the local parties. These two processes are: (i) obtaining the raw data distribution statistics and (ii) fusing together the statistics to derive the optimal feature partition of the decision tree. This pattern of decomposing a centralized implementation of the algorithm into different components and assigning specific routines to the aggregator and the party is a recurring theme found in both the ID3 Decision Tree algorithm and the Gradient Boosted Decision Tree algorithm, as we will later demonstrate.

At a high level, this entails the aggregator querying each of the individual parties for its statistical data, to which each of the respective party responds and collected by the aggregator. The aggregator then takes this data and fuses the statistics into a single statistic, where it uses this fused statistic to perform a branch partition of the model. It then repeats these two processes until some termination criteria have been met.

> During the response process of these queries from each individual party, either of the two (or both) privacy-preserving techniques may be applied to ensure further guarantees of privacy.

We now describe the ID3 decision tree algorithm as defined in Algorithm 2.1. In Sect. 2.3.2, we have outlined the three key steps in order to learn a decision tree structure for the ID3 decision tree algorithm. First, we defined the number of colluding parties (line 1), $\bar{t}$, our evenly allocated privacy budget for determining the counts, assigned to ϵ_1, (line 2), respectively, and finally our root node (line 3). To privately train a decision tree model in our proposed Federated Learning system, we determine the best possible feature to perform our split on, such that we can maximize the total information gain. For this, we first query each of the parties for their respective counts, which is assigned to f (line 10).

> Differentially private noise and/or encryption techniques such as threshold homomorphic encryption may be applied on the raw data distribution during every query transaction that occurs between a party and the aggregator. If a trusted federation scenario is in place, this step may be omitted.

We determine whether or not the aggregate values of the counts fall into any of the criteria for generating a leaf node (line 11). Based on the class counts, we perform another query process to determine which of the class the leaf should be classified as (lines 12–14). Otherwise, we specify a new privacy budget based on the initial privacy budget, ϵ_1, divided by 2 times the size of the feature of the dataset (line 16). We subsequently obtain the counts for each attribute set and correspondingly compute the entropy value, V_F for that given feature (lines 17–

Algorithm 2.1 Private ID3 Decision Tree—Federated Learning

Input: $\mathcal{D}$, Input Dataset; A, Aggregator; t, Minimum Number of Honest, Non-Colluding Parties; ϵ, Privacy Guarantee; $\mathcal{F}$, Attribute Set; C, Class Attribute; d, Max Tree Depth; pk, Public Key

Output: $\mathcal{M}$, Trained Global ID3 Decision Tree Model

1:   $\bar{t} \leftarrow n - t + 1$
2:   $\epsilon_1 \leftarrow \frac{\epsilon}{2(d+1)}$
3:   Define current splits, $S \leftarrow \emptyset$, for root node
4:   $\mathcal{M} \leftarrow$ BuildTree(S, $\mathcal{D}$, t, ϵ_1, $\mathcal{F}$, C, d, pk)
5:   **return** $\mathcal{M}$
6:
7:   **Function** *BuildTree*(S, $\mathcal{D}$, t, ϵ_1, $\mathcal{F}$, C, d, pk):
8:     $f \leftarrow max_{F \in \mathcal{F}} |F|$
9:     Asynchronously query $\mathcal{P}$: *counts*(S, ϵ_1, t)
10:    $N \leftarrow$ decrypted aggregate of noisy counts
11:    **if** $\mathcal{F} \leftarrow \emptyset$ or $d \leftarrow 0$ or $\frac{N}{f|C|} < \frac{\sqrt{2}}{\epsilon_1}$ **then**
12:      Asynchronously query $\mathcal{P}$: *class_counts*(S, ϵ_1, t)
13:      $N_c \leftarrow$ vector of decrypted, noisy class counts
14:      **return** node labeled with $argmax_c N_c$
15:    **else**
16:      $\epsilon_2 \leftarrow \frac{\epsilon_1}{2|\mathcal{F}|}$
17:      **for** $F \in \mathcal{F}$ **do**
18:        **for** $f_i \in F$ **do**
19:          Update set of split values to send to child node: $S_i \leftarrow S + \{F = f_i\}$
20:          Asynchronously query $\mathcal{P}$: *counts*(S_i, ϵ_2, t) & *class_counts*(S_i, ϵ_2, t)
21:          $N'^F_i \leftarrow$ aggregate of *counts*
22:          $N'^F_{i,c} \leftarrow$ element-wise aggregate of *class_counts*
23:          Recover N^F_i from $\bar{t}$ partial decryptions of N'^F_i
24:          Recover $N^F_{i,c}$ from $\bar{t}$ partial decryptions of $N'^F_{i,c}$
25:        **end for**
26:        $V_F \leftarrow \sum_{i=1}^{|F|} \sum_{c=1}^{|C|} N^F_{i,c} \cdot log \frac{N^F_{i,c}}{N^F_i}$
27:      **end for**
28:      $\bar{F} \leftarrow argmax_F V_F$
29:      Create root node M with label $\bar{F}$
30:      **for** $f_i \in \bar{F}$ **do**
31:        $S_i \leftarrow S + \{F = f_i\}$
32:        $M_i \leftarrow BuildTree(S_i, \mathcal{P}, t, \epsilon_1, \mathcal{F} \setminus \bar{F}, C, d, pk)$
33:        Set M_i as child of M with edge f_i
34:      **end for**
35:      **return** $\mathcal{M}$
36:    **end if**
37: **end**

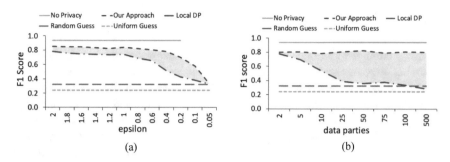

Fig. 2.3 Comparison of local models versus Federated-Learning-based implementation of decision trees. (**a**) Comparison over ranges of epsilon. (**b**) Comparison over the number of parties

27). Prior to computing the entropy values for each feature, we then determine the best split point of the tree (line 26) and update our model M by creating a root node with label $\bar{F}$ (line 28).

We then split the data subset correspondingly (line 31) and perform the same operation on that given data subset recursively (line 32). We perform this process until the termination criteria, defined in Sect. 2.3.2, have been met. Alternatively, within a Federated Learning setting, the algorithm can also terminate when the counts are small relative to the degree of noise that prevents the process of obtaining any useful information, which is evaluated during the step in line 11. This yields the final model, M, which is used as the final global model trained within the Federated Learning system.

We now briefly demonstrate some performance evaluations from Truex et al.'s [38] paper. In this experimental setup, the Nursery Dataset from the UCI Machine Learning repository [51] was used to conduct experiments on various different comparisons of the ID3 Decision Tree algorithm. In particular, the comparison was focused primarily between training a decision tree model locally versus within a Federated Learning setting. As shown in Fig. 2.3, the results in both plots demonstrate the efficacy of training a Federated-Learning-based ID3 Decision Tree (outlined in blue) is better than that of the local DP approaches (i.e., training model independently from a federated setting). In particular, for the comparison of using different values of ϵ as shown in Fig. 2.3a, we see that despite decreasing the overall privacy budget, the Federated Learning approach still outperforms the local DP method consistently. In Fig. 2.3b, we see that as we increase the number of parties, the Federated Learning model's performance is consistent despite the varying number of samples per party. However for the local DP setting, the F1 score decreases as we increase the number of parties since the overall number of samples per party decreases, therefore reducing the performance of the models themselves. For further details on the individual experimental setup and evaluation, we refer readers to the original paper.

2.5 XGBoost for Federated Learning

In this section, we introduce a XGBoost-based implementation within a Federated Learning system, also known as the *Party-Adaptive XGBoost* (PAX), proposed by Ong et al. [31]. As noted in Sect. 2.3.3, one of the major challenges in training a gradient boosted decision tree is to find the optimal feature and value to split using the computed gain score. Similar to the method of decomposing the routines of the algorithm as described in the previous section, the same principles of the aggregator querying each party for its distribution statistics, fusing those statistics, and finding the optimal partition based on those fused statistics hold for the Gradient Boosted Decision Tree. In this case, instead of simple data count statistics, we deal with histograms of data distributions instead.

Optimized GBDT methods such as XGBoost [4] and LightGBM [20] utilize a *quantile-based approximation* to efficiently reduce the overall search space of the split finding process by approximating the raw data distribution as a surrogate histogram representation. Empirically, it has been shown that quantile approximations work just as well as the exact greedy solutions [4, 20, 21]. This method for data approximations can serve as a semi-secure approach for training Gradient Boosted Decision Trees under a *Trusted Federations* security policy. By quantizing or reducing the resolution of the raw distribution, we can effectively generate a surrogate representation of the raw data containing relatively lower fidelity of information than the original data distribution. Therefore, this does not directly reveal the raw data distribution of the original data source directly.

> This method can further be supplemented with additional layers of security such as adding differential private noise and/or applying encryption techniques to satisfy the *Protected Federations* scenario of Federated Learning.

Many methods for building distributed quantile sketches exist, including GKMethod [16] and its extended variants [45]. However, each comes with its own trade-offs in performance, speed, and reconstruction accuracy. For the proposed method of PAX, this algorithm implements the Distributed Distribution Sketch or *DDSketch* [28], an efficient and robust method that constructs highly accurate quantile sketch approximations of data distributions with the ability to merge multiple quantile sketches together. Furthermore, another advantage of using *DDSketch* enables training of XGBoost where parties are able to join during intermediate steps of the boosting process when a new party joins the federation. Due to its ability to merge quantile sketch histograms efficiently and accurately, this method enables dynamic adaptations to new data distributions in the data as new parties join the federation.

Just as privacy policies for encryption and differential privacy methods exist, approximation-based methods for fidelity reduction policies can be employed. To determine the most optimal approximation error bounds or thresholds for the quantile sketch process, or effectively the bin size of the histogram approximation for the data, various heuristics can be applied to mitigate against potential data leaks. As opposed to equally approximating the data distribution in the same manner, PAX considers the relative sample size of each party's data with respect to those of the other parties' sample sizes to determine the most optimal bin size for discretizing the data. This is achieved by defining an approximation error parameter ϵ,[5] used by XGBoost for *each individual party*. Intuitively, the inverse of the ϵ parameter, or $1/\epsilon$, is roughly equivalent to the *bin size of the histogram*. Hence, using an approximated representation tailored within the context of the party and the other participants in the network can help to limit any "raw" data leakage to a defined bound.

To train a gradient boosted decision tree within a FL setting, the aggregator first initializes a global null model, $f_\emptyset^{(\mathcal{A})}$ (line 1). The aggregator also defines the global hyperparameter $\epsilon^{(A)}$, which denotes the error tolerance budget for the training process for XGBoost. This parameter sets the upper-bound histogram approximation error and equivalently the number of maximum bins used in training.

Given the defined global $\epsilon^{(A)}$ parameter, we then determine the appropriate policies for how we construct the local party's surrogate histogram by computing the party's local ϵ_i parameter (line 2). The aggregator first queries each party for their dataset size, $|d_i|$, and maintains a list of counts, which we denote here as S. For each ith party, we compute the local party's ϵ_i parameter (line 36):

$$\epsilon_i = \epsilon^{(A)} \left(\frac{|d_i|}{\sum_{d \in \mathcal{D}} |d|} \right).$$

Equivalently, rewriting the inverse term of ϵ_i can give us the corresponding number of bins that each ith party will use to construct their surrogate histogram representation of the local dataset:

$$\frac{1}{\epsilon_i} = \frac{1}{\epsilon^{(A)} \left(\frac{|d_i|}{\sum_{d \in \mathcal{D}} |d|} \right)} = \frac{\sum_{d \in \mathcal{D}} |d|}{\epsilon^{(A)} |d_i|}.$$

The aggregator replies to each party with their respective ϵ_i parameter and correspondingly assigns the local party epsilon (lines 5–6). For this purpose, the party uses the given ϵ_i and their respective local data distribution, $\mathcal{D}^{(p_i)}$, to compute a surrogate histogram representation of the local party's data distribution, $\tilde{\mathcal{D}}^{(p_i)}$ (line 7). Each party then subsequently constructs their own surrogate histogram

---

[5] Note: While ϵ is used here as a notation for the histogram approximation error, this term does not collude with the notation used in differential privacy (DP).

Algorithm 2.2 Party-Adaptive XGBoost (PAX)

Input: $\mathcal{D}$, Input Dataset; A, Aggregator; P, Participating Parties in FL Training; $\epsilon^{(\mathcal{A})}$, Global Error Tolerance Budget; T, Maximum Number of Training Rounds; l, Model Loss Function

Output: $f^{(\mathcal{A})}$, Trained Global XGBoost Model

1:   $\mathcal{A}$ Initialize Global Null Model: $f_{\emptyset}^{(A)} \leftarrow 0$

2:   $\{\epsilon_1, \ldots, \epsilon_{|\mathcal{P}|}\} \leftarrow compute\_local\_epsilon(\epsilon^{(\mathcal{A})})$

3:

4:   **for** $i = 1, \ldots, |\mathcal{P}|$ **do**

5:      $\mathcal{A}$ Transmits ϵ_i to p_i

6:      Assign Local Epsilon ϵ_i to p_i

7:      $\tilde{\mathcal{D}}^{(p_i)} \leftarrow compute\_histogram(\mathcal{D}^{(p_i)}, \epsilon_i)$

8:      p_i Transmits $\tilde{\mathcal{D}}_X^{(p_i)}$ to $\mathcal{A}$

9:      $\bar{\mathcal{D}}^{(\mathcal{A})} \leftarrow \bar{\mathcal{D}}^{(\mathcal{A})} \cup \tilde{\mathcal{D}}_X^{(p_i)}$

10:   **end for**

11:

12:   **repeat**

13:      $(\bar{\mathcal{D}}^{(\mathcal{A})}, G^{(\mathcal{A})}, H^{(\mathcal{A})}) \leftarrow (\emptyset, \emptyset, \emptyset)$

14:      **for** $i = 1, \ldots, |\mathcal{P}|$ **do**

15:         $\mathcal{A}$ Transmits $f_t^{(\mathcal{A})}$ to Party: $f_t^{(p_i)} \leftarrow f_t^{(\mathcal{A})}$

16:         p_i Generate Predictions: $\hat{y}_t^{(p_i)} = f_t^{(p_i)}(\tilde{\mathcal{D}}_X^{(p_i)})$

17:         p_i Computes $g^{(p_i)}$ and $h^{(p_i)}$

18:         p_i Transmits $g^{(p_i)}$ and $h^{(p_i)}$ to $\mathcal{A}$

19:         $G^{(\mathcal{A})} \leftarrow G^{(\mathcal{A})} \cup g^{(p_i)}$

20:         $H^{(\mathcal{A})} \leftarrow H^{(\mathcal{A})} \cup h^{(p_i)}$

21:      **end for**

22:      $\bar{\mathcal{D}}_m^{(\mathcal{A})}, G_m^{(\mathcal{A})}, H_m^{(\mathcal{A})} \leftarrow merge\_hist(\bar{\mathcal{D}}^{(\mathcal{A})}, G^{(\mathcal{A})}, H^{(\mathcal{A})}, \epsilon_m^{(\mathcal{A})})$

23:      $f_t^{(\mathcal{A})} \leftarrow grow\_tree(\bar{\mathcal{D}}_m^{(\mathcal{A})}, G_m^{(\mathcal{A})}, H_m^{(\mathcal{A})})$

24:   **until** $t \leq T$ or other termination criteria.

25:

26:   **Function** $compute\_local\_epsilon(\epsilon^{(\mathcal{A})})$:

27:      $S \leftarrow \emptyset$

28:      **for** $p_i \in \mathcal{P}$ **do**

29:         $\mathcal{A}$ Queries Data Count: $S \leftarrow S \cup |\mathcal{D}^{(p_i)}|$

30:         $E \leftarrow \emptyset$

31:         **for** $i = 1, \ldots, |\mathcal{P}|$ **do**

32:            Compute i^{th} Party ϵ: $\epsilon_i \leftarrow \epsilon^{(A)} \left(\frac{s_i}{\sum_{q \in S} q} \right)$

33:            $E \leftarrow E \cup \epsilon_i$

34:         **end for**

35:      **end for**

36:   **end**

representation, $\mathcal{D}^{(p_i)}$, of the i-th local party's p_i raw data distribution and transmits the resulting sketch to the aggregator, where the merge process can fuse together the entire federations' distributions as a single view of the whole dataset for the model to train. Note that this process can occur once before the training starts with the initial set of parties or alternatively as new parties join the federation during intermediate phases of the training process.

After computing the surrogate histogram representation of the data, we then initiate the iterative Federated Learning process. First, the aggregator $\mathcal{A}$ transmits their global model, $f_t^{(\mathcal{A})}$ to each party, p_i, which is assigned to each party's respective local model $f_t^{(p_i)}$ (line 13). We then evaluate $f_t^{(p_i)}$ on $\tilde{\mathcal{D}}_X^{(p_i)}$ to obtain the model's predictions, $\hat{y}_t^{(p_i)}$ (line 14). Afterward, given the predictions, we compute the loss function that is used to compute the gradient, $g^{(p_i)}$, and Hessians, $h^{(p_i)}$, for each of the corresponding surrogate input feature-value split candidates (line 15). Gradient and Hessian statistics that fall under a certain bin interval are grouped together within their respective value buckets [4]. The gradient and Hessian for each party are then sent back as replies to the aggregator and then collected until some quorum criterion has been met (line 16). Given the collected results from each party, we perform a fusion operation to merge the final histogram representation used toward boosting the decision tree model, as formulated in the method of DDSketch [28].

The final ϵ value based on this heuristic is denoted as $\epsilon_m^{(\mathcal{A})}$, where m denotes the variables pertaining to the merging process. Using the derived $\epsilon_m^{(\mathcal{A})}$, we can utilize the existing methods such as [5] and [1] for implementing merging histogram routines with error guarantees based on some defined error parameter. This produces the final outputs for the merged gradients, $G_m^{(\mathcal{A})}$, Hessians, $H_m^{(\mathcal{A})}$, and feature-value split candidates, $\tilde{\mathcal{D}}_m^{(\mathcal{A})}$, which are used for the boosting process (lines 25–26). With a new $f_t^{(\mathcal{A})}$ generated, we repeat our training process for T rounds, or until some other stopping criteria depending on whether early stopping or other heuristics are considered (line 27).

We now briefly demonstrate some performance evaluations from Ong et al.'s [31] paper. In this experimental setup, the airline delay causes dataset from the U.S. Department of Transportation's (DOT) Bureau of Transportation Statistics (BTS) [52]. However, a slightly preprocessed version of the dataset from Kaggle[6] was used to conduct experiments on various different comparisons of the different Federated Learning algorithms. We evaluate our model against different variants of Federated Learning algorithms including Logistic Regression and SecureBoost [6], which utilizes a form of encryption to protect user's data (Table 2.2). Based on these results, we see that the proposed XGBoost method outperforms both the Logistic Regression model and the SecureBoost model. Further descriptions on the experimental setup and analysis can be found in the corresponding paper.

---

[6] https://www.kaggle.com/giovamata/airlinedelaycauses

Table 2.2 Results for FL XGBoost on Testing Samples from the Airline Dataset

	Airline (Random)					Airline (Balanced)				
Model	ACC	PRE	REC	AUC	F1	ACC	PRE	REC	AUC	F1
PAX (Ours)	**0.88**	**0.88**	**0.88**	**0.87**	**0.88**	**0.87**	**0.85**	**0.90**	**0.87**	**0.88**
Homo SecureBoost	0.74	0.71	0.82	0.81	0.72	0.81	0.84	0.79	0.86	0.81
Logistic regression	0.58	0.72	0.49	0.58	0.45	0.58	0.72	0.49	0.58	0.45

2.6 Open Problems and Future Research Directions

In this section, we outline some of the potential research directions that stem from the development of tree-based models for Federated Learning. Here, we specifically look at various new problems and open research challenges that emerged as a result of the development of these new class of models for Federated Learning. Key challenges include data representation policies for reduction-based data obfuscation techniques, mitigation of bias and fairness, and considering training methodologies for alternative Federated Learning network topologies.

2.6.1 Data Fidelity Threshold Policies

In the previous section, we introduced one possible policy that can be employed to derive party-specific approximation thresholds based on some underlying statistics of the individual party's dataset. However, there exist many other policies and methods for finding the most optimal approximation thresholds based on different design objectives or mitigation against specific types of attacks that an adversary might deploy to target information about the party's raw data distribution. We consider the risks and trade-offs (*in particular model performance*) that must be accounted for when training a model depending on the different security scenarios, as described in Sect. 2.2.3.

Another dimension to evaluate these surrogate representation methods is to juxtapose their respective risks, trade-offs, and error bounds against additive obfuscation techniques such as differential privacy. For methods such as quantile sketch and clustering-based approaches, these methods attempt to obfuscate the raw data distribution using reductive techniques to reduce or limit the amount of information present in the dataset. This presents researchers to consider alternative methods for how to jointly optimize these additive and reductive techniques for data obfuscation methods toward building a robust statistical approach for better data privacy protection.

2.6.2   Fairness and Bias Mitigation Methods for Tree-Based FL Models

One of the key benefits of Federated Learning enables the amalgamation of disparate data sources to come together and jointly train models. This, however, introduces new challenges and problems as parties of different data sources often contribute widely different data distribution characteristics and statistical properties that may significantly differ from one party to another. In particular, the work of Abbay et al. [47] explores these different challenges in the context of Federated Learning through different bias mitigation techniques and fairness evaluation metrics to evaluate the effects of different models and biased datasets. Hence for tree-based methods, we can study the various effects of different bias mitigation techniques such as local reweighing, global reweighing, and in-processing methods that impose a regularization component [48] to discourage the model from deriving biased decisions in its predictions. Future work in this space would aim to investigate the full effects of bias of tree-based models in a Federated Learning setting and correspondingly devising methods to resolve such issues from emerging during training.

2.6.3   Training Tree-Based FL Models on Alternative Network Topologies

In this chapter, we introduced a tree-based model training algorithm based on a single-point aggregator topology orchestrating the overall training procedure. However, there exist other topologies for Federated Learning, such as a tier-based Federated Learning system [49], which divides the corresponding parties of a Federation to be divided among different groups based on their training performance to mitigate against stragglers during a model training process in Federated Learning. Given these newly emerging Federated Learning network topologies, devising new ways to also train models for these new architectures will be another possible direction of research areas to explore. Specifically, since one of the main challenges in tree-based models entails figuring out how to perform fusion across different party sources, a tier-based topology presents us with a non-trivial challenge of determining how to perform fusion in a hierarchical manner and determine what information to propagate and at what frequency to pass information across the different hierarchies of the network.

2.7   Conclusion

In this chapter, we introduced an emerging paradigm of tree-based models for Federated Learning systems. We outlined some of the major challenges of implementing such models within a Federated Learning setting and also evaluated how these models complement, and even strengthen the core objective of Federated Learning systems through some of the key advantages tree-based models provide. Furthermore, we surveyed some of the various state-of-the-art methods and generalized some key trends and observations for the different variations in implementations. Finally, we introduced two different implementations of tree-based models for Federated Learning: the ID3 Decision Tree and XGBoost implementation with different security measures.

References

1. Blomer J, Ganis G (2015) Large-scale merging of histograms using distributed in-memory computing. J Phys Conf Ser 664:092003. IOP Publishing
2. Bonawitz K, Ivanov V, Kreuter B, Marcedone A, McMahan HB, Patel S, Ramage D, Segal A, Seth K (2017) Practical secure aggregation for privacy-preserving machine learning. In: Proceedings of the 2017 ACM SIGSAC conference on computer and communications security, pp 1175–1191
3. Breiman L (1996) Bagging predictors. Mach Learn 24(2):123–140
4. Chen T, Guestrin C (2016) XGBoost: a scalable tree boosting system. In: Proceedings of the 22nd ACM SIGKDD international conference on knowledge discovery and data mining, pp 785–794
5. Chen T, Guestrin C (2016) XGBoost: a scalable tree boosting system supplementary material
6. Cheng K, Fan T, Jin Y, Liu Y, Chen T, Yang Q (2019) SecureBoost: a lossless federated learning framework. arXiv preprint arXiv:1901.08755
7. Dang Z, Gu B, Huang H (2020) Large-scale kernel method for vertical federated learning. In: Federated learning. Springer, Cham, pp 66–80
8. de Souza LAC, Rebello GAF, Camilo GF, Guimarães LCB, Duarte OCMB (2020) DFedForest: decentralized federated forest. In: 2020 IEEE international conference on blockchain (Blockchain). IEEE, pp 90–97
9. Dimitrakopoulos GN, Vrahatis AG, Plagianakos V, Sgarbas K (2018) Pathway analysis using XGBoost classification in biomedical data. In: Proceedings of the 10th Hellenic conference on artificial intelligence, pp 1–6
10. Dwork C, McSherry F, Nissim K, Smith A (2006) Calibrating noise to sensitivity in private data analysis. In: Theory of cryptography conference. Springer, pp 265–284
11. Fang W, Chen C, Tan J, Yu C, Lu Y, Wang L, Zhou J, Alex X (2020) A hybrid-domain framework for secure gradient tree boosting. ArXiv, abs/2005.08479
12. Feng Z, Xiong H, Song C, Yang S, Zhao B, Wang L, Chen Z, Yang S, Liu L, Huan J (2019) SecureGBM: secure multi-party gradient boosting. In: 2019 IEEE international conference on Big Data (Big Data). IEEE, pp 1312–1321
13. Friedman JH (2001) Greedy function approximation: a gradient boosting machine. Ann Stat 29:1189–1232
14. Friedman JH (2002) Stochastic gradient boosting. Comput Stat Data Anal 38(4):367–378
15. Giacomelli I, Jha S, Kleiman R, Page D, Yoon K (2019) Privacy-preserving collaborative prediction using random forests. AMIA Summits Transl Sci Proc 2019:248

16. Greenwald M, Khanna S (2001) Space-efficient online computation of quantile summaries. ACM SIGMOD Rec 30(2):58–66
17. Hard A, Rao K, Mathews R, Ramaswamy S, Beaufays F, Augenstein S, Eichner H, Kiddon C, Ramage D (2018) Federated learning for mobile keyboard prediction. arXiv preprint arXiv:1811.03604
18. Hastie T, Tibshirani R, Friedman J (2009) The elements of statistical learning: data mining, inference, and prediction. Springer Science & Business Media, New York
19. Kairouz P, McMahan HB, Avent B, Bellet A, Bennis M, Bhagoji AN, Bonawitz K, Charles Z, Cormode G, Cummings R et al (2019) Advances and open problems in federated learning. arXiv preprint arXiv:1912.04977
20. Ke G, Meng Q, Finley T, Wang T, Chen W, Ma W, Ye Q, Liu T-Y (2017) LightGBM: a highly efficient gradient boosting decision tree. In: Advances in neural information processing systems, pp 3146–3154
21. Keck T (2017) FastBDT: a speed-optimized multivariate classification algorithm for the belle II experiment. Comput Softw Big Sci 1(1):2
22. Leung C (2020) Towards privacy-preserving collaborative gradient boosted decision tree learning
23. Li Q, Wen Z, He B (2020) Practical federated gradient boosting decision trees. In: Proceedings of the AAAI conference on artificial intelligence, vol 34, pp 4642–4649
24. Li S, Zhang X (2019) Research on orthopedic auxiliary classification and prediction model based on XGBoost algorithm. Neural Comput Appl 32(7):1971–1979
25. Liu Y, Liu Y, Liu Z, Liang Y, Meng C, Zhang J, Zheng Y (2020) Federated forest. IEEE Trans Big Data
26. Liu Y, Ma Z, Liu X, Ma S, Nepal S, Deng R (2019) Boosting privately: privacy-preserving federated extreme boosting for mobile crowdsensing. arXiv preprint arXiv:1907.10218
27. Ludwig H, Baracaldo N, Thomas G, Zhou Y, Anwar A, Rajamoni S, Ong Y, Radhakrishnan J, Verma A, Sinn M et al (2020) IBM federated learning: an enterprise framework white paper v0. 1. arXiv preprint arXiv:2007.10987
28. Masson C, Rim JE, Lee HK (2019) DDSketch: a fast and fully-mergeable quantile sketch with relative-error guarantees. arXiv preprint arXiv:1908.10693
29. McMahan HB, Moore E, Ramage D, Hampson S et al (2016) Communication-efficient learning of deep networks from decentralized data. arXiv preprint arXiv:1602.05629
30. Nobre J, Neves RF (2019) Combining principal component analysis, discrete wavelet transform and XGBoost to trade in the financial markets. Expert Syst Appl 125:181–194
31. Ong YJ, Zhou Y, Baracaldo N, Ludwig H (2020) Adaptive histogram-based gradient boosted trees for federated learning. arXiv preprint arXiv:2012.06670
32. Pelttari H et al (2020) Federated learning for mortality prediction in intensive care units
33. Quinlan JR (1986) Induction of decision trees. Mach Learn 1(1):81–106
34. Salzberg SL (1993, 1994) C4.5: programs for machine learning by J. Ross Quinlan. Morgan Kaufmann Publishers, Inc., San Mateo
35. Sjöberg A, Gustavsson E, Koppisetty AC, Jirstrand M (2019) Federated learning of deep neural decision forests. In: International conference on machine learning, optimization, and data science. Springer, pp 700–710
36. Sweeney L (2002) k-anonymity: a model for protecting privacy. Int J Uncertainty Fuzziness Knowl-Based Syst 10(05):557–570
37. Tian Z, Zhang R, Hou X, Liu J, Ren K (2020) FederBoost: private federated learning for GBDT. arXiv preprint arXiv:2011.02796
38. Truex S, Baracaldo N, Anwar A, Steinke T, Ludwig H, Zhang R (2018) A hybrid approach to privacy-preserving federated learning
39. Wang Z, Yang Y, Liu Y, Liu X, Gupta BB, Ma J (2020) Cloud-based federated boosting for mobile crowdsensing. arXiv preprint arXiv:2005.05304
40. Wu Y, Cai S, Xiao X, Chen G, Ooi BC (2020) Privacy preserving vertical federated learning for tree-based models. arXiv preprint arXiv:2008.06170

41. XingFen W, Xiangbin Y, Yangchun M (2018) Research on user consumption behavior prediction based on improved XGBoost algorithm. In: 2018 IEEE international conference on Big Data (Big Data). IEEE, pp 4169–4175
42. Yamamoto F, Wang L, Ozawa S (2020) New approaches to federated XGBoost learning for privacy-preserving data analysis. In: International conference on neural information processing. Springer, pp 558–569
43. Yang M, Song L, Xu J, Li C, Tan G (2019) The tradeoff between privacy and accuracy in anomaly detection using federated XGBoost. arXiv preprint arXiv:1907.07157
44. Zhang J, Zhao X, Yuan P (2020) Federated security tree algorithm for user privacy protection. J Comput Appl 40(10):2980–2985
45. Zhang Q, Wang W (2007) A fast algorithm for approximate quantiles in high speed data streams. In: 19th international conference on scientific and statistical database management (SSDBM 2007). IEEE, p 29
46. Xie L, Liu J, Lu S, Chang T-H, Shi Q (2021) An efficient learning framework for federated XGBoost using secret sharing and distributed optimization. arXiv preprint arXiv:2105.05717
47. Abay A, Zhou Y, Baracaldo N, Rajamoni S, Chuba E, Ludwig H (2020) Mitigating Bias in Federated Learning. arXiv preprint arXiv:2012.02447
48. Ravichandran S, Khurana D, Venkatesh B, Edakunni NU (2020) FairXGBoost: fairness-aware classification in XGBoost arXiv preprint arXiv:2009.01442
49. Chai Z, Ali A, Zawad S, Truex S, Anwar A, Baracaldo N, Zhou Y, Ludwig H, Yan F, Cheng Y (2020) TiFL: a tier-based federated learning system. arXiv preprint arXiv:2001.09249
50. Chen X, Zhou S, Yang K, Fan H, Feng Z, Chen Z, Wang H, Wang Y (2021) Fed-EINI: an efficient and interpretable inference framework for decision tree ensembles in federated learning. arXiv preprint arXiv:2105.09540
51. Dua D, Graff C. UCI Machine Learning Repository. School of Information and Computer Science, University of California, Irvine. http://archive.ics.uci.edu/ml
52. U.S. Department of Transportation (2009) Airline On-Time Statistics and Delay Causes. https://www.transtats.bts.gov/OT_Delay/OT_DelayCause1.asp

Chapter 3
Semantic Vectorization: Text- and Graph-Based Models

Shalisha Witherspoon, Dean Steuer, and Nirmit Desai

Abstract Semantic vector embedding techniques have proven useful in developing mathematical relationships of non-numeric data such as text. A key application enabled by such techniques is the ability to measure semantic similarity between given data samples and find similar data points via encoding comparison. State-of-the-art embedding approaches assume all data are available at a centralized location. However, in many scenarios, data are distributed across multiple edge locations and cannot be aggregated due to a variety of constraints. Hence, the applicability of state-of-the-art embedding approaches is limited to freely shared datasets, leaving out applications with sensitive or mission-critical data.

In this chapter, we address this gap by reviewing novel unsupervised algorithms for learning and applying semantic vector embeddings in a variety of distributed settings. Specifically, for scenarios where multiple edge locations can engage in joint learning, we adapt the proposed federated learning techniques for semantic vector embedding. Where joint learning is not possible, we propose novel semantic vector translation algorithms to enable semantic query across multiple edge locations, each with its own semantic vector space. Experimental results on natural language as well as graph datasets show that this may be a promising new direction.

3.1 Introduction

Exponential growth of IoT devices and the need to analyze the vast amounts of data they generate closer to its origin have led to an emergence of the *edge computing* paradigm [15]. The factors driving such a paradigm shift are fundamental: (a) costs involved in transporting large amounts of data to Cloud, (b) regulatory constraints in moving data across sites, and (c) latency in placing all data analytics in Cloud.

S. Witherspoon (✉) · D. Steuer · N. Desai
IBM Research – Yorktown Heights, Yorktown Heights, NY, USA
e-mail: shalisha.witherspoon@ibm.com; dean.steuer@ibm.com; nirmit.desai@us.ibm.com

© The Author(s), under exclusive license to Springer Nature Switzerland AG 2022 53
H. Ludwig, N. Baracaldo (eds.), *Federated Learning*,
https://doi.org/10.1007/978-3-030-96896-0_3

Further, deployments of applications enabled by 5G network architecture rely on edge computing for meeting the low-latency requirements [4].

A critical application of edge computing is the extraction of insights from the edge data by running machine learning computations at the edge agent, without needing to export the data to a central location such as the Cloud [18]. However, most of the recent advances in machine learning have focused on performance improvements while assuming all data are aggregated in a central location with massive computational capacity. Recently proposed federated learning techniques have charted a new direction by enabling model training from data residing locally across many edge locations [8, 17].

However, previous work on federated learning has not been primarily focused on machine learning tasks beyond classification and prediction. Specifically, representation learning and semantic vector embedding techniques have proven effective across a variety of machine learning tasks across multiple domains. For text data, sentence and paragraph embedding techniques such as doc2vec [7], GloVe [11], and BERT [1] have led to highly accurate language models for a variety of Natural Language Processing tasks. Similar results have been achieved in graph learning tasks [3, 16] and image recognition tasks [2, 10]. Key reasons behind the effectiveness of semantic embedding techniques include their ability to numerically represent rich features in low-dimensional vectors and their ability to preserve semantic similarity among such rich features. Further, little or no labeled data is needed in learning the semantic vector embedding models. Clearly, semantic vector embedding will remain a fundamental tool in addressing many machine learning problems in the future.

This chapter addresses the challenge of representation learning when data cannot reside in a centralized location. Two new research problems are introduced that generalize federated learning. First, we introduce the problem of learning semantic vector embedding wherein each edge site with data participates in an iterative joint-learning process. However, unlike the previous work on federated learning, the edge sites must agree on the vector-space encoding. Second, we address a different setting where the separate parties are unable to participate in an iterative joint-learning process. Instead, each edge site maintains a semantic vector embedding model of its own. Such scenarios are quite common where edge sites may not have continuous connectivity and may join and leave dynamically.

It is important to note that while the edge scenario motivated the study and development of the aforementioned research problems, they are not limited to edge scenarios and can be applied in many settings where data cannot be aggregated centrally, such as in mobile or enterprise computing use cases. In the edge environment, an edge device, be it a mobile phone, computer, sensor, etc., can be treated as a party in the traditional federated learning scenario and thus can be utilized in any setting federated learning is carried out.

3.2 Background

Before discussing the topic of semantic vector federation, it is necessary to define several terms and techniques used in the approach. The first is a brief overview of Natural Language Processing and natural language embedding. It is also necessary to define the algorithms that utilize components of natural language embedding that allow for the federated semantic search that follows.

3.2.1 Natural Language Processing

Natural Language Processing (NLP) is a broad field covering computer interpretation of human speech and text. NLP has a long history of study within computer science, with the first explorations going back to the 1950s. During that time, work focused on breaking speech and text into its formative components and interpreting language as ontologies from which computers could more easily reason [13, 14].

In the 1990s, compute power and new algorithms in the field had advanced sufficiently for research to move away from complex rules and toward utilizing machine learning algorithms [5, 6, 12] to identify patterns. Researchers moved to focus on unsupervised algorithms, as the abundance of information was difficult or impossible to classify. The proliferation of more complex algorithms such as neural networks in recent years has served as the backbone for continued research into natural language understanding.

Natural language embedding is a technique by which human speech and text are converted into numeric vectors on which a computer can make calculations. This conversion of words into a numeric representation is referred to as vectorization and enables tasks such as finding semantically similar words, clustering documents, classifying text, extracting text features, etc. Techniques such as stemming, lemmatization, and stopword removal can also be used to reduce the size of the corpus into a smaller, but more valuable set of data by removing text that has little information. Once the text has been converted into vectors, similarity functions can be applied. One such example is cosine similarity, which works by projecting two vectors into a two-dimensional space. The cosine angle between these two vectors is then determined where the smaller the angle, the higher the similarity between two vectors. This process is done for the entire corpus of text to generate cosine similarity vector scores for all words.

Representation learning and semantic vector embedding techniques have proven effective across a variety of machine learning tasks across multiple domains. For text data, sentence and paragraph embedding techniques such as Doc2Vec, GloVe, and BERT have led to highly accurate language models for a variety of NLP tasks. Key reasons behind the effectiveness of semantic embedding techniques include their ability to numerically represent rich features in low-dimensional vectors and their ability to preserve semantic similarity among such rich features. Furthermore, little or no labeled data is needed in training semantic vector embedding models as they rely on unsupervised learning.

3.2.2   Text Vectorizers

Text vectorizers are a series of algorithms used to embed text data. These algorithms attempt to identify and categorize text into more machine interpretable forms. One of the prominent algorithms to be developed in this area is Word2Vec. Word2Vec was first developed by Tomas Mikolov and team in 2013 [9]. This algorithm was able to solve the challenge of maintaining semantic meaning in text space and allow for words in similar contexts to be correlated. Typically, the corpus is many thousands, or millions of words. Underlying the Word2Vec model is a neural network that takes as input the vectorized words and creates a mapping of the vast input data. Two addition techniques are a part of the Word2Vec algorithm. These are the continuous bag-of-words (CBOW) model and the Skip-Gram model. CBOW works by creating a vector projection of the words that surround a word w in order to predict the word. The number of words to include is defined by a "window." A window describes the number of words before and after the query word in a sentence to include in the vector projection. In a sample sentence such as "the dog jumped over the lazy fox," suppose we wanted to find the vector space of the word "over," the algorithm would look at the context surrounding the word. If we additionally provide a window size of two, our vector projection would include the words "dog" (w-2), "jumped" (w-1), "the" (w+1), and "lazy" (w+2). Skip-Gram works in a reverse fashion where we attempt to predict the words surrounding some particular word. Using the sample sentence again, and using the same word, "over," the approach would attempt to learn that the word "jumped" and "the" are contextually close to "over." See Fig. 3.1 to compare the approaches.

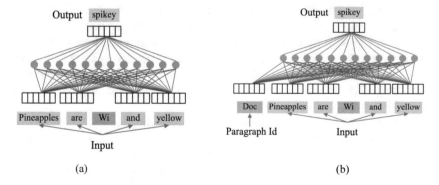

(a)   (b)

Fig. 3.1 Text vectorizers. (**a**) Word2vec. (**b**) Doc2vec, based on the CBOW algorithm (continuous bag-of-words)

In Word2Vec, vectors of fixed dimensions representing each word in the vocabulary are initialized randomly. The learning task is defined as predicting a given word based on the preceding N words and following N words. The loss function is defined as the error in predicting the given word. By iterating through many sentences during training, the word vectors are optimized using gradient descent and updated to minimize the loss and accurately represent the semantic concept. Interestingly, such semantic vectors also exhibit algebraic properties, e.g., vector representing "Queen" is similar to the one corresponding to subtracting "Man" from "King" and adding "Woman." Doc2Vec is a simple yet clever tweak of Word2Vec where a vector representing an entire document, e.g., a paragraph, is learned along with the words in it.

3.2.3 Graph Vectorizers

Another area of valuable semantic meaning is that of graphs. Graphs are generally structured as a series of nodes linked by edges. A neighborhood defines a portion of the nodes in the overall graph that are connected together. These graphs can be complex or small depending on the datasets. Examples of graph datasets include social network graphs where individuals are nodes and friendships are edges; author collaboration networks where authors are nodes and co-authorships are edges; and road networks where cities are nodes and roads are edges. Given these scenarios, it becomes valuable to find patterns in these potentially massive graphs. Node2Vec is an algorithm for representation learning on graph data that was first proposed by Grover and Leskovec in 2016 [3].

The paper's effort is two-fold; first, by using graphs and node neighborhoods, the algorithm can generate nodes of similar semantic meaning; second, by using graphs where some subsets of the links are missing in an attempt to predict where links should exist. Our work focuses on the first technique of identifying semantically similar nodes. Semantically similar nodes can be described in two ways: homophily and structural equivalence. Homophily describes a scenario where nodes are highly interconnected and, therefore, similar to each other. Structural equivalence describes a scenario where nodes that are similarly connected or fulfill a similar role within the graph are similar to each other. These nodes need not be highly connected or even connected to each other.

As an example, consider a grade school population that consists of all students and staff. Suppose a node represents a single individual and an edge represents individuals attending the same class. A cohort of students of a particular grade are likely to appear in several classes together and act as a neighborhood. A teacher may teach this cohort but may also teach other classes of entirely different students at different times. Students who appear in the same classes would be homophilic and considered a highly interconnected group of nodes. The teacher is structurally equivalent to other teachers who are a point of a single connection to other large

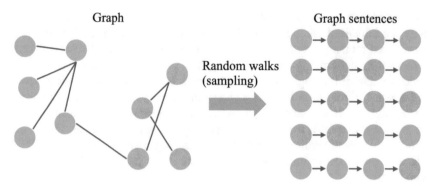

Fig. 3.2 Node2vec: random walks on graphs

groups of students. Structurally equivalent nodes need not be transition nodes into larger neighborhoods.

Nodes that exist on the periphery of a graph with single connections or no connections at all may also be treated as structurally equivalent. The preference to identify nodes via homophily or structural equivalence is treated as a parameter during Node2Vec model training. Given a graph, Node2Vec can learn vector representations for the nodes, which can then be used for node comparison and link prediction. Unlike text sentences where each word is preceded or followed by at most one word, graphs have a complex structure. It is necessary to convert this potentially complex and interconnected graph in a sequence of elements much like a sentence. One of the key innovations behind Node2Vec is mapping a graph to node sequences, a.k.a. graph sentences, by generating random walks and then using Word2Vec to learn the vector representation of the nodes in these sequences. Hyperparameters control the number of walks to generate, walk length, as well as the preference of the walk to keep close to its starting node and potentially revisit nodes it has already seen, or explore further out away from the starting node. Once the walk sequences have been generated, the walks are provided as sentences to a text vectorizer as described above (Fig. 3.2).

3.3 Problem Formulation

With the background on semantic embedding covered, we are ready to formally define the problem of semantic vector federation for edge environments.

In the introduction, we presented two problem scenarios for semantic vector federation in edge environments: the first being when iterative joint learning is possible and the second when edge sites were unable to participate in joint learning. Joint learning is the process by which multiple parties collaborate and share some form of information. In cases where the data is not sensitive or all parties are controlled by a single organization, the raw information could be shared. However, in many scenarios, it may be necessary to minimize what data is being shared with others. The latter scenario was used to inform and design new algorithms for the joint-learning process.

To address the conflicting challenges, we developed novel algorithms for each scenario. In the case of joint learning, prior to beginning the iterative distributed gradient descent, edge sites collaborate to compute an aggregate feature set so that the semantic vector spaces across edge sites are aligned. In the case where joint learning is not possible, edge sites learn their own semantic vector embedding models from local data. As a result, the semantic vector spaces across edge sites are not aligned, and semantic similarity across edge sites is not preserved. To address this problem, we propose a novel approach for learning a mapping function between the vector spaces such that vectors of one edge site can be mapped to semantically similar vectors on another edge site.

3.3.1 Joint Learning

Joint learning can be described as scenarios where synchronous learning is able to transpire, i.e., all parties are able to participate in federated learning at the same time to train a global model. The global model could then be used in performing semantic similarity searches across all edges sites for new data.

The joint-learning algorithm adapts the federated averaging algorithm, which achieves model fusion by averaging the learned weights during iterative rounds of training, to a semantic vector embedding setting. The main challenge in applying federated learning in semantic embedding models is in ensuring that concepts across edge sites are aligned. For example, in the case of text data, if the vocabulary of words is different across edge sites, federated averaging cannot be readily applied because the learned weights of the embedding model are what eventually ends up as the word embeddings, and if they are not aligned, the correct embeddings would not be updated properly during averaging. Hence, a key innovation in the joint-learning adaption is to align the vocabulary of concepts as a prerequisite step in the iterative synchronous training process to ensure consistent embedding across sites.

Figure 3.3 depicts an illustration wherein EDGE1 wants to perform a global search for top-3 experts most similar to person X. Assuming that a Doc2Vec model m_1 has been distributed to all edge sites via joint learning, EDGE1 uses m_1 to vectorize person X's document as vector v_1 and sends v_1 to other sites. Other sites apply a similarity metric, e.g., cosine similarity, to find top-3 nearest-neighbor vectors to v_1 and return the corresponding person identities and cosine similarity score back to EDGE1. After receiving the results from all edge sites, EDGE1 can select the top-3 results having the highest cosine similarity.

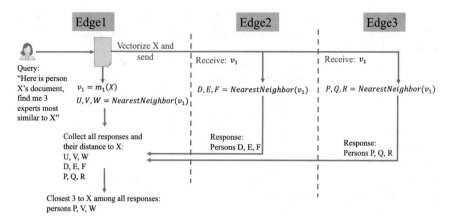

Fig. 3.3 Semantic search for the motivating example: joint learning

3.3.2   Vector-Space Mapping

The problem of *vector-space mapping* of semantic vector embedding is defined as having N edge sites, each edge site i with local dataset D_i and a pre-trained semantic vector embedding model m_i trained on D_i. Each of the edge sites wants to collaborate in performing global similarity search for a new example d across all edge sites but do not want to share their data with each other and are not able to participate in jointly training a common model.

A key property of semantic embedding models is that each has what is known as its own vector space. This means that the real-valued vectors produced for semantic representations are initialized randomly. As an example, because of this property, even if two word embedding models were trained on the exact same corpus, their semantic vectors would be different, and the semantic meaning would not be preserved across the other embedding models. This introduces a concept known as *vector-space mapping*, which is the ability to translate the vector space of one embedding model into another independently trained embedding model's vector space in order to retain the semantic meaning learned across both models and enable queries of similarity among their vectors.

One of the main challenges is to identify a training set of semantically similar words in the different vector spaces and use the corresponding vectors as reference vectors that can be used to generate a function that is capable of transforming any vector from one vector space to the other. Given this, our algorithm makes use of the properties of multi-layer perceptron (MLP) neural networks to potentially learn universal functions and, therefore, the possibility to train such a network to learn the mapping. However, training a MLP model requires a training set that is commonly available to all edge sites. Availability of such training data is highly constrained, especially given that the sites do not wish to share their proprietary datasets with each other.

Hence, another key innovation of our vector-space mapping algorithm is the idea of leveraging any publicly available corpus, regardless of its domain, as a training dataset generator for the mapper MLP model. The formal algorithm definition can be defined as outlined in Algorithm 3.1 and illustrated in Fig. 3.4.

Algorithm 3.1 Vector-Space Mapping Algorithm

Input: Local Dataset D_i, Public dataset D_p, Loss Function F_i, Epochs T, learning rate η

Function $Main(D_i, D_p, F_i, \eta)$:

  $m_i \leftarrow TrainDoc2Vec(D_i, D_p, F_i, \eta)$

  store m_i

Function $Map_j(m_j, D_p, F_i)$:

  $W_{i \rightarrow j} \leftarrow RandomNN()$

  for all $b \in D_p$ **do**

    $v_i \leftarrow predict(m_i, b)$

    $v_j \leftarrow predict(m_j, b)$

    $L \leftarrow F_i(v_i, v_j)$

    $\nabla \leftarrow Gradient(L, F_i, W_{i \rightarrow j})$

    $W_{i \rightarrow j} \leftarrow W_{i \rightarrow j} - \eta \nabla L$

  end for

  $m_{i \rightarrow j} \leftarrow Model(W_{i \rightarrow j})$

  store $m_{i \rightarrow j}$

Function $GlobalSearch(d)$:

  for all $Edge_j \in Edges$ **do**

    $v_i \leftarrow predict(m_i, d)$

    $v_j \leftarrow m_{i \rightarrow j}(v_i)$

    Send query v_j to $Edge_j$

    $V_{sim} \leftarrow$ Receive result vectors from $Edge_j$

  end for

  return V_{sim}

As shown in Fig. 3.4, consider that a semantic vector embedding model m_1 is trained from local data on EDGE1 and another semantic vector embedding model m_2 is trained from local data on EDGE2. The objective is to train a mapper MLP model that can map vectors produced by vector space of m_1 to the vector space of m_2. An auxiliary dataset D_p that is accessible to both edge sites can serve as the training samples generator and facilitate the training of MLP mapper model. Input to the MLP model are the vectors produced by m_1 on samples of D_p, and the ground-truth labels are the vectors produced by m_2 on the same samples of D_p. Since the input and the output of the MLP mapper model can have a different dimensionality, this approach works even when EDGE1 and EDGE2 choose a different dimensionality for their semantic vectors.

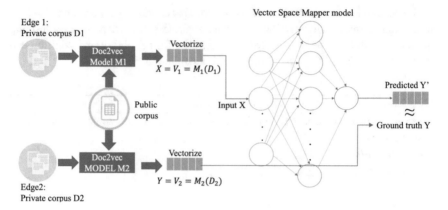

Fig. 3.4 Learning to map vector space of Edge1 to that of Edge2

3.4 Experimentation and Setup

We evaluate the two algorithms of joint learning and vector-space mapping via extensive experiments on two data modalities: natural language and graph. The experiments are anchored on the motivating example of performing a global semantic search for individuals with expertise. The evaluation metric is dependent on the algorithm. For joint learning, we perform an objective evaluation of how well the federated semantic vector embedding model performs relative to the baseline of a centralized model; comparing with a baseline is a standard practice for unsupervised algorithms since there is no ground truth on semantic similarity between samples. And for vector-space mapping, we perform an objective analysis comparing cosine similarity of reference vectors with and without our mapping algorithm being performed.

3.4.1 Datasets

For the natural language modality, we leverage three different datasets: (a) an internal dataset consisting of Slack collaboration conversations, (b) the 2017 Wikipedia dataset with 10K samples, and (c) the 20-newsgroup public datasets with 18,846 samples. For joint-learning experiments, (a) is used for both the centralized and federated experiments. For vector-space mapping experiments, (b) is used as the private datasets, with (c) acting as the public dataset accessible by all edge sites. For the graph modality, we leverage (a) above for the joint-learning experiments but instead of looking at the text content of the posts, we construct a collaboration graph between users.

The Slack dataset (a) consists of natural language conversations across 7367 Slack channels among 14,208 unique users. Of these, only 1576 users having sufficient activity (more than 100 posts) are used in the experiments. All Slack posts of a user are treated as a single document in training the Doc2vec models. For the centralized case, Slack posts of all users are used for training a single Doc2vec model, whereas for the federated case (joint learning), the users are uniformly distributed across two edge sites. No additional knowledge of the organization hierarchy, projects, or teams is included, leaving the models to rely solely on the content of the Slack posts as a basis of semantic vector embedding representing each user.

In constructing a graph from the Slack dataset, each user is treated as a node in the graph, and other users who participate in the same Slack channel as the user are treated as the edges. For avoiding noisy edges due to having channels with a large number of users, a pair of users participating together, i.e., co-occurring, in less than 10 channels do not have an edge between them. Another approach would have been to assign weights to edges; however, Node2vec does not take advantage of edge weight information. The entire graph is used for training the centralized Nodes2vec model. For the federated case, users are randomly assigned to one of the edge sites. When doing so, the cross-site edges are handled in two alternative ways: (1) the cross-site edges are not retained, so each edge site has edges only among the users assigned to the site, called *no retention*, and (2) the nodes involved on cross-site edges are retained on both sites, called *retention*.

3.4.2 Implementation

For the natural language dataset, we use the Doc2vec model architecture with the Skip-Gram PV-DM algorithm with 40 epochs and a learning rate of 0.025. Doc2vec semantic vectors are 50-dimensional real-valued vectors. For the graph dataset, we use the Node2vec architecture with 40 epochs and a learning rate of 0.025. Node2vec semantic vectors are 124-dimensional real-valued vectors. The hyperparameters of the Node2vec favor homophily approach where the return parameter p is favored over the in–out parameter q. We set $p = 0.6$ and $q = 0.1$. The walk length parameter, the number of hops to other nodes from the start node, is set to 20, and the number of walks, the number of iterations of node hopping to perform, is also set to 20.

In the case of vector-space mapping, the mapper model is an MLP model with a single hidden layer with 1200 neurons and a dropout ratio of 0.2. We use the cosine embedding loss in training the MLP as the semantic similarity is based on cosine similarity. ADAM optimizer with a learning rate of 0.00001 and 20 epochs of training with the batch size of 64 was applied.

It is worth emphasizing that these details are provided for completeness and these parameters are quite commonly used in the literature. The objective here is not to produce the best-performing semantic vector embedding models. Instead, we are primarily interested in evaluating the *relative* performance of the federated algorithms compared to the traditional centralized ones. Hence, all of the above parameters are kept the same for the centralized and federated cases.

3.5   Results: Joint Learning

3.5.1   Metrics

For objectively measuring how well the federated algorithms perform relative to the centralized case, the degree of overlap is computed as follows. For a given document d in the dataset, the centralized model is used to vectorize the document, and the set of top-k most similar documents from the dataset is found based on cosine similarity, called d_c^k. Then, using the respective federated algorithm, the set of top-k most similar documents is found for the same document d, called d_f^k. The degree of overlap sim_k then is the ratio of cardinality of the intersecting set and k, denoted as $sim_k = \frac{|d_c^k \cap d_f^k|}{k}$. For multiple documents in the dataset, a simple mean of sim_k is computed over all documents. When $sim_k = 1$, the centralized and federated models produce identical results on semantic search. The idea behind the measure is simple: the higher the sim_k, the closer the federated case performance is to the centralized case. In evaluating the federated algorithms relative to the centralized case, we set $k = 10$.

3.5.1.1   Natural Language

Figure 3.5 shows the distribution of the number of overlaps between the centralized case and the joint-learning case ($sim_{10} \times 10$) when the joint-learning algorithm is applied to the Slack dataset. As indicated by the sim_k of 0.609, for a majority of the users, the joint-learning model found about 6 of the same users found by the centralized model. It is important to note that an average degree of 6 out of

Fig. 3.5 Performance of Doc2vec joint learning relative to centralized learning, $sim_k = 0.609$

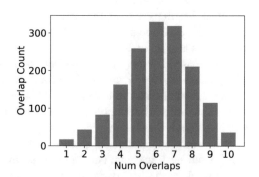

10 overlaps is an adequate result, because we found that when retrieving the top 10 results from cosine similarity, the bottom half results are usually inconsistent between experiments, even in the centralized case, due to the fact that the latter results are lower and closer in score, often only separated by trailing decimal digits. Thus, the fact that the federated model was able to overlap with more than half of the results produced from the centralized model demonstrates similar performance from the models.

Based on the above, we can conclude that there is not a significant loss in performance introduced by the joint-learning algorithm when compared to the centralized model, making the joint-learning algorithm a viable alternative to the centralized case.

3.5.1.2 Graph

Figure 3.6 shows the distribution of the number of overlaps between the centralized case and the joint-learning case ($sim_{10} \times 10$) when the joint-learning algorithm is applied to the graph dataset with no retention of cross-site collaborators. As seen in the distribution as well as indicated by the sim_k of 0.138, for a majority of the users, the joint-learning model found almost no users returned by the centralized model. This is not an encouraging result by itself. However, since the cross-site edges are dropped from the graph corresponding to the joint-learning case, valuable information about those users' collaboration behavior is lost compared to the centralized case having the entire graph. Although this explanation is intuitive to validate it, we need to examine the result when the cross-site collaborators are retained and discussed next.

Figure 3.7 shows the distribution of the number of overlaps between the centralized case and the joint-learning case ($sim_{10} \times 10$) when the joint-learning algorithm is applied to the graph dataset with all cross-site collaborators retained across both sites. As seen in the distribution as well as indicated by the sim_k of 0.253, for a majority of the users, the joint-learning model found more than 2 of the same users returned by the centralized model. Compared to the no retention

Fig. 3.6 Performance of Node2vec joint learning relative to centralized learning, no retention, $sim_k = 0.138$

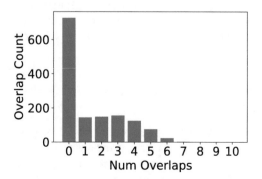

Fig. 3.7 Performance of Node2vec joint learning with collaborator retention relative to centralized learning, $sim_k = 0.253$

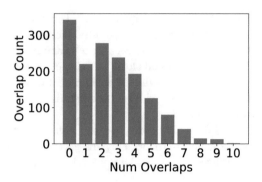

result, this is a significantly better result. Thus, the explanation above is validated as retaining the cross-site collaborators clearly helps the joint-learning model achieve more accurate user embedding.

Although the difference between the Node2vec joint-learning results above can be explained by the difference in retention policy, the inferior results of Node2vec when compared with Doc2vec require further investigation. One hypothesis is that the random assignment of users to edge sites can have an adverse effect on the joint-learning performance because such an assignment can have an uneven effect on the collaborative user clusters in the graph. For example, one edge site may end up having most of its collaborative clusters unaffected, whereas another may have its collaborative clusters split into two sites. Although the immediate collaborators may be preserved via cross-site retention, the higher-order collaborations are still affected.

3.6 Results: Vector-Space Mapping

To construct the required vector spaces, we used 10,000 randomly shuffled subsamples from the 2017 Wikipedia dataset as the private data on two edge sites to train two Doc2vec models using different initial random weights. For our public dataset used to generate input and ground-truth vectors for training the MLP mapper model, we leveraged the 20-newsgroup data consisting of 18,886 samples. Our experiments focused on mapping the vector space of EDGE1 to EDGE2.

3.6.1 Cosine Distance

To illustrate the impact of not having mapping across vector spaces, we measured the cosine similarity between vectors for the same documents in both vector spaces. Without mapping, the resulting cosine distance distribution was shown to have a similar distribution to orthogonal vectors, which is essentially akin to

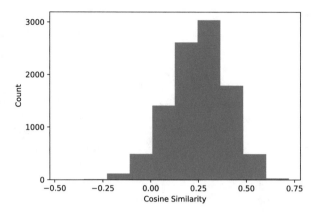

Fig. 3.8 Distribution of cosine distance, no mapping

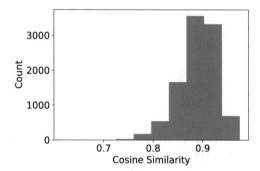

Fig. 3.9 Distribution of cosine distance, mapping performed

comparing random vectors, as shown in Fig. 3.8. For comparison, Fig. 3.9 shows the distribution of cosine distance after mapping the vector spaces, which shows a significant shift in the mean and variance of the distribution away from the random distribution and toward a similarity around 1.0.

3.6.2   Rank Similarity

To further determine the quality of the vector-space mapping, we measured the rank similarity of comparable vectors in both vector spaces. To do this, we vectorized documents in EDGE1's vector space and performed the mapping into EDGE2's vector space to find its 20 nearest matching vectors. If the nearest matching vector was the same as the test document, we gave it a rank of 0; otherwise, we assigned it the rank that it appeared in the similarity result. In cases where the test document was not returned in the similarity result, we assigned it a rank of 20. As shown in Fig. 3.10, we achieve a 0.95 percent accuracy of a perfect match between the

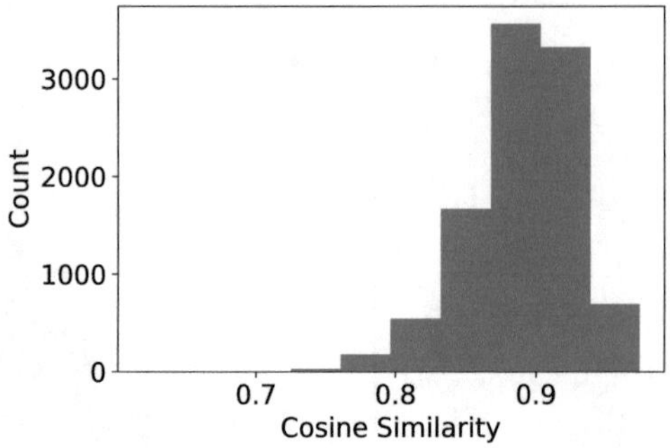

Fig. 3.10 Distribution of
rank similarity, mapping
performed

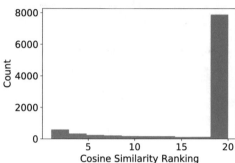

Fig. 3.11 Distribution of
rank similarity, no mapping

document vectors after mapping. We also performed the rank similarity experiment
without mapping, shown in Fig. 3.11, which resulted in a mere 0.03 percent accuracy
for matching the appropriate vector, and the majority of the results gives a rank
of 20. Thus both results illustrate the effectiveness of our vector-space mapping
algorithm for semantic search across independently trained local models.

3.7 Conclusions and Future Work

With the increasing regulation and the growth in data originating at the edge, edge
computing is poised to be a critical area of research with significant impact on
how IT systems are developed, deployed, and managed. This chapter introduced
the novel research direction of federated semantic vector embedding, building on
the unique combination of the well-known techniques of federated learning and
semantic vector embedding. Specifically, two research problems were formulated to
cater to two separate settings in which edge sites want to collaborate in performing
global semantic search across sites without sharing any raw data.

The first setting, called joint learning, is when the edge sites have a tightly coupled collaboration to participate in a synchronous joint-learning process and have an
agreement on the model architecture, training algorithm, vector dimensionality, and
data format. A novel algorithm to address the joint-learning problem is presented

with the novel idea of vocabulary aggregation before starting the iterative federated learning process.

The second setting, called vector-space mapping, is when the edge sites do not agree on the various parameters of joint learning or cannot participate in a synchronous process as they may need to join and leave dynamically. This is clearly a challenging setting and one of great significance in practice. Based on the novel idea of training another model to learn the mapping between vector spaces based on a public dataset from any domain, an algorithm for addressing the vector-space mapping problem was presented.

Experimental evaluation using multiple natural languages as well as graph datasets shows that these algorithms show promising results for both algorithms compared to the baseline centralized case where all data can be aggregated on one site. Several important research questions remain open. How do these algorithms scale in the number of edge sites, differences in data distributions, and the amount of data at edge site? How do we interpret such semantic vectors and explain the similarity results they produce? The work covered here is one of the first in the area of federated semantic vector embedding and has unlocked several key challenges for future research.

References

1. Devlin J, Chang M, Lee K, Toutanova K (2018) BERT: pre-training of deep bidirectional transformers for language understanding. CoRR abs/1810.04805, http://arxiv.org/abs/1810.04805,1810.04805
2. Frome A, Corrado GS, Shlens J, Bengio S, Dean J, Ranzato M, Mikolov T (2013) Devise: a deep visual-semantic embedding model. In: Advances in neural information processing systems, pp 2121–2129
3. Grover A, Leskovec J (2016) Node2vec: scalable feature learning for networks. In: Proceedings of the 22nd ACM SIGKDD international conference on knowledge discovery and data mining, KDD'16. Association for Computing Machinery, New York, pp 855–864
4. Hu YC, Patel M, Sabella D, Sprecher N, Young V (2015) Mobile edge computing—a key technology towards 5G. ETSI White Pap 11(11):1–16
5. Kanerva P, Kristofersson J, Holst A (2000) Random indexing of text samples for latent semantic analysis. In: Proceedings of the 22nd annual conference of the cognitive science society, vol 1036. Erlbaum, New Jersey
6. Uesaka Y, Kanerva P, Asoh H, Karlgren J, Sahlgren M (2001) From words to understanding. In: Foundations of real-world intelligence. CSLI Publications, p 294). chapter 26
7. Le Q, Mikolov T (2014) Distributed representations of sentences and documents. In: International conference on machine learning, pp 1188–1196
8. McMahan HB, Moore E, Ramage D, y Arcas BA (2016) Federated learning of deep networks using model averaging. CoRR abs/1602.05629. http://arxiv.org/abs/1602.05629,1602.05629
9. Mikolov T, Chen K, Corrado G, Dean J (2013) Efficient estimation of word representations in vector space. 1301.3781
10. Norouzi M, Mikolov T, Bengio S, Singer Y, Shlens J, Frome A, Corrado G, Dean J (2014) Zero-shot learning by convex combination of semantic embeddings. In: Proceedings of 2nd international conference on learning representations

11. Pennington J, Socher R, Manning CD (2014) GloVe: global vectors for word representation. In: Proceedings of the 2014 conference on empirical methods in natural language processing (EMNLP), pp 1532–1543
12. Sahlgren M, Kanerva P (2008) Permutations as a means to encode order in word space. In: Cognitive science—COGSCI
13. Salton G (1962) Some experiments in the generation of word and document associations. In: Proceedings of the fall joint computer conference, AFIPS'62 (Fall), 4–6 Dec 1962. Association for Computing Machinery, New York, pp 234–250. https://doi.org/10.1145/1461518.1461544
14. Salton G, Wong A, Yang CS (1975) A vector space model for automatic indexing. Commun ACM 18(11):613–620
15. Satyanarayanan M (2017) The emergence of edge computing. Computer 50(1):30–39
16. Wang Q, Mao Z, Wang B, Guo L (2017) Knowledge graph embedding: a survey of approaches and applications. IEEE Trans Knowl Data Eng 29(12):2724–2743
17. Yang Q, Liu Y, Chen T, Tong Y (2019) Federated machine learning: concept and applications. ACM Trans Intell Syst Technol (TIST) 10(2):1–19
18. Zhou Z, Chen X, Li E, Zeng L, Luo K, Zhang J (2019) Edge intelligence: paving the last mile of artificial intelligence with edge computing. Proc IEEE 107(8):1738–1762

Chapter 4
Personalization in Federated Learning

Mayank Agarwal, Mikhail Yurochkin, and Yuekai Sun

Abstract Typical federated learning (FL) problem formulation requires learning a single model suitable for all parties while prohibiting parties from sharing their data with the aggregator. However, it may not be possible to learn a common single model that is suitable for all parties. For example, consider a sentence completion problem: "I live in the state of . . ." The answer clearly depends on the party, and no single model is appropriate here. To handle such situations, various personalization strategies have been proposed in the recent literature. In particular, the problem appears to have a close connection to meta-learning. We review recent FL personalization techniques categorizing them into eight groups and summarize three strategies and corresponding datasets for benchmarking personalization in federated learning. We provide an overview of the statistical challenges of personalization in federated learning. At a high level, personalization leads to an increase in the model complexity, which in turn increases the hardness of the federated learning task. We study when too much personalization can prevent standard approaches to personalized federated learning from learning the common parts of the parties and present alternative approaches that overcome such issues.

M. Agarwal (✉)
IBM Research, Cambridge, MA, USA
e-mail: mayank.agarwal@ibm.com

M. Yurochkin
MIT-IBM Watson AI Lab, IBM Research, Cambridge, MA, USA
e-mail: mikhail.yurochkin@ibm.com

Y. Sun
Department of Statistics, University of Michigan, Ann Arbor, MI, USA
e-mail: yuekai@umich.edu

© The Author(s), under exclusive license to Springer Nature Switzerland AG 2022
H. Ludwig, N. Baracaldo (eds.), *Federated Learning*,
https://doi.org/10.1007/978-3-030-96896-0_4

4.1   Introduction

Centralized federated learning aims to learn a global model from individual parties' data while keeping their local data private and localized to their individual machines. This global model has the advantage of utilizing data from all the parties and thus generalizes better to test data across parties. In practical scenarios, however, datasets on individual parties are often heterogeneous (non-IID), thus rendering one global model performance sub-optimal for some parties. On the other hand, if each party trains a local model on their local data, they train on a data distribution similar to what is expected at test time but might fail to generalize due to the paucity of data available on a local party. Personalized federated learning aims to learn a model that has the generalization capabilities of the global model but can also perform well on the specific data distribution of each party.

To illustrate the need for personalization, consider the case of a language model learned in a federated learning setting [10]: if we use a global model to predict the next word for the prompt: "*I live in the state of ...,*" the global model will predict the same token (name of the state) for every party irrespective of their local data distribution. Thus, while a global model might be able to learn the general semantics of language well, it fails to personalize to the individual party.

In addition to the aforementioned qualitative example, we can also quantitatively demonstrate the need for personalization. We set up our experiment using the MNIST dataset[1] divided among 100 parties. We distribute the data among these parties in a heterogeneous manner using a Dirichlet distribution with different concentration parameters (α) [64]. We train a 2-layer fully connected network in two settings to measure the benefits parties gain from participating in federated learning. In the first setting, we train an individual network for each of the 100 parties for 10 epochs using solely the parties' own data and measure the performance of these individual networks on their respective parties' test data ($\text{Acc}_i^{\text{local}}$). In the second setting, all 100 parties participate in training a global model using Federated Averaging (FedAvg) [39] for 100 communication rounds, and we measure the performance of this global model on each of the party's test data ($\text{Acc}_i^{\text{global}}$). Figure 4.1 shows histograms of the differences in the performance of the global model and the local model ($\text{Acc}_i^{\text{global}} - \text{Acc}_i^{\text{local}}$) for each of the parties, under different levels of data heterogeneity. As is evident in the plots, the global model does not benefit each party participating in its training, and this phenomenon is more pronounced when the non-IID characteristics of the data are more severe (smaller α values). This experiment emphasizes the need for personalization of the global model on the parties' local data distribution to ensure that every party benefits from its participation in the learning setup.

In this chapter, we review different personalization techniques proposed in federated learning literature and discuss the connections between federated learning and

---

[1] http://yann.lecun.com/exdb/mnist/

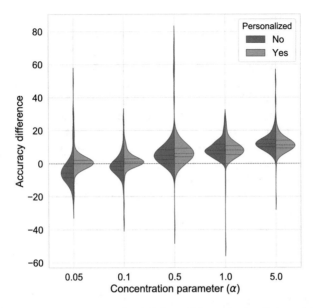

Fig. 4.1 Difference in the accuracies of a global model learned using Federated Averaging and the local models trained solely on the parties' local datasets. Without personalization, under more severe cases of heterogeneity (smaller α), the global model underperforms on a significant number of parties as compared to their local models. With a naive method of personalization by fine-tuning, this effect is attenuated, with the performance improving even in the extreme cases of heterogeneity

first-order meta-learning [18]. We also study the statistical limits of personalization in federated learning. In particular, we show that personalization improves party-specific performance up to a point. After this point, adding more parties to the problem does not lead to improvements in performance.

4.2 First Steps Toward Personalization

In this section, we look at a basic technique that combines federated learning and personalization and explore why this technique is a strong baseline for the personalization task.

4.2.1 Fine-Tuning Global Model for Personalization

A straightforward method to personalize a global model learned using federated learning is to train it further on the local data. This method allows us to control the level of personalization through the number of local updates performed on the global model—zero local updates retain the global model, while as the number of local updates increases, the model becomes more personalized to the local data.

While this technique might look simple, it is a strong baseline for the personalization task. We study this fine-tuning approach in the experiment presented in Sect. 4.1 and Fig. 4.1. We personalize the global model learned over 100 parties with data distributed in a heterogeneous manner by fine-tuning it for 1 epoch on the local data and then measure the performance of this fine-tuned model on the parties' local test data. As is evident through the results of this experiment, this simple fine-tuning technique considerably improves the performance of the global model as compared to the local models. For the extreme cases of heterogeneity, this method improves the performance for a significant number of parties and also does not negatively impact the performance in less severe cases of heterogeneity. We now aim to understand the reason behind the strong performance of this fine-tuning approach.

4.2.2 Federated Averaging as a First-Order Meta-learning Method

In this section, we try to understand the reason behind the effectiveness of fine-tuning the global model learned using federated averaging. We replicate the derivations of Jiang et al. [29] to show that the updates in Federated Averaging are a combination of federated SGD updates and the first-order MAML (FOMAML) updates.

What is meta-learning and MAML?
While conventional machine learning approaches aim to learn parameters that perform best on a given task, meta-learning or learning to learn [55–57, 59] aims to learn parameters that can be quickly adapted to new tasks. Model-Agnostic Meta-Learning (MAML) [18] is among the most popular meta-learning approaches: its goal is to find model parameters that can be adapted to a new task in few gradient updates. However, to achieve this, the MAML objective requires computing the second-order derivatives, which are computationally expensive. First-order MAML (FOMAML) [43] approximates the MAML objective by considering only the first-order derivatives, thereby reducing the computation demand of MAML. See Sect. 4.3.5 for further discussion of meta-learning and related federated learning personalization strategies.

We start the analysis by defining the update of FedSGD (equation (4.1)). FedSGD operates by taking a single gradient step on each of the N parties, communicating these gradients back to the aggregator, and then aggregating these gradients to update the global model. We use ∇_k^i to denote the kth-step gradient on party i.

$$\nabla_{\text{FedSGD}} = \frac{1}{N} \sum_{i=1}^{N} \frac{\partial \mathcal{L}_i(\theta)}{\partial \theta} = \frac{1}{N} \sum_{i=1}^{N} \nabla_1^i, \tag{4.1}$$

where θ are the model parameters (e.g., neural network weights) and $\mathcal{L}_i(\theta)$ is the loss of party i. Next, we derive the update of MAML and first-order MAML [18] in similar terms. Assume θ_K^i is the personalized model of party i obtained after taking K steps of the gradient of loss with β as the learning rate:

$$\theta_K^i = \theta - \beta \sum_{j=1}^{K} \frac{\partial \mathcal{L}_i(\theta_j^i)}{\partial \theta}. \tag{4.2}$$

The MAML update is then defined as the gradient of the personalized model θ_K^i with respect to the initial parameters θ, averaged across N parties. Unfortunately, this computation requires higher-order derivatives and is expensive even for $K = 1$. FOMAML ignores the higher-order derivatives and only uses the first-order gradients:

$$\nabla_{\text{FOMAML}}(K) = \frac{1}{N} \sum_{i=1}^{N} \frac{\partial \mathcal{L}_i(\theta_K^i)}{\partial \theta} = \frac{1}{N} \sum_{i=1}^{N} \nabla_{K+1}^i. \tag{4.3}$$

Having computed the updates for FedSGD and FOMAML, we now look at the update of Federated Averaging (FedAvg). The update for FedAvg is the average of party updates, which are the sums of local gradient updates ∇_j^i:

$$\nabla_{\text{FedAvg}} = \frac{1}{N} \sum_{i=1}^{N} \sum_{j=1}^{K} \nabla_j^i = \frac{1}{N} \sum_{i=1}^{N} \left(\nabla_1^i + \sum_{j=1}^{K-1} \nabla_{j+1}^i\right) \tag{4.4}$$

$$= \frac{1}{N} \sum_{i=1}^{N} \nabla_1^i + \sum_{j=1}^{K-1} \frac{1}{N} \sum_{i=1}^{N} \nabla_{j+1}^i. \tag{4.5}$$

Rearranging these terms allows us to derive a relation between the updates of FedAvg, FedSGD, and FOMAML:

$$\nabla_{\text{FedAvg}} = \nabla_{\text{FedSGD}} + \sum_{j=1}^{K-1} \nabla_{\text{FOMAML}}(j). \tag{4.6}$$

The FedAvg update for 1-gradient update ($K = 1$) in FedAvg before every communication reduces to the FedSGD setting according to equation (4.6). Increasing the number of gradient updates progressively increases the FOMAML part in the update. According to Jiang et al. [29], models trained with $K = 1$ are hard to personalize, while increasing K increases the personalization capabilities of the models up to a certain point, beyond which the performance of the initial model becomes unstable.

4.3   Personalization Strategies

Personalization in a federated learning setting has gained considerable interest in the research community in recent years. In this section, we look at the various techniques proposed for this problem and classify them into 8 main categories. The classification criteria along with the methods that fall under the respective criteria are summarized in Table 4.1. In the following subsections, we delve deeper into each criterion defined in the table and look at its strengths and weaknesses.

4.3.1   Client (Party) Clustering

The central premise of personalization in federated learning is that one global model might not work for all parties due to the non-IID heterogeneous distribution of data on the parties. Client (party) clustering techniques for personalization operate under a common assumption: among the N parties present in the system, there are $K < N$ distinct data distributions. This assumption enables the techniques to cluster parties into K clusters to alleviate the non-IID data distribution conditions and learns a common global model for each of the K clusters. Thus, under this formulation, the personalization problem is then sub-divided into two sub-problems: (1) Defining a clustering hypothesis to cluster parties together and (2) Aggregating and learning a model for each defined cluster.

Clustered federated learning (CFL) [47] assumes that there exists a partitioning $C = \{c_1, \ldots, c_K\}, \bigcup_{k=1}^{K} c_k = \{1, \ldots, N\}$, such that every subset of parties $c_k \in C$ satisfies the conventional federated learning assumption of a global model minimizing the risk on all the parties' data distributions at the same time. However, instead of identifying the complete clustering C of parties at once, CFL recursively bi-partitions parties into clusters until all the clusters are identified. The algorithm proceeds by training local models until convergence up to a certain limit. These individual party models are then aggregated, and the global model is checked for its congruence, i.e., how well does the global model minimize the risk for each party. If the global model fits a certain stopping criterion for the parties, CFL is terminated. Otherwise, parties are partitioned into two sub-clusters, and CFL is recursively executed on each. Since the bi-partitioning approach works recursively,

Table 4.1 Classification of different methods of personalization in federated learning setting. For each classification, we briefly describe the core idea of this classification along with the methods that fall under this classification criterion

Personalization strategy	Description and methods
Client clustering[a]	Cluster similar parties together to learn models for similar data distributions
	Methods: CFL [47], 3S-Clustering[b] [19], IFCA [20], HypCluster [36]
Client contextualization[a]	Learn contextual features private to the parties to add contextual information to the models along with input features
	Methods: FURL [5], FCF [2]
Data augmentation	Augment local data with data from other parties or global data to increase its diversity and size
	Methods: DAPPER [36], XorMixup [51], global-data-sharing[b] [66]
Distillation	Distill information between local and global models
	Methods: FML [50], FedMD [34]
Meta-learning approach	Formulate the personalization problem as a meta-learning [25, 57] problem
	Methods: FedMeta [9], Per-FedAvg [17], ARUBA [31], FedPer [3]
Mixture of models	Maintain a local model along with a global model and use a combination of the two
	Methods: APFL [15], LG-FedAvg [35], FL+DE [44], MAPPER [36]
Model regularization	Optimize a regularized version of the loss function to balance local model with global model
	Methods: L2GD [23], FedAMP [26], pFedMe [16], Fed+ [61]
Multi-task learning	Use multi-task learning framework [46, 65] for federated learning setting
	Methods: MOCHA [53], VIRTUAL [14]

[a]We use the terms "Client" and "Party" interchangeably here. While the term "Client" is used in some research papers, its analogous term "Party" is the one used in this book
[b]The authors of these methods have not assigned a specific name to their proposed algorithm/technique. We choose to call them so for brevity purposes

the number of clusters K is not required to be known a priori. Additionally, since the clustering mechanism is implemented on the aggregator, parties do not bear the computational burden of the approach. Instead, the aggregator, with its generally greater compute power than parties, can reduce the overhead of clustering.

3S-Clustering [19] also formulates the problem similar to CFL [47], but instead of recursive bi-partitioning of parties, it aims to find the K clusters at once on the aggregator. Once the local models are trained and communicated to the aggregator,

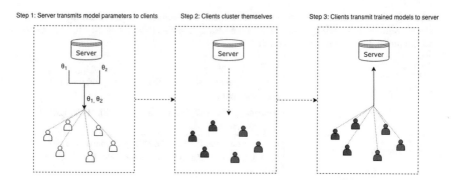

Fig. 4.2 Client (party) clustering strategy for personalization. In this particular instance, the aggregator maintains K separate model parameters and sends these to each individual party, where the parties decide which of the K model parameters they should use

3S-Clustering executes a clustering method—typically, KMeans works, but other clustering methods can also be employed as shown in the original paper where they study this method primarily for byzantine-robust distributed optimization—on the aggregator to find the K clusters. This method, however, is restricted to convex objectives and, hence, does not apply to non-convex objectives such as in the case of deep neural networks.

The aforementioned two methods use the aggregator for party clustering, while the other two methods in this category, IFCA [20] and HypCluster [36], utilize the parties to identify their own cluster memberships (Fig. 4.2). These two methods are pretty similar to each other and operate by the aggregator maintaining K cluster centers and the associated model parameters. At each round, the aggregator broadcasts the cluster parameters to each of the parties, which in turn estimates its cluster identity by choosing the parameters that achieve the lowest loss value. These cluster centers are then used as initializers of the local model, fine-tuned on the local data, and sent back to the aggregator along with the cluster identity for aggregation. The aggregator then aggregates models according to their cluster membership, and the entire process repeats.

4.3.2   Client Contextualization

Learning user-specific contextual features or embeddings has been widely used to improve the personalization of models in problems unrelated to federated learning [1, 22, 27, 38, 58]. Client contextualization utilizes the same approach of learning user embeddings to the task of personalization in federated learning. The rationale behind this approach is that the embeddings for each party capture characteristics specific to the particular party and act as indicators to the global model to utilize this context to adapt its predictions to the specific party.

Federated collaborative filtering (FCF) [2] proposes a collaborative filtering [48]-based recommender system learned in a federated manner. Collaborative filtering models the interaction between N users and M items through a user–item interaction matrix $\boldsymbol{R} \in \mathbb{R}^{N \times M}$ as a linear combination of the user-factor matrix $\boldsymbol{X} \in \mathbb{R}^{K \times N}$ and the item-factor matrix $\boldsymbol{Y} \in \mathbb{R}^{K \times M}$ as

$$\boldsymbol{R} = \boldsymbol{X}^T \boldsymbol{Y}. \tag{4.7}$$

In each iteration of the FCF algorithm, the aggregator sends the item-factor matrix $\boldsymbol{Y}$ to each of the parties, which in turn use their local data to update the user-factor matrix $\boldsymbol{X}$ and the item-factor matrix $\boldsymbol{Y}$. The updated item-factor matrices are sent back to the aggregator for aggregation, while the user-factor matrices are kept private on the individual parties. This allows each party to learn its own set of user-factor matrices while utilizing the item-factor information from across parties. In experiments comparing FCF with standard collaborative filtering, FCF is shown to closely match the performance of standard collaborative filtering on multiple recommendation performance metrics and for multiple recommendation datasets.

While FCF is specific to collaborative filtering, FURL [5] generalizes this approach by (a) defining private and federated parameters and (b) specifying the independent local training constraint and the independent aggregation constraint. The independent local training constraint specifies that the loss function used by local parties is independent of the private parameters of the other parties, while the independent aggregation constraint specifies that the global aggregation step is independent of the private parameters of the parties. When these conditions are met, FURL guarantees no model quality loss from splitting the parameters into private and federated.

An application of FURL to the task of document classification is shown in Fig. 4.3a. Here, the user embeddings are private to each party, while the parameters of the BiLSTM and the MLP are federated parameters that are shared across all users. Each party trains the private and federated parameters jointly but only shares the federated parameters with the aggregator for aggregation. In experiments, personalization through FURL is shown to significantly improve the performance on the document classification task.

Personalization through FURL, however, has several drawbacks. First, using FURL requires incorporating private parameters into the modeling that might require making changes to the network architecture. Subsequently, incorporating the private parameters into the model increases the number of parameters to be learned on the party, which, given the paucity of data on parties, might make the task more difficult. Finally, the FURL technique suffers from a cold-start problem for new parties. Since the private parameters are specific to each party, new parties joining the framework need to first train their private parameters before they can utilize the power of personalization and might suffer from a degraded performance before that.

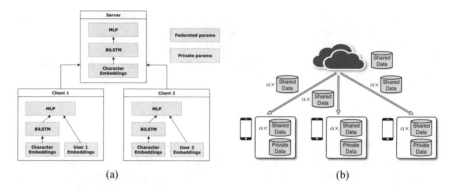

(a)   (b)

Fig. 4.3 Client contextualization and data augmentation strategies for personalization in federated learning. (**a**) FURL document classification model. Federated parameters (character embeddings) are used along with private parameters (user embeddings) on each party. (**b**) Illustration of data-sharing strategy. Each party uses its private data along with a subset of the global shared data to train its local model. (Image source: original paper [66])

4.3.3 Data Augmentation

Data augmentation techniques have been utilized in standard machine learning problems either to alleviate the problems of class imbalance, non-IID datasets, or to artificially inflate the otherwise lower-sized datasets. Techniques for these range from oversampling under-represented class samples [8] to training GANs to generate augmenting data samples [37]. Readers interested in this area should refer to surveys on data augmentation techniques to gain an understanding of the landscape of the field [40, 52].

Since federated learning also suffers from a paucity of data on the parties, while a significantly large amount of data is available globally, it is natural to ask if the global data (data across all parties) can be used to improve the performance of a particular party. In the same spirit, methods have been proposed either to share a small amount of data globally to help improve performance on parties [66] or to train a Generative Adversarial Network (GAN) in addition to the local model to augment data samples [28].

One straightforward way to augment data is to collect a subset of data from all parties to create a global shared dataset that each party can use to augment their local datasets. DAPPER [36] and global-data-sharing [66] methods fall under this category. Both these methods utilize a global dataset D_G that is indicative of the global data distribution. The global-data-sharing method proposes to initialize the federated learning process by sharing a warm-up model trained on the global dataset, along with a random subset of the dataset (αD_G) to each party. Each party augments its local dataset with the dataset provided by the aggregator to train its local model, which is then transmitted back to the aggregator for aggregation. An illustration of this process is shown in Fig. 4.3b.

DAPPER [36], on other hand, instead of directly augmenting the local dataset with the global dataset, optimizes the objective:

$$\lambda D_{\text{party}} + (1 - \lambda) D_G. \tag{4.8}$$

Each party, at each optimization step, selects the local dataset D_{party} with probability λ and the global dataset D_G with probability $(1 - \lambda)$ for optimization. The rest of the optimization and aggregation steps remain unchanged.

Both DAPPER and global-data-sharing methods show significant improvements over models trained with no personalization, but they require the transfer of parties' data to the global aggregator and to other parties as well. Moving party's data outside their machines violates the privacy guarantees of federated learning, and thus these methods might not be directly implementable in practice.

XorMixup [51] aims to circumvent the privacy concerns associated with transferring party data while still utilizing the personalization power of data augmentation. It proposes to use an XOR encoding scheme to obfuscate data samples to upload to the aggregator. Each party implementing XorMixup first selects data samples from two different class labels and then creates an encoded data sample using an XOR of the two. Each party uploads such encoded samples to the aggregator, which decodes them using a separate data sample from the specified class label, and uses these decoded samples to train a model. These encoded samples are shown to have a high dissimilarity with the original data samples and also show improvement in the performance of models under non-IID conditions.

4.3.4 Distillation

Under the general formulations of federated learning, when the aggregator sends a model to the party, it uses that model as the starting point to train on the local data. Personalization through distillation takes a different approach to the problem. Rather than using the central model parameters as a starting point, distillation-based approaches use knowledge distillation [21, 24] to transfer knowledge between models without explicitly copying parameters. A key advantage of utilizing distillation instead of copying model parameters is that the model architectures need not be the same across parties and the aggregator for the framework to work. For example, parties can choose model architectures more suited to their data and/or hardware constraints. Federated Mutual Learning (FML) [50] and FedMD [34] are two main methods that follow this approach.

What is knowledge distillation?
Model compression [11] is the task of reducing the model size, thereby reducing the memory required to store the model and increasing the speed of inference, while preserving the information in the original neural network. Knowledge distillation [21, 24] is a type of model compression technique that aims to effectively transfer information or knowledge from a larger network to a smaller network. There are 3 main components in knowledge distillation— teacher network, student network, and knowledge. The teacher network is the bigger model that encodes the knowledge, and this knowledge needs to be transferred into the student network that is typically smaller in size. There are several ways to define the knowledge to distill [21]—it can either be outputs of certain layers of the network (such as response-based and feature-based knowledge) or it can be relationships between the different layers or data samples (such as relation-based knowledge).

FML adopts a two-way distillation between the local model and the global model. Parties implementing FML maintain a local model that is continuously trained on their data without sharing the data with the aggregator. At each communication round, the aggregator sends the global model to each party that is updated by the party through a two-way knowledge distillation between the global and the local model. The corresponding objective functions are as follows:

$$\mathcal{L}_{\text{local}} = \alpha \mathcal{L}_{\text{local}} + (1 - \alpha) D_{KL}(p_{\text{global}} \| p_{\text{local}}). \tag{4.9}$$

$$\mathcal{L}_{\text{global}} = \beta \mathcal{L}_{\text{global}} + (1 - \beta) D_{KL}(p_{\text{local}} \| p_{\text{global}}). \tag{4.10}$$

Since the connection between the local and global models in FML is through the KL divergence of the output probabilities unlike through parameter copying in other federated learning methods, the local and global model architectures can be different. The original work for FML [50] conducts experiments to demonstrate this effect. Using different network architectures on different parties, the authors show improvement over independent training of a global model on the complete dataset under this setting as well.

FedMD [34] also proposes a similar formulation of personalization using distillation as FML. The FedMD framework requires a public dataset that is shared across parties and the aggregator, along with the private datasets that each party maintains. The framework proceeds by parties first training models on the public dataset followed by training on their respective private dataset and then communicating the class scores for each sample in the public dataset to the central aggregator. An aggregation of these class scores across all parties is used as the target distribution that each party learns from using distillation. Similar to FML, FedMD has the advantage of supporting different model architectures across parties. However,

FedMD requires a large public dataset that needs to be shared between parties, thereby increasing the communication cost between parties and aggregator.

4.3.5 Meta-learning Approach

Contemporary machine learning models are trained to perform well on a single task. Meta-learning or "learning to learn" [25, 57, 59], on the other hand, aims to learn models that can be rapidly adapted to new tasks with only a handful of examples. There are multiple ways of achieving this—metric-based, model-based, and optimization-based methods [59]. In this section, we focus on optimization-based methods that are better suited for our purposes. Optimization-based techniques for meta-learning aim to learn model parameters that can be quickly modified to new tasks given a handful of examples and within a few gradient updates. Model-Agnostic Meta-Learning (MAML) [18] is a fairly popular method that is applicable to any model that can be learned with gradient-based methods. Instead of training model parameters to minimize the loss on a given task, MAML trains model parameters to minimize the loss on tasks after a few parameter adaptation steps. If we consider each task to be a party in federated learning setting, we can draw a parallel between personalized federated learning and meta-learning. We want to train a global model to act as a good initializer for party models such that it is able to adapt, i.e., personalize, quickly to party data distributions. In Sect. 4.2, we reviewed the connections between the naive personalization baseline, i.e., fine-tuning of FedAvg, and meta-learning. We now review other recent methods based on meta-learning.

ARUBA [31] is a framework that combines meta-learning with multi-task learning techniques to enable meta-learning methods to learn and take advantage of task similarities to improve their performance. One of the motivations behind ARUBA is that in meta-learning models, certain model weights act as feature extractors and are transferable across tasks without much modification, while other weights vary greatly. Having a per-coordinate learning rate allows parameters to be adapted at different rates depending on their transferability across tasks. ARUBA, when tested on the next-character prediction task in a federated learning setting, matches the performance of a fine-tuned FedAvg baseline, but without additional hyperparameter optimization over the fine-tuning learning rates.

FedMeta [9]—proposed concurrently with ARUBA—incorporates standard meta-learning algorithms into federated learning setting. Under this setting, the aggregator aims to maintain an initialization that a party can quickly adapt to its local data distribution. The party trains by executing the inner loop (adaptation steps on the support data) of the meta-learning algorithm locally and returns the gradient of the outer loop (query data) back to the aggregator, which uses this information to update its initialization. While FedMeta incorporates meta-learning into personalization by running the meta-learning step on the parties while aggregating the model initialization on the aggregator, Per-FedAvg [17] shows that this formulation may not perform well in some cases. Instead, Per-FedAvg assumes that each party takes the global model as initialization and updates it for one gradient step with respect to its own loss function, changing the problem formulation as follows (4.11):

$$\min_{\theta} F(\theta) := \frac{1}{N} \sum_{i=1}^{N} \mathcal{L}_i(\theta - \alpha \nabla \mathcal{L}_i(\theta)). \tag{4.11}$$

Finally, FedPer [3] proposes to separate the global model into a base network that acts as a feature extractor, and a personalization layer. The parties jointly train the base and the personalization layers but only share the base network with the aggregator for aggregation. This allows the system to learn the representation extraction network using information from multiple parties, while learning a personalization layer specific to each party. The connection between this method and meta-learning is not explicitly explored in the original paper, and there is related work in the meta-learning literature, Almost No Inner Loop (ANIL) [45], which proposes separating the network into a body and head network and only adapting the head to a new task in the inner loop of meta-learning.

4.3.6 Mixture of Models

In the standard formulation of federated learning, the local model is trained on local data, while the global model aggregates information from parties to build a global model. The key incentive for a party to participate in federated learning is to utilize the information on other parties to reduce its generalization error as compared to training a model locally. However, there can be cases where the global model performs worse for certain parties in the federated learning system than the individual models that these parties could train locally [62], e.g., see experiment in Fig. 4.1. This motivates the idea of mixing the global and local models by learning a parameter to optimally combine the two models.

FL+DE [44] learns to combine the predicted class probabilities of the local and global models using a mixture of experts technique [63]. Each party maintains a local domain expert (DE) model trained on the local data, while also collaborating with other parties to build a global model. The gating function $(\alpha_i(x))$—parameterized as a logistic regression model in the original work [44]—is learned along with the federated learning setup to optimally combine the predicted class probabilities of the global model $(\hat{y}_G)$ and the local domain expert $(\hat{y}_i)$. The gating function thus learns regions of preferences between the two models conditioned on the input. The final prediction for a given data sample x is then a convex combination of the predicted class probabilities of the two models:

$$\hat{y}_i = \alpha_i(x)\hat{y}_G(x) + (1 - \alpha_i(x))\hat{y}_{\text{local}}(x). \tag{4.12}$$

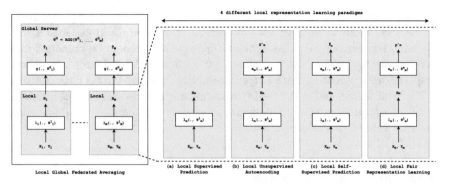

Fig. 4.4 LG-FedAvg—the local models $\ell_i(\,\cdot\,;\theta_i^\ell)$ learn to extract high-level representations H_i given the local data (X_i, Y_i), while the global model $g(\,\cdot\,;\theta^g)$ operates solely on the learned representations H_i. Owing to this bifurcation, the local model can be trained using specialized techniques, 4 of which are shown in this figure

$$\alpha_i(x) = \sigma(w_i^T x + b_i). \tag{4.13}$$

Instead of using a mixture of experts technique to combine the output probabilities, the MAPPER [36] and APFL [15] methods learn a mixing parameter (α) to optimally combine the local and global models. In APFL, while the global model is still trained to minimize the empirical risk on the aggregated domain as in traditional federated learning, the local model (h_{local}) is trained to also incorporate part of the global model (h_g) using α (equation (4.14)). The personalized model for the i-th party is a convex combination of the global model (h_g) and the local model (h_{local}) (equation (4.15)).

$$h_{\text{local}} = \arg\min_h \hat{\mathcal{L}}_{D_i}(\alpha_i h + (1 - \alpha_i)h_g). \tag{4.14}$$

$$h_{\alpha_i} = \alpha_i h_{\text{local}} + (1 - \alpha_i)h_g. \tag{4.15}$$

The three methods we have looked at so far maintain both the local and global models. However, both these models are trained for the same task. LG-FedAvg [35] instead proposes to bifurcate the learning task between the local and global models—each party learns to extract high-level representations from raw data, and the global model operates on these representations (rather than the raw data) from all devices. We depict the LG-FedAvg process for an image classification task in Fig. 4.4. Here, the local models are trained to extract high-level representations from raw images, and while the global model is trained using supervised learning, the local models are free to choose the technique to learn these representations. As shown in Fig. 4.4, these can be learned using supervised prediction task (using an auxiliary model to map representations to predictions) or unsupervised or semi-supervised techniques (sub-figures (a) to (c)). The local models can also be trained to learn fair representation through adversarial training against protected attributes (sub-figure d). This bifurcation has several benefits: (a) Operating the global model on representations rather than raw data reduces the size of the global model, thus reducing the number of parameters and updates that need to be communicated between the aggregator and the parties, (b) it allows the local parties to choose a specialized encoder to extract representations depending on the characteristics of its local dataset rather than using a common global model, and (c) it allows local models to learn fair representations that obfuscate protected attributes thus enhancing the privacy of the local data.

4.3.7 Model Regularization

In conventional supervised federated learning, the system is optimized for the following:

$$\min_{\theta} \left\{ \mathcal{L}(\theta) := \frac{1}{N} \sum_{i=1}^{N} \mathcal{L}_i(\theta) \right\}. \tag{4.16}$$

Here, N is the number of parties, θ are the model parameters, and $\mathcal{L}_i(\theta)$ denotes the loss over the ith party data distribution.

Regularization-based personalization techniques, on the other hand, optimize for a regularized version of the standard objective. Loopless Local Gradient Descent (L2GD) [23], Federated Attentive Message Passing (FedAMP) [26], pFedME [16], and Fed+ [61] are all instantiations of the regularization technique, differing primarily in their definition of the regularization objective.

L2GD [23] defines the regularization objective to be the L2 norm of the difference between the local model parameters (θ_i) and the average parameters across parties ($\bar{\theta}$), and the entire system optimizes the objective defined in equations (4.17) and (4.18). To optimize for this objective, L2GD proposes a non-uniform SGD method with convergence analysis over the required number of communication rounds between the parties and the aggregator. The method views the objective as a 2-sum problem, sampling either $\nabla \mathcal{L}$ or $\nabla \psi$ to estimate ∇F, and defines an unbiased estimator of the gradient as in equation (4.19). At each time step, either the local models take a local gradient step with probability $1 - p$ or the aggregator shifts the local models toward the average with probability p.

$$\text{L2GD}: \min_{\theta_1, \dots, \theta_N} \left\{ F(w) := \mathcal{L}(\theta) + \lambda \psi(\theta) \right\}. \tag{4.17}$$

$$\mathcal{L}(\theta) := \frac{1}{N} \sum_{i=1}^{N} \mathcal{L}_i(\theta_i), \quad \psi(\theta) := \frac{1}{2N} \sum_{i=1}^{N} \|\theta_i - \bar{\theta}\|^2. \tag{4.18}$$

$$G(\theta) := \begin{cases} \frac{\nabla \mathcal{L}(\theta)}{1-p} & \text{with probability } 1 - p, \\ \frac{\lambda \nabla \psi(\theta)}{p} & \text{with probability } p. \end{cases} \tag{4.19}$$

L2GD is applicable to and provides guarantees for convex loss functions and thus is not directly applicable to non-convex loss functions typically encountered in neural network models.

pFedMe [16] models the personalization problem as a bi-level optimization problem through the objective defined as follows:

$$\texttt{pFedMe} : \min_{\theta} \left\{ F(\theta) = \frac{1}{N} \sum_{i=1}^{N} F_i(\theta) \right\}. \qquad (4.20)$$

$$F_i(\theta) = \min_{\theta_i} \left\{ \mathcal{L}_i(\theta_i) + \frac{\lambda}{2} \|\theta_i - \theta\|^2 \right\}. \qquad (4.21)$$

Here, θ_i is the ith party's personalized model, trained on its local data distribution while maintaining a bounded distance from the global model parameters θ at the inner level. The optimal personalized model for a party is then defined as

$$\hat{\theta}_i(\theta) := \text{prox}_{\mathcal{L}_i/\lambda}(\theta) = \arg\min_{\theta_i} \left\{ \mathcal{L}_i(\theta_i) + \frac{\lambda}{2} \|\theta_i - \theta\|^2 \right\}. \qquad (4.22)$$

Similar to FedAvg, a system implementing pFedMe sends the global model weights to the parties at each communication round and performs model aggregation using weights returned by parties after a certain number of local rounds. Unlike FedAvg, party locally minimizes equation (4.21), which is a bi-level optimization problem. At each local round, the party first solves equation (4.22) to find the optimal personalized party parameters $\hat{\theta}_i(\theta_{i,r}^t)$. Here, $\theta_{i,r}^t$ is the local model of party i at global round t and local round r, where $\theta_{i,0}^t = \theta^t$. Thereafter, at the outer level, the party updates the local model $\theta_{i,r}^t$ using gradients with respect to F_i in equation (4.21).

FedAMP [26] proposes the following objective for personalization:

$$\min_{\theta} \left\{ F(\theta) = \sum_{i=1}^{N} \mathcal{L}_i(\theta_i) + \lambda \sum_{i<j}^{N} A(\|\theta_i - \theta_j\|^2) \right\}. \qquad (4.23)$$

The second part of the objective defines an attention-inducing function $A(\|\theta_i - \theta_j\|^2)$, which measures the similarity between party parameters in a non-linear manner, and aims to improve the collaboration between parties. The attention-inducing function can take any form; however, in the proposed work, the authors use the negative exponential function $A(\|\theta_i - \theta_j\|^2) = 1 - e^{-\|\theta_i-\theta_j\|^2/\sigma}$. To optimize for the objective, FedAMP adopts an alternate-optimization strategy, first optimizing $\sum_{i<j}^{N} A(\|\theta_i - \theta_j\|^2)$ on the aggregator through weights collected from all parties, and then optimizing $\mathcal{L}_i(\theta_i)$ on the corresponding parties using their local datasets.

Fed+ [61] argues that robust aggregation allows better handling of the heterogeneity of data across parties and accommodates it through a model regularization approach to personalization. Fed+ introduces a convex penalty function ϕ and constants α, μ as follows:

$$\min_{\theta, z, \bar{\theta}} \frac{1}{N} \left\{ F_{\mu, \alpha}(\theta, z, \bar{\theta}) = \sum_{i=1}^{N} \mathcal{L}_i(\theta_i) + \frac{\alpha}{2} \|\theta_i - z_i\|_2^2 + \mu \phi(z_i - \bar{\theta}) \right\} \quad (4.24)$$

and proposes a robust combination of the current local and global models obtained by minimizing (4.24) with respect to z_i, keeping θ_i and $\bar{\theta}$ fixed. Setting $\phi(\cdot) = \| \cdot \|_2$ gives a ρ-smoothed approximation of the Geometric Median, a form of robust aggregation:

$$z_i \leftarrow \bar{\theta} + \text{prox}_{\phi/\rho}(\theta_i - \bar{\theta}), \quad \text{where} \quad \rho := \mu/\alpha, \text{ hence}$$

$$z_i = (1 - \lambda_i) \theta_i + \lambda_i \bar{\theta}, \quad \text{where} \quad \lambda_i := \min\left\{1, \rho/\|\theta_i - \bar{\theta}\|_2\right\}.$$

To compute the robust $\bar{\theta}$ from $\{\theta_i\}$, the aggregator runs the following two-step iterative procedure initialized with $\bar{\theta} = \theta_{\text{mean}} := \text{Mean}\{\theta_i\}$ until $\bar{\theta}$ converges:

$$v_i \leftarrow \max\left\{0, 1 - (\rho/\|\theta_i - \bar{\theta}\|_2)\right\} (\theta_i - \bar{\theta}),$$

$$\bar{\theta} \leftarrow \theta_{\text{mean}} - \text{Mean}\{v_i\}.$$

4.3.8 Multi-task Learning

Traditional machine learning approaches typically optimize a model for a single task. Multi-task learning (MTL) [4, 7, 54] extends this traditional approach to learn multiple tasks jointly, thus exploiting the commonalities and differences between the tasks to potentially enhance the performance on individual tasks. Since these methods can learn relationships between non-IID and unbalanced datasets, they are well suited to apply to the federated learning setting [53]. Readers interested in multi-task learning should refer to the following survey papers [46, 65] to gain an overview of the field.

While MTL methods are appealing in the federated learning context, they do not account for the communication challenges such as fault tolerance and stragglers in the framework. MOCHA [53] was the first framework for federated learning using multi-task learning that factored in fault tolerance and stragglers during training. MTL approaches generally formulate the problem as follows:

$$\min_{w, \Omega} \left\{ \sum_{i=1}^{N} \mathcal{L}_i(\theta_i) + R(\Omega) \right\}. \quad (4.25)$$

Here, N is the total number of tasks, and $\mathcal{L}_i(\theta_i)$ and θ_i are the loss function and parameters for task i. The matrix $\Omega \in \mathbb{R}^{N \times N}$ models relationships among tasks and is either known a priori or is estimated while simultaneously learning task models. MTL approaches differ in their formulation of R that promotes suitable structure among the tasks through the Ω matrix. MOCHA optimizes this objective in the federated learning setting using the objective's distributed primal–dual optimization formulation. This allows it to separate computation across nodes by only requiring data available on the party to update the local model parameters. MOCHA showed the applicability of multi-task learning approaches to federated learning setting and showed improved performance as compared to global and local models trained on the experimental datasets. It, however, is designed for only convex models and is therefore inapplicable to non-convex deep learning models.

VIRTUAL (variational federated multi-task learning) [14] extends the multi-task learning framework to non-convex models through the usage of variational inference methods. Given N parties, each with datasets D_i, local model parameters θ_i, and the central model parameters θ, VIRTUAL computes the posterior distribution

$$p(\theta, \theta_1, \ldots, \theta_N | D_{1:N}) \propto \frac{\prod_{i=1}^{N} p(\theta, \theta_i | D_i)}{p(\theta)^{N-1}}. \tag{4.26}$$

This posterior distribution assumes: (1) that party data is conditionally independent given aggregator and party parameters, $p(D_{1:N}|\theta, \theta_1, \ldots, \theta_N) = \prod_{i=1}^{N} p(D_i|\theta, \theta_i)$, and (2) a factorization of the prior as $p(\theta, \theta_1, \ldots, \theta_N) = p(\theta) \prod_{i=1}^{N} p(\theta_i)$. Since the posterior distribution defined in equation (4.26) is intractable, the algorithm proposes an expectation propagation like algorithm [41] to approximate the posterior.

4.4 Benchmarks for Personalization Techniques

In this section, we review datasets suitable for benchmarking methods for personalization in federated learning. We consider datasets with non-IID party data distributions, i.e., each party's data is sampled from a different distribution. We review datasets used in the prior works, as well as other datasets that might fit the personalization problem setting.

Broadly classified, prior works in this domain have used one of the following types of datasets: (a) Synthetic datasets, where a generative process for the dataset is defined to generate data samples for parties, (b) Simulating federated datasets by partitioning commonly available datasets such as MNIST or CIFAR-10 [32] according to some hypothesis, and (c) Utilizing datasets that have a natural partitioning such as data collected from multiple parties. We now look at each of these types in detail.

4.4.1 Synthetic Federated Datasets

A fairly common way of generating synthetic federated dataset is to follow the method proposed by Shamir et al. [49], with some added modifications to inject heterogeneity among parties. While the exact way of generating the dataset varies between proposed methods, the underlying process is as follows: for each device k, samples (X_k, Y_k) are generated according to the model $y = \arg\max(\text{softmax}(Wx + b))$. The model parameters W_k and b_k are controlled by a parameter α and are sampled as: $u_k \sim \mathcal{N}(0, \alpha)$, $W_k \sim \mathcal{N}(u_k, 1)$, and $b_k \sim \mathcal{N}(u_k, 1)$. The generation of X_k is controlled by the second parameter β and is sampled as: $B_k \sim \mathcal{N}(0, \beta)$, $v_k \sim \mathcal{N}(B_k, 1)$, Σ is a diagonal covariance matrix with $\Sigma_{j,j} = j^{-1.2}$, and $x_k \sim \mathcal{N}(v_k, \Sigma)$.

This synthetic dataset `Synthetic`(α, β) has two parameters: α and β. Here, α controls how much local models differ from each other, and β controls how much the local data on each device differs from data on other parties.

4.4.2 Simulating Federated Datasets

A common way of simulating federated datasets is to use commonly available datasets and partitioning them across parties according to a hypothesis. Prior works have generally utilized datasets such as MNIST,[2] CIFAR-10, and CIFAR-100 [32] for the task. Since these datasets have no specific natural feature that can be used to partition them across parties, we need to partition them according to a hypothesis. One way to partition is to sample data points for each party from a particular subset of classes, to ensure that parties do not see data from all classes and thus do not have features representative of all the classes in the dataset. Another way of partitioning includes using a probabilistic allocation of data samples across parties—such as sampling $p_k \sim \text{Dir}_N(\alpha)$, and allocating $p_{k,i}$ proportion of instances of class k to party i [64].

Synthetically created federated datasets have the advantage of allowing control over the amount of heterogeneity in the dataset; however, they are limited by the number of parties they can support. Prior works have set up their experiments with the number of parties in the order of tens. While this is suitable for federated learning in an enterprise setting where typically the number of parties does not scale too much, this setting does not consider the scale typically encountered in smartphones or IoT type of applications.

---

[2] http://yann.lecun.com/exdb/mnist/

Table 4.2 Federated learning datasets

Dataset	Task	No. of parties	Total samples	Samples per device	
				Mean	Std
FEMNIST	Image classification	3,550	805,263	226.83	88.94
CelebA	Image classification	9,343	200,288	21.44	7.63
Shakespeare	Language modeling	1,129	4,226,158	3,743.28	6, 212.26
Reddit	Language modeling	1,660,820	56,587,343	34.07	62.95
Sent140	Sentiment analysis	660,120	1,600,498	2.42	4.71

4.4.3 Public Federated Datasets

Besides synthetic and simulated federated datasets, there are datasets available, which support natural partitioning of data among parties. LEAF [6] is a popular benchmark for federated learning methods that provides multiple datasets for image classification, language modeling, and sentiment analysis tasks. These datasets are the preferred datasets for benchmarking personalization techniques due to their proximity to real-life non-IID characteristics of data, and the scale of parties they support. Some of the datasets included in LEAF are:

- **FEMNIST:** Extended MNIST (EMNIST) [13] is a dataset containing hand-written samples of digits, uppercase, and lowercase characters for a total of 62 class labels. The EMNIST dataset is partitioned by the original writers of the handwritten samples to create the FEMNIST dataset with over 3,500 parties.
- **Shakespeare dataset:** Intended for language modeling task, this dataset is built from *The Complete Works of William Shakespeare*,[3] by considering each speaking character in each play as a separate party.

Details about the datasets available in LEAF along with their aggregate and party-level statistics can be found in Table 4.2.

In addition to the datasets provided in LEAF, there are other datasets that have the characteristics required for personalization in federated learning task. Some of these datasets are:

- **Google Landmarks Dataset v2 (GLDv2):** The GLDv2[4] is a large-scale fine-grained dataset intended for instance recognition and image retrieval tasks [60]. It is composed of approximately 5 million images of human-made and natural landmarks across 246 countries, with approximately 200,000 distinct labels for these images. This dataset can be utilized in the federated learning setting by either partitioning according to the geographical location of the landmarks or the

---

[3] http://www.gutenberg.org/ebooks/100

[4] https://github.com/cvdfoundation/google-landmark

landmark categories, or the author. Given its scale and diversity, this dataset is a strong test bed for the personalization task.

- **MIMIC-III:** The Medical Information Mart for Intensive Care III (MIMIC-III) [30] is a large-scale de-identified health-related data of over 40,000 patients who stayed at the critical care units of a hospital in Boston, Massachusetts between 2001 and 2012. It includes information such as demographics, vital sign measurements, laboratory test results, procedures, medications, caregiver notes, imaging reports, and mortality (both in and out of hospital), for over 60,000 critical care unit stays. While the scale of this dataset is limited in comparison to the GLDv2, it is one of the largest available medical datasets and thus provides an important benchmark for evaluating personalization in federated learning.

4.5   Personalization as the Incidental Parameters Problem

There is a possible theoretical explanation for the limitations of personalization in federated learning: the incidental parameters problem. We consider a general model of personalization in federated learning: the aggregator and the parties aim to solve an optimization problem of the form

$$\min_{\theta, \theta_1, \ldots, \theta_N} \frac{1}{N} \sum_{i=1}^{N} \mathcal{L}_i(\theta, \theta_i), \tag{4.27}$$

where θ are the shared model parameters, the θ_i's are the party-specific parameters, and $\mathcal{L}_i$ is the empirical risk on the i-th party. In most federated learning settings, the sample size per party is limited, so the party-specific parameters θ_i are only estimated up to a certain accuracy that depends on the sample size per party. We may hope that it is possible to estimate the shared parameters θ more accurately, but this is only possible up to a point.

The population version of the problem is

$$\min_{\theta, \theta_1, \ldots, \theta_N} \frac{1}{N} \sum_{i=1}^{N} \mathcal{R}_i(\theta, \theta_i), \tag{4.28}$$

where $\mathcal{R}_i$ is the (population) risk on the i-th party: $\mathcal{R}_i(\cdot) \triangleq E[\mathcal{L}_i(\cdot)]$. Let $(\widehat{\theta}, \widehat{\theta}_1, \ldots, \widehat{\theta}_N)$ and $(\theta^*, \theta_1^*, \ldots, \theta_N^*)$ be the argmin of (4.27) and (4.28), respectively. The estimates of the shared parameters satisfy the *score equations*:

$$0 = \frac{1}{N} \sum_{i=1}^{N} \partial_\theta \mathcal{L}_i(\widehat{\theta}, \widehat{\theta}_1, \ldots, \widehat{\theta}_N). \tag{4.29}$$

Expanding the score equations around $(\theta^*, \theta_1^*, \ldots, \theta_N^*)$ (and dropping the higher-order terms), we have

$$0 = \frac{1}{N} \sum_{i=1}^{N} \partial_\theta \mathcal{L}_i(\theta^*, \theta_1^*, \ldots, \theta_N^*) + \partial_\theta^2 \mathcal{L}_i(\theta^*, \theta_1^*, \ldots, \theta_N^*)(\widehat{\theta} - \theta^*)$$

$$+ \partial_{\theta_i} \partial_\theta \mathcal{L}_i(\theta^*, \theta_1^*, \ldots, \theta_N^*)(\widehat{\theta}_i - \theta_i^*).$$

We rearrange to isolate the estimation error in the shared parameters

$$\widehat{\theta} - \theta^* = \left(\frac{1}{N} \sum_{i=1}^{n} \partial_\theta^2 \mathcal{L}_i(\theta^*, \theta_1^*, \ldots, \theta_N^*) \right)^{-1} \left(\begin{array}{c} \frac{1}{N} \sum_{i=1}^{N} \partial_\theta \mathcal{L}_i(\theta^*, \theta_1^*, \ldots, \theta_N^*) \\ + \partial_{\theta_i} \partial_\theta \mathcal{L}_i(\theta^*, \theta_1^*, \ldots, \theta_N^*)(\widehat{\theta}_i - \theta_i^*) \end{array} \right).$$

We see that the estimation error in the party-specific parameters affects the estimation error of the shared parameters through the average of the terms $\partial_{\theta_i} \partial_\theta \mathcal{L}_i(\theta^*, \theta_1^*, \ldots, \theta_N^*)(\widehat{\theta}_i - \theta_i^*)$. This average is generally not mean zero, so it does not converge to zero even as the number of parties grows. For this average to converge to zero, one of two things must occur:

1. $\widehat{\theta}_i - \theta_i^* \xrightarrow{p} 0$: the estimation errors of the personalized parameters converge to zero. This is only possible if the sample size per party grows. Unfortunately, computational and storage constraints on the parties preclude this scenario in most federated learning problems.
2. $\partial_{\theta_i} \partial_\theta \mathcal{L}_i(\theta^*, \theta_1^*, \ldots, \theta_N^*)$ is the mean zero. This is equivalent to the score equations (4.29) satisfying a certain orthogonality property [12, 42]. If the score equations satisfy this property, then the estimation errors of the party-specific parameters do not affect (to first-order) the estimates of the shared parameters. Although this is highly desirable, orthogonality only occurs in certain special cases.

In summary, the estimation error in the shared parameters is generally affected by the estimation errors in the party-specific parameters, and it does not converge to zero in realistic federated learning settings in which the number of parties grows but the samples size per party remains bounded. Taking a step back, this is to be expected from a degree-of-freedom point of view. As we grow the number of parties, although the total sample size increases, the total number of parameters we must learn also increases. In other words, personalization in federated learning is a high-dimensional problem. Such problems are known to be challenging, and estimation errors generally do not converge to zero unless the parameters exhibit special structure (e.g., sparsity, low rank). Unfortunately, this is typically not the case in federated learning.

Practically, this means that if we incorporate personalization in a federated learning problem, it is profligate to increase the number of parties beyond a certain point without increasing the sample size per party. Whether we are beyond this point can be ascertained by checking whether the quality of the shared parameters estimates is improving as more parties are added. If we are beyond this point, the estimation error in the party-specific parameters is dominating the estimation error in the shared parameters, so there is no benefit to parameter sharing. This is known as the *incidental parameters* problem, and it has a long history in statistics. We refer to [33] for a review of the problem.

4.6   Conclusion

In this chapter, we motivate the need for personalization by demonstrating that native federated learning does not necessarily help all parties train a better model as compared to them training the models locally. To alleviate this problem, different techniques for personalization have been proposed. We group these techniques into eight major categories based on the type of personalization strategy they employ. In addition to a review of personalization strategies, we also provide an overview of the statistical challenges of personalization in federated learning. We conclude this chapter by providing recommendations for practical considerations while implementing or utilizing personalization strategies, and future research directions, and open problems in theoretical understanding of personalization in federated learning.

Practical considerations. Choosing a personalization strategy for your application is closely tied to the properties of the parties and the aggregator participating in the federated learning setup. Specifically, the questions that will help in making an informed choice are: (1) Do all parties have the same model architecture? (2) Is there a data-sharing mechanism available to augment local data? (3) What are the compute capabilities available on the participating parties and the aggregator? and (4) How much data do you expect to be present on each party?

If not all parties have the same model architecture, or if it is preferable to have different architectures for different parties, then either distillation-based approaches (Sect. 4.3.4) or the LG-FedAvg [35] method should be explored. These techniques support and have been experimentally shown to work with different party and global model architectures. Another important consideration is if there is global data available to augment the local data. While sharing party data violates the core tenet of federated learning, if it is possible to collect a shared dataset, data augmentation techniques for personalization (Sect. 4.3.3) can be powerful candidates in these scenarios.

The party and aggregator compute capabilities also play an important role in selecting a personalization strategy. Specifically for parties, if the compute and memory capabilities are sufficiently available, then a mixture of models approach (Sect. 4.3.6) can be explored. Because the mixture of models approach maintains a local and a global model on the parties and uses a combination of the two for inference, it significantly increases the memory and compute requirements for the participating parties. On similar lines, if the aggregator has enough memory capacity to allow for maintaining multiple model parameters, then client (party) clustering approaches (Sect. 4.3.1) might be helpful.

The final question that will help in making an informed decision is regarding the amount of data available on each party. This is an important consideration to apply contextualization approaches. Client (party) contextualization (Sect. 4.3.2) increases the number of parameters to learn from the local data, and if sufficient data is available on parties, then it might be possible to learn these contextual parameters to help in personalization.

Lastly, meta-learning (Sect. 4.3.5) and model regularization (Sect. 4.3.7) approaches are applicable irrespective of whether the aforementioned conditions are met or not and should always be considered when choosing a personalization strategy.

Advancing personalization in federated learning. With the growing number of personalization algorithms proposed in the literature, an important next step in our opinion is to establish benchmarks and performance metrics to effectively and reliably measure the performance of the proposed techniques. This requires the availability of datasets that mimic the conditions typically encountered in practical deployments. While some datasets exist for this purpose, there is a need for more datasets in a broader range of application domains. Benchmarking on standardized datasets will allow for better interpretation of the capabilities and limitations of the proposed techniques and will also enable easy comparison across the techniques. In addition to datasets, there is a need for a standardized evaluation setup for personalization. The typical way of evaluating federated learning techniques is to measure the performance of the global model, and this technique has been ported over to the personalization problem as well. However, as we saw in the motivating example for personalization, measuring the global model accuracy does not necessarily provide a complete picture of the performance of each party. Thus, defining an evaluation setup that considers the performance of each party will serve as an important contribution for effectively evaluating personalization techniques.

Theoretical understanding of personalization. As we saw, there is a scalability issue with personalization due to the incidental parameters problem. This issue is distinguished from most scalability issues in machine learning by its statistical nature. Overcoming this issue is a requirement for scaling personalization to large party clouds. Unfortunately, a general solution to the underlying incidental parameters problems has eluded the statistics community for the better part of a century, so it is unlikely that there is a general way to perform personalization at scale. However, it may be possible to develop solutions tailored to the particular model/application, and this is a rich area of future work.

References

1. Amir S, Wallace BC, Lyu H, Silva PCMJ (2016) Modelling context with user embeddings for sarcasm detection in social media. arXiv preprint arXiv:160700976
2. Ammad-Ud-Din M, Ivannikova E, Khan SA, Oyomno W, Fu Q, Tan KE, Flanagan A (2019) Federated collaborative filtering for privacy-preserving personalized recommendation system. arXiv preprint arXiv:190109888
3. Arivazhagan MG, Aggarwal V, Singh AK, Choudhary S (2019) Federated learning with personalization layers. arXiv preprint arXiv:191200818
4. Baxter J (2000) A model of inductive bias learning. J Artif Intell Res 12:149–198
5. Bui D, Malik K, Goetz J, Liu H, Moon S, Kumar A, Shin KG (2019) Federated user representation learning. arXiv preprint arXiv:190912535
6. Caldas S, Duddu SMK, Wu P, Li T, Konečnỳ J, McMahan HB, Smith V, Talwalkar A (2018) Leaf: a benchmark for federated settings. arXiv preprint arXiv:181201097
7. Caruana R (1997) Multitask learning. Mach Learn 28(1):41–75
8. Chawla NV, Bowyer KW, Hall LO, Kegelmeyer WP (2002) Smote: synthetic minority over-sampling technique. J Artif Intell Res 16:321–357
9. Chen F, Luo M, Dong Z, Li Z, He X (2018) Federated meta-learning with fast convergence and efficient communication. arXiv preprint arXiv:180207876
10. Chen M, Suresh AT, Mathews R, Wong A, Allauzen C, Beaufays F, Riley M (2019) Federated learning of n-gram language models. In: Proceedings of the 23rd conference on computational natural language learning (CoNLL), pp 121–130
11. Cheng Y, Wang D, Zhou P, Zhang T (2017) A survey of model compression and acceleration for deep neural networks. arXiv preprint arXiv:171009282
12. Chernozhukov V, Chetverikov D, Demirer M, Duflo E, Hansen C, Newey W, Robins J (2017) Double/debiased machine learning for treatment and causal parameters. arXiv:160800060 [econ, stat] 1608.00060
13. Cohen G, Afshar S, Tapson J, Van Schaik A (2017) EMNIST: extending MNIST to handwritten letters. In: 2017 international joint conference on neural networks (IJCNN). IEEE, pp 2921–2926
14. Corinzia L, Beuret, A, Buhmann JM (2019) Variational federated multi-task learning. arXiv preprint arXiv:190606268
15. Deng Y, Kamani MM, Mahdavi M (2020) Adaptive personalized federated learning. arXiv preprint arXiv:200313461
16. Dinh CT, Tran NH, Nguyen TD (2020) Personalized federated learning with Moreau envelopes. arXiv preprint arXiv:200608848
17. Fallah A, Mokhtari A, Ozdaglar A (2020) Personalized federated learning: a meta-learning approach. arXiv preprint arXiv:200207948
18. Finn C, Abbeel P, Levine S (2017) Model-agnostic meta-learning for fast adaptation of deep networks. In: International conference on machine learning. PMLR, pp 1126–1135
19. Ghosh A, Hong J, Yin D, Ramchandran K (2019) Robust federated learning in a heterogeneous environment. arXiv preprint arXiv:190606629
20. Ghosh A, Chung J, Yin D, Ramchandran K (2020) An efficient framework for clustered federated learning. arXiv preprint arXiv:200604088
21. Gou J, Yu B, Maybank SJ, Tao D (2020) Knowledge distillation: a survey. arXiv preprint arXiv:200605525
22. Grbovic M, Cheng H (2018) Real-time personalization using embeddings for search ranking at Airbnb. In: Proceedings of the 24th ACM SIGKDD international conference on knowledge discovery & data mining, pp 311–320
23. Hanzely F, Richtárik P (2020) Federated learning of a mixture of global and local models. arXiv preprint arXiv:200205516
24. Hinton G, Vinyals O, Dean J (2015) Distilling the knowledge in a neural network. arXiv preprint arXiv:150302531

25. Hospedales T, Antoniou A, Micaelli P, Storkey A (2020) Meta-learning in neural networks: a survey. arXiv preprint arXiv:200405439
26. Huang Y, Chu L, Zhou Z, Wang L, Liu J, Pei J, Zhang Y (2021) Personalized cross-silo federated learning on non-IID data. In: Proceedings of the AAAI conference on artificial intelligence, vol 35, pp 7865–7873
27. Jaech A, Ostendorf M (2018) Personalized language model for query auto-completion. arXiv preprint arXiv:180409661
28. Jeong E, Oh S, Kim H, Park J, Bennis M, Kim SL (2018) Communication-efficient on-device machine learning: federated distillation and augmentation under non-IID private data. arXiv preprint arXiv:181111479
29. Jiang Y, Konečnỳ J, Rush K, Kannan S (2019) Improving federated learning personalization via model agnostic meta learning. arXiv preprint arXiv:190912488
30. Johnson AE, Pollard TJ, Shen L, Li-Wei HL, Feng M, Ghassemi M, Moody B, Szolovits P, Celi LA, Mark RG (2016) MIMIC-III, a freely accessible critical care database. Sci Data 3(1):1–9
31. Khodak M, Balcan MF, Talwalkar A (2019) Adaptive gradient-based meta-learning methods. arXiv preprint arXiv:190602717
32. Krizhevsky A, Hinton G et al (2009) Learning multiple layers of features from tiny images
33. Lancaster T (2000) The incidental parameter problem since 1948. J Econ 95(2):391–413. https://doi.org/10.1016/S0304-4076(99)00044-5
34. Li D, Wang J (2019) FedMD: Heterogenous federated learning via model distillation. arXiv preprint arXiv:191003581
35. Liang PP, Liu T, Ziyin L, Allen NB, Auerbach RP, Brent D, Salakhutdinov R, Morency LP (2020) Think locally, act globally: federated learning with local and global representations. arXiv preprint arXiv:200101523
36. Mansour Y, Mohri M, Ro J, Suresh AT (2020) Three approaches for personalization with applications to federated learning. arXiv preprint arXiv:200210619
37. Mariani G, Scheidegger F, Istrate R, Bekas C, Malossi C (2018) BAGAN: data augmentation with balancing GAN. arXiv preprint arXiv:180309655
38. McGraw I, Prabhavalkar R, Alvarez R, Arenas MG, Rao K, Rybach D, Alsharif O, Sak H, Gruenstein A, Beaufays F et al (2016) Personalized speech recognition on mobile devices. In: 2016 IEEE international conference on acoustics, speech and signal processing (ICASSP). IEEE, pp 5955–5959
39. McMahan B, Moore E, Ramage D, Hampson S, y Arcas BA (2017) Communication-efficient learning of deep networks from decentralized data. In: Artificial intelligence and statistics. PMLR, pp 1273–1282
40. Mikołajczyk A, Grochowski M (2018) Data augmentation for improving deep learning in image classification problem. In: 2018 international interdisciplinary PhD workshop (IIPhDW). IEEE, pp 117–122
41. Minka TP (2013) Expectation propagation for approximate Bayesian inference. arXiv preprint arXiv:13012294
42. Neyman J (1979) $C(\alpha)$ tests and their use. Sankhyā: Indian J Stat Ser A (1961–2002) 41(1/2):1–21
43. Nichol A, Achiam J, Schulman J (2018) On first-order meta-learning algorithms. arXiv:180302999 [cs] 1803.02999
44. Peterson D, Kanani P, Marathe VJ (2019) Private federated learning with domain adaptation. arXiv preprint arXiv:191206733
45. Raghu A, Raghu M, Bengio S, Vinyals O (2019) Rapid learning or feature reuse? Towards understanding the effectiveness of MAML. In: International conference on learning representations
46. Ruder S (2017) An overview of multi-task learning in deep neural networks. arXiv preprint arXiv:170605098
47. Sattler F, Müller KR, Samek W (2020) Clustered federated learning: model-agnostic distributed multitask optimization under privacy constraints. IEEE Trans Neural Netw Learn Syst 1–13. https://doi.org/10.1109/TNNLS.2020.3015958

48. Schafer JB, Frankowski D, Herlocker J, Sen S (2007) Collaborative filtering recommender systems. In: Brusilovsky P, Kobsa A, Nejdl W (eds) The adaptive web: methods and strategies of web personalization. Springer, pp 291–324. https://doi.org/10.1007/978-3-540-72079-9_9
49. Shamir O, Srebro N, Zhang T (2014) Communication-efficient distributed optimization using an approximate Newton-type method. In: International conference on machine learning. PMLR, pp 1000–1008
50. Shen T, Zhang J, Jia X, Zhang F, Huang G, Zhou P, Wu F, Wu C (2020) Federated mutual learning. arXiv preprint arXiv:200616765
51. Shin M, Hwang C, Kim J, Park J, Bennis M, Kim SL (2020) XOR mixup: privacy-preserving data augmentation for one-shot federated learning. arXiv preprint arXiv:200605148
52. Shorten C, Khoshgoftaar TM (2019) A survey on image data augmentation for deep learning. J Big Data 6(1):1–48
53. Smith V, Chiang CK, Sanjabi M, Talwalkar A (2017) Federated multi-task learning. arXiv preprint arXiv:170510467
54. Thrun S (1996) Is learning the n-th thing any easier than learning the first? In: Advances in neural information processing systems. Morgan Kaufmann Publishers, San Mateo, pp 640–646
55. Thrun S (1998) Lifelong learning algorithms. In: Learning to learn. Springer, Boston pp 181–209
56. Vanschoren J (2018) Meta-learning: a survey. arXiv preprint arXiv:181003548
57. Vilalta R, Drissi Y (2002) A perspective view and survey of meta-learning. Artif Intell Rev 18(2):77–95
58. Vu T, Nguyen DQ, Johnson M, Song D, Willis A (2017) Search personalization with embeddings. In: European conference on information retrieval. Springer, pp 598–604
59. Weng L (2018) Meta-learning: learning to learn fast. lilianwenggithubio/lil-log. http://lilianweng.github.io/lil-log/2018/11/29/meta-learning.html
60. Weyand T, Araujo A, Cao B, Sim J (2020) Google Landmarks Dataset v2-a large-scale benchmark for instance-level recognition and retrieval. In: Proceedings of the IEEE/CVF conference on computer vision and pattern recognition, pp 2575–2584
61. Yu P, Kundu A, Wynter L, Lim SH (2021) Fed+: a unified approach to robust personalized federated learning. 2009.06303
62. Yu T, Bagdasaryan E, Shmatikov V (2020) Salvaging federated learning by local adaptation. arXiv preprint arXiv:200204758
63. Yuksel SE, Wilson JN, Gader PD (2012) Twenty years of mixture of experts. IEEE Trans Neural Netw Learn Syst 23(8):1177–1193
64. Yurochkin M, Agarwal M, Ghosh S, Greenewald K, Hoang N, Khazaeni Y (2019) Bayesian nonparametric federated learning of neural networks. In: International conference on machine learning. PMLR, pp 7252–7261
65. Zhang Y, Yang Q (2017) A survey on multi-task learning. arXiv preprint arXiv:170708114
66. Zhao Y, Li M, Lai L, Suda N, Civin D, Chandra V (2018) Federated learning with non-IID data. arXiv preprint arXiv:180600582

Chapter 5
Personalized, Robust Federated Learning with Fed+

Pengqian Yu, Achintya Kundu, Laura Wynter, and Shiau Hong Lim

Abstract Fed+ is a unified family of methods designed to better accommodate the real-world characteristics found in federated learning training, such as the lack of IID data across parties and the need for robustness to outliers. Fed+ does not require all parties to reach a consensus, allowing each party to train local, personalized models through a form of regularization while benefiting from the federation to improve accuracy and performance. The methods included in the Fed+ family are shown to be provably convergent. Experiments indicate that Fed+ outperform other methods when data is not IID, and the robust versions of Fed+ outperform other methods in the presence of outliers.

5.1 Introduction

Enabling parties to train large machine learning models without sharing data through federated learning allows satisfying privacy concerns, and, when data available at any one party would be insufficient, it allows increasing accuracy and reducing training times. Parallel distributed stochastic gradient descent (SGD) bears similarities to federated learning, but in federated learning communication is minimized by parties performing a number of iterations locally before sending their parameters to the aggregator. Contrary to the usual distributed SGD use cases, federations tend to be diverse. This can result in non-Independently-and-Identically-Distributed (i.e., non-IID) data across parties, which in turn has a significant impact on the algorithms performance. In addition, FL settings often involve parties whose data are outliers with respect to the other parties. Indeed, the most likely real-world settings will involve heterogeneous, non-IID data, and outliers. Many federated aggregation procedures, however, can result in failure of the training process itself, in that the training does not converge or converges to a poor solution, when

P. Yu · A. Kundu · L. Wynter (✉) · S. H. Lim
IBM Research, Singapore, Singapore
e-mail: achintya.k@ibm.com; lwynter@sg.ibm.com; shonglim@sg.ibm.com

© The Author(s), under exclusive license to Springer Nature Switzerland AG 2022   99
H. Ludwig, N. Baracaldo (eds.), *Federated Learning*,
https://doi.org/10.1007/978-3-030-96896-0_5

datasets are too heterogeneous. The fact that this can occur then in precisely those applications where federated learning would have the greatest benefit implies that an approach designed for heterogeneous FL is required in practice.

Personalization of federated learning model training, when judiciously performed, is one means of avoiding such training failure. In addition, personalization of federated training allows for greater accuracy on the data that matters to each party. As we shall see in this chapter, personalization means that each party solves its own problem, but benefits from the federation as a whole through a distinct component of the problem they each solve.

Real-world federated learning settings also often require the party models to perform well on data specific to each party. A single common model across all parties may not perform optimally for each party. As remarked by [15], an application of sentence completion for a mobile user should be optimized for that user's context, and not be identical across all users. While it is possible with e.g. neural network models to do fine tuning locally upon a pre-trained model, the resulting solution is unlikely to be as good in performance as a fully personalized federated model.

We discuss in this chapter personalized methods for federated learning. Personalized federated learning is designed to address precisely these issues of avoiding training failure, increasing robustness to outliers and stragglers, and improving performance on the applications of interest, when party-level evaluation data need not be common across parties.

Another aspect important to the success of federated learning in practice is the ability to produce robust solutions. Robustness in a federation means that the aggregate solution will not be overly skewed by the outliers that may be a party whose data is significantly different from the other parties, or a party that experiences corruption when its update is sent to the aggregator. Most federated learning algorithms aggregate all the active parties using the mean. However, other measures, such as the median, are more robust to outliers than the mean.

In this chapter we shall discuss personalized federated learning models that allow the central server to employ both mean and robust methods of aggregating local models while keeping the structure of local computation intact. These personalized methods also relax the requirement that the parties must reach a full consensus.

It is important in federated learning not to make explicit assumptions on the distributions of the local data, which are assumed private to each party and so for which assumptions can be neither verified nor enforced. Hence, we discuss forms of personalized federated learning that assumes a global shared parameter space with locally computed loss functions.

We call the approach described in this chapter Fed+. The Fed+ theory is equipped to handle heterogeneous data and heterogeneous computing environments including stragglers. The convergence results presented here do not make assumptions on the homogeneity of the party-level data, and they cover the general setting where the number of local update steps across parties can vary. Through Fed+, we define a unified framework for personalized federated learning that handles both average and robust aggregation methods and provide a comprehensive convergence theory for this class of methods, which includes non-smooth convex and non-convex loss functions. We also illustrate the methods through a set of numerical experiments comparing personalized versus non-personalized federated learning algorithms.

5.2 Literature Review

The original and most intuitive federated learning algorithm is FederatedAveraging (FedAvg; [16]). However, its element-wise mean update is susceptible to corrupted values by faulty parties [21]. Furthermore, it has been observed in practice that FedAvg can lead to failure of the training process. The work by [13] showed that FedAvg as initially defined can converge to a point, which is not a solution to the original problem; the authors proposed to add a learning rate that decreases at each federated round, and with that modification they provide a theoretical convergence guarantee for the algorithm, even when the data is not IID. On the other hand, the resulting algorithm is very slow to converge and hence not efficient in practice.

Robustness of federated learning to corrupted updates from the parties, or, more generally, to outliers has been explored in some recent work. We focus on two such works in particular. The authors of [19] propose Robust Federated Aggregation (RFA) and argue that federated learning can be made robust to corrupted updates by replacing the weighted arithmetic mean aggregation by an approximate geometric median. The authors of [22] propose a Byzantine-robust distributed statistical learning algorithm based on the coordinate-wise median of model weights, with a focus on achieving optimal statistical performance. Both robust federated learning algorithms, RFA [19] and coordinate-wise median [22], involve training a single global model on distributed data with assumptions on the maximum percentage of adversarial parties, usually that it is less than one half. As we will see, such methods are not robust to non-IID data and can, as FedAvg, lead to failure of the learning process.

In an effort to better handle non-IID data, in [11] a regularization term is introduced in the form of a proximal distance from the FedAvg solution in their FedProx algorithm. The work of [9, 13] seeks to explain the non-convergence of FedAvg while proposing new algorithms: The works of [3, 14, 17] propose new algorithms, called FedSplit and LocalUpdate, and Local Fixed Point, resp., and obtain tight bounds on the number of communication rounds required to achieve an ϵ accuracy. These algorithms all require the convergence of all parties to a common model.

The works by [5, 7, 8, 12, 15, 20, 23] all advocate, as we do, for a fully personalized approach whereby each client trains a local model while contributing to a global model. In [15], the authors propose explicitly clustering parties and solving the aggregate model only within each cluster. While this approach would likely eliminate the training failure we observe in practice, it adds considerable overhead and reduces the benefit of the federation by decreasing the likely size of each federation. The authors of [5] propose a method that has similarity to our

FedPlus+. In [20], it is proposed to use a more complex procedure where not only the local parameters are optimized but each party also optimizes a (local version of) the global parameters. The work of [8] provides a unification of a number of personalized federated aggregation methods in the smooth and convex settings. The authors of [12] propose a bilevel programming framework that alternates between solving for an average aggregate solution and a local solution. Bilevel programs however are non-convex and the problem could be more readily solved in two phases, first obtaining the average aggregate solution and then using it for each local party. In [23] it is suggested that the average aggregate solution can be personalized for each as a set of weighted average aggregate solutions. The most important difference between Fed+ and the above methods is that only Fed+ covers the case of robust aggregation, both in the definition of the algorithm and in the convergence theory, which allows for the non-smooth loss functions that result.

5.3   Illustration of Federated Learning Training Failure

Next, we illustrate the training failure that can occur in real-world federated learning settings. The study involves a federated reinforcement learning-based financial portfolio management problem, presented in detail in a separate chapter in this volume. The key observation, shown here in Fig. 5.1, is that replacing at each round the local party models with a common, central (e.g., average) model can lead to dramatic spikes in model changes, which can trigger training failure for the federation as a whole. Specifically, the figure shows the mean and standard deviation of the change in neural network parameter values before and after a

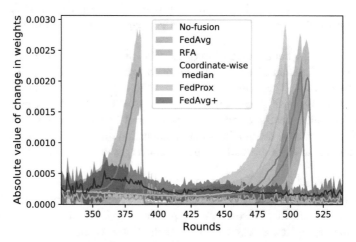

Fig. 5.1 Illustration of federated learning training failure on a real-world application when parties are forced to converge to a single, common aggregate solution

federated learning aggregation step. The federated aggregation methods illustrated are: FedAvg, RFA using the geometric median, coordinate-wise median, and FedProx, as well as the no fusion case where each party trains independently on its own data, and the FedAvg+ version of our personalized Fed+ approach. FedAvg, RFA, coordinate-wise median, and FedProx all cause large spikes in the parameter change that do not occur without federated learning or when using personalized federated learning.

Such dramatic model change can lead to failure of the training process. In particular, the large spikes coincide precisely with training failure, as shown in Fig. 5.2 (bottom four figures). Note that this example *does not involve adversarial parties or party failure in any way*, as evidenced from the fact that the single-party training on the same dataset (top curve) does not suffer any failure. Rather, this is an example of federated learning on a real-world application where parties' data are not drawn IID from a single dataset. As such, it is conceivable that federated model failure would be a relatively common occurrence in real-world applications using the vast majority of algorithms. A deeper understanding of this collapse can be gleaned from Fig. 5.3, which shows what occurs before and after an aggregation step.

Figure 5.3 motivates the Fed+ approach to personalized federated learning by illustrating the behavior of the algorithms as a function of how close the local party moves from a purely local model toward a common, central model. A local party update occurs in each subplot on the left side, at $\lambda = 0$. Observe that the local updates improve the performance from the previous aggregation indicated by the dashed lines. However, performance degrades after the subsequent aggregation, corresponding to the right-hand side of each subplot, where $\lambda = 1$. In fact, for FedAvg, RFA, and FedProx, performance of the subsequent aggregation is worse

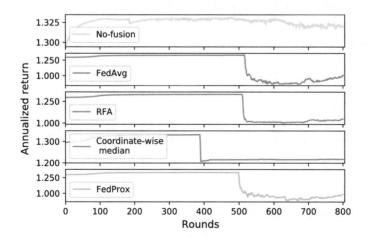

Fig. 5.2 Illustration of federated training collapse on the financial portfolio management problem. There is no training collapse without federated learning (top sub-figure), but all of the non-personalized federated learning algorithms exhibit the collapse

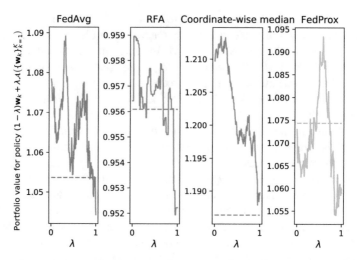

Fig. 5.3 Before and after aggregation, along the line given by varying $\lambda \in [0, 1]$ using a convex combination of local update and the common model on the financial portfolio management problem. Dashed lines represent the aggregate model at the previous round. A right-hand side lower than the left-hand side means that a full step toward averaging (or median) all parties, i.e., $\lambda = 1$, degrades local performance. This is the case with the standard FedAvg as well as with the non-personalized robust methods

than the previous value (dashed line). Intermediate values of λ correspond to moving toward, but not reaching, the common, central model.

5.4 Personalized Federated Learning

Personalization is aimed at better handling federated learning in real-world settings including non-IID data across parties, parties having outlier data with respect to other parties, stragglers, in that updates are transmitted late, and an implicit requirement for the final trained model(s) to perform well on each party's own datasets. Fed+ in particular accomplishes these goals using a robust approach, and, importantly, does not require all parties to converge to a single central point. This requires generalizing the objective of the federated learning training process, as follows.

5.4.1 Problem Formulation

Consider a federation of K parties with local loss functions $f_k : \mathbb{R}^d \to \mathbb{R}, k = 1, 2, \ldots, K$. The original FedAvg formulation [16] involves training a central model

$\tilde{\mathbf{w}} \in \mathbb{R}^d$ by minimizing the average local loss over the K parties:

$$\min_{\tilde{\mathbf{w}} \in \mathbb{R}^d, \ \mathbf{W} \in \mathbb{R}^{d \times K}} \left[F(\mathbf{W}) := \frac{1}{K} \sum_{k=1}^{K} f_k(\mathbf{w}_k) \right] \quad \text{subject to} \quad \mathbf{w}_k = \tilde{\mathbf{w}}, \ k = 1, \dots, K,$$

(5.1)

where we use the notation $\mathbf{W} := (\mathbf{w}_1, \mathbf{w}_2, \dots, \mathbf{w}_K) \in \mathbb{R}^{d \times K}$ with $\mathbf{w}_k \in \mathbb{R}^d$ denoting the local model of party k. In place of the hard equality constraints in (5.1), we advocate for a penalization-based approach and propose the following objective for the overall federated training process:

$$\min_{\mathbf{W} \in \mathbb{R}^{d \times K}} \quad F_\mu(\mathbf{W}) := \frac{1}{K} \sum_{k=1}^{K} \left[f_k(\mathbf{w}_k) + \mu \, \mathcal{B}\big(\mathbf{w}_k, \mathcal{A}(\mathbf{W})\big) \right], \quad (5.2)$$

where $\mu > 0$ is a user-chosen penalty constant, $\mathcal{A}$ is an aggregation function that outputs a central aggregate $\tilde{\mathbf{w}} \in \mathbb{R}^d$ of $\mathbf{w}_1, \dots, \mathbf{w}_K$, and $\mathcal{B}(\cdot, \cdot)$ is a distance function that penalizes the deviation of a local model $\mathbf{w}_k$ from the central aggregate $\tilde{\mathbf{w}} = \mathcal{A}(\mathbf{W})$. Note that $\tilde{\mathbf{w}}$ may be the mean, median, etc. When $\mu = 0$, problem (5.2) reduces to the non-federated setting where every party independently minimizes its local objective function. On the other hand, for $\mu > 0$ and $\mathcal{A}(\mathbf{W}) = \frac{1}{K} \sum_k \mathbf{w}_k$, if one sets $\mathcal{B}$ such that $\mathcal{B}(\mathbf{w}, \tilde{\mathbf{w}}) = \infty$ if $\mathbf{w} \neq \tilde{\mathbf{w}}$ and $\mathcal{B}(\mathbf{w}, \tilde{\mathbf{w}}) = 0$ otherwise, then (5.2) is equivalent to problem (5.1). More generally, the distance function $\mathcal{B}$ may be the usual squared Euclidean distance, the metric induced by the ℓ_p norm (denoted as $\|\cdot\|_p$) with $p \in [1, \infty]$, or other Bregman divergence measures such as

$$\mathcal{B}(\mathbf{w}_k, \tilde{\mathbf{w}}) = \frac{1}{2} \|\mathbf{w}_k - \tilde{\mathbf{w}}\|_Q^2, \quad (5.3)$$

where $Q \in \mathbb{R}^{d \times d}$ is a symmetric positive semi-definite matrix and $\|\mathbf{w}\|_Q := \sqrt{\mathbf{w}^\top Q \mathbf{w}}$. By setting Q to be a diagonal matrix with non-negative entries, (5.3) can serve to weight each component's contribution differently depending on the application and information available during model training.

5.4.2 Handling Robust Aggregation

We exploit the machinery of the function $\mathcal{B}$ to define a family of aggregation functions that includes the mean, geometric median, coordinate-wise median, etc. That is, the global model $\tilde{\mathbf{w}}$ is computed by aggregating the current local models

$\{\mathbf{w}_1, \ldots, \mathbf{w}_K\}$ via

$$\tilde{\mathbf{w}} \leftarrow \mathcal{A}(\mathbf{W}) := \arg\min_{\mathbf{w} \in \mathbb{R}^d} \frac{1}{K} \sum_{k=1}^{K} \mathcal{B}(\mathbf{w}_k, \mathbf{w}). \tag{5.4}$$

The mean aggregation function $\mathcal{A}$ is recovered by choosing $\mathcal{B}(\mathbf{w}, \mathbf{w}') = \|\mathbf{w} - \mathbf{w}'\|_2^2$ in (5.4). The geometric median and coordinate-wise median aggregation are obtained by setting $\mathcal{B}(\mathbf{w}, \mathbf{w}') = \|\mathbf{w} - \mathbf{w}'\|_2$ and $\mathcal{B}(\mathbf{w}, \mathbf{w}') = \|\mathbf{w} - \mathbf{w}'\|_1$, resp. To unify the aggregation methods that include non-smooth functions $\mathcal{B}$ in (5.2), we introduce the following general class of parameterized functions $\mathcal{B}$ where we choose a convex function $\phi : \mathbb{R}^d \rightarrow [0, \infty]$ and a smoothness-robustness parameter $\rho > 0$:

$$\mathcal{B}(\mathbf{w}_k, \tilde{\mathbf{w}}) = \Phi_\rho(\mathbf{w}_k - \tilde{\mathbf{w}}), \quad \Phi_\rho(\mathbf{w}) := \min_{\mathbf{w}' \in \mathbb{R}^d} \left[\phi(\mathbf{w}') + \frac{1}{2\rho} \|\mathbf{w} - \mathbf{w}'\|_2^2 \right].$$
$$\tag{5.5}$$

We call the minimizer in the above problem the proximal operator of ϕ and denote it as $\mathrm{prox}_\phi^\rho(\mathbf{w})$. Note that Φ_ρ is a smooth function known as the Moreau envelope of ϕ. By choosing ϕ to be the ℓ_2 norm, we obtain a $(1/\rho)$-smooth approximation of the geometric median aggregation as in [19]. Setting ϕ to the ℓ_1 norm gives a smooth approximation of coordinate-wise median aggregation. The usual mean aggregation is naturally recovered in both of these cases: (i) $\phi(\mathbf{w}) = \frac{1}{2}\|\mathbf{w}\|_2^2$, and (ii) $\phi(\mathbf{w}) = 0$ if $\mathbf{w} = \mathbf{0}$ and $+\infty$ otherwise.

5.4.3 Personalization

While problem (5.2) can be solved centrally, we are interested in the federated setting where at every round each active party k solves its own version of the problem, running E_k iterations of the following update, with learning rate $\eta > 0$:

$$\mathbf{w}_k \leftarrow \theta[\mathbf{w}_k - \eta \nabla f_k(\mathbf{w}_k)] + (1 - \theta)\mathbf{z}_k, \quad \text{for} \quad i = 1, \ldots, E_k. \tag{5.6}$$

Note that personalization occurs via the party-specific regularization term $\mathbf{z}_k$ and the constant $\theta \in (0, 1]$ controls the degree of regularization while training the local model using (5.6). In practice, the exact gradient $\nabla f_k(\mathbf{w}_k)$ in (5.6) is replaced by an unbiased random estimate. In standard methods, $\mathbf{z}_k$ is set to the current global model $\tilde{\mathbf{w}}$. Fed+ proposes however robust personalization as we shall see below.

5.4.4   Reformulation and Unification of Mean and Robust Aggregation

To obtain a unified framework that covers personalization along with robustness, consider again the original FedAvg (5.1), expressed equivalently as

$$
\min_{\mathbf{W}, \mathbf{Z} \in \mathbb{R}^{d \times K}, \tilde{\mathbf{w}} \in \mathbb{R}^d} \frac{1}{K} \sum_{k=1}^{K} f_k(\mathbf{w}_k) \quad \text{subject to} \quad \mathbf{w}_k = \mathbf{z}_k, \ \mathbf{z}_k = \tilde{\mathbf{w}}, \ k = 1, \dots, K,
$$

$$(5.7)$$

where $\mathbf{Z} := (\mathbf{z}_1, \dots, \mathbf{z}_K)$ with $\mathbf{z}_k \in \mathbb{R}^d$. While ADMM can be used to solve equality-constrained problems of the above form, we take a penalty-based approach, as in [24], suitable for both convex and non-convex settings. Then, to handle heterogeneous data and computing environments across parties, we replace the equality constraints in (5.7) with penalty functions:

$$
\min_{\mathbf{W}, \mathbf{Z} \in \mathbb{R}^{d \times K}, \tilde{\mathbf{w}} \in \mathbb{R}^d} H_{\mu, \alpha}(\mathbf{W}, \mathbf{Z}, \tilde{\mathbf{w}}) :=
$$

$$
\frac{1}{K} \sum_{k=1}^{K} \left[f_k(\mathbf{w}_k) + \frac{\alpha}{2} \|\mathbf{w}_k - \mathbf{z}_k\|_2^2 + \mu \phi(\mathbf{z}_k - \tilde{\mathbf{w}}) \right],
$$

$$(5.8)$$

where $\alpha > 0$ is a user-chosen penalty constant and $\phi : \mathbb{R}^d \to [0, \infty]$ is a convex penalty function. The following proposition connects this new formulation (5.8) to our original objective (5.2).

Proposition 5.1 *Problem* (5.8) *is a special case of* (5.2), *where the $\mathcal{A}$ & $\mathcal{B}$ functions are as defined in* (5.4) & (5.5) *respectively (with $\rho = \mu/\alpha$), and the following relation between the two optimization objectives holds*

$$
F_\mu(\mathbf{W}) = \min_{\mathbf{Z} \in \mathbb{R}^{d \times K}, \tilde{\mathbf{w}} \in \mathbb{R}^d} H_{\mu, \alpha}(\mathbf{W}, \mathbf{Z}, \tilde{\mathbf{w}}).
$$

$$(5.9)$$

Proof Using $\rho = \mu/\alpha$, we have from (5.8)

$$
\min_{\mathbf{Z} \in \mathbb{R}^{d \times K}} H_{\mu, \alpha}(\mathbf{W}, \mathbf{Z}, \tilde{\mathbf{w}}) = \frac{1}{K} \sum_{k=1}^{K} \left[f_k(\mathbf{w}_k) + \mu \min_{\mathbf{z}_k \in \mathbb{R}^d} \left\{ \frac{1}{2\rho} \|\mathbf{w}_k - \mathbf{z}_k\|_2^2 + \phi(\mathbf{z}_k - \tilde{\mathbf{w}}) \right\} \right]
$$

$$
= \frac{1}{K} \sum_{k=1}^{K} \left[f_k(\mathbf{w}_k) + \mu \min_{\mathbf{v}_k \in \mathbb{R}^d} \left\{ \frac{1}{2\rho} \|(\mathbf{w}_k - \tilde{\mathbf{w}}) - \mathbf{v}_k\|_2^2 + \phi(\mathbf{v}_k) \right\} \right]
$$

$$= \frac{1}{K} \sum_{k=1}^{K} \left[f_k(\mathbf{w}_k) + \mu \, \Phi_\rho(\mathbf{w}_k - \tilde{\mathbf{w}}) \right],$$

$$= \frac{1}{K} \sum_{k=1}^{K} \left[f_k(\mathbf{w}_k) + \mu \, \mathcal{B}(\mathbf{w}_k, \tilde{\mathbf{w}}) \right], \tag{5.10}$$

where the second equality is obtained by the change of variable $\mathbf{z}_k \to \tilde{\mathbf{w}} + \mathbf{v}_k$ and the last two equalities are due to (5.5). Now, further minimizing (5.10) w.r.t. $\tilde{\mathbf{w}}$ and using the definition (5.4), we arrive at (5.9). $\qquad \square$

The Fed+ formulation (5.8) suggests a natural choice for the personalization function $\mathcal{R} : \mathbb{R}^d \times \mathbb{R}^d \to \mathbb{R}^d$, which computes $\mathbf{z}_k := \mathcal{R}(\tilde{\mathbf{w}}, \mathbf{w}_k)$ as a (robust) combination of the current local and central model. To be precise, Fed+ proposes setting $\mathcal{R}(\tilde{\mathbf{w}}, \mathbf{w}_k)$ by minimizing (5.8) w.r.t. $\mathbf{z}_k$ while keeping $\mathbf{w}_k$ and $\tilde{\mathbf{w}}$ fixed. Thus, we have the following closed form update:

$$\mathbf{z}_k \leftarrow \mathcal{R}(\tilde{\mathbf{w}}, \mathbf{w}_k) = \tilde{\mathbf{w}} + \mathrm{prox}_\phi^\rho(\mathbf{w}_k - \tilde{\mathbf{w}}), \quad \rho = \mu/\alpha. \tag{5.11}$$

Next, we connect the local party's update (5.6) to our new formulation (5.8).

Proposition 5.2 Let $\theta := \frac{1}{1+\alpha\eta}$. Then, the local update (5.6) (line 14 in Fed+ algorithm) is a gradient descent iteration with learning rate $\eta' := \frac{\eta}{1+\alpha\eta}$ applied to the following sub-problem:

$$\min_{\mathbf{w}_k \in \mathbb{R}^d} F_k(\mathbf{w}_k; \mathbf{z}_k, \tilde{\mathbf{w}}) := f_k(\mathbf{w}_k) + \frac{\alpha}{2} \|\mathbf{w}_k - \mathbf{z}_k\|_2^2 + \mu\phi(\mathbf{z}_k - \tilde{\mathbf{w}}), \tag{5.12}$$

where $\mathbf{z}_k$ & $\tilde{\mathbf{w}}$ are kept fixed by setting them to $\mathbf{z}_k^{t-1}$ and $\tilde{\mathbf{w}}^{t-1}$, respectively.

Proof The gradient descent iteration for the function $F_k(\cdot\,; \mathbf{z}_k^{t-1}, \tilde{\mathbf{w}}^{t-1})$ with stepsize $\eta' := \frac{\eta}{1+\alpha\eta}$ is given by

$$\begin{aligned}
\mathbf{w}_k^t \leftarrow{}& \mathbf{w}_k^t - \eta' [\nabla f_k(\mathbf{w}_k^t) + \alpha(\mathbf{w}_k^t - \mathbf{z}_k^{t-1})] \\
={}& (1 - \alpha\eta') \left[\mathbf{w}_k^t - \frac{\eta'}{1 - \alpha\eta'} \nabla f_k(\mathbf{w}_k^t) \right] + (\alpha\eta') \mathbf{z}_k^{t-1} \\
={}& \left(\frac{1}{1 + \alpha\eta} \right) [\mathbf{w}_k^t - \eta \nabla f_k(\mathbf{w}_k^t)] + \left(\frac{\alpha\eta}{1 + \alpha\eta} \right) \mathbf{z}_k^{t-1}.
\end{aligned}$$

Thus, we have the local update of the form (5.6) (line 14 in Fed+ algorithm) where $\theta := \frac{1}{1+\alpha\eta}$. $\qquad \square$

5.4.5 The Fed+ Algorithm

Fed+ is defined in Algorithm 5.1 as a family of federated learning methods for solving problem (5.2) with $\mathcal{B}$ as in (5.5) and $\mathcal{A}$ as defined in (5.4). Fed+ is designed to allow for robust aggregation functions $\mathcal{A}$, where local copies of shared parameters are aggregated. Importantly, Fed+ does not require all parties to agree on a single common model. We argue that this offers the benefits of federation without the pitfall of training failure that can occur in real-world implementations of federated learning.

So as to encompass important special cases, Algorithm 5.1 introduces a number of parameters. Specifically, we define $\lambda \in [0, 1]$, $\theta \in (0, 1]$, and $\mathcal{R} : \mathbb{R}^d \times \mathbb{R}^d \to \mathbb{R}^d$. A main difference between our approach and other federated algorithms is that the parties do not set the aggregated central model as their starting point when performing the local update step (5.6); i.e., parties need not set $\lambda = 1$ in line 12 of Algorithm 5.1. Instead, Fed+ advocates initializing local models at each round with the value from the previous round (i.e., $\lambda = 0$). This mitigates the dramatic changes in local models seen in Fig. 5.1.

Algorithm 5.1 Fed+: Parties $k = 1, \ldots, K$; aggregation fcn. $\mathcal{A}$; local iterations per round at party k, E_k; learning rate η; and $\theta \in (0, 1]$; $\lambda \in [0, 1]$; and robust personalization fcn. $\mathcal{R} : \mathbb{R}^d \times \mathbb{R}^d \to \mathbb{R}^d$

Initialization:
1:   Each party k sends its initial local model $\mathbf{w}_k^0$ to the Aggregator, which computes the central value $\tilde{\mathbf{w}}^0 \leftarrow \mathcal{A}(\mathbf{W}^0)$.

Aggregator:
2: **for** round $t = 1, \ldots, T$ **do**
3:       Sample parties to obtain $S_t \subseteq \{1, \ldots, K\}$.
4:       Send the current global model $\tilde{\mathbf{w}}^{t-1}$ to each party $k \in S_t$.
5:       **for** each party $k \in S_t$ **in parallel do**
6:             $\mathbf{w}_k^t \leftarrow$ Local-Solve$(k, t, \tilde{\mathbf{w}}^{t-1}, \mathbf{w}_k^{t-1})$.   // Each party $k \notin S_t$ sets $\mathbf{w}_k^t \leftarrow \mathbf{w}_k^{t-1}$.
7:             Party sends $\mathbf{w}_k^t$ to Aggregator.
8:       **end for**
9:       Compute the aggregated central model: $\tilde{\mathbf{w}}^t \leftarrow \mathcal{A}(\mathbf{W}^t)$.
10: **end for**
        Local-Solve $(k, t, \tilde{\mathbf{w}}^{t-1}, \mathbf{w}_k^{t-1})$:   // Run on each active party $k \in S_t$
11: Compute a robust local model: $\mathbf{z}_k^{t-1} := \mathcal{R}(\tilde{\mathbf{w}}^{t-1}, \mathbf{w}_k^{t-1})$.
12: Initialize the current local model: $\mathbf{w}_k^t \leftarrow (1 - \lambda)\mathbf{w}_k^{t-1} + \lambda\tilde{\mathbf{w}}^{t-1}$.
13: **for** $i = 1, \ldots, E_k$ **do**
14:     $\mathbf{w}_k^t \leftarrow \theta[\mathbf{w}_k^t - \eta\nabla f_k(\mathbf{w}_k^t)] + (1 - \theta)\mathbf{z}_k^{t-1}$.
15: **end for**

5.4.6   Mean and Robust Variants of Fed+

We introduce three variants of interest of Fed+, unified through their choice of function ϕ, and discuss how they can be combined. We propose setting $\lambda = 1$ in the initialization of the Local-Solve for each party, and $\theta := \frac{1}{1+\alpha\eta}$ with $\alpha > 0$ tuned as a hyper-parameter. We recommend tuning the hyper-parameter ρ to control the amount of robust personalization.

5.4.6.1   FedAvg+

A mean aggregation-based method with better training performance than FedAvg via personalization. Choose $\phi(\mathbf{w}) = \frac{1}{2}\|\mathbf{w}\|_2^2$. $\mathcal{A}$ is the mean, i.e., $\tilde{\mathbf{w}}^t := \texttt{Mean}\{\mathbf{w}_k^t : k \in S_t\}$ and (11) takes the form: $\mathcal{R}(\tilde{\mathbf{w}}^t, \mathbf{w}_k^t) = (1 - \lambda_k^t)\,\mathbf{w}_k^t + \lambda_k^t\,\tilde{\mathbf{w}}^t$, where $\lambda_k^t := \rho/(1+\rho)$.

5.4.6.2   FedGeoMed+

A robust aggregation-based method that offers stability in training in the presence of outliers/adversaries. Set $\phi(\mathbf{w}) = \|\mathbf{w}\|_2$. $\mathcal{A}$ is a ρ-smoothed approximation of the Geometric Median, and (11) becomes

$$\mathcal{R}(\tilde{\mathbf{w}}^t, \mathbf{w}_k^t) = (1 - \lambda_k^t)\,\mathbf{w}_k^t + \lambda_k^t\,\tilde{\mathbf{w}}^t, \quad \text{where } \lambda_k^t := \min\left\{1,\ \rho/\|\mathbf{w}_k^t - \tilde{\mathbf{w}}^t\|_2\right\}.$$

To compute $\tilde{\mathbf{w}}^t$ from $\{\mathbf{w}_k^t : k \in S_t\}$ the aggregator runs the following two step iterative procedure initialized with $\tilde{\mathbf{w}} = \mathbf{w}_{\text{mean}} := \texttt{Mean}\{\mathbf{w}_k^t : k \in S_t\}$ until $\tilde{\mathbf{w}}$ converges

$$\mathbf{v}_k \leftarrow \max\left\{0,\ 1 - (\rho/\|\mathbf{w}_k^t - \tilde{\mathbf{w}}\|_2)\right\}(\mathbf{w}_k^t - \tilde{\mathbf{w}}), \quad \forall k \in S_t,$$

$$\tilde{\mathbf{w}} \leftarrow \mathbf{w}_{\text{mean}} - \texttt{Mean}\{\mathbf{v}_k : k \in S_t\}.$$

5.4.6.3   FedCoMed+

FedCoMed+ offers the benefits of robust aggregation via the median with added flexibility in allowing different parameters for each coordinate. Choose $\phi(\mathbf{w}) = \|\mathbf{w}\|_1$. $\mathcal{A}$ is a ρ-smoothed approximation of the Coordinate-wise Median and (11) takes the following form:

$$\mathcal{R}(\tilde{\mathbf{w}}^t, \mathbf{w}_k^t) = (I - \Lambda_k^t)\,\mathbf{w}_k^t + \Lambda_k^t\,\tilde{\mathbf{w}}^t, \quad \text{where } \Lambda_k^t \text{ is a diagonal matrix with}$$
$$\Lambda_k^t(i, i) := \min\left\{1,\ \rho/|\mathbf{w}_k^t(i) - \tilde{\mathbf{w}}^t(i)|\right\}, \quad i = 1, \ldots, d.$$

To compute $\tilde{\mathbf{w}}^t$ from $\{\mathbf{w}_k^t \ : \ k \in S_t\}$ the aggregator starts with $\tilde{\mathbf{w}} = \mathbf{w}_{\text{mean}} :=$ $\text{Mean}\{\mathbf{w}_k^t \ : \ k \in S_t\}$ and runs the following two step iterative procedure until $\tilde{\mathbf{w}}$ converges

$$\mathbf{v}_k \leftarrow \max\left\{\mathbf{0}, \ \mathbf{w}_k^t - \tilde{\mathbf{w}}^t - \rho\, sign(\mathbf{w}_k^t - \tilde{\mathbf{w}}^t)\right\}, \quad \forall k \in S_t,$$

$$\tilde{\mathbf{w}} \leftarrow \mathbf{w}_{\text{mean}} - \text{Mean}\{\mathbf{v}_k : k \in S_t\}.$$

5.4.6.4 Hybridization via the Unified Fed+ Framework with Layer-Specific ϕ

The unification of aggregation methods through a single formulation allows for seamlessly combining different methods of aggregation and personalization applied to different layers in training deep neural networks. For example, initial layers may use FedAvg+, while final layers may benefit from FedCoMed+ and greater personalization via ρ.

Specifically, Fed+ provides a framework where one can employ robust personalization functions specific to different layers in a neural network as follows.

Let $\mathbf{w} = (\mathbf{w}_{[1]}, \cdots, \mathbf{w}_{[L]})$, where $\mathbf{w}_{[l]}$ denotes the weights of the l-th layer. In this case, we define

$$\phi(\mathbf{w}) := \sum_{l=1}^{L} \phi_l(\mathbf{w}_{[l]}),$$

where ϕ_l, $l = 1, \ldots, L$, can be chosen independently giving rise to different methods of robust personalization for each layer.

5.4.7 Deriving Existing Algorithms from Fed+

Many non-personalized federated learning methods fit into the Fed+ framework and can be obtained by setting the parameters in Algorithm 5.1 appropriately.

This is useful when a single code is to be used in a federated learning system, and for different applications, a personalized or non-personalized method would be preferred, and a mean aggregation-based or a robust aggregation-based method may be preferred. To handle all of these applications from a common code, it is helpful when they can be obtained by simply setting parameters of the foundational method appropriately.

This is the case with the Fed+ formulation above. To obtain pure Local (Stochastic) Gradient Descent without Federation, one need only set $\lambda = 0$, $\theta = 1$. The seminal federated learning method, based on the mean and common solution for all parties, FedAvg, can be obtained by setting the parameters as $\lambda = 1$,

$\theta = 1$, $E_k = E$ for all k, $\tilde{\mathbf{w}}^t = \text{Mean}\{\mathbf{w}_k^t : k \in S_t\}$. A regularized version of FedAvg, which bears some similarity to personalized federated learning in that the local problem is augmented by a proximal term containing the aggregator solution is FedProx. FedProx does not, however, start the local training of each federated round from the last local solution, and so to obtain FedProx from the above algorithm for Fed+, one must set the parameters as follows: $\lambda = 1$, $\theta = \frac{1}{1+\mu\eta}$, $\mathcal{R}(\tilde{\mathbf{w}}^t, \mathbf{w}_k^t) = \tilde{\mathbf{w}}^t = \text{Mean}\{\mathbf{w}_k^t : k \in S_t\}$.

Non-personalized robust federated learning can also be obtained by appropriately setting the parameters of Fed+. For RFA, it involves setting $\lambda = 1$, $\theta = 1$, $\tilde{\mathbf{w}}^t = \text{Geometric Median}\{\mathbf{w}_k^t : k \in S_t\}$. For Coordinate-wise median, one would set $\lambda = 1$, $\theta = 1$, and $\tilde{\mathbf{w}}^t = \text{Coordinatewise Median}\{\mathbf{w}_k^t : k \in S_t\}$.

5.5 Fixed Points of Fed+

Here, we characterize the fixed points of the Fed+ algorithm to gain insights on the kind of personalized solution it offers. Before proceeding further we make the following assumption:

Assumption 5.3 For each $k = 1, \ldots, K$, f_k is convex, all the parties actively participate in every round of the federating learning process, and the Local-Solve subroutine in Fed+ returns $\mathbf{w}_k^t$ as the exact minimizer of $F_k(\cdot; \mathbf{z}_k^{t-1}, \tilde{\mathbf{w}}^{t-1})$, i.e.,

$$\mathbf{w}_k^t = \text{prox}_{f_k}^{\frac{1}{\alpha}}(\mathbf{z}_k^{t-1}) := \underset{\mathbf{w}_k}{\arg\min} f_k(\mathbf{w}_k) + \frac{\alpha}{2}\|\mathbf{w}_k - \mathbf{z}_k^{t-1}\|_2^2, \ \forall t \geq 1, \ k = 1, \ldots, K.$$

$$(5.13)$$

We define $\tilde{f}_k : \mathbb{R}^d \to \mathbb{R}$ to be the Moreau envelope of f_k with smoothing parameter $(1/\alpha)$, i.e.,

$$\tilde{f}_k(\mathbf{z}) := \underset{\mathbf{w}_k \in \mathbb{R}^d}{\min} f_k(\mathbf{w}_k) + \frac{\alpha}{2}\|\mathbf{w}_k - \mathbf{z}\|_2^2, \ \forall \mathbf{z} \in \mathbb{R}^d.$$

Now, we present the fixed-point characterization of Fed+ under Assumption 5.3:

Theorem 5.4 Consider the Fed+ algorithm for solving problem (5.9) under Assumption 5.3. Let $\mathcal{R}$ be as in (5.11) and $(\mathbf{W}^*, \mathbf{Z}^*, \tilde{\mathbf{w}}^*)$ be a fixed point of Fed+. Then, the following conditions are satisfied:

$$\frac{1}{K}\sum_{k=1}^{K}\tilde{f}_k(\mathbf{z}_k^*) = 0, \ \mathbf{w}_k^* = \mathbf{z}_k^* - \frac{1}{\alpha}\nabla\tilde{f}_k(\mathbf{z}_k^*), \ \mathbf{z}_k^* = \tilde{\mathbf{w}}^* + \text{prox}_\phi^\rho(\mathbf{w}_k^* - \tilde{\mathbf{w}}^*),$$

$$k = 1, \ldots, K.$$

$$(5.14)$$

Proof We start with the following observations about Fed+: $\forall\, t \geq 1,:$

$$\mathbf{w}_k^t = \mathbf{z}_k^{t-1} - \frac{1}{\alpha}\nabla \tilde{f}_k(\mathbf{z}_k^{t-1}),\ k = 1,\dots,K,$$

$$\tilde{\mathbf{w}}^t = \frac{1}{K}\sum_{k=1}^K \mathbf{w}_k^t - \frac{1}{K}\sum_{k=1}^K \mathrm{prox}_\phi^\rho(\mathbf{w}_k^t - \tilde{\mathbf{w}}^t), \tag{5.15}$$

$$\mathbf{z}_k^t = \tilde{\mathbf{w}}^t + \mathrm{prox}_\phi^\rho(\mathbf{w}_k^t - \tilde{\mathbf{w}}^t),\ k = 1,\dots,K,$$

where the first equation is a direct consequence of (5.13), the second one comes from (5.28), and the last one is by choice of $\mathcal{R}$. Therefore, for a fixed point, the second & third equations in (5.14) obviously hold. Now, for a fixed point, we also have the following from (5.15):

$$\tilde{\mathbf{w}}^* = \frac{1}{K}\sum_{k=1}^K \mathbf{w}_k^* - \frac{1}{K}\sum_{k=1}^K \mathrm{prox}_\phi^\rho(\mathbf{w}_k^* - \tilde{\mathbf{w}}^*). \tag{5.16}$$

Replacing the first $\mathbf{w}_k^*$ in (5.16) with $\mathbf{z}_k^* - \frac{1}{\alpha}\nabla \tilde{f}_k(\mathbf{z}_k^*)$ and subsequently $\mathbf{z}_k^*$ by $\tilde{\mathbf{w}}^* + \mathrm{prox}_\phi^\rho(\mathbf{w}_k^* - \tilde{\mathbf{w}}^*)$, we get

$$
\begin{aligned}
\tilde{\mathbf{w}}^* &= \frac{1}{K}\sum_{k=1}^K \Big[\mathbf{z}_k^* - \frac{1}{\alpha}\nabla \tilde{f}_k(\mathbf{z}_k^*)\Big] - \frac{1}{K}\sum_{k=1}^K \mathrm{prox}_\phi^\rho(\mathbf{w}_k^* - \tilde{\mathbf{w}}^*)\\
&= \frac{1}{K}\sum_{k=1}^K \Big[\tilde{\mathbf{w}}^* + \mathrm{prox}_\phi^\rho(\mathbf{w}_k^* - \tilde{\mathbf{w}}^*)\Big] - \frac{1}{\alpha K}\sum_{k=1}^K \nabla \tilde{f}_k(\mathbf{z}_k^*) - \frac{1}{K}\sum_{k=1}^K \mathrm{prox}_\phi^\rho(\mathbf{w}_k^* - \tilde{\mathbf{w}}^*)\\
&= \tilde{\mathbf{w}}^* - \frac{1}{\alpha K}\sum_{k=1}^K \nabla \tilde{f}_k(\mathbf{z}_k^*).
\end{aligned}
$$

Thus, we have the first equation in (5.14).   □

With the help of above Theorem, we now analyze two extreme choices of $\mathcal{R}$ in part (a) & (b) of the following Corollary:

Corollary 5.5 *Consider the Fed+ algorithm under Assumption 5.3. Let* $(\mathbf{W}^*, \mathbf{Z}^*, \tilde{\mathbf{w}}^*)$ *be a fixed point of Fed+. Then, the following are true:*

(a) If Fed+ sets $\mathcal{R}(\tilde{\mathbf{w}}, \mathbf{w}_k) = \mathbf{w}_k$, *then*

$$\mathbf{w}_k^* \in \underset{\mathbf{w}}{\arg\min}\ f_k(\mathbf{w}),\ k = 1,\dots,K. \tag{5.17}$$

(b) If Fed+ uses $\mathcal{R}(\tilde{\mathbf{w}}, \mathbf{w}_k) = \tilde{\mathbf{w}}$ along with $\mathcal{A}(\mathbf{W}) = \frac{1}{K}\sum_k \mathbf{w}_k$, then

$$\frac{1}{K}\sum_{k=1}^{K} \tilde{f}_k(\tilde{\mathbf{w}}^*) = 0, \quad \mathbf{w}_k^* = \tilde{\mathbf{w}}^* - \frac{1}{\alpha}\nabla \tilde{f}_k(\tilde{\mathbf{w}}^*), \ k = 1, \ldots, K. \quad (5.18)$$

(c) If Fed+ employs $\mathcal{R}(\tilde{\mathbf{w}}, \mathbf{w}_k) = (1-\gamma)\mathbf{w}_k + \gamma\tilde{\mathbf{w}}$, $\gamma \in (0,1)$ with $\mathcal{A}(\mathbf{W}) = \frac{1}{K}\sum_k \mathbf{w}_k$, then

$$\mathbf{w}_k^* = \tilde{\mathbf{w}}^* - \frac{1}{\alpha\gamma}\nabla \tilde{f}_k\big((1-\gamma)\mathbf{w}_k^* + \gamma\tilde{\mathbf{w}}^*\big), \ k = 1, \ldots, K, \ where \ \tilde{\mathbf{w}}^* = \frac{1}{K}\sum_{k=1}^{K}\mathbf{w}_k^*.$$
$$(5.19)$$

Proof To prove (a), we apply Theorem 5.4 with $\phi = 0$. This choice of ϕ leads to $\mathbf{z}_k = \mathcal{R}(\tilde{\mathbf{w}}, \mathbf{w}_k) = \mathbf{w}_k$ from (5.11). Therefore, Fed+ boils to applying the proximal point algorithm $\mathbf{w}_k^t = \text{prox}_{\tilde{f}_k}^{\frac{1}{\alpha}}(\mathbf{w}_k^{t-1})$, $t \geq 1$, at each local party $k = 1, \ldots, K$. Therefore, we obtain the result (5.17) as $\mathbf{w}_k^* = \text{prox}_{\tilde{f}_k}^{\frac{1}{\alpha}}(\mathbf{w}_k^*)$. Alternatively, setting $\mathbf{z}_k^* = \mathbf{w}_k^*$ in (5.14), we get

$$\mathbf{w}_k^* = \mathbf{w}_k^* - \frac{1}{\alpha}\nabla \tilde{f}_k(\mathbf{w}_k^*) \implies \mathbf{w}_k^* \in \underset{\mathbf{w}}{\arg\min} f_k(\mathbf{w}). \quad (5.20)$$

Next, we prove part (b) by applying Theorem 5.4 with the following choice of ϕ: $\phi(\mathbf{w}) = 0$ iff $\mathbf{w} = \mathbf{0}$ and $+\infty$ otherwise. This particular ϕ corresponds to the choice $\mathbf{z}_k = \mathcal{R}(\tilde{\mathbf{w}}, \mathbf{w}_k) = \tilde{\mathbf{w}}$ from (5.11). Also, the aggregation function $\mathcal{A}$ becomes the mean as from (5.4) and (5.5) we get

$$\mathcal{A}(\mathbf{W}) = \underset{\tilde{\mathbf{w}}}{\arg\min} \frac{1}{K}\sum_{k=1}^{K} \Phi_\rho(\mathbf{w}_k - \tilde{\mathbf{w}}), \ \text{where} \ \Phi_\rho(\mathbf{w}) = \frac{1}{2\rho}\|\mathbf{w}\|_2^2.$$

Now, putting $\mathbf{z}_k^* = \tilde{\mathbf{w}}^*$ in (5.14), we arrive at (5.18). Finally, we show part (c) by setting $\phi(\mathbf{w}) = \frac{1}{2}\|\mathbf{w}\|_2^2$, $\mathbf{w} \in \mathbb{R}^d$ in Theorem 5.4. Let the constant μ (or ρ) be set in such a way that $(\mu/\alpha) = \rho = \gamma/(1-\gamma)$. Then, (5.11) becomes $\mathbf{z}_k = \mathcal{R}(\tilde{\mathbf{w}}, \mathbf{w}_k) = (1-\gamma)\mathbf{w}_k + \gamma\tilde{\mathbf{w}}$. Also, like in part (b), the aggregation function $\mathcal{A}$ becomes the mean here as well. Now, we complete the proof by using $\mathbf{z}_k^* = (1-\gamma)\mathbf{w}_k^* + \gamma\tilde{\mathbf{w}}^*$ in (5.14):

$$\mathbf{w}_k^* = (1-\gamma)\mathbf{w}_k^* + \gamma\tilde{\mathbf{w}}^* - \frac{1}{\alpha}\nabla \tilde{f}_k(\mathbf{z}_k^*) \implies \mathbf{w}_k^* = \tilde{\mathbf{w}}^* - \frac{1}{\alpha\gamma}\nabla \tilde{f}_k(\mathbf{z}_k^*).$$

Note that part (b) of the Corollary recovers the fixed-point result of FedProx given in [17]. In the next Proposition, we characterize the fixed points of Fed+ for the general case where $\mathcal{R}$ and $\mathcal{A}$ need not be defined through a common ϕ function.

Proposition 5.6 *Consider the Fed+ algorithm with an arbitrary aggregation function $\mathcal{A}$ and the following personalization function: $\mathbf{z}_k = \mathcal{R}(\tilde{\mathbf{w}}, \mathbf{w}_k) = (1 - \gamma_k)\mathbf{w}_k + \gamma_k\tilde{\mathbf{w}}$, $\gamma_k \in (0, 1]$, $k = 1, \ldots, K$. Then, under Assumption 5.3, the following holds for any fixed point $(\mathbf{W}^*, \mathbf{Z}^*, \tilde{\mathbf{w}}^*)$ of Fed+:*

$$\mathbf{w}_k^* = \tilde{\mathbf{w}}^* - \frac{1}{\alpha\gamma_k}\nabla\tilde{f}_k\big((1 - \gamma_k)\mathbf{w}_k^* + \gamma_k\tilde{\mathbf{w}}^*\big), \ k = 1, \ldots, K,$$

$$\text{where } \tilde{\mathbf{w}}^* = \mathcal{A}(\mathbf{w}_1^*, \ldots, \mathbf{w}_K^*).$$

(5.21)

Moreover, if the aggregation function $\mathcal{A}$ (such as Mean, Geometric Median, Coordinate-wise Median) in Fed+ satisfy the following translation & sign invariance property

$$\forall \ \mathbf{w}, \mathbf{w}_1, \ldots, \mathbf{w}_K \in \mathbb{R}^d, \quad \mathcal{A}(\mathbf{w} - \mathbf{w}_1, \ldots, \mathbf{w} - \mathbf{w}_K) = \mathbf{w} - \mathcal{A}(\mathbf{w}_1, \ldots, \mathbf{w}_K),$$

(5.22)

then the following holds

$$\mathcal{A}\left(\frac{1}{\alpha\gamma_1}\nabla\tilde{f}_1(\mathbf{z}_1^*), \ \ldots, \ \frac{1}{\alpha\gamma_K}\nabla\tilde{f}_K(\mathbf{z}_K^*)\right) = 0,$$

$$\text{where } \mathbf{z}_k^* = (1 - \gamma_k)\mathbf{w}_k^* + \gamma_k\tilde{\mathbf{w}}^*, \ k = 1, \ldots, K.$$

(5.23)

Proof Similar to the proof of Theorem 5.4, we have the following from Fed+ algorithm: $\forall t \geq 1$,

$$\mathbf{w}_k^t = \mathbf{z}_k^{t-1} - \frac{1}{\alpha}\nabla\tilde{f}_k(\mathbf{z}_k^{t-1}), \ k = 1, \ldots, K,$$

$$\tilde{\mathbf{w}}^t = \mathcal{A}(\mathbf{w}_1^t, \ldots, \mathbf{w}_K^t),$$

$$\mathbf{z}_k^t = (1 - \gamma_k)\mathbf{w}_k^t + \gamma_k\tilde{\mathbf{w}}^t, \ k = 1, \ldots, K.$$

For a fixed point, we thus have

$$\mathbf{w}_k^* = \mathbf{z}_k^* - \frac{1}{\alpha}\nabla\tilde{f}_k(\mathbf{z}_k^*), \ k = 1, \ldots, K,$$

(5.24)

$$\tilde{\mathbf{w}}^* = \mathcal{A}(\mathbf{w}_1^*, \ldots, \mathbf{w}_K^*),$$

(5.25)

$$\mathbf{z}_k^* = (1 - \gamma_k)\mathbf{w}_k^* + \gamma_k\tilde{\mathbf{w}}^*, \ k = 1, \ldots, K.$$

(5.26)

Now, replacing the first $\mathbf{z}_k^*$ in (5.24) by (5.26), we arrive at

$$\mathbf{w}_k^* = \tilde{\mathbf{w}}^* - \frac{1}{\alpha\gamma_k}\nabla\tilde{f}_k(\mathbf{z}_k^*), \ k = 1, \ldots, K.$$

(5.27)

Thus, we have (5.21). Further, using (5.27) in (5.25) and utilizing the property (5.22) gives us (5.23). □

5.6 Convergence Analysis

Consider Algorithm 5.1 and the following assumption concerning the values of the parameters used in the algorithm.

Assumption 5.7 Let the parameters be set as follows. (i) $\phi : \mathbb{R}^d \to [0, \infty]$ is any convex function with easy to compute proximal operator, (ii) the function $\mathcal{R}$ is set as in the reference (5.11), (iii) the initialization parameter λ in line 12 is set to 0, (iv) set $\theta := \frac{1}{1+\alpha\eta}$ in line 14, and (v) compute the aggregation step in line 9 via (5.4), where $\mathcal{B}$ is given by (5.5). The parameters $\alpha > 0$, $\rho > 0$, and $\eta > 0$ are considered to be tunable unless specified otherwise.

To implement the aggregation step $\tilde{\mathbf{w}} \leftarrow \mathcal{A}(\mathbf{w}_1, \ldots, \mathbf{w}_K)$ for a general choice of ϕ, we propose the following iterative procedure, which is initialized with $\tilde{\mathbf{w}} = \mathbf{w}_{\mathrm{mean}} := \mathrm{Mean}\{\mathbf{w}_1, \ldots, \mathbf{w}_K\}$:

$$\tilde{\mathbf{w}} \leftarrow \mathbf{w}_{\mathrm{mean}} - \mathrm{Mean}\left\{ \mathrm{prox}_\phi^\rho(\mathbf{w}_1 - \tilde{\mathbf{w}}), \cdots, \mathrm{prox}_\phi^\rho(\mathbf{w}_K - \tilde{\mathbf{w}}) \right\}. \quad (5.28)$$

The above setting then gives rise to the following property:

$$(\mathbf{z}_1^t, \ldots, \mathbf{z}_K^t, \tilde{\mathbf{w}}^t) = \underset{\mathbf{Z}\in\mathbb{R}^{d\times K}, \tilde{\mathbf{w}}\in\mathbb{R}^d}{\arg\min} \; H_{\mu,\alpha}(\mathbf{W}^t, \mathbf{Z}, \tilde{\mathbf{w}}), \; t = 1, 2, \ldots \quad (5.29)$$

To analyze Fed+, we make the following smoothness assumption:

Assumption 5.8 For each $k = 1, 2, \ldots, K$, $f_k : \mathbb{R}^d \to \mathbb{R}$ is differentiable and the gradient ∇f_k is Lipschitz continuous with constant L_f, i.e., $\|\nabla f_k(\mathbf{w}) - \nabla f_k(\mathbf{w}')\|_2 \leq L_f \|\mathbf{w} - \mathbf{w}'\|_2$, $\forall \mathbf{w}, \mathbf{w}' \in \mathbb{R}^d$.

Proposition 5.9 *Under Assumption 5.8 and the stepsize choice $\eta = 1/L_f$, the following holds for Fed+: $\forall k \in S_t$,:*

$$F_k(\mathbf{w}_k^t; \mathbf{z}_k^{t-1}, \tilde{\mathbf{w}}^{t-1}) \leq F_k(\mathbf{w}_k^{t-1}; \mathbf{z}_k^{t-1}, \tilde{\mathbf{w}}^{t-1}) - \frac{1}{2(L_f + \alpha)}\|\nabla F_k(\mathbf{w}_k^{t-1}; \mathbf{z}_k^{t-1}, \tilde{\mathbf{w}}^{t-1})\|_2^2,$$
$$(5.30)$$

where F_k is defined in (5.12) and the gradient is w.r.t. $\mathbf{w}_k$.

Proof Let us first recall the following well-known descent lemma [1] for functions with Lipschitz continuous gradient.

Lemma 5.10 *Let $f : \mathbb{R}^d \to \mathbb{R}$ be continuously differentiable and ∇f be Lipschitz continuous with constant $L > 0$. Then, the following holds*

$$f\left(\mathbf{w} - \frac{1}{L}\nabla f(\mathbf{w})\right) \leq f(\mathbf{w}) - \frac{1}{2L}\|\nabla f(\mathbf{w})\|_2^2, \ \forall \mathbf{w} \in \mathbb{R}^d.$$

From Proposition 5.2, we know that the local update (5.6) (line 14 in Fed+ algorithm) is a gradient descent iteration with learning rate $\eta' = \frac{\eta}{1+\alpha\eta} = \frac{1}{L_f+\alpha}$ applied to the function $F_k(\cdot\,;\mathbf{z}_k^{t-1}, \tilde{\mathbf{w}}^{t-1})$. Clearly, $\nabla F_k(\cdot\,;\mathbf{z}_k^{t-1}, \tilde{\mathbf{w}}^{t-1})$ is Lipschitz continuous with constant $L = (L_f + \alpha)$. Therefore, applying the above Lemma, we have the following after one gradient descent iteration (starting with $\mathbf{w}_k^{t-1}$) at the Local-Solve subroutine: $\forall k \in S_t$,

$$F_k(\mathbf{w}_k^t; \mathbf{z}_k^{t-1}, \tilde{\mathbf{w}}^{t-1}) \leq F_k(\mathbf{w}_k^{t-1}; \mathbf{z}_k^{t-1}, \tilde{\mathbf{w}}^{t-1}) - \frac{1}{2L}\|\nabla F_k(\mathbf{w}_k^{t-1}; \mathbf{z}_k^{t-1}, \tilde{\mathbf{w}}^{t-1})\|_2^2.$$
$$(5.31)$$

Now, note the fact that $F_k(\mathbf{w}_k^t; \mathbf{z}_k^{t-1}, \tilde{\mathbf{w}}^{t-1})$ remains non-increasing after each gradient descent step. This completes the proof.   □

Combining the relation (5.29) with (5.30) we derive the following convergence result for Fed+.

Theorem 5.11 Assume that $H_{\mu,\alpha}$ in (5.8) is bounded from below, and parties are sampled with equal probability. Then, under Assumption 5.8 and the stepsize choice $\eta = 1/L_f$, the following holds for Fed+:

$$\lim_{t\to\infty} \mathbb{E}\left[\sum_{k=1}^{K}\|\nabla F_k(\mathbf{w}_k^{t-1}; \mathbf{z}_k^{t-1}, \tilde{\mathbf{w}}^{t-1})\|_2^2\right] = 0, \qquad (5.32)$$

where the expectation is with respect to the random subsets S_t, $t \geq 1$.

Moreover, the federated objective $F_\mu(\mathbf{W}^t)$ monotonically decreases with round t and converges to a value $\hat{F}_\mu \geq \min_{\mathbf{W}} F_\mu(\mathbf{W})$. Additionally, if the f_ks are convex, all parties are active in every round, and the level set $\{(\mathbf{W}, \mathbf{Z}, \tilde{\mathbf{w}}) : H_{\mu,\alpha}(\mathbf{W}, \mathbf{Z}, \tilde{\mathbf{w}}) \leq H_{\mu,\alpha}(\mathbf{W}^0, \mathbf{Z}^0, \tilde{\mathbf{w}}^0)\}$ is compact, then $\lim_{t\to\infty} F_\mu(\mathbf{W}^t) = \min_{\mathbf{W}} F_\mu(\mathbf{W})$ and the rate of convergence is $O(1/t)$.

Proof We start the proof with following observations from the Proposition 5.1 and the definition (5.11):

$$\tilde{\mathbf{w}}^t = \arg\min_{\tilde{\mathbf{w}} \in \mathbb{R}^d}\left[\min_{\mathbf{Z} \in \mathbb{R}^{d \times K}} H_{\mu,\alpha}(\mathbf{W}^t, \mathbf{Z}, \tilde{\mathbf{w}})\right],$$

$$(\mathbf{z}_1^t, \ldots, \mathbf{z}_K^t) = \arg\min_{\mathbf{Z} \in \mathbb{R}^{d \times K}} H_{\mu,\alpha}(\mathbf{W}^t, \mathbf{Z}, \tilde{\mathbf{w}}^t), \ t = 1, 2, \ldots$$

Combining the above, we have the following:

$$(\mathbf{z}_1^t, \ldots, \mathbf{z}_K^t, \tilde{\mathbf{w}}^t) = \underset{\mathbf{Z} \in \mathbb{R}^{d \times K}, \tilde{\mathbf{w}} \in \mathbb{R}^d}{\arg\min} H_{\mu,\alpha}(\mathbf{W}^t, \mathbf{Z}, \tilde{\mathbf{w}}), \ t = 1, 2, \ldots \quad (5.33)$$

This implies

$$H_{\mu,\alpha}(\mathbf{W}^t, \mathbf{Z}^t, \tilde{\mathbf{w}}^t) \leq H_{\mu,\alpha}(\mathbf{W}^t, \mathbf{Z}^{t-1}, \tilde{\mathbf{w}}^{t-1}), \ t = 1, 2, \ldots \quad (5.34)$$

Before moving further, we introduce the following notation $F_k^t(\mathbf{w}) := F_k(\mathbf{w}; \mathbf{z}_k^{t-1}, \tilde{\mathbf{w}}^{t-1})$. Now, we have the following from Proposition 5.9:

$$F_k^t(\mathbf{w}_k^t) \leq F_k^t(\mathbf{w}_k^{t-1}) - \frac{1}{2L} \|\nabla F_k^t(\mathbf{w}_k^{t-1})\|_2^2, \ \forall k \in S_t, \quad (5.35)$$

where $L := (L_f + \alpha)$. Moreover, $\mathbf{w}_k^t = \mathbf{w}_k^{t-1}$ for all $k \notin S_t$ implies that

$$F_k^t(\mathbf{w}_k^t) \leq F_k^t(\mathbf{w}_k^{t-1}), \ \forall k \notin S_t. \quad (5.36)$$

Summing (5.35) and (5.36), we get $\forall t = 1, 2, \ldots$,

$$H_{\mu,\alpha}(\mathbf{W}^t, \mathbf{Z}^{t-1}, \tilde{\mathbf{w}}^{t-1}) \leq H_{\mu,\alpha}(\mathbf{W}^{t-1}, \mathbf{Z}^{t-1}, \tilde{\mathbf{w}}^{t-1}) - \frac{1}{2KL} \sum_{k \in S_t} \|\nabla F_k^t(\mathbf{w}_k^{t-1})\|_2^2. \quad (5.37)$$

We can also express (5.37) in expectation form:

$$\mathbb{E}[H_{\mu,\alpha}(\mathbf{W}^t, \mathbf{Z}^{t-1}, \tilde{\mathbf{w}}^{t-1})] \leq H_{\mu,\alpha}(\mathbf{W}^{t-1}, \mathbf{Z}^{t-1}, \tilde{\mathbf{w}}^{t-1}) - \frac{p}{2KL} \sum_{k=1}^{K} \|\nabla F_k^t(\mathbf{w}_k^{t-1})\|_2^2, \quad (5.38)$$

where the expectation is w.r.t the random subset S_t and $p \in (0, 1]$ is the probability of $k \in S_t$. Taking, expectations w.r.t $S_1, S_2, \ldots, S_t$ (i.e., all the randomness), we get $\forall t = 1, 2, \ldots$,

$$\mathbb{E}[H_{\mu,\alpha}(\mathbf{W}^t, \mathbf{Z}^{t-1}, \tilde{\mathbf{w}}^{t-1})] \leq \mathbb{E}[H_{\mu,\alpha}(\mathbf{W}^{t-1}, \mathbf{Z}^{t-1}, \tilde{\mathbf{w}}^{t-1})] - \frac{p}{2KL} \sum_{k=1}^{K} \mathbb{E}[\|\nabla F_k^t(\mathbf{w}_k^{t-1})\|_2^2]. \quad (5.39)$$

Combining (5.39) and (5.34), we have $\forall t = 1, 2, \ldots$,

$$\mathbb{E}[H_{\mu,\alpha}(\mathbf{W}^t, \mathbf{Z}^t, \tilde{\mathbf{w}}^t)] \leq \mathbb{E}[H_{\mu,\alpha}(\mathbf{W}^{t-1}, \mathbf{Z}^{t-1}, \tilde{\mathbf{w}}^{t-1})] - \frac{p}{2KL} \sum_{k=1}^{K} \mathbb{E}[\|\nabla F_k^t(\mathbf{w}_k^{t-1})\|_2^2].$$

Summing over all t and using the fact $H_{\mu,\alpha}$ is bounded below, we arrive at (5.32). On the other hand, combining (5.37) and (5.34), we get $\forall t = 1, 2, \ldots,$

$$H_{\mu,\alpha}(\mathbf{W}^t, \mathbf{Z}^t, \tilde{\mathbf{w}}^t) \leq H_{\mu,\alpha}(\mathbf{W}^{t-1}, \mathbf{Z}^{t-1}, \tilde{\mathbf{w}}^{t-1}) - \frac{1}{2KL} \sum_{k \in S_t} \|\nabla F_k^t(\mathbf{w}_k^{t-1})\|_2^2.$$

(5.40)

Now, from (5.33) and (5.9) we see that $F_\mu(\mathbf{W}^t) = H_{\mu,\alpha}(\mathbf{W}^t, \mathbf{Z}^t, \tilde{\mathbf{w}}^t)$. Thus, from (5.40) we have that $\{F_\mu(\mathbf{W}^t)\}_{t=0}^\infty$ is monotonically non-decreasing; therefore, also converges to some real value say $\hat{F}_\mu$ because $H_{\mu,\alpha}$ is bounded below. The rest of the proof, when f_ks are convex, follows from Theorem 3.7 in [1] as (5.40) and (5.33) together suggest that Fed+ is basically an (approximate) alternating minimization approach for solving (5.9). □

5.7 Experiments

5.7.1 Datasets

We first provide an overview of the standard federated datasets and the way in which heterogeneity was imposed as well as the models used in our experiments. We curated a diverse set of synthetic and non-synthetic datasets, including those used in prior work on federated learning [11], and some proposed in LEAF, a benchmark for federated settings [2]. We then report and discuss the numerical results for the baseline algorithms together with our proposed Fed+ algorithms.

We evaluate Fed+ on standard federated learning benchmarks including the non-identical synthetic dataset of [11], a convex classification problem with MNIST [10] and FEMNIST [2, 4, 11], and a non-convex text sentiment analysis task called Sentiment140 (Sent140; [6]). Hyperparameters are the same as those of [11] and use the best μ reported in FedProx. Data is randomly split for each local party into an 80% training set and a 20% testing set. For all experiments, the number of local iterations per round $E = 20$, the number of selected parties per round $K = 10$, and the batch size is 10. In addition, the neural network models for all datasets are the same as those of [11]. Learning rates are 0.01, 0.03, 0.003, and 0.3 for synthetic, MNIST and FEMNIST and Sent140 datasets, respectively. The experiments used a fixed regularization parameter $\alpha = 0.01$ for each party's Local-Solve and the penalty parameters ρ for the mixture model are 1000, 10, and 10 for FedAvg+, FedGeoMed+, and FedCoMed+ methods, respectively. On the Sent140 dataset, we found that initializing the local model to a mixture model (i.e., setting $\lambda = 0.001$ instead of the default $\lambda = 0$) at the beginning of every Local-Solve subroutine for each party gives the best performance. We simulate the federated learning setup (1 aggregator N parties) on a commodity-hardware machine with 16 Intel® Xeon® E5-2690 v4 CPU and 2 NVIDIA® Tesla P100 PCIe GPU.

To generate non-identical synthetic data, we follow a similar setup to that of [11], additionally imposing heterogeneity among parties. In particular, for each party k, we generate samples (X_k, Y_k) according to the model $y = \arg\max(\text{softmax}(Wx + b))$, $x \in \mathbb{R}^{60}$, $W \in \mathbb{R}^{10 \times 60}$, $b \in \mathbb{R}^{10}$. We model $W_k \sim \mathcal{N}(u_k, 1)$, $b_k \sim \mathcal{N}(u_k, 1)$, $u_k \sim \mathcal{N}(0, \zeta)$; $x_k \sim \mathcal{N}(v_k, \Sigma)$, where the covariance matrix Σ is diagonal with $\Sigma_{j,j} = j^{-1.2}$. Each element in the mean vector v_k is drawn from $\mathcal{N}(B_k, 1)$, $B_k \sim \mathcal{N}(0, \beta)$. Therefore, ζ controls how much the local models differ from each other and β controls how much the local data at each party differs from that of other parties. In order to better characterize statistical heterogeneity and study its effect on convergence, we choose $\zeta = 1000$ and $\beta = 10$. There are $K = 30$ parties in total, and the number of samples on each party follows a power law.

Three datasets curated from prior work in federated learning are tested [2, 16]. First, we consider a convex classification problem using MNIST [10] with multinomial logistic regression. To impose statistical heterogeneity, we distribute the data among $K = 1000$ parties such that each party has samples of only one digit and the number of samples per party follows a power law. The input of the model is a flattened 784-dimensional (28×28) image, and the output is a class label between 0 and 9.

We then study a more complex 62-class Federated Extended MNIST [2, 4] (FEMNIST) dataset proposed in [11] using the same model. The heterogeneous data partitions in FEMNIST are generated by subsampling 10 lower case characters ("a"-"j") from EMNIST dataset [4] and distributing only 5 classes to each party. There are $K = 200$ parties in total. The input of the model is a flattened 784-dimensional (28×28) image, and the output is a class label between 0 and 9.

To address non-convex settings, we consider a text sentiment analysis task on tweets from Sentiment140 [6] (Sent140) with a two layer LSTM binary classifier containing 256 hidden units with pre-trained 300D GloVe embedding [18]. There are $K = 772$ parties in total. Each twitter account corresponds to a party. The model takes as input a sequence of 25 characters, embeds each of the characters into a 300-dimensional space by looking up Glove and outputs one character per training sample after 2 LSTM layers and a densely connected layer. We consider the highly heterogeneous setting where there are 90% stragglers (see [11] for more details).

5.7.2   Results

In Fig. 5.4, we report the testing performance for the baseline algorithms FedAvg, FedProx, RFA, and coordinate-wise median together with our proposed Fed+ family of algorithms. Overall, on the benchmark problems, the robust federated learning algorithms perform the worst on these non-IID datasets. FedProx uses a proximal term that, as we know from [11], is beneficial in heterogeneous settings as compared with FedAvg. However, all of the baselines produce a single global model not specific to the parties. Not only does Fed+ improve performance, it often also

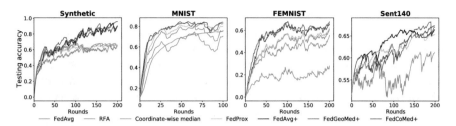

Fig. 5.4 Performance of Fed+, i.e., FedAvg+, FedGeoMed+, and FedCoMed+, is superior to that of the baselines

speeds up the learning convergence, as shown in Fig. 5.4. The Fed+ methods can improve the baseline algorithms' performance on these four datasets by 28.72%, 6.24%, 11.32%, and 13.89%, respectively. In particular, the best Fed+ algorithm can improve the most competitive implementation of the baseline algorithm FedProx's performance on these four datasets by 9.90% on average. In addition, the FedAvg+ method can achieve similar performances as FedGeoMed+ on the MNIST and FEMNIST datasets, but it fails to outperform the robust variants of Fed+, namely FedGeoMed+ and FedCoMed+, on the synthetic and Sent140 datasets. This shows the benefits of incorporating robust statistics such as the geometric median and coordinate-wise median rather than using the average as an aggregation statistic.

We also evaluate the impact of increasing the number of parties in training on the test accuracy. On the synthetic dataset, average test accuracy improves from 70.22% to 90.73% to 98.03% when the number of parties participating in training goes from $K = 3$ to $K = 15$ to $K = 30$. The average is taken over FedAvg+, FedGeoMed+, and FedCoMed+. On the MNIST dataset, average accuracies over the Fed+ algorithms are 69.80%, 81.34%, and 83.36% when the number of parties in training goes from $K = 100$ to $K = 500$ to $K = 1000$, respectively. On the FEMNIST dataset, average accuracies over the Fed+ algorithms are 25.16%, 68.71%, and 78.66% when the number of parties in training goes from $K = 20$ to $K = 100$ to $K = 200$, respectively. On the Sent140 dataset, average accuracies over the Fed+ algorithms are 57.13%, 60.77%, and 65.43% when the number of parties goes from $K = 77$ to $K = 386$ to $K = 772$, respectively. This shows the benefit of using Fed+ increases as the number of parties increases.

5.8 Conclusion

Personalized federated learning is designed to better handle the statistical heterogeneity inherent in federated settings, specifically, the lack of IID data across parties. Personalized federated learning results in more stable learning and achieves significantly better performance.

Heterogeneous data across parties is often accompanied by outliers, in that some parties' data is significantly different from the others. Robust aggregation via the median in place of the mean is one way to mitigate the impact of such outliers. The Fed+ family of personalized federated learning methods allows for seamless integration of both robust aggregation and mean aggregation. Fed+ unifies numerous algorithms, personalized and non-personalized, robust and average-based, while keeping the structure of the local computation intact.

This chapter provided an illustration of the pitfalls of non-personalized federated learning in very heterogeneous settings, introduced the Fed+ framework, provided convergence guarantees for this class of methods for non-smooth convex and non-convex loss functions and for the case of stragglers, and included a number of experiments comparing personalized and non-personalized, robust and average-based aggregation.

Additional experiments using personalized federated learning with Fed+ on a financial portfolio management problem are provided in Chap. 21 of this volume.

References

1. Beck A (2015) On the convergence of alternating minimization for convex programming with applications to iteratively reweighted least squares and decomposition schemes. SIAM J Optim 25(1):185–209. https://doi.org/10.1137/13094829X
2. Caldas S, Wu P, Li T, Konečný J, McMahan HB, Smith V, Talwalkar A (2018) LEAF: a benchmark for federated settings. arXiv preprint arXiv:181201097
3. Charles Z, Konecný J (2020) On the outsized importance of learning rates in local update methods. ArXiv abs/2007.00878
4. Cohen G, Afshar S, Tapson J, Van Schaik A (2017) EMNIST: extending MNIST to handwritten letters. In: 2017 international joint conference on neural networks (IJCNN). IEEE, pp 2921–2926
5. Deng Y, Kamani MM, Mahdavi M (2020) Adaptive personalized federated learning. arXiv preprint arXiv:200313461
6. Go A, Bhayani R, Huang L (2009) Twitter sentiment classification using distant supervision. CS224N project report, Stanford 1(12):2009
7. Hanzely F, Richtárik P (2020) Federated learning of a mixture of global and local models. ArXiv abs/2002.05516
8. Hanzely F, Zhao B, Kolar M (2021) Personalized federated learning: a unified framework and universal optimization techniques. 2102.09743
9. Karimireddy SP, Kale S, Mohri M, Reddi S, Stich S, Suresh AT (2019) SCAFFOLD: stochastic controlled averaging for on-device federated learning. ArXiv abs/1910.06378
10. LeCun Y, Bottou L, Bengio Y, Haffner P (1998) Gradient-based learning applied to document recognition. Proc IEEE 86(11):2278–2324
11. Li T, Sahu AK, Zaheer M, Sanjabi M, Talwalkar A, Smith V (2020) Federated optimization in heterogeneous networks. Proc Mach Learn Syst 2:429–450
12. Li T, Hu S, Beirami A, Smith V (2021) Ditto: fair and robust federated learning through personalization. 2012.04221
13. Li X, Huang K, Yang W, Wang S, Zhang Z (2020) On the convergence of FedAvg on non-IID data. ICLR, Arxiv, abs/1907.02189
14. Malinovsky G, Kovalev D, Gasanov E, Condat L, Richtárik P (2020) From local SGD to local fixed point methods for federated learning. ICML Arxiv, abs/2004.01442

15. Mansour Y, Mohri M, Ro J, Theertha Suresh A (2020) Three approaches for personalization with applications to federated learning. arXiv e-prints arXiv:2002.10619, 2002.10619
16. McMahan B, Moore E, Ramage D, Hampson S, y Arcas BA (2017) Communication-efficient learning of deep networks from decentralized data. In: Artificial intelligence and statistics. PMLR, pp 1273–1282
17. Pathak R, Wainwright M (2020) FedSplit: an algorithmic framework for fast federated optimization. ArXiv abs/2005.05238
18. Pennington J, Socher R, Manning CD (2014) Glove: global vectors for word representation. In: Proceedings of the 2014 conference on empirical methods in natural language processing (EMNLP), pp 1532–1543
19. Pillutla K, Kakade SM, Harchaoui Z (2019) Robust aggregation for federated learning. arXiv preprint arXiv:191213445
20. Dinh CT, Tran N, Nguyen J (2020) Personalized federated learning with Moreau envelopes. In: Larochelle H, Ranzato M, Hadsell R, Balcan MF, Lin H (eds) Advances in neural information processing systems, vol 33. Curran Associates, Inc., pp 21394–21405. https://proceedings. neurips.cc/paper/2020/file/f4f1f13c8289ac1b1ee0ff176b56fc60-Paper.pdf
21. Tyler DE (2008) Robust statistics: theory and methods. J Am Stat Assoc 103(482):888–889. https://doi.org/10.1198/jasa.2008.s239
22. Yin D, Chen Y, Kannan R, Bartlett P (2018) Byzantine-robust distributed learning: towards optimal statistical rates. PMLR, Stockholmsmässan, Stockholm, vol 80. Proceedings of Machine Learning Research, pp 5650–5659. http://proceedings.mlr.press/v80/yin18a.html
23. Zhang M, Sapra K, Fidler S, Yeung S, Alvarez JM (2021) Personalized federated learning with first order model optimization. In: International conference on learning representations. https:// openreview.net/forum?id=ehJqJQk9cw
24. Zhang S, Choromanska AE, LeCun Y (2015) Deep learning with elastic averaging SGD. In: Cortes C, Lawrence N, Lee D, Sugiyama M, Garnett R (eds) Advances in neural information processing systems, vol 28. Curran Associates, Inc. https://proceedings.neurips.cc/paper/2015/ file/d18f655c3fce66ca401d5f38b48c89af-Paper.pdf

Chapter 6
Communication-Efficient Distributed Optimization Algorithms

Gauri Joshi and Shiqiang Wang

Abstract In federated learning, the communication link connecting the edge parties with the central aggregator is sometimes bandwidth-limited and can have high network latency. Therefore, there is a critical need to design and deploy communication-efficient distributed training algorithms. In this chapter, we will review two orthogonal communication-efficient distributed stochastic gradient descent (SGD) methods: (1) local-update stochastic gradient descent (SGD), where clients make multiple local model updates that are periodically aggregated, and (2) gradient compression and sparsification methods to reduce the number of bits transmitted per update. In both these methods, there is a trade-off between the error convergence with respect to the number of iterations and the communication efficiency.

6.1 Introduction

Stochastic Gradient Descent in ML training. A majority of supervised learning problems are solved using the empirical risk minimization framework [5, 41], where the goal is to minimize the empirical risk objective function $F(x) = \sum_{j=1}^{n} f(x, \xi_j)/n$. Here, n is the size of the training dataset, ξ_j is the j−th labeled training sample, and $f(x; \xi_j)$ is the (generally non-convex) loss function. A ubiquitous algorithm to optimize $F(x)$ is *stochastic gradient descent (SGD)*, where we compute the gradient of $f(x; \xi_n)$ over small, randomly chosen subsets $\mathcal{B}$ (called mini-batches) of b samples each [4, 12, 25, 35, 37, 53] and update x according to $x_{k+1} = x_k - \eta \sum_{i \in \mathcal{B}} \nabla f(x_k, \xi_i)/b$, where η is referred to as the learning rate or step size. Although designed for convex objectives, mini-batch SGD has been

G. Joshi
Carnegie Mellon University, Pittsburgh, PA, USA
e-mail: gaurij@andrew.cmu.edu

S. Wang (✉)
IBM T. J. Watson Research Center, Yorktown Heights, NY, USA
e-mail: wangshiq@us.ibm.com

© The Author(s), under exclusive license to Springer Nature Switzerland AG 2022   125
H. Ludwig, N. Baracaldo (eds.), *Federated Learning*,
https://doi.org/10.1007/978-3-030-96896-0_6

shown to perform well even on non-convex loss surfaces due to its ability to escape saddle points and local minima [7, 31, 42, 55]. Therefore, it is the dominant training algorithm in state-of-the-art machine learning.

For massive datasets such as Imagenet [38], running mini-batch SGD on a single node can be prohibitively slow. A standard way to parallelize gradient computation is the parameter server (PS) framework [11], consisting of a central server and multiple worker nodes. Straggling workers and communication delays can become a bottleneck in scaling this framework to a large number of worker nodes. Several methods such as asynchronous [10, 13, 15, 56] and periodic gradient aggregation [43, 46, 54] have been proposed to improve the scalability of data-center based ML training.

Motivation for Federated Learning. In spite of the algorithmic and systems advances that improve the efficiency and scalability, there is one main limitation of data-center based training. It requires the training dataset to be centrally available at the parameter server, which shuffles and splits it across the worker nodes. The rapid proliferation of edge parties such as phones, IoT sensors, and cameras with on-device computation capabilities has led to a major shift in this data partitioning paradigm. Edge parties collect rich information from their environment, which can be used for data-driven decision-making. Due to limited communication capabilities as well as privacy concerns, the data cannot be directly sent over to the cloud for centralized processing or shared with other nodes. The federated learning framework proposes to keep the data at the edge party, and instead bring model training to the edge. In federated learning, data is kept at the edge parties and the model is trained in a distributed manner. Only gradients or model updates are being exchanged between the edge parties and the aggregator.

System Model and Notation. A typical federated learning setting consists of a central aggregator connected to K edge parties as shown in Fig. 6.1, where K can be of the order of thousands or even millions. Each party i has a local dataset $\mathcal{D}_i$ consisting of n_i samples, which cannot be transferred over to the central aggregator

Fig. 6.1 In federated optimization, the goal of the aggregator is to minimize a weighted average of the local objective functions $F_i(x)$ at the edge parties

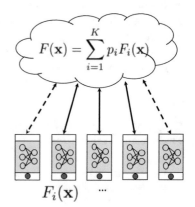

$$F(\mathbf{x}) = \sum_{i=1}^{K} p_i F_i(\mathbf{x})$$

$$F_i(\mathbf{x})$$

or shared with other edge parties. We use $p_i = n_i/n$ to denote the fraction of data at the i-th party, where $n = \sum_{i=1}^{K} n_i$. The aggregator seeks to train a machine learning model $x \in \mathbb{R}^d$ using the union of local datasets $\mathcal{D} = \cup_{i=1}^{K} \mathcal{D}_i$. The model vector x contains the parameters of the model, for example, the weights and biases of a neural network. In order to train the model x, the aggregator seeks to minimize the following empirical risk objective function:

$$F(x) := \sum_{i=1}^{K} p_i F_i(x) \tag{6.1}$$

where $F_i(x) = \frac{1}{n_i} \sum_{\xi \in \mathcal{D}_i} f(x; \xi)$ is the local objective function at the i-th party. Here, f is the loss function (possibly non-convex) defined by the model x and ξ represents a data sample from local dataset $\mathcal{D}_i$. Observe that we assign the weight p_i proportional to the data fraction for the i-th party. This is because, we would like to emulate a centralized training scenario with all the training data transferred to a central parameter server. Thus, parties with more data will get a higher weight in the global objective function.

Due to the resource-limitations at the edge parties and the large number of parties, the federated training algorithms have to operate under strict communication constraints and cope with data and computational heterogeneity. For example, the wireless communication link connecting each edge party with the central aggregator can be bandwidth-limited and have high network latency. Also, due to limited network connectivity and battery constraints, the edge parties may only be intermittently available. Thus, only a subset of m out of the K edge parties are available participating in training the model x at a given time. To operate within these communication constraints, the federated learning framework requires new distributed training algorithms, going beyond the algorithms used in the data-center setting. In Sect. 6.2 we review the local-update SGD algorithm and its variants, which reduce the frequency of communication of the edge parties with the aggregator. In Sect. 6.3, we review compressed and quantized distributed training algorithms that reduce the number of bits communicated for every update sent by the edge parties to the aggregator.

6.2   Local-Update SGD and FedAvg

In this section we first discuss local-update SGD and its variants. The FedAvg algorithm, which is at the core of federated learning, is an extension of local-update SGD. We discuss how FedAvg builds on local-update SGD, and various strategies that are used to handle data and computational heterogeneity in federated learning.

6.2.1 Local-Update SGD and Its Variants

Synchronous Distributed SGD. In the data-center setting, the training dataset $\mathcal{D}$ is shuffled and equally distributed across m worker nodes. The standard method to train machine learning models is to use synchronous distributed SGD [11], where the gradients are computed by the workers and then aggregated by a central parameter server. In every iteration t of synchronous SGD, the workers pull the current version of the model x_t from the parameter server. Each worker i computes a mini-batch stochastic gradient $g_i(x) = \sum_{\xi \in \mathcal{B}} f(x; \xi)$ using a mini-batch $\mathcal{B}$ of B samples drawn from this local dataset $\mathcal{D}_i$. The parameter server then collects the gradients from all the workers and updates the model parameters as per

$$x_{t+1} = x_t - \frac{\eta}{m} \sum_{i=1}^{m} g_i(x). \tag{6.2}$$

As the number of workers m increases, the error versus iterations convergence of synchronous SGD improves. However, due to variabilities in the local gradient computation times at the workers, the time taken to wait for all workers to finish their gradient computations increases. To improve the scalability with the number of workers, straggler-resilient variants of synchronous SGD that perform asynchronous gradient aggregation have been proposed in [13, 15, 28, 29, 56, 58].

Local-Update SGD. In spite of the effectiveness of asynchronous aggregation methods to improve the scalability of distributed SGD, in many distributed systems, the communication time to exchange gradients and model updates between the workers and the parameter server can dominate the variabilities in the local gradient computation times. Thus, constant inter-node communication after every iteration can be prohibitively expensive and slow. Local-update SGD is a communication-efficient distributed SGD algorithm that overcomes this issue by having the workers nodes perform multiple local SGD updates instead of just computing a mini-batch gradient.

Local-update SGD divides the training into communication rounds, as illustrated in Fig. 6.2. In a communication round, each worker locally optimizes its objective function $F_i(x)$ using SGD. Each worker i starts from the current global model, denoted by x_t and performs τ SGD iterations to obtain the model $x_{t+\tau}^{(i)}$. The resulting models are then sent by the m workers to the parameter server, which averages them to update the global model as follows:

$$x_{t+\tau} = \frac{1}{m} \sum_{i=1}^{m} x_{t+\tau}^{(i)}. \tag{6.3}$$

Runtime Per Iteration of Local-update SGD. By performing τ local updates at each worker before communicating with the parameter server, local-update SGD

Fig. 6.2 In local-update SGD, each worker makes τ local SGD updates, after which the resulting models are aggregated by the parameter server

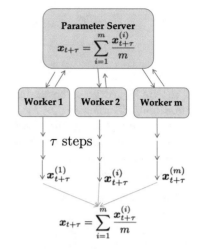

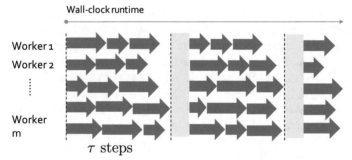

Fig. 6.3 In local-update SGD, by reducing the frequency of communication, it helps amortize the communication delay (shown in yellow) across τ iterations

reduces the expected runtime per iteration. Let us quantify this runtime saving by considering the following delay model. The time taken by the ith worker to compute a mini-batch gradient at the kth local-step is modeled a random variable $Y_{i,k} \sim F_Y$, assumed to be independent and identically distributed (i.i.d.) across workers and mini-batches. The communication delay is represented by a constant D, and it includes the time taken to send the local model to the parameter server and receive the averaged global model from the parameter server. Since each worker i makes τ local updates, its average local computation time (time taken to complete one sequence of 3 blue arrows in Fig. 6.3) is given by

$$\overline{Y}_i = \frac{Y_{i,1} + Y_{i,2} + \dots Y_{i,\tau}}{\tau}. \tag{6.4}$$

If $\tau = 1$, in which case local-update SGD reduces to synchronous SGD, the random variable $\overline{Y}$ is identical to Y. Since the communication delay D is amortized over τ iterations, the runtime time per iteration (also illustrated in Fig. 6.3) is given by

$$\mathbb{E}[T_{\text{Local-update}}] = \mathbb{E}[\max(\overline{Y}_1, \overline{Y}_2, \ldots, \overline{Y}_m)] + \frac{D}{\tau} \qquad (6.5)$$

$$= \mathbb{E}[\overline{Y}_{m:m}] + \frac{D}{\tau}. \qquad (6.6)$$

The term $Y_{m:m}$ denotes the maximum order statistic of m i.i.d. random variables with probability distribution $Y \sim F_Y$. From (6.6) we observe that performing more local updates reduces the runtime per iteration in two ways. Firstly, the communication delay gets amortized across τ iterations and reduces by a factor of τ. Secondly, performing local updates also provides a straggler mitigation benefit because the tail of $\overline{Y}_{m:m}$ is lighter than $Y_{m:m}$ and thus the first term in (6.6) reduces with τ.

Error Convergence of Local-update SGD. As we see above, reducing the frequency of communication between the workers and the parameter server to only once in τ iterations can give a significant reduction in the runtime per iteration. However, setting a large value of τ, the number of local updates results in inferior error convergence. This is because, the worker nodes' models $x_{t+\tau}^{(i)}$ diverge from each other as τ increases. The papers [43, 46, 47] give an error convergence analysis of local-update SGD in terms of number of local updates τ. Suppose the objective function $F(x)$ is L-Lipschitz-smooth, and the learning rate η satisfies $\eta L + \eta^2 L^2 \tau(\tau - 1) \leq 1$. The stochastic gradient $g(x; \xi)$ is an unbiased estimate of $\nabla F(x)$, that is, $\mathbb{E}_\xi[g(x; \xi)] = \nabla F(x)$. The stochastic gradient $g(x; \xi)$ is assumed to have bounded variance, that is, $\text{Var}(g(x; \xi)) \leq \sigma^2$. If the starting point is x_1, then $F(x_T)$ after T iterations of local-update SGD is bounded as

$$\mathbb{E}\left[\frac{1}{T}\sum_{t=1}^{T} \|\nabla F(x_t)\|^2\right] \leq \frac{2[F(x_1) - F_{\text{inf}}]}{\eta t} + \frac{\eta L \sigma^2}{m} + \eta^2 L^2 \sigma^2 (\tau - 1) \qquad (6.7)$$

where x_t denotes the averaged model at the tth iteration. Setting $\tau = 1$ makes local-update SGD and its error convergence bound identical to that of synchronous distributed SGD. As τ increases, the last term of the bound increases, thus increasing the error floor at convergence.

Adaptive Communication Strategies. From the runtime and error analyses above we can see that there is a trade-off between error and communication delay per iteration as we vary τ. A larger τ reduces the expected communication delay, but yields worse error convergence. In order to get fast convergence as well as a low error floor, [47, 52] propose a strategy to adapt τ during the training process. For a fixed learning rate η, the following strategy in [47] gradually reduces τ:

$$\tau_l = \left\lceil \sqrt{\frac{F(x_{t=lT_0})}{F(x_{t=0})}} \tau_0 \right\rceil \qquad (6.8)$$

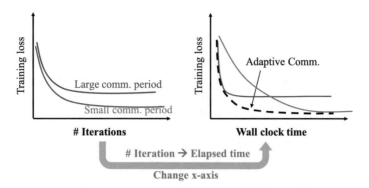

Fig. 6.4 Motivation for adapting the number of local update τ during the course of training

where τ_l is the number of local updates in the l-th interval of T_0 seconds in the training. This update rule can also be modified to account for an underlying variable learning rate schedule (Fig. 6.4).

Elastic Averaging and Overlap SGD. In local-update SGD, the updated global model needs to be communicated to the nodes before the next set of τ updates can commence. Moreover, the global model cannot be updated until the slowest of the m nodes finishes its τ local updates. This communication barrier can become a bottleneck in the global model being updated and increase the expected runtime per training round. Since this communication barrier is imposed by the algorithm rather than the systems implementation, we need an algorithmic approach to remove it and allow communication to overlap with local computation. Works such as [9, 11, 13, 15, 19, 33, 56] use asynchronous gradient aggregation to remove the synchronization barrier. However, asynchronous aggregation causes model staleness, that is, slow nodes can have arbitrarily outdated versions of the global model. Some recent works [48, 57] propose variants of local-update SGD that allow an overlap of communication and computation. In these algorithms, the worker nodes start their local updates from an anchor model, which is available even before the slowest nodes finish the previous round of local updates. This approach is inspired by the elastic-averaging SGD (EASGD) algorithm proposed in [57], which adds a proximal term to the objective function. Proximal methods such as [6, 32, 57], although not designed for this purpose, naturally allow overlapping communication and computation.

6.2.2 Federated Averaging (FedAvg) Algorithm and Its Variants

The FedAvg Algorithm. Due to limited communication capabilities of edge parties in federated learning, local-update SGD is especially suitable for the federated learning framework, where it is referred to as the FedAvg algorithm. The main differences are as follows. Firstly, the worker nodes that are servers in the cloud are replaced by edge parties such as mobile and IoT devices. Due to the intermittent availability of the edge parties, unlike the data-center setting, only a subset of m out of K parties participate in each training round. Secondly, the datasets $\mathcal{D}_i$ can be highly heterogeneous across the edge parties both in their size and composition, unlike the data-center setting where the dataset $\mathcal{D}$ is shuffled and evenly partitioned across the worker nodes.

The federated averaging algorithm (FedAvg) [30] also divides the training into communication rounds. In a communication round, the aggregator selects m edge parties uniformly at random from among the available parties. Each edge party locally optimizes its objective function $F_i(x)$ using SGD similar to local-update SGD. Unlike basic local-update SGD where each worker performs the same number of local updates τ, in FedAvg the number of local updates τ_i may vary across edge parties and communication rounds. A common implementation practice is that the parties run for the same local epochs E. Thus, $\tau_i = \lfloor En_i/B \rfloor$, where B is the mini-batch size. Alternately, if each communication round has a fixed length in terms of wall-clock time, then τ_i represents the local iterations completed by party i within the time window and may change across clients (depending on their computation speeds and availability) and across communication rounds. In the r-th communication round, edge parties start from the global model $x_{r,0}$ and perform τ_i local updates each. Suppose their resulting models are denoted by $x_{r,\tau_i}^{(i)}$. The shared global model x_r is updated as follows:

$$x_{r+1,0} = \sum_{i=1}^{m} p_i x_{r,\tau_i}^{(i)} \tag{6.9}$$

where $p_i = |\mathcal{D}_i|/|\mathcal{D}|$, the fraction of data at the i-th edge party.

Strategies to Handle Data Heterogeneity. Since the datasets $\mathcal{D}_i$ are highly heterogeneous across the nodes, the locally trained models at the edge parties can be different significantly from each other. And as the number of local updates increase, the models may become overfitted to the local datasets. As a result, the FedAvg algorithm may converge to an incorrect point that is not a stationary point of the global objective function $F(x)$. For example, suppose each edge party performs a large number of local updates and the i-th party's local model converges to $x_*^{(i)} = \min F_i(x)$. Then the weighted average of these local models will converge to $x = \sum_{i=1}^{K} p_i x_*^{(i)}$, which can be arbitrarily different from the true global minimum $x_* = \min F(x)$. One solution to reduce this solution bias caused by data heterogeneity is to choose a small or decaying learning rate η and/or keep the number of local updates τ small. Other techniques used to overcome the solution bias include proximal local-update methods such as [39, 40] that add a regularization term to the global objective and methods that aim to minimize cross-party model

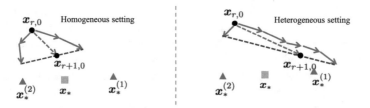

Fig. 6.5 Model updates in the parameter space. Green squares and blue triangles denote the minima of global and local objectives, respectively. In the heterogeneous update setting, the solution will be biased toward the parties with more local updates

drift [23] by exchanging control variates. At a high level, these techniques deter the edge parties' models from drifting away from the global model.

Strategies to Handle Computational Heterogeneity The effect of data heterogeneity can be exacerbated by computational heterogeneity across the edge parties. Even if the edge parties make different number of local updates τ_i, the standard FedAvg algorithm proposes that the resulting models are simply aggregated in proportion of the data fractions p_i. However, this can result in an inconsistent solution that is mismatching with the intended global objective, as shown in [49] and illustrated in Fig. 6.5. The final solution becomes biased toward the local optimum $x_*^{(i)} = \min F_i(x)$, and it can be arbitrarily far away from the global minimum $x_*^{(i)} = \min F(x)$. The paper [49] fixes this inconsistency by normalizing the accumulated local update $(x_{r,\tau_i}^{(i)} - x_{r,0}^{(i)})$ by the number local updates τ_i, before sending it to the central aggregator. This normalized federated averaging algorithm called FedNova results in a consistent solution while preserving the fast convergence rate.

Besides variability in the number of local updates τ_i, computational heterogeneity and solution inconsistency may also occur due to the edge parties using local momentum, adaptive local optimizers such as AdaGrad, or different learning rate schedules. A generalized version of FedNova [51] is required to fix the inconsistency in these cases.

Strategies to Handle Intermittent Availability of Edge Parties. The total number of edge parties in a federated learning setup can be of the order of thousands or even millions of devices. Due to their local computing resource constraints and bandwidth limitations, the edge parties are only intermittently available to participate in training. For example, cell phones are currently used for federated training only when they are plugged in for charging in order to conserve the battery. Therefore, in each communication round, only a small subset of the edge parties participate in the FedAvg algorithm.

Most work on designing and analyzing federated learning algorithms assume that the subset of edge parties is selected uniformly at random from the entire set of edge parties [27]. Such partial and intermittent participation amplifies the adverse effects of data heterogeneity by adding a variance term to the error. Some recent works [8, 21, 36] propose client selection methods that cope with such heterogeneity and improve the convergence speed. These strategies assign a higher selection probability to edge parties with higher local losses and show that it can speed-up progress of the global model. However, this speed-up comes at the cost of a higher non-vanishing bias, which increases with the degree of data heterogeneity. The paper [8] proposes an adaptive strategy that gradually reduces the selection skew in order to achieve the best trade-off between convergence speed and error floor.

6.3   Model Compression

In addition to performing multiple local updates, the models can be also compressed during communication and computation. One approach is to use standard loss-less compression techniques, which, however, can only reduce the model size by a limited degree and requires decompression at the receiver side. In this section, we discuss a specific class of lossy compression techniques that is designed for improving the communication efficiency in federated learning and distributed SGD in general. These techniques do not require decompression at the receiver and can guarantee training convergence. We focus on approaches to improve the communication efficiency in Sects. 6.3.1 and 6.3.2 and to improve both communication and computation efficiency in Sect. 6.3.3.

6.3.1   SGD with Compressed Updates

A widely used approach is to compress the model updates transmitted between parties and the aggregator [2, 24]. In particular, we define a compressor $C(z)$ that produces a compressed version of an arbitrary vector z. Popular compressors include those that implement quantization [2] and sparsification [44]. Based on their characteristics, compressors can be categorized into unbiased and general (i.e., possibly biased). We discuss these two variants of compressors as follows, where we consider a technique called error feedback for general compressors, which is useful for avoiding variance blow-up and guaranteeing convergence. Note that our notion of bias in this section is in the context of probabilistic modeling, where an unbiased compressor means that the expected value of a compressed vector (obtained from this compressor) is equal to the original vector.

6.3.1.1   Unbiased Compressor Without Error Feedback

An *unbiased compressor* $C(z)$ satisfies both of the following characteristics:

$$\mathbb{E}\left[C(z)|z\right] = z \tag{6.10}$$

$$\mathbb{E}\left[\|C(z) - z\|^2 \,|z\right] \le q\|z\|^2 \tag{6.11}$$

where $q \ge 0$ is a constant capturing the relative approximation gap achieved by the compressor. Intuitively, the relative approximation gap means the relative error of the compressed vector compared to that of the original vector. We can easily see that $q = 0$ is a necessary condition of $C(z) = z$ (i.e., no compression) and $q = 1$ is a necessary condition of $C(z) = \mathbf{0}$ (i.e., no transmission). In general, a larger q corresponds to a more compressed vector produced by $C(z)$. As we will see in the "random-k" example presented next, in some cases we may amplify the compression result to guarantee unbiasedness, which can produce a value of q that is greater than one.

Examples An example of an unbiased compressor is a randomized quantizer that gives

$$[C(z)]_i = \begin{cases} \lfloor z_i \rfloor, & \text{with probability } \lceil z_i \rceil - z_i \\ \lceil z_i \rceil, & \text{with probability } z_i - \lfloor z_i \rfloor \end{cases} \tag{6.12}$$

for the i-th component of the vector, where $\lfloor \cdot \rfloor$ and $\lceil \cdot \rceil$ denote the floor (rounding down to integer) and ceiling (rounding up to integer) operators, respectively. We note that the integer here can be the base in the case of floating point representation. It can be easily seen that this quantization operation satisfies the unbiasedness property (6.10). Noting that the quantization operation gives $q = \max_{y \in [0,1]}(1 - y)^2 y + y^2(1 - y)$, we have $q = 1/4$.

Another example is to randomly select k components from the original vector z with equal probability of k/d and amplifying the result by d/k, i.e.,

$$[C(z)]_i = \begin{cases} \frac{d}{k} z_i, & \text{with probability } \frac{k}{d} \\ 0, & \text{with probability } 1 - \frac{k}{d} \end{cases} \tag{6.13}$$

for the i-th component of the vector. This is often known as the random-k sparsification technique. It is apparent that this operation is also unbiased. The left-hand side of (6.11) is the sum of $\left[(d/k - 1)^2 \cdot k/d + 1 \cdot (1 - k/d)\right] z_i^2$ over all components i. Hence, we have $q = d/k - 1$.

Local-Update SGD with Compressed Updates. When using compression with local-update SGD, each party computes its local updates as usual. The updates are compressed before they are sent to the aggregator, and the aggregator then averages the compressed updates to obtain the next global model parameter. Assuming that a round with τ iterations starts at iteration t, this gives the following recurrence relation:

$$x_{t+\tau} = x_t + \frac{1}{m} \sum_{i=1}^{m} C\left(x_{t+\tau}^{(i)} - x_t\right). \tag{6.14}$$

In a different implementation, another compression operation can be applied at the server to maintain the same compression level (e.g., quantization precision or number of components to transmit). This gives

$$x_{t+\tau} = x_t + C\left(\frac{1}{m} \sum_{i=1}^{m} C\left(x_{t+\tau}^{(i)} - x_t\right)\right). \tag{6.15}$$

The operations in (6.14) and (6.15) are similar, possibly with a different overall approximation gap q.

Convergence Bound. With an appropriately chosen learning rate, the optimality (expressed as the squared norm of gradient) after T iterations using (6.14) can be bounded as [34]:

$$\mathbb{E}\left[\frac{1}{T} \sum_{t=1}^{T} \|\nabla F(x_t)\|^2\right] = O\left(\frac{1+q}{\sqrt{T}} + \frac{\tau}{T}\right) \tag{6.16}$$

where $x_t := \frac{1}{m} \sum_{i=1}^{m} x_t^{(i)}$ for all t, even if no compression/aggregation occurs in iteration t, and constants other than q, τ, and T are absorbed in $O(\cdot)$, where $O(\cdot)$ is the big-O notation that stands for an upper bound while ignoring constants.

Variance Blow-Up. From (6.16), we can see that when T is sufficiently large, the error is dominated by the first term $O\left(\frac{1+q}{\sqrt{T}}\right)$. This error is related to the value of q. When q is large, we need to increase the number of iterations T by q^2 times to eliminate the effect of q and reach the same error, which is problematic because the advantage of compression would be canceled out by the increased amount of computation, particularly for compressors such as random-k where $1+q$ is inversely proportional to k, as we discussed earlier. Since the first term of (6.16) is also proportional to the variance of the stochastic gradient, which we absorbed into the $O(\cdot)$ notation for simplicity, this phenomenon is also known as variance blow-up in the literature [44].

Next, we will see that error feedback can fix the variance blow-up problem by accumulating the difference between the compressed and actual parameter vectors locally so that it can be transmitted in a future communication round.

6.3.1.2 General Compressor with Error Feedback

We first proceed with the introduction of a general (possibly biased) compressor. A *general compressor* $C(z)$ satisfies the following property:

$$\mathbb{E}\left[\|C(z) - z\|^2 \,|z\right] \le \alpha\|z\|^2 \tag{6.17}$$

where α is a constant with $0 \le \alpha < 1$ capturing the relative approximation gap achieved by the compressor. Compared to the properties of the unbiased compressor in (6.10) and (6.11), the key difference is that the general compressor does not guarantee unbiasedness. Equations (6.11) and (6.17) are essentially the same when we let $\alpha = q$, except that we require $\alpha < 1$ for the purpose of convergence analysis. Another reason for keeping α different from q is to distinguish between the two types of compressors. Compressors satisfying (6.17) are also known as α-contractive compressors [1]. There is also a stricter version of (6.17) where the inequality holds without expectation.

Example A typical example of a general compressor is the top-k sparsification technique that selects k components with the largest magnitude. This can be expressed as follows:

$$[C(z)]_i = \begin{cases} z_i, & \text{if } |z_i| \text{ is among the } k \text{ largest elements of } \{|z_j| : \forall j \in \{1, 2, \ldots, d\}\} \\ 0, & \text{otherwise} \end{cases} \tag{6.18}$$

for the i-th component of the vector. As this operation is deterministic for given z, it is biased. We can obtain $\alpha = 1 - \frac{k}{d}$ because the square of the remaining components in z cannot be larger than the k components with the largest magnitudes.

Local-Update SGD with Compressed Updates and Error Feedback. When using error feedback, in addition to exchanging compressed updates between clients and the server, the portion that has not been communicated (referred to as the "error" here) will be accumulated locally. In the next round, the accumulated error will be added to the latest updates in that round, and this sum vector will be used by the compressor to compute the compressed vector. Each party i keeps an error vector $e^{(i)}$ that is initialized as $e_0^{(i)} = \mathbf{0}$. In every round r, the following steps are executed.

1. For each party $i \in \{1, 2, \ldots, m\}$ in parallel:

   a. Compute τ steps of local gradient descent to obtain $x_{r,\tau}^{(i)}$, starting from the global parameter x_r.
   b. Sum up the accumulated error with the current update: $z_r^{(i)} := e_r^{(i)} + x_{r,\tau}^{(i)} - x_r$.
   c. Compute the compression result $\Delta_r^{(i)} := C\left(z_r^{(i)}\right)$ (this is what will be sent to the aggregator).
   d. Subtract the compression result to obtain the remaining error for the next round $e_{r+1}^{(i)} = z_r^{(i)} - \Delta_r^{(i)}$.

2. The aggregator updates the global parameter for the next round based on the compressed updates received from parties, i.e.,

$$x_{r+1} = x_r + \frac{1}{m}\sum_{i=1}^{m}\Delta_r^{(i)} = x_r + \frac{1}{m}\sum_{i=1}^{m}C\left(z_r^{(i)}\right). \tag{6.19}$$

We can see that the only difference between (6.14) and (6.19) is that we now compress on $z_r^{(i)}$, which includes the accumulated error from previous rounds. Note that we use the round r index here for convenience, instead of the iteration index t in (6.14). Similar to (6.15), the above procedure can also be extended to compressing and accumulating errors at both the parties and the aggregator [45].

Convergence Bound. Similar to (6.16), we present the optimality bound for the error-feedback mechanism. With an appropriately chosen learning rate, we have [3]

$$\mathbb{E}\left[\frac{1}{Tm}\sum_{t=1}^{T}\sum_{i=1}^{m}\|\nabla F(x_t^{(i)})\|^2\right] = O\left(\frac{1}{\sqrt{T}} + \frac{\tau^2}{(1-\alpha)^2 T}\right). \tag{6.20}$$

We note that although the left-hand sides of (6.16) and (6.20) are slightly different, their physical meanings are the same. The slight difference is due to the different techniques used in deriving these bounds. Compared to (6.16), we see that the approximation gap due to compression, captured by α, is now in the second term in (6.20). When T is sufficiently large, we now have a convergence rate of $O\left(\frac{1}{\sqrt{T}}\right)$, which avoids the variance blow-up problem.

Note that as we require $0 \le \alpha < 1$, our analysis here does not hold for the random-k compressor in (6.13), but we can modify (6.13) by removing the amplification coefficient d/k since we do not require unbiasedness anymore. The resulting compressor satisfies $\alpha = 1 - k/d$, which is the same as for top-k. However, in practice, top-k usually works better than random-k because its actual approximation gap is usually much less than the upper bound of $1 - k/d$.

These results suggest that error-feedback mechanisms generally perform better than non-error-feedback mechanisms. However, there is recent work [20] suggesting that, by transforming biased compressors into unbiased ones in a systematic manner, we may actually obtain better performance. This is an active area of research, and practitioners may need to experiment with different compression techniques to see which one works the best for the problem at hand.

6.3.2 Adaptive Compression Rate

A question in SGD with compressed updates is how to determine the compression rate (i.e., the quantities q and α in (6.11) and (6.17)) to minimize the training time for reaching some target value of the objective function. The optimal compression rate in this case depends on the physical time incurred by computation in every iteration and communication in every round. This problem is similar to determining the optimal number of local updates τ as discussed in Sect. 6.2.1, but here the control variable is the compression rate instead. A similar approach where the method of compression rate adaptation is derived from the convergence bound, as in Sect. 6.2.1, can be applied to solve this problem. To overcome the difficulty of estimating or eliminating unknown parameters in the convergence bound, model-free approaches such as those based on online learning [16] can also be used. In essence, the online learning based approach uses exploration–exploitation, which explores different choices of compression rates in initial rounds and gradually switches to exploiting those rates that have been beneficial before. A challenge is that the exploration needs to have minimal overhead, because otherwise it will prolong the training time even compared to the case without optimization.

To facilitate efficient exploration, a problem can be formulated to find the best compression rate that minimizes the training time for reducing the empirical risk by a unit amount [16]. The exact objective of this problem is unknown, because it is difficult to predict how training will progress when using different compression rates. However, empirical evidence shows that for a given (current) empirical risk, we can assume that the compression used before is independent of the progression of future empirical risk. Together with some other assumptions, we can cast this problem in an online convex optimization (OCO) framework [18], which can be solved using online gradient descent with the gradient being the derivative of the training time for unit risk reduction with respect to the compression rate. Note that this gradient here is different from the gradient of the learning problem. The online gradient decent procedure is then to update the compression rate using gradient descent on the training time objective in *each* round, where different rounds can have different objectives that are unknown beforehand. It can be theoretically proven that, although we perform gradient descent only on each round's objective, the accumulated optimality gap (known as the *regret*) grows sublinearly in time, so that the time-averaged regret goes to zero as time goes to infinity. However, this approach requires a gradient oracle that gives the exact derivative at the compression rate chosen in each round, which is difficult to obtain in practice.

To overcome this issue, a sign-based online gradient descent method is used in [16], which updates the compression rate only based on the sign, instead of the actual value, of the derivative. It is relatively easy to estimate the sign of the derivative, and as long as the probability of estimating the correct sign is higher than that of estimating the wrong sign, a similar sublinear regret is guaranteed. Empirical results have shown that this algorithm converges to a near-optimal compression rate quickly and improves the performance over choosing an arbitrarily fixed compression rate.

6.3.3   Model Pruning

In addition to compressing the parameter updates, the models themselves can be compressed by pruning (removing) some insignificant weights in neural networks,

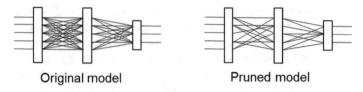

Fig. 6.6 Illustration of model pruning

which speeds up both computation and communication while maintaining a similar accuracy of the final model [14, 17]. An illustration of pruning is shown in Fig. 6.6. A well-known pruning method is to iteratively train and prune the model, by removing a certain percentage of weights that have small magnitudes at an interval that includes multiple SGD iterations.

When combining pruning with federated learning, a two-stage procedure can be used, where the model is trained and pruned on a single party in the first stage, and then pruned further during the regular federated learning process that involves multiple parties [22]. The initial pruning stage allows federated learning to start with a small model to save both computation and communication compared to starting with the full model, while still converging to the global optimum as the model and its weights are adjusted in the further pruning stage. To determine which weights should be pruned (or added back in the second stage), an objective can be formulated so that the pruned model approximates the original model and also maintains the "trainability" in future rounds. To approximate the original model, standard magnitude-based pruning with a properly chosen pruning rate can be applied, so that only those weights with a small enough magnitude can be pruned. The trainability can be captured using a first-order approximation of the empirical risk reduction when performing one step of SGD from the pruned model. Based on this approximation, we can solve for the set of weights that should be pruned (or added back if they are already pruned before) to maintain trainability [22]. Overall, this approach adapts the model size over time to (approximately) maximize the training efficiency.

6.4 Discussion

In this chapter we reviewed communication-efficient distributed optimization algorithms that are used in federated learning, in particular, local-update SGD algorithms that reduce the communication frequency and compression methods that reduce the number of bits communicated. These methods could be combined with other algorithms that improve the convergence speed and efficiency of federated learning. For example, instead of using classic SGD as the local solver, the edge parties may use acceleration [50], variance reduction [23, 26], or adaptive optimization methods.

References

1. Albasyoni A, Safaryan M, Condat L, Richtárik P (2020) Optimal gradient compression for distributed and federated learning. arXiv preprint arXiv:2010.03246
2. Alistarh D, Grubic D, Li J, Tomioka R, Vojnovic M (2017) QSGD: communication-efficient SGD via gradient quantization and encoding. In: Advances in neural information processing systems, pp 1709–1720
3. Basu D, Data D, Karakus C, Diggavi SN (2020) Qsparse-local-SGD: distributed SGD with quantization, sparsification, and local computations. IEEE J Sel Areas Inf Theory 1(1):217–226
4. Bottou L, Curtis FE, Nocedal J (2018) Optimization methods for large-scale machine learning. arXiv preprint arXiv:1606.04838
5. Boyd S, Vandenberghe L (2004) Convex optimization. Cambridge University Press, Cambridge
6. Boyd S, Parikh N, Chu E, Peleato B, Eckstein J (2011) Distributed optimization and statistical learning via the alternating direction method of multipliers. Found Trends Mach Learn 3(1): 1–122
7. Chaudhari P, Soatto S (2017) Stochastic gradient descent performs variational inference, converges to limit cycles for deep networks. CoRR, abs/1710.11029. http://arxiv.org/abs/1710.11029
8. Cho YJ, Wang J, Joshi G (2020) Client selection in federated learning: convergence analysis and power-of-choice selection strategies
9. Cipar J, Ho Q, Kim JK, Lee S, Ganger GR, Gibson G, Keeton K, Xing E (2013) Solving the straggler problem with bounded staleness. In: Proceedings of the workshop on hot topics in operating systems
10. Cui H, Cipar J, Ho Q, Kim JK, Lee S, Kumar A, Wei J, Dai W, Ganger GR, Gibbons PB, Gibson GA, Xing EP (2014) Exploiting bounded staleness to speed up big data analytics. In: Proceedings of the USENIX annual technical conference, pp 37–48
11. Dean J, Corrado GS, Monga R, Chen K, Devin M, Le QV, Mao MZ, Ranzato M, Senior A, Tucker P, Yang K, Ng AY (2012) Large scale distributed deep networks. In: Proceedings of the international conference on neural information processing systems, pp 1223–1231
12. Dekel O, Gilad-Bachrach R, Shamir O, Xiao L (2012) Optimal distributed online prediction using mini-batches. J Mach Learn Res 13(1):165–202
13. Dutta S, Joshi G, Ghosh S, Dube P, Nagpurkar P (2018) Slow and stale gradients can win the race: error-runtime trade-offs in distributed SGD. In: International conference on artificial intelligence and statistics (AISTATS). https://arxiv.org/abs/1803.01113
14. Frankle J, Carbin M (2019) The lottery ticket hypothesis: finding sparse, trainable neural networks. In: International conference on learning representations
15. Gupta S, Zhang W, Wang F (2016) Model accuracy and runtime tradeoff in distributed deep learning: a systematic study. In: IEEE international conference on data mining (ICDM). IEEE, pp 171–180
16. Han P, Wang S, Leung KK (2020) Adaptive gradient sparsification for efficient federated learning: an online learning approach. In: 2020 IEEE 40th international conference on distributed computing systems (ICDCS), pp 300–310
17. Han S, Mao H, Dally WJ (2015) Deep compression: compressing deep neural networks with pruning, trained quantization and Huffman coding. arXiv preprint arXiv:1510.00149
18. Hazan E (2016) Introduction to online convex optimization. Found Trends Optim 2(3–4): 157–325. ISSN 2167-3888
19. Ho Q, Cipar J, Cui H, Kim JK, Lee S, Gibbons PB, Gibson GA, Ganger GR, Xing EP (2013) More effective distributed ml via a stale synchronous parallel parameter server. In: Proceedings of the international conference on neural information processing systems, pp 1223–1231
20. Horváth S, Richtarik P (2021) A better alternative to error feedback for communication-efficient distributed learning. In: International conference on learning representations

21. Jee Cho Y, Gupta S, Joshi G, Yagan O (2020) Bandit-based communication-efficient client selection strategies for federated learning. In: Proceedings of the asilomar conference on signals, systems, and computers, pp 1066–1069. https://doi.org/10.1109/IEEECONF51394.2020.9443523

22. Jiang Y, Wang S, Valls V, Ko BJ, Lee W-H, Leung KK, Tassiulas L (2019) Model pruning enables efficient federated learning on edge devices. arXiv preprint arXiv:1909.12326

23. Karimireddy SP, Kale S, Mohri M, Reddi SJ, Stich SU, Suresh AT (2019) SCAF-FOLD: stochastic controlled averaging for on-device federated learning. arXiv preprint arXiv:1910.06378

24. Karimireddy SP, Rebjock Q, Stich S, Jaggi M (2019) Error feedback fixes SignSGD and other gradient compression schemes. In: International conference on machine learning. PMLR, pp 3252–3261

25. Li M, Zhang T, Chen Y, Smola AJ (2014) Efficient mini-batch training for stochastic optimization. In: Proceedings of the ACM SIGKDD international conference on knowledge discovery and data mining, pp 661–670

26. Li T, Sahu AK, Zaheer M, Sanjabi M, Talwalkar A, Smith V (2020) FedDANE: a federated newton-type method

27. Li X, Huang K, Yang W, Wang S, Zhang Z (2020) On the convergence of FedAvg on non-IID data. In: International conference on learning representations (ICLR). https://arxiv.org/abs/1907.02189

28. Lian X, Huang Y, Li Y, Liu J (2015) Asynchronous parallel stochastic gradient for nonconvex optimization. In: Proceedings of the international conference on neural information processing systems, pp 2737–2745

29. Lian X, Zhang W, Zhang C, Liu J (2018) Asynchronous decentralized parallel stochastic gradient descent. In: Proceedings of the 35th international conference on machine learning. Proceedings of machine learning research, vol 80. PMLR, pp 3043–3052. http://proceedings.mlr.press/v80/lian18a.html

30. McMahan HB, Moore E, Ramage D, Hampson S, y Arcas BA (2017) Communication-efficient learning of deep networks from decentralized data. In: International conference on artificial intelligence and statistics (AISTATS). https://arxiv.org/abs/1602.05629

31. Neyshabur B, Tomioka R, Salakhutdinov R, Srebro N (2017) Geometry of optimization and implicit regularization in deep learning. CoRR, abs/1705.03071. http://arxiv.org/abs/1705.03071

32. Parikh N, Boyd S (2014) Proximal algorithms. Found Trends Optim 1(3):127–239

33. Recht B, Re C, Wright S, Niu F (2011) Hogwild: a lock-free approach to parallelizing stochastic gradient descent. In: Proceedings of the international conference on neural information processing systems, pp 693–701

34. Reisizadeh A, Mokhtari A, Hassani H, Jadbabaie A, Pedarsani R (2020) FedPAQ: a communication-efficient federated learning method with periodic averaging and quantization. In: International conference on artificial intelligence and statistics. PMLR, pp 2021–2031

35. Robbins H, Monro S (1951) A stochastic approximation method. In: The annals of mathematical statistics, pp 400–407

36. Ruan Y, Zhang X, Liang S-C, Joe-Wong C (2021) Towards flexible device participation in federated learning. In: Banerjee A, Fukumizu K (eds) Proceedings of the 24th international conference on artificial intelligence and statistics. Proceedings of machine learning research, vol 130. PMLR, pp 3403–3411. http://proceedings.mlr.press/v130/ruan21a.html

37. Ruder S (2016) An overview of gradient descent optimization algorithms. arXiv preprint arXiv:1609.04747

38. Russakovsky O, Deng J, Su H, Krause J, Satheesh S, Ma S, Huang Z, Karpathy A, Khosla A, Bernstein M, Berg AC, Fei-Fei L (2015) ImageNet large scale visual recognition challenge. Int J Comput Vis 115(3):211–252

39. Sahu AK, Li T, Sanjabi M, Zaheer M, Talwalkar A, Smith V (2019) Federated optimization in heterogeneous networks. In: Proceedings of the machine learning and systems (MLSys) conference

40. Sahu AK, Li T, Sanjabi M, Zaheer M, Talwalkar A, Smith V (2019) Federated optimization for heterogeneous networks. https://arxiv.org/abs/1812.06127
41. Shalev-Shwartz S, Ben-David S (2014) Understanding machine learning: from theory to algorithms. Cambridge University Press, New York
42. Shwartz-Ziv R, Tishby N (2017) Opening the black box of deep neural networks via information. CoRR, abs/1703.00810. http://arxiv.org/abs/1703.00810
43. Stich SU (2018) Local SGD converges fast and communicates little. arXiv preprint arXiv:1805.09767
44. Stich SU, Cordonnier J-B, Jaggi M (2018) Sparsified SGD with memory. In: Advances in neural information processing systems, pp 4447–4458
45. Tang H, Yu C, Lian X, Zhang T, Liu J (2019) DoubleSqueeze: parallel stochastic gradient descent with double-pass error-compensated compression. In: International conference on machine learning. PMLR, pp 6155–6165
46. Wang J, Joshi G (2018) Cooperative SGD: unifying temporal and spatial strategies for communication-efficient distributed SGD, preprint. https://arxiv.org/abs/1808.07576
47. Wang J, Joshi G (2019) Adaptive communication strategies for best error-runtime trade-offs in communication-efficient distributed SGD. In: Proceedings of the SysML conference. https://arxiv.org/abs/1810.08313
48. Wang J, Liang H, Joshi G (2020) Overlap local-SGD: an algorithmic approach to hide communication delays in distributed SGD. In: Proceedings of international conference on acoustics, speech, and signal processing (ICASSP)
49. Wang J, Liu Q, Liang H, Joshi G, Poor HV (2020) Tackling the objective inconsistency problem in heterogeneous federated optimization. In: Proceedings on neural information processing systems (NeurIPS). https://arxiv.org/abs/2007.07481
50. Wang J, Tantia V, Ballas N, Rabbat M (2020) SlowMo: improving communication-efficient distributed SGD with slow momentum. In: International conference on learning representations. https://openreview.net/forum?id=SkxJ8REYPH
51. Wang J, Xu Z, Garrett Z, Charles Z, Liu L, Joshi G (2021) Local adaptivity in federated learning: convergence and consistency
52. Wang S, Tuor T, Salonidis T, Leung KK, Makaya C, He T, Chan K (2019) Adaptive federated learning in resource constrained edge computing systems. IEEE J Sel Areas Commun 37(6): 1205–1221
53. Yin D, Pananjady A, Lam M, Papailiopoulos D, Ramchandran K, Bartlett P (2018) Gradient diversity: a key ingredient for scalable distributed learning. In: Proceedings of the twenty-first international conference on artificial intelligence and statistics. Proceedings of machine learning research, vol 84. pp 1998–2007. http://proceedings.mlr.press/v84/yin18a.html
54. Yu H, Yang S, Zhu S (2018) Parallel restarted SGD for non-convex optimization with faster convergence and less communication. arXiv preprint arXiv:1807.06629
55. Zhang C, Bengio S, Hardt M, Recht B, Vinyals O (2017) Understanding deep learning requires rethinking generalization. In: International conference on learning representations
56. Zhang J, Mitliagkas I, Re C (2017) Yellowfin and the art of momentum tuning. CoRR, arXiv:1706.03471. http://arxiv.org/abs/1706.03471
57. Zhang S, Choromanska AE, LeCun Y (2015) Deep learning with elastic averaging SGD. In: NIPS'15 proceedings of the 28th international conference on neural information processing systems, pp 685–693
58. Zhang W, Gupta S, Lian X, Liu J (2015) Staleness-aware Async-SGD for distributed deep learning. arXiv preprint arXiv:1511.05950

Chapter 7
Communication-Efficient Model Fusion

Mikhail Yurochkin and Yuekai Sun

Abstract We consider the problem of learning a federated model where the number of communication rounds is severely limited. We discuss recent works on model fusion, a special case of Federated Learning where only a single communication round is allowed. This setting has a unique feature where it is sufficient for clients to have a pre-trained model, but not the data. Data storage regulations such as GDPR make this setting appealing as the data can be immediately deleted after updating the local model before FL starts. However, model fusion methods are limited to relatively shallow neural network architectures. We discuss extensions of model fusion applicable to deep learning models that require more than one communication round, but remain very efficient in terms of communication budget, i.e., number of communication rounds and size of the messages exchanged between the clients and the server. We consider both homogeneous and heterogeneous client data scenarios, including scenarios where training on the aggregated data is suboptimal due to biases in the data. In addition to deep learning methods, we cover unsupervised settings such as mixture models, topic models, and hidden Markov models.

We compare the statistical efficiency of model fusion to that of a hypothetical centralized approach in which a learner with unlimited compute and storage capacity simply aggregates data from all clients and trains a model in a non-federated way. As we shall see, although the model fusion approach generally matches the convergence rate of the (hypothetical) centralized approach, it may not have the same efficiency. Further, this discrepancy between the centralized and federated approaches is amplified when client data is heterogeneous.

M. Yurochkin (✉)
MIT-IBM Watson AI Lab, IBM Research, Cambridge, MA, USA
e-mail: mikhail.yurochkin@ibm.com

Y. Sun
Department of Statistics, University of Michigan, Ann Arbor, MI, USA
e-mail: yuekai@umich.edu

© The Author(s), under exclusive license to Springer Nature Switzerland AG 2022   145
H. Ludwig, N. Baracaldo (eds.), *Federated Learning*,
https://doi.org/10.1007/978-3-030-96896-0_7

7.1 Introduction

Standard federated learning algorithms, e.g., Federated Averaging [34], rely on simple parameter averaging when aggregating model parameters from the clients. Due to its simplicity, this approach is compatible with most models and deep learning architectures. However, it also has some drawbacks. In particular, the number of communication rounds required to train a high-performance model is often on the order of hundreds in many application areas. This communication cost may be prohibitive, especially in settings in which the overhead of a communication round is high. For example, in applications in which the clients are mobile devices, a communication round may correspond to a daily synchronization task with a server. In other applications, initiating a round of communication may require human approval (e.g., hospitals forming a data/model sharing coalition). In such applications, it is challenging to set up the infrastructure to support frequent communications.

The aforementioned challenges motivate us to consider federated learning algorithms that only require a few rounds of communication. We review the recent *model fusion* [48, 49] techniques that only require a single round of communication. The model fusion approach has the additional benefit of model fusion that clients do not need to store data, i.e., they are only required to provide their local models for the federated learning of a more powerful global model. Data storage is regulated by GDPR [19], making this feature of model fusion practically appealing. As we shall see, model fusion techniques rely on bipartite *matching* [49] via variants of the Hungarian algorithm [28] or Wasserstein barycenters [1, 42]. These approaches take into account similarity between model components (e.g., neuron weights) when averaging parameters, which allow them to produce good global models in as little as a single communication round. The drawback is that the corresponding optimization problem aligning model components becomes intractable for deep neural networks such as VGG architectures [41].

In order to handle deep neural networks, we also consider the layer-wise matching strategy of [47], which can train a powerful federated model in a fixed number of communication rounds (depending on the depth of the neural network). However, this approach requires clients to store data in order to perform model updates locally as in the other federated learning algorithms.

Finally, we also explore the statistical properties of model fusion. In particular, we compare the model fusion approach to a hypothetical centralized approach in which a server aggregates all the data from the clients and trains an ML model without any communication constraints. This is impractical in many applications of federated learning, but it is a gold-standard to which we can compare the statistical performance of model fusion. Ideally, we hope that model fusion matches the statistical efficiency of this (hypothetical) centralized approach. As we shall see, this is almost the case.

In this chapter we review model fusion techniques and demonstrate their applicability to federated learning in a single communication round with simpler neural network architectures [49]; unsupervised learning with mixture models, hidden Markov models, and topic models [48, 50]. We present extensions for posterior fusion [14] to support federated learning of Bayesian neural networks [36]. We review the method of [47] suitable for federated learning of deep neural networks under the limited communication budget. Finally, we study the statistical properties of model fusion, and we conclude with a discussion of open challenges and promising future work directions.

7.2 Permutation-Invariant Structure of Models

Many machine learning models can be described by *sets* of parameter vectors rather than single vector of parameters. For example, a mixture model for clustering is characterized by a set of cluster centroids. Any permutation of the order of the centroids in the set yields equivalent clustering quality, i.e., same data likelihood. In the federated learning context, it is now evident why simply averaging two sets of cluster centroids obtained from different clients may be detrimental: even in the simplest case of two clients fitting a clustering model on homogeneous datasets and recovering identical centroids, the ordering of centroids in their solution is arbitrary and elementwise averaging is likely to result in poor federated global model. Other prominent examples of permutation-invariant unsupervised models that we will cover in this chapter are topic models and hidden Markov models.

Permutation invariance of model parameters is also present in supervised learning, specifically in neural networks. Consider a simple one-hidden layer fully connected neural network with L hidden units

$$f(x) = \sigma(x W_1) W_2, \tag{7.1}$$

where $\sigma(\cdot)$ is a non-linearity applied entry-wise, $x \in \mathbb{R}^D$, $W_1 \in \mathbb{R}^{D \times L}$, $W_2 \in \mathbb{R}^{L \times K}$. D and K are the input and output dimensions, and we omit bias terms without loss of generality. Let $W_{1,\cdot l}$ denote lth column of W_1 and $W_{2,l\cdot}$ denote lth row of W_2, then we can write (7.1) as

$$f(x) = \sum_{l=1}^{L} W_{2,l\cdot} \sigma(\langle x, W_{1,\cdot l} \rangle). \tag{7.2}$$

Sum is a permutation-invariant operation, therefore any reordering of the neurons, i.e., columns of W_1 and correspondingly rows of W_2, result in a neural network with the same prediction rule. To account for the permutation invariance, we re-write (7.1) as

$$f(x) = \sigma(x W_1 \Pi) \Pi^T W_2, \tag{7.3}$$

where Π is one of the $L!$ possible permutation matrices. Recall that a permutation matrix is an orthogonal matrix that acts on rows when applied on the left and on columns when applied on the right. Suppose $\{W_1, W_2\}$ are optimal weights, then, according to equation (7.3), training on two homogeneous datasets X_j and X'_j will result in two sets of weights $\{W_1\Pi_j, \Pi_j^T W_2\}$ and $\{W_1\Pi_{j'}, \Pi_{j'}^T W_2\}$. Naive averaging of these two sets of parameters is suboptimal, i.e., $\Pi_j \neq \Pi_{j'}$ with high probability, therefore $\frac{1}{2}(W_1\Pi_j + W_1\Pi_{j'}) \neq W_1\Pi$ for any Π. To optimally average neural network weights we should first undo the permutation $(W_1\Pi_j\Pi_j^T + W_1\Pi_{j'}\Pi_{j'}^T)/2 = W_1$.

Neural network invariances beyond permutations

The model fusion techniques that we are about to present in this chapter are designed to take into account permutation invariance of client models when performing the fusion. However, in the case of neural networks, other invariances might exist. Neural networks are typically vastly overparametrized and learning the corresponding weights is a non-convex optimization problem with possibly many equivalent (in a sense of training loss) local optima. For a single hidden layer neural network with L neurons, as in equation (7.1), for any solution there are at least $L!$ equivalent ones due to permutation invariance. It is possible that there are other invariant (or otherwise equivalent) solutions. This is an unresolved question in the literature. A potentially fruitful perspective is to study the loss landscape of neural networks—there is no complete theoretical understanding of this problem; however, some progress has been made [15, 18, 21, 30]. Considering the loss landscape perspective for developing new federated learning and model fusion algorithms is an interesting future work direction.

7.2.1 General Formulation of Matched Averaging

We now formalize the idea of averaging parameters of models with inherent permutation invariance following the perspective of [47]. We continue the example of a one-hidden layer neural network; however, the idea easily generalizes to other models as we will show in the subsequent sections.

Let w_{jl} be the lth neuron weights learned on jth client data. We consider w_{jl} to be the concatenation of the lth column of W_1 and lth row of W_2 from the preceding discussion. Let θ_i denote the (unknown) ith neuron weights in the global model and $c(\cdot, \cdot)$ be an appropriate similarity function, e.g., squared Euclidean distance. The *matched averaging* optimization problem is as follows:

$$\min_{\{\pi_{il}^j \in \{0,1\}\}} \sum_{i=1}^{L} \sum_{j,l} \min_{\theta_i} \pi_{il}^j c(w_{jl}, \theta_i) \text{ s.t. } \sum_{i} \pi_{il}^j = 1 \,\forall\, j, l; \; \sum_{l} \pi_{il}^j = 1 \,\forall\, i, j.$$

(7.4)

The inner optimization of θ_i is trivial when similarity $c(\cdot, \cdot)$ is a squared Euclidean distance, i.e., it is the average of the matched client neuron weights $\theta_i = \frac{\sum_{j,l} \pi_{il}^j w_{jl}}{\sum_{j,l} \pi_{il}^j}$. The name matched averaging is due to the relation of equation (7.4) to the maximum bipartite matching problem. This optimization problem is also related to the Wasserstein barycenter [1] that was utilized by [42] in their model fusion approach based on optimal transport.

7.2.2 Solving Matched Averaging

We discuss a general perspective of solving equation (7.4) and present a concrete algorithm in the subsequent section.

Optimization problem (7.4) can be solved using the following iterative algorithm: fix all but one $\pi^{j'}$, then find $\pi^{j'}$ using Hungarian matching algorithm [28] and iterate over j' until convergence. The limitation of equation (7.4) is the implicit assumption of homogeneity of client datasets. Specifically, it assumes that each client has the same model architecture, that is also equivalent to the global model architecture. Although this is typical in federated learning, e.g., same assumption is made by the Federated Averaging [34] algorithm, it is not always practical. In the matched averaging perspective, it is equivalent to saying that every neuron of a given client has a matching neuron in the neural networks of all other clients. When thinking of neurons as feature extractors and taking into account potential heterogeneity of the client datasets, this seems unrealistic. Different datasets require different feature extractors that may overlap only partially. To better account for data heterogeneity and allow partial matching (overlap) of the client neurons, we treat global model size L as an unknown, possibly larger than individual model sizes, i.e., $\max_j L_j \le L \le \sum_j L_j$, where L_j is the number of neurons in jth client's model.

The resulting objective remains amendable to the iterative optimization with Hungarian algorithm. At each iteration, fixing all but one $\pi^{j'}$, we compute current global model parameter estimates $\{\theta_i = \arg\min_{\theta_i} \sum_{j \ne j', l} \pi_{il}^j c(w_{jl}, \theta_i)\}_{i=1}^{L}$ (e.g., via taking a mean for the squared Euclidean similarity) and solve the following problem to update $\pi^{j'}$:

$$\min_{\{\pi_{il}^{j'} \in \{0,1\}\}} \sum_{i=1}^{L+L_{j'}} \sum_{j=1}^{L_{j'}} \pi_{il}^{j'} C_{il}^{j'} \text{ s.t. } \sum_i \pi_{il}^{j'} = 1 \ \forall \ l; \ \sum_l \pi_{il}^{j} \in \{0, 1\} \ \forall \ i, \text{ where}$$

$$C_{il}^{j'} = \begin{cases} c(w_{j'l}, \theta_i), & i \leq L \\ \epsilon + \lambda(i), & L < i \leq L + L_{j'}. \end{cases}$$

(7.5)

Parameter ϵ is interpreted as a maximum dissimilarity between a pair of neurons before we declare them to have different functionality, i.e., keeping them as separate neurons in the global model. To control the global model growth we introduce an additional penalty function $\lambda(i)$ increasing in i. After updating $\pi^{j'}$, the new global model size is $L = \max\{i : \pi_{il}^{j'} = 1, l = 1, \ldots, L_{j'}\}$. This formulation can be directly utilized given a user-specified similarity $c(\cdot, \cdot)$, threshold ϵ, and penalty function $\lambda(\cdot)$. In the following section we describe a Bayesian nonparametric approach where these choices arise naturally from the model.

7.3   Probabilistic Federated Neural Matching

In federated learning via model fusion the goal is to aggregate model parameters learned from different datasets into a more powerful global model. The input to a model fusion algorithm is a collection of local model parameters, and the output is the global model parameters. Bayesian hierarchical modeling is a natural choice for modeling such inputs and unknowns. Bayesian hierarchical model typically describes a generating process starting from the parameters of the global model, which in turn generate local model parameters, from which the data arises. The inference process reverses the generating process, i.e., infers the unknown global model parameters from the data, or, in the model fusion case, from the local parameters estimated from the corresponding datasets. Probabilistic Federated Neural Matching (PFNM) [49] is one such Bayesian hierarchical approach specialized to fusion of neural networks that we review in this section.

7.3.1   PFNM Generative Process

Following the notations of Sect. 7.2.1, we observe neural network weights of J clients $\{\{w_{jl} \in \mathbb{R}^{D+K+1}\}_{l=1}^{L_j}\}_{j=1}^{J}$. The dimension of each w_{jl} can be understood as follows: D is the data dimension and correspondingly the number of weights in-going into the neuron (column of W_1), 1 is for the bias term of the neuron, and

K is the output dimension and correspondingly the number of weights out-going from the neuron (row of W_2). In accordance with the data heterogeneity discussion that motivated equation (7.5), we want a nonparametric prior for the global model parameters, i.e., one that allows for unknown global model size. Yurochkin et al. [49] utilize Beta-Bernoulli process [45] to achieve this. First generate the collection of the global model parameters from the Beta process:

$$Q := \sum_i q_i \delta_{\theta_i} \sim \text{BetaProc}(\alpha, \gamma_0 H), \text{ where}$$

(7.6)

$$H = \mathcal{N}(\mu_0, \Sigma_0) \text{ is the base measure, i.e. } \theta_i \sim H, \ i = 1, \ldots$$

Parameters $\mu_0 \in \mathbb{R}^{D+1+K}$, covariance Σ_0, and $\gamma_0, \alpha \in \mathbb{R}_+$ are the prior hyperparameters. For simplicity we set $\mu_0 = 0$, suggesting that neural network weights are small in magnitude, and isotropic diagonal covariance $\Sigma_0 = \sigma_0^2 I$. The Beta process concentration parameter α controls the degree of sharing across local models (we will set $\alpha = 1$ for simplicity), and the mass parameter γ_0 controls our prior beliefs of the global model size, i.e., how heterogeneous we expect client datasets to be (larger γ_0 suggests larger global model sizes a priori).

Next step considers the heterogeneity of the client datasets: clients need only parts of the global model feature extraction capabilities to model their data. We select a subset of the global model neurons for each client $j = 1, \ldots, J$ via the Bernoulli process:

$$\mathcal{T}_j := \sum_i b_{ji} \delta_{\theta_i}, \text{ where } b_{ji} | q_i \sim \text{Bern}(q_i) \ \forall i.$$

(7.7)

$\mathcal{T}_j$ is then a set of global model weights selected for client j, i.e., $\mathcal{T}_j = \{\theta_i : b_{ji} = 1 \ i = 1, \ldots\}$. Finally we model the local model weights that we observe accounting for local data noise:

$$w_{jl} | \mathcal{T}_j \sim \mathcal{N}(\mathcal{T}_{jl}, \Sigma_j) \text{ for } l = 1, \ldots, L_j; \ \ L_j := \text{card}(\mathcal{T}_j).$$

(7.8)

For simplicity we assume diagonal isotropic covariances $\Sigma_j = \sigma_j^2 I$.

Indian Buffet Process and the Beta-Bernoulli Process

The Indian buffet process (IBP) is a Bayesian nonparametric prior over sparse binary matrices with infinitely many columns [22]. The name is due to the following culinary metaphor: suppose J customers are arriving sequentially to a buffet and choose dishes to sample. The first customer samples Poisson(γ_0) dishes. The j-th customer then tries each of the dishes selected by previous customers with probability proportional to the dish's popularity, and addi-

(continued)

tionally samples Poisson(γ_0/j) new dishes. Thibaux and Jordan [45] showed that the de Finetti mixing distribution corresponding to the IBP is a Beta-Bernoulli Process. Let Q be a random measure drawn from a Beta process, $Q \mid \alpha, \gamma_0, H \sim \text{BP}(\alpha, \gamma_0 H)$, with concentration parameter α, mass parameter γ_0, and base measure H over Ω with $H(\Omega) = 1$. Then Q is a discrete measure $Q = \sum_i q_i \delta_{\theta_i}$ formed by an infinitely countable set of (weight, atom) pairs $(q_i, \theta_i) \in [0, 1] \times \Omega$. The weights $\{q_i\}_{i=1}^{\infty}$ can be shown to have a "stick-breaking" distribution [44]: $\nu_1 \sim \text{Beta}(\gamma_0, 1)$, $q_i = \prod_{k=1}^{i} \nu_k$, and the atoms θ_i are drawn i.i.d. from H. Then a subset of atoms in Q is selected via a Bernoulli process, i.e., each subset $\mathcal{T}_j$ for $j = 1, \ldots, J$ is characterized by a Bernoulli process with base measure Q: $\mathcal{T}_j \mid Q \sim \text{BeP}(Q)$. Each subset $\mathcal{T}_j$ is also a discrete measure formed by pairs $(b_{ji}, \theta_i) \in \{0, 1\} \times \Omega$, i.e., $\mathcal{T}_j := \sum_i b_{ji} \delta_{\theta_i}$, where $b_{ji} \mid q_i \sim \text{Bern}(q_i) \,\forall i$ is a binary random variable indicating whether atom θ_i belongs to subset $\mathcal{T}_j$. The collection of such subsets is then said to be distributed by a Beta-Bernoulli process. Marginalizing over the Beta Process Q yields the predictive distribution $\mathcal{T}_J \mid \mathcal{T}_1, \ldots, \mathcal{T}_{J-1} \sim \text{BeP}\left(\frac{\alpha \gamma_0}{J+\alpha-1} H + \sum_i \frac{m_i}{J+\alpha-1} \delta_{\theta_i}\right)$, where $m_i = \sum_{j=1}^{J-1} b_{ji}$, which is equivalent to the IBP.

7.3.2 PFNM Inference

To estimate the unknown latent variables $\{b_{ji}\}$ and $\{\theta_i\}$ we appeal to maximum a posteriori (MAP) estimation, i.e., maximizing posterior probability. First we note that there is a one-to-one correspondence between $\{b_{ji}\}$ and matching variables $\{\pi_{il}^j\}$ from equation (7.5), i.e., $\pi_{il}^j = 1$ if $\mathcal{T}_{jl} = \theta_i$ and 0 otherwise.

To derive MAP estimates we write the posterior probability of the latent variables:

$$\underset{\{\theta_i\}, \{\pi^j\}}{\arg\max} \, P(\{\theta_i\}, \{\pi^j\} | \{w_{jl}\}) \propto P(\{w_{jl}\} | \{\theta_i\}, \{\pi^j\}) P(\{\pi^j\}) P(\{\theta_i\}). \tag{7.9}$$

Optimal values of $\{\theta_i\}$ can be expressed as functions of $\{\pi^j\}$ due to Gaussian–Gaussian conjugacy as formalized in the following proposition of [49]:

Proposition 7.1 *Given $\{\pi^j\}$, the MAP estimate of $\{\theta_i\}$ is given by*

$$\theta_i = \frac{\mu_0/\sigma_0^2 + \sum_{j,l} \pi_{il}^j w_{jl}/\sigma_j^2}{1/\sigma_0^2 + \sum_{j,l} \pi_{il}^j/\sigma_j^2} \, \textit{for } i = 1, \ldots, L. \tag{7.10}$$

Using the above proposition and taking the natural logarithm we reformulate (7.9) as a simpler optimization problem in $\{\pi^j\}$ only:

$$\arg\max_{\{\pi^j\}} \frac{1}{2} \sum_i \frac{\left\| \frac{\mu_0}{\sigma_0^2} + \sum_{j,l} \pi_{il}^j \frac{w_{jl}}{\sigma_j} \right\|^2}{1/\sigma_0^2 + \sum_{j,l} \pi_{il}^j/\sigma_j^2} + \log P(\{\pi^j\})$$

$$\text{s.t. } \pi_{il}^j \in \{0, 1\} \; \forall \, j, i, l; \; \sum_i \pi_{il}^j = 1 \; \forall \, j, l; \; \sum_l \pi_{il}^j \in \{0, 1\} \; \forall \, j, i. \tag{7.11}$$

This optimization problem can be solved with the strategy outlined in Sect. 7.2.2: fix all but one π^j, re-write the objective in a form of a matching problem to solve for π^j, and iterate over j. We will write $-j$ to denote "all but j," and let $L_{-j} = \max\{i \, : \, \pi_{il}^{-j} = 1\}$ denote global model size omitting client j. To arrive at an objective function analogous to (7.5), we first expand the *first term* of (7.11) into cases when $i = 1, \ldots, L_{-j}$ and when $i = L_{-j} + 1, \ldots, L_{-j} + L_j$. We use the following *substraction trick* of [48]:

Proposition 7.2 (Substraction trick) *When $\sum_l \pi_{il} \in \{0, 1\}$ and $\pi_{il} \in \{0, 1\}$ for $\forall i, l$, optimizing $\sum_i f(\sum_l \pi_{il} x_l + C)$ for π is equivalent to optimizing $\sum_{i,l} \pi_{il}(f(x_l + C) - f(C))$ for any function f, $\{x_l\}$ and C independent of π.*

Algorithm 7.1 Single Layer Neural Matching

1: Collect weights and biases from the J clients and form w_{jl}.
2: Form assignment cost matrix per (7.15).
3: Compute matching assignments π^j using the Hungarian algorithm.
4: Enumerate all resulting unique global neurons and use (7.10) to infer the associated global weight vectors from all instances of the global model neurons across the J clients.
5: Concatenate the global neurons and the inferred weights and biases to form the new global hidden layer.

With the help of the above proposition we re-write the first term of (7.11):

$$\frac{1}{2} \sum_i \frac{\| \mu_0/\sigma_0^2 + \sum_{j,l} \pi_{il}^j w_{jl}/\sigma_j^2 \|^2}{1/\sigma_0^2 + \sum_{j,l} \pi_{il}^j/\sigma_j^2} =$$

$$\sum_{i=1}^{L_{-j}+L_j} \sum_{l=1}^{L_j} \pi_{il}^j \left(\frac{\| \mu_0/\sigma_0^2 + w_{jl}/\sigma_j^2 + \sum_{-j,l} \pi_{il}^j w_{jl}/\sigma_j^2 \|^2}{1/\sigma_0^2 + 1/\sigma_j^2 + \sum_{-j,l} \pi_{il}^j/\sigma_j^2} - \frac{\| \mu_0/\sigma_0^2 + \sum_{-j,l} \pi_{il}^j w_{jl}/\sigma_j^2 \|^2}{1/\sigma_0^2 + \sum_{-j,l} \pi_{il}^j/\sigma_j^2} \right).$$
$$\tag{7.12}$$

Next we consider the *second* term of (7.11):

$$\log P(\{\pi^j\}) = \log P(\pi^j|\pi^{-j}) + \log P(\pi^{-j}). \tag{7.13}$$

To expand this term we appeal to the exchangeability property of the Indian Buffet Process (IBP) [22, 45]: we can always consider j to be the last "customer" in the IBP. Denote $m_i^{-j} = \sum_{-j,l} \pi_{il}^j$ to be the number of times client weights were matched with the global model weight i outside of client j, i.e., the "dish" popularity in the IBP. Now we expand (7.13):

$$\log P(\{\pi^j\}) = \sum_{i=1}^{L_{-j}} \sum_{l=1}^{L_j} \pi_{il}^j \log \frac{m_i^{-j}}{J - m_i^{-j}} + \sum_{i=L_{-j}+1}^{L_{-j}+L_j} \sum_{l=1}^{L_j} \pi_{il}^j \left(\log \frac{\gamma_0}{J} - \log(i - L_{-j}) \right). \tag{7.14}$$

Combining (7.12) and (7.14) we obtain a cost expression analogous to equation (7.5):

$$C_{il}^j = - \begin{cases} \dfrac{\left\| \frac{\mu_0}{\sigma_0^2} + \frac{w_{jl}}{\sigma_j^2} + \sum_{-j,l} \pi_{il}^j \frac{w_{jl}}{\sigma_j^2} \right\|^2}{\frac{1}{\sigma_0^2} + \frac{1}{\sigma_j^2} + \sum_{-j,l} \pi_{il}^j/\sigma_j^2} - \dfrac{\left\| \frac{\mu_0}{\sigma_0^2} + \sum_{-j,l} \pi_{il}^j \frac{w_{jl}}{\sigma_j^2} \right\|^2}{\frac{1}{\sigma_0^2} + \sum_{-j,l} \pi_{il}^j/\sigma_j^2} + 2\log \frac{m_i^{-j}}{J - m_i^{-j}}, i \le L_{-j} \\[2em] \dfrac{\left\| \frac{\mu_0}{\sigma_0^2} + \frac{w_{jl}}{\sigma_j^2} \right\|^2}{\frac{1}{\sigma_0^2} + \frac{1}{\sigma_j^2}} - \dfrac{\left\| \frac{\mu_0}{\sigma_0^2} \right\|^2}{\frac{1}{\sigma_0^2}} - 2\log \frac{i - L_{-j}}{\gamma_0/J}, \qquad L_{-j} < i \le L_{-j} + L_j. \end{cases} \tag{7.15}$$

Now we can simply use Hungarian algorithm to minimize $\sum_i \sum_l \pi_{il}^j C_{il}^j$ to update π^j and iterate over j until convergence. We illustrate and summarize the resulting PFNM algorithm in Fig. 7.1.

Fusion of multilayer neural networks with PFNM

PFNM can be extended to fuse deep neural networks in a single communication round; however, its efficacy reduces for deeper networks. We refer interested reader to Section 3.2 of [49] for the precise extension of the PFNM model to multiple layers. In Sect. 7.6.1 we discuss the high level idea, as well as details regarding applying PFNM to convolutional and recurrent architectures. Figure 7.6 shows experiments where PFNM can successfully fuse 4-layer LeNets [29], but underperforms on a deeper 9-layer VGG-9 [41] architecture.

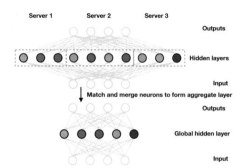

Fig. 7.1 Single layer Probabilistic Federated Neural Matching algorithm showing matching of three MLPs. Nodes in the graphs indicate neurons, neurons of the same color have been matched. PFNM approach consists of using the corresponding neurons in the output layer to convert the neurons in each of the J clients to weight vectors referencing the output layer. These weight vectors are then used to form a cost matrix, which the Hungarian algorithm then uses to do the matching. The matched neurons are then aggregated via Proposition 7.2 to form the global model

7.3.3 PFNM in Practice

To illustrate practical advantages of PFNM we discuss a sample of the results presented in [49]. They simulate federated learning scenario on MNIST dataset as follows: (1) to obtain homogeneous data partitioning, each client is assigned a random sample of equal size from each of the $K = 10$ classes; (2) for heterogeneous client data, they draw $p_k \sim \text{Dir}_J(0.5)$ and allocate p_{kj} proportion of instances of class k to client j. Due to small Dirichlet concentration parameter (0.5), it is likely that some clients receive very little amount of examples of certain classes (or none at all). This heterogeneous partitioning strategy was further explored in the work of [26].

After partitioning the data, [49] train a neural network with $L_j = 100$ hidden units independently for each client and perform fusion of the resulting models in a *single* communication round. In this setting clients do not need to have the data available, i.e., they can safely delete it after training their models and still benefit from federated learning with PFNM. This is the most communication restrictive scenario that is the easiest to implement in practice, i.e., clients simply need to send their model weights to the server once and then download the fused global model. This can be performed with standard data sharing tools and does not require setting up a specific compute infrastructure between the clients and the server.

Yurochkin et al. [49] studied the following baselines: to consider model fusion beneficial, the resulting global model should outperform client local models. Ensemble [8, 17], i.e., averaging model predictions, is a standard way to benefit from multiple models; however, in the context of neural networks it makes prediction computationally expensive (all models need to be stored and propagated through) and is often impractical. Besides, clients may not want to explicitly share their model parameters with each other. They also consider other model fusion strategies:

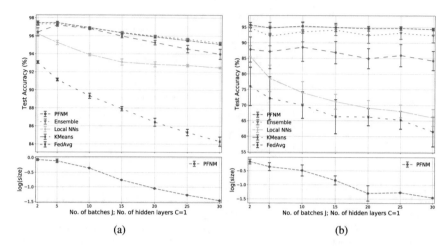

Fig. 7.2 Single communication federated learning. Test accuracy and normalized model size ($\log \frac{L}{\sum_j L_j}$) as a function of varying number of batches (J). PFNM consistently outperforms local models and federated averaging while performing comparably to ensembles at a fraction of the storage and computational costs. (**a**) MNIST homogeneous. (**b**) MNIST heterogeneous

Federated Averaging [34] corresponds to a naive elementwise parameter averaging of the client models. As [34] recommended, the client models were trained from the same random initialization, although this requirement may be impractical if clients trained their models before deciding to participate in federated learning. They also proposed a k-means [31] based fusion strategy as a baseline. The key difference with PFNM is that clustering, unlike matching, allows neurons of the same model to be averaged together. Considering that each neuron of a given model has dedicated feature extraction functionality, this is an undesirable property.

We present the results in Fig. 7.2. The top of each plot shows test accuracy as a function of the number of clients J and the bottom emphasizes model compression in comparison to ensemble by plotting $\log \frac{L}{\sum_j L_j}$, i.e., the log ratio of the PFNM global model size to the total number of neurons across client models (equivalent to the ensemble model size). PFNM performs comparable to ensemble, while producing a much smaller model, outperforms local client models and other model fusion baselines, demonstrating its utility in the context of single-round federated learning.

Single-round fusion of deep neural networks

As the disadvantage of PFNM, we notice a relatively low performance on the CIFAR10 dataset. The neural networks considered in this experiment are too shallow to be efficient on this dataset. Although PFNM can be applied

(continued)

to deeper neural networks, its performance is not as good when fusing deep architectures such as VGG [41]. The fusion problem becomes significantly more complicated for deeper architectures and remains an open problem. In Sect. 7.6 we present techniques based on PFNM that can efficiently fuse deep neural networks when more than one communication round is allowed.

7.4 Unsupervised FL with SPAHM

In this section we review Statistical Parameter Aggregation via Heterogeneous Matching (SPAHM) of [48] that extends the modeling framework of PFNM to a single-round federated learning of a variety of popular unsupervised models that exhibit permutation invariance of their parameters, such as Gaussian mixture models (GMM), hidden Markov models (HMM), and topic models. Recall that the base measure H in the PFNM construction in equation (7.6) was set to be Gaussian. This is a reasonable base measure for modeling neural network weights, but is not appropriate to models with latent quantities of different nature (e.g., known to be positive). Additionally, recall that PFNM had hyperparameters $\mu_0, \gamma_0, \sigma_0$, and σ_j that needed to be set by the user. SPAHM extends PFNM framework with the functionality to estimate its own hyperparameters via empirical Bayes (except γ_0).

7.4.1 SPAHM Model

Similar to PFNM, SPAHM uses Beta-Bernoulli process construction, but with a more general base measure. As before we will denote ith global model parameters as θ_i and lth local model parameters of the jth client as w_{jl}. However note that these parameters can now be cluster centers in a GMM, hidden states of an HMM, etc.

As in PFNM, start with the Beta Process prior on the global model parameters:

$$Q := \sum_i q_i \delta_{\theta_i} \sim \text{BetaProc}(\alpha, \gamma_0 H), \ \theta_i \sim H, \ i = 1, \ldots \quad (7.16)$$

Base measure H is allowed to be any exponentially family distribution that is application appropriate. Exponential family density can be written in a general form as follows:

$$p_\theta(\theta \mid \tau, n_0) = \mathcal{H}(\tau, n_0)\exp(\tau^T \theta - n_0 \mathcal{A}(\theta)), \quad (7.17)$$

where τ, n_0 are the hyperparameters and $\mathcal{H}(\cdot, \cdot)$ and $\mathcal{A}(\cdot)$ are functions that define a specific distribution within the exponential family.

Then select a subset of global model parameters for each client with the Bernoulli process as in (7.7):

$$\mathcal{T}_j := \sum_i b_{ji} \delta_{\theta_i}, \text{ where } b_{ji} | q_i \sim \text{Bern}(q_i) \ \forall i. \tag{7.18}$$

This step is the same as in PFNM, and we define $\mathcal{T}_j$ as the set of global model parameters selected for client j, i.e., $\mathcal{T}_j = \{\theta_i : b_{ji} = 1 \ i = 1, \ldots\}$. To model the observed local model parameters we again generalize PFNM construction to exponential family distributions:

$$w_{jl} | \mathcal{T}_j \sim F(\cdot | \mathcal{T}_{jl}) \text{ for } l = 1, \ldots, L_j; \quad L_j := \text{card}(\mathcal{T}_j), \tag{7.19}$$

where the probability density of F is

$$p_w(w \mid \theta) = h(w) \exp(\theta^T T(w) - \mathcal{A}(\theta)), \tag{7.20}$$

with $T(\cdot)$ being the sufficient statistics function.

7.4.2 SPAHM Inference

To estimate the unknown matching variables $\pi = \{\pi^j\}$, similar to PFNM, we will maximize the posterior probability, which we should now derive in a general (for any exponential family distribution) form. Let $Z_i = \{(j, l) \mid \pi_{il}^j = 1\}$ be the index set of the local parameters assigned to the ith global parameter, then we have

$$P(\pi \mid w) \propto P(\pi) \int p_w(w \mid \pi, \theta) p_\theta(\theta) \, d\theta = P(\pi) \prod_i \int \prod_{z \in Z_i} p_w(w_z \mid \theta_i) p_\theta(\theta_i) \, d\theta_i$$

$$= P(\pi) \prod_i \mathcal{H}(\tau, n_0) \int \left(\prod_{z \in Z_i} h(w_z) \right) \exp \left((\tau + \sum_{z \in Z_i} T(w_z))^T \theta_i - (\text{card}(Z_i) + n_0) \mathcal{A}(\theta) \right) d\theta_i$$

$$= P(\pi) \prod_i \frac{\mathcal{H}(\tau, n_0) \prod_{z \in Z_i} h(w_z)}{\mathcal{H}(\tau + \sum_{z \in Z_i} T(w_z), \text{card}(Z_i) + n_0)}. \tag{7.21}$$

Taking the logarithm and noticing that $\sum_i \sum_{j,l} \pi_{il}^j \log h(w_{jl})$ is constant in π, we arrive at the objective function for π generalizing (7.10)

$$\underset{\pi}{\arg\max} \log P(\pi) - \sum_i \log \mathcal{H} \left(\tau + \sum_{j,l} \pi_{il}^j T(w_{jl}), \sum_{j,l} \pi_{il}^j + n_0 \right) \tag{7.22}$$

$$\text{s.t. } \pi_{il}^j \in \{0, 1\} \ \forall \ j, i, l; \ \sum_i \pi_{il}^j = 1 \ \forall \ j, l; \ \sum_l \pi_{il}^j \in \{0, 1\} \ \forall \ j, i.$$

Now we follow the strategy similar to the PFNM derivations, i.e., fix all but one π^j and utilize substraction trick from Proposition 7.2 to formulate the problem in a form amendable to the Hungarian algorithm to update π^j. The resulting generalized expression for the cost is as follows:

$$
C_{il}^j = - \begin{cases} \log \dfrac{m_i^{-j}}{\alpha+J-1-m_i^{-j}} - \log \dfrac{\mathcal{H}\left(\tau+T(w_{jl})+\sum_{-j,l}\pi_{il}^j T(w_{jl}), 1+m_i^{-j}+n_0\right)}{\mathcal{H}\left(\tau+\sum_{-j,l}\pi_{il}^j T(w_{jl}), m_i^{-j}+n_0\right)}, & i \le L_{-j} \\ \log \dfrac{\alpha\gamma_0}{(\alpha+J-1)(i-L_{-j})} - \log \dfrac{\mathcal{H}(\tau+T(w_{jl}), 1+n_0)}{\mathcal{H}(\tau, n_0)}, & L_{-j} < i \le L_{-j}+L_j. \end{cases} \quad (7.23)
$$

Besides generalization, we mentioned that SPAHM also provides hyperparameter estimation procedure. Hyperparameters can be updated at each iteration as follows:

$$
\arg\max_{\tau, n_0} \sum_{i=1}^{L} \left(\log \mathcal{H}(\tau, n_0) - \log \mathcal{H}\left(\tau + \sum_{j,l} \pi_{il}^j T(w_{jl}), \sum_{j,l} \pi_{il}^j + n_0 \right) \right). \quad (7.24)
$$

This optimization problem can be solved with gradient based methods in general; however, it also admits a closed form update for certain exponential family distributions. Notably, for Gaussian distribution, assuming $\sigma^2 = \sigma_j^2 \; \forall \; j$, updates are as follows:

$$
\mu_0 = \frac{1}{L} \sum_{i=1}^{L} \frac{1}{m_i} \sum_{j,l} \pi_{il}^j w_{jl}, \quad \sigma^2 = \frac{1}{N-L} \sum_{i=1}^{L} \left(\sum_{j,l} \pi_{il}^j w_{jl}^2 - \frac{(\sum_{j,l} \pi_{il}^j w_{jl})^2}{m_i} \right),
$$

$$
\sigma_0^2 = \frac{1}{L} \sum_{i=1}^{L} \left(\frac{\sum_{j,l} \pi_{il}^j w_{jl}}{m_i} - \mu_0 \right)^2 - \sum_{i=1}^{L} \frac{\sigma^2}{m_i},
$$

$$(7.25)$$

where $N = \sum_j L_j$ and $m_i = \sum_{j,l} \pi_{il}^j$. This result can be obtained by setting corresponding derivatives of Eq. (7.24) $+ \sum_{j,l} \log h_\sigma(w_{jl})$ to 0 and solving the system of equations.

7.4.3 SPAHM in Practice

To illustrate SPAHM applications we highlight two experiments from [48]: simulations with Gaussian mixture models and Gaussian topic models [16] on the Gutenberg dataset of 40 books considered as clients for federated learning.

Gaussian mixture models. In the GMM experiment the data is generated using $L = 50$ true global model centroids $\theta_i \in \mathbb{R}^{50}$ from a Gaussian distribution $\theta_i \sim \mathcal{N}(\mu_0, \sigma_0^2 \mathbf{I})$. To generate heterogeneous client datasets, for $j = 1, \dots, J$

they picked a random subset of global centroids for each client and added noise with variance σ^2 to obtain the "true" local centroids, $\{w_{jl}\}_{l=1}^{L_j}$. This data simulation process is based on the model description in Sect. 7.4.1 with Gaussian densities. Then each dataset is sampled from a GMM with the corresponding set of local centroids.

Each client fits a k-means model locally and provides the cluster centroids as inputs to SPAHM. This can be viewed as a single communication round federated learning of a clustering model. To quantify the effect of estimation error endured by clients, they compared to SPAHM that uses true data generating local parameters. Additionally they considered two clustering-based fusion methods—as in the PFNM case, clustering is inferior to matching in the model fusion context. Different methods are compared based on the Hausdorff distance between the estimates and true data generating *global* centroids. Results of their experiments are presented in Fig. 7.3. Increasing the noise variance σ increases heterogeneity among client datasets making the problem harder; however, SPAHM degrades gracefully in Fig. 7.3a. Increasing the number of clients J does not degrade the performance of SPAHM in Fig. 7.3b.

Federated topic modeling. In this experiment [48] considered federated Gaussian LDA topic modeling [16] on 40 books from the Gutenberg dataset. Each client is viewed as a single book and they estimate their own topics, providing inputs to SPAHM that fuses them into a single topic model encompassing all 40 books. Illustration of the fusion procedure is given in Fig. 7.4.

We refer the reader to the SPAHM paper [48] for additional examples of federated unsupervised learning.

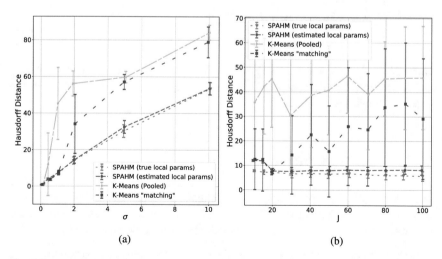

(a)   (b)

Fig. 7.3 Federated learning of a clustering model. Simulated experiment with Gaussian mixture models. SPAHM has the lowest estimation error measured with the Hausdorff distance. (**a**) Increasing heterogeneity via noise σ. (**b**) Increasing number of clients J

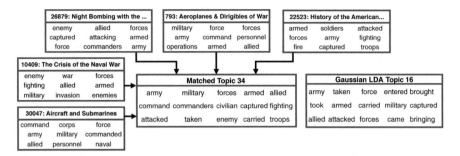

Fig. 7.4 Topic related to war found by SPAHM and Gaussian LDA. The five boxes pointing to the Matched topic represent local topics that SPAHM fused into the global one. The headers of these five boxes state the book names along with their Gutenberg IDs

Extensions to clients with time stamps
The idea of approaching model fusion using the Beta-Bernoulli process as in PFNM and SPAHM was first proposed in the context of topic modeling in [50]. Their goal was to fuse topic models learned from (possibly heterogeneous) text corpora partitioned by time and source, e.g., articles published in different conferences over multiple years. Partitioning by source is analogous to the client partitioning in federated learning and inspired PFNM and SPAHM. However, their modeling tools are also applicable to data with time-stamps, i.e., they proposed a method to fuse topics estimated from articles published in different years to study evolution of topics in time. This functionality has not been applied in the federated learning context so far.

7.5   Model Fusion of Posterior Distributions

We have discussed various model fusion strategies; however, so far they have all been targeting fusion of "frequentist" models, i.e., clients learn estimates of their parameters that are being fused. In situations where uncertainty quantification is important, clients may wish to train Bayesian models instead, e.g., Bayesian neural networks [36] or Gaussian mixture models where the full posterior is of interests rather than only the cluster centroids. In such scenario we need model fusion techniques that are able to ingest posterior distributions from the clients and produce a global fused posterior as the output. In this section we present one such technique that enables single-round federated learning of Bayesian models.

Distributed posterior estimation has been actively studied in the literature [3, 9, 10, 24, 43]; however, as in the case of Federated Averaging [34], they typically require many communication rounds to converge, are limited to homogeneous client datasets, or do not handle permutation invariance nature of the many popular models. The only earlier method that accounts for permutation invariance is by [11]; however, it assumes data homogeneity and is computationally too expensive.

7.5.1 Model Fusion with KL Divergence

We review the KL-fusion method of [14] for fusing posterior distributions learned from heterogeneous datasets. Their method assumes that each client performed mean-field variational inference (VI) [27] locally, i.e., client j obtains posterior of the form:

$$p_j(z_1, \ldots, z_{L_j}) = \prod_{l=1}^{L} q(z_l | w_{jl}), \tag{7.26}$$

where $q(z_l | w_{jl})$ is the approximate posterior of component z_l parameterized by w_{jl}. We note that KL-fusion could be also applied to other approximate parametric posterior inference techniques, e.g., Laplace approximations [5], assumed density filtering [38], and expectation propagation [35]; however, here we focus on one of the most popular methods, i.e., mean-field VI.

Variational inference

Variational inference [7, 27, 46] is a technique for approximating the true posterior distribution by a parametric approximate distribution minimizing the KL divergence between the variational approximation and the true posterior. Comparing to Markov chain Monte Carlo methods, VI is an optimization problem and can benefit from the modern stochastic gradient methods allowing VI based algorithms to scale to large data and models with a large number of parameters, such as Bayesian Neural Networks (BNNs) [36].

The goal of KL-fusion is to infer global posterior of a similar mean-field form:

$$\bar{p}(z_1, \ldots, z_L) = \prod_{i=1}^{L} q(z_i | \theta_i), \tag{7.27}$$

where $\{\theta_i\}$ are the parameters of the global posterior. The global posterior should approximate the client local posteriors. Claici et al. [14] propose the following optimization problem to achieve this:

$$
\min_{\{\theta_i\},\{\pi^j\}} \sum_{j=1}^{J} \mathcal{D}\left(\prod_{l=1}^{L_j} q\left(z_i \middle| \sum_{i=1}^{L} \pi_{il}^j \theta_i \right) \,\middle\|\, \prod_{l=1}^{L_j} q(z_l | w_{jl}) \right)
$$
(7.28)

$$
\text{subject to} \quad \sum_{l=1}^{L_j} \pi_{il}^j \leq 1, \ \sum_{i=1}^{L} \pi_{il}^j = 1, \ P_{il}^j \in \{0, 1\}.
$$

Here we use $\{\pi^j\}$ to denote matching variables as before. Tractability of this problem depends on the divergence $\mathcal{D}(\cdot \,\|\, \cdot)$. A convenient choice is the Kullback–Leibler (KL) divergence due to its property of factorizing over the product distributions. With KL (7.28) can be simplified as follows:

$$
\min_{\{\theta_i\},\{\pi^j\}} \sum_{j=1}^{J} \sum_{l=1}^{L_j} \sum_{i=1}^{L} \pi_{il}^j \mathrm{KL}\left(q(z_i | \theta_i) \,\|\, q(z_l | w_{jl}) \right)
$$
(7.29)

$$
\text{subject to} \quad \sum_{l=1}^{L_j} \pi_{il}^j \leq 1, \ \sum_{i=1}^{L} \pi_{il}^j = 1, \ \pi_{il}^j \in \{0, 1\}.
$$

It remains to address the problem of estimating the number of global posterior components L. Comparing (7.28) to (7.5), (7.11) and (7.22) we notice similarities; however there is no term regularizing L, which previously stemmed from the IBP prior. Claici et al. [14] propose the following $L_{2,2,1}$ regularization term inspired by the $L_{2,1}$ matrix norm utilized by the approach of [12] for clustering using optimal transport:

$$
\sum_{i=1}^{L} \left(\sum_{j=1}^{J} \left(\sum_{l=1}^{L_j} (\pi_{il}^j)^2 \right) \right)^{1/2}.
$$
(7.30)

This quantity is the $L_{2,1}$ norm of the $L \times J$ matrix whose element at position (i, j) is the norm of row i in π^j.

We state the finalized KL-fusion objective function for completeness:

$$
\min_{\{\theta_i\},\{\pi^j\}} \sum_{j=1}^{J} \sum_{l=1}^{L_j} \sum_{i=1}^{L} \pi_{il}^j \mathrm{KL}\left(q(z_i | \theta_i) \,\|\, q(z_l | w_{jl}) \right) + \lambda \sum_{i=1}^{L} \left(\sum_{j=1}^{J} \left(\sum_{l=1}^{L_j} (\pi_{il}^j)^2 \right) \right)^{1/2}
$$

$$
\text{subject to} \quad \sum_{l=1}^{L_j} \pi_{il}^j \leq 1, \ \sum_{i=1}^{L} \pi_{il}^j = 1, \ \pi_{il}^j \in \{0, 1\}.
$$
(7.31)

KL-fusion algorithm alternates between updating $\{\pi^j\}$ and θ_i at every iteration. Due to the regularizer tying all $\{\pi^j\}$, it is no longer amendable to the Hungarian algorithm; however, relaxing the binary constraint $\pi_{il}^j \in \{0, 1\}$ turns (7.31) into a convex problem in $\{\pi^j\}$ amendable to convex problem solvers.

Updating θ_i requires solving a KL barycenter problem. Banerjee et al. [2] studied this problem and showed that if the distributions are in the same exponential family, then the natural parameter of the barycenter is equal to the average of the input distributions natural parameters. Specifically, let $\{q_i\}$ be the input distributions in the same exponential family Q with natural parameters $\{\eta_i\}$, and let $\lambda_i \geq 0$ be a set of weights with $\sum_i \lambda_i = 1$, then the solution to $\min_{q \in Q} \sum_{i=1}^n \lambda_i \mathrm{KL}(q \parallel q_i)$ is a distribution $q^* \in Q$ with natural parameter $\eta^* = \sum_{i=1}^n \lambda_i \eta_i$. In the context of KL-fusion, given matching variables $\{\pi^j\}$, this result allows us to update $\{\theta_i\}$ for each i by solving analogous KL barycenter problem:

$$\min_{q_i \in Q} \sum_{j=1}^{J} \sum_{l=1}^{L} \pi_{il}^j \mathrm{KL}\left(q_i \parallel q(z_l | w_{jl})\right). \tag{7.32}$$

7.5.2   KL-Fusion in Practice

We present a Gaussian mixture model simulated experiment analogous to the one presented in Sect. 7.4.3. The key difference is that local datasets are now generated with an arbitrary covariance matrix instead of an isotropic one. Instead of k-means, clients now learn their local mixture model parameters with variational inference, estimating both means and covariances. SPAHM can only utilize the means when fusing the local models, while KL-Fusion can benefit from the covariance information in the local posteriors. DP-clustering [6, 20] is a nonparametric clustering-based baseline; VI-oracle is an idealized non-federated scenario where VI is performed on the combined data of all clients. We present results in Fig. 7.5.

We emphasize Fig. 7.5b: when means of the client data generating mixture components are similar, i.e., lower x-axis values, covariance information is crucial for effective fusion. KL-fusion, as opposed to SPAHM, can utilize covariance information and achieves lower estimation error as a result.

Another important application of KL-Fusion is federated learning of *Bayesian* neural networks (BNN). Such neural networks excel at quantifying prediction uncertainty and can be used to identify out-of-distribution examples at test time instead of making wrongful predictions. We refer the reader to the experiments in Section 5.4 of [14] for examples of federated learning of BNNs with KL-Fusion.

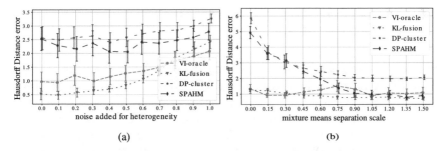

Fig. 7.5 Federated learning of a Bayesian clustering model. Simulated experiment with Gaussian mixture models. KL-fusion benefits from the posterior information improving upon SPAHM estimation error, especially when means of all mixture components are hardly distinguishable (**b**). (**a**) Heterogeneity in local datasets. (**b**) Separation between data generating means

Going beyond mean-field VI and KL

KL-fusion is specialized to posterior approximations that factorize the posterior into a product of distributions of the quantities of interest, e.g., mean-field VI. Such factorization is popular due to its scalability and computational tractability, which is crucial especially in the context of BNNs. However it enforces independence of the quantities of interest in the posterior, potentially introducing large errors when approximating the true posterior. Another major line of posterior learning techniques in Bayesian literature are the sampling based methods, such as Markov Chain Monte Carlo (MCMC) or Hamiltonian Monte Carlo (HMC) [4, 37]. These techniques are known to be less scalable, but provide more accurate posteriors. Although there are dedicated packages, e.g., Stan [13], providing efficient HMC implementations, they remain limited to smaller models, i.e., inapplicable to BNNs.

The key challenge in extending KL-fusion to sampling based methods is that posterior is now represented as a collection of samples, rather than a product of distributions. Such extension would be non-trivial as it requires new objective function formulation to perform posterior fusion. We also note that KL-fusion considers KL divergence as a measure of closeness between the fused global posterior and the client posteriors (recall the general form in equation (7.28))—this choice was made out of convenience, i.e., KL divergence of product distributions is equal to the sum of KLs between the corresponding terms yielding (7.29). To accommodate sampling based posteriors it is likely more appropriate to consider Wasserstein distance to take into account the geometry of the posterior samples. We note the related work of [43] that studied posterior aggregation using Wasserstein barycenters; however, it is limited to homogeneous data and models without permutation invariances.

7.6 Fusion of Deep Neural Networks with Low-Communication Budget

We have presented a series of model fusion algorithms capable of a *single* round federated learning in various contexts. Despite the ideal communication efficiency, these approaches have limitations when it comes to federated learning of *deep* neural networks (DNN) as discussed toward the end of Sect. 7.3.3. In this section we review the Federated Matched Averaging (FedMA) [47] algorithm that was designed to address this limitation, however at the cost of additional communication rounds. On a high level, FedMA capitalizes on the strength of PFNM in fusing shallow neural networks: it fuses one layer at a time and asks the clients to fine-tune the remaining layers at every communication round. FedMA is able to perform federated learning of DNNs, requiring number of communication rounds equal to the number of layers. In some federated learning applications communication constraints are guided by the size of the model parameters exchanged between clients and the server, as opposed to the number of times any amount of information is exchanged. FedMA is particularly appealing in such scenarios as it only exchanges parameters of a single layer at a time, resulting in the total communication cost of a single model size. From this perspective, it could also be considered a single-round federated learning algorithm.

7.6.1 Extending PFNM to Deep Neural Networks

Recall our discussion in Sect. 7.2 illustrating the permutation invariance nature of a single hidden layer fully connected (FC) neural network. The idea behind equation (7.3) can be extended to deep networks:

$$x_n = \sigma(x_{n-1} \Pi_{n-1}^T W_n \Pi_n), \tag{7.33}$$

where $n = 1, \ldots, N$ is the layer index and Π_0 is identity indicating non-ambiguity in the ordering of the input features $x = x_0$. Π_N is also an identity assuming clients assign same output indices to the same classes. As before, $\sigma(\cdot)$ is a non-linearity function, except for $f(x) = x_N$ where it could be an identity function or a softmax if we want probabilities instead of logits. Notice that for $N = 2$ we recover (7.3).

Note that permutations of any consecutive pair of intermediate layers are dependent, leading to a NP-hard combinatorial optimization problem. However we can utilize PFNM by solving the problem recursively: suppose we have $\{\Pi_{j,n-1}\}$ for all J clients, then taking $\{\Pi_{j,n-1}^T W_{j,n}\}$ to be PFNM inputs, we estimate $\{\Pi_{j,n}\}$ and proceed to the next layer. The base of recursion is $\{\Pi_{j,0}\}$, which is known to be an identity permutation for any j. While this is a feasible solution, the problem remains NP-hard and the quality of the approximation obtained with the aforementioned procedure is likely to deteriorate for larger number of layers. The *main idea of*

FedMA is to have clients to locally fine-tune layers $n + 1$ and above after fusing layer n with PFNM. We summarize FedMA in Algorithm 7.2 [47] and proceed to discuss treatment of convolutional and long short-term memory (LSTM) [25] layers.

Algorithm 7.2 Federated Matched Averaging (FedMA)

Input: local weights of N-layer architectures $\{W_{j,1}, \ldots, W_{j,N}\}_{j=1}^{J}$ from J clients.
Output: global model weights $\{W_1, \ldots, W_N\}$

1: $n = 1$
2: **while** $n \leq N$ **do**
3:     **if** $n < N$ **then**
4:       $\{\Pi_j\}_{j=1}^{J} = \text{PFNM}\big(\{W_{j,n}\}_{j=1}^{J}\big)$ (find permutation matrices matching neurons)
5:       $W_n = \frac{1}{J}\sum_j W_{j,n}\Pi_j^T$ (compute global model layer n weights)
6:     **else**
7:       $W_n = \sum_{k=1}^{K}\sum_j p_{jk}W_{jl,n}$ where p_k is fraction of data points with label k on client j (perform basic weighted averaging for last layer)
8:     **end if**
9:     **for** $j \in \{1, \ldots, J\}$ **do**
10:      $W_{j,n+1} \leftarrow \Pi_j W_{j,n+1}$ (permute the next-layer weights)
11:      Train $\{W_{j,n+1}, \ldots, W_{j,L}\}$ with $\{W_{j,1}, \ldots, W_{j,n}\}$ frozen
12:    **end for**
13:    $n = n + 1$
14: **end while**

Convolutional layers. In convolutional neural networks (CNN), the permutation invariance is with respect to the channels instead of neurons. Let $\text{Conv}(x, W)$ denote the convolutional operation on input x with weights $W \in \mathbb{R}^{C^{in} \times w \times h \times C^{out}}$, where C^{in}, C^{out} are the numbers of input/output channels and w, h are the width and height of the filters. Applying any permutation to the output channel dimension of a layer and the same permutation to the input channel dimension of the subsequent layer does not change the CNN. Analogous to (7.33) we obtain

$$x_n = \sigma\big(\text{Conv}(x_{n-1}, \Pi_{n-1}^T W_n \Pi_n)\big). \tag{7.34}$$

We note that pooling operations do not affect the formulation as they act within channels. To use PFNM or FedMA with CNNs, client j forms inputs to PFNM matching procedure as $\{w_{jl} \in \mathbb{R}^D\}_{l=1}^{C_n^{out}}$, $j = 1, \ldots, J$, where D is the flattened $C_n^{in} \times w \times h$ dimension of $\Pi_{j,n-1}^T W_{j,n}$.

LSTM layers

Permutation invariance of recurrent neural networks (RNN) such as LSTMs is due to the invariant ordering of the hidden states. At a first glance, it appears similar to the already familiar FCs and CNNs; however, the problem is more nuanced in the RNN case. The subtlety is due to the hidden-to-hidden weights: let L be the hidden state dimension and $H \in \mathbb{R}^{L \times L}$ hidden-to-hidden weights, then we can notice that any permutation of the hidden states affects *both* rows and columns of H. To illustrate, suppose we want to match H_j and $H_{j'}$ of two clients, then to find optimal matching we need to minimize $\|\Pi^T H_j \Pi - H_{j'}\|_2^2$ over permutations Π. This is known as a quadratic assignment problem that is NP-hard [32] even in this simple case.

To complete the specification of PFNM to RNNs, we recall a basic RNN cell

$$h_t = \sigma(h_{t-1}\Pi^T H \Pi + x_t W \Pi), \qquad (7.35)$$

where t is indexing the input at position t. PFNM can be applied to the input-to-hidden weights $\{W_j\}$ as in the case of fully connected networks, then we can compute global hidden-to-hidden weights utilizing the estimated matchings $H = \frac{1}{J}\sum_j \Pi_j H_j \Pi_j^T$. Extending this to (multilayer) LSTM cells is straightforward, see [47].

We note that this PFNM-based approach ignores the information in the hidden-to-hidden weights when estimating the matchings. It may be possible to improve this procedure by considering approximate optimization tools from the optimal transport literature developed for the Gromov–Wasserstein metric [23], which is a similar quadratic assignment problem; for example, approximate algorithms for computing Gromov–Wasserstein barycenters [39].

7.6.2 FedMA in Practice

We summarize the key experimental findings of [47]. In Fig. 7.6 we present performance of PFNM and FedMA with convolutional neural networks: on MNIST with LeNet [29] (4 layers) and on CIFAR10 with a more sophisticated VGG-9 [41] (9 layers). PFNM attempts to perform federated learning in a single communication round, while FedMA utilizes number of rounds equal to the number of layers in the respective architectures. We see that while PFNM performs well on MNIST with LeNet, i.e., it can handle moderate number of layers, its performance drops significantly for a deeper network VGG-9. FedMA, on the other hand, successfully executes federated learning with VGG-9 client models and achieves strong test

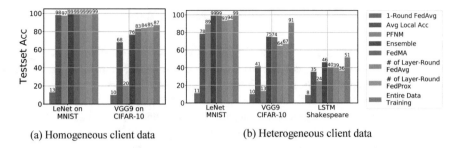

(a) Homogeneous client data   (b) Heterogeneous client data

Fig. 7.6 Comparison of federated learning methods with limited number of communications on LeNet trained on MNIST; VGG-9 trained on CIFAR10; LSTM trained on Shakespeare dataset over: (**a**) homogeneous and (**b**) heterogeneous data partitions

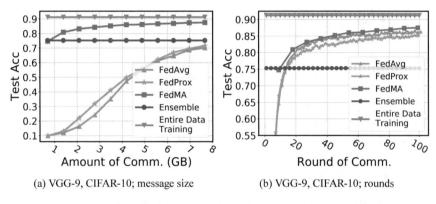

(a) VGG-9, CIFAR-10; message size   (b) VGG-9, CIFAR-10; rounds

Fig. 7.7 Convergence rates of federated learning methods as a function of communication cost: training VGG-9 on CIFAR10 with $J = 16$ clients

performance on CIFAR10 outperforming other federated learning baselines, i.e., FedAvg [34] and FedProx [40] restricted to the same number of communication rounds.

FedMA can also be easily extended to benefit from additional communication rounds by simply iterating over layers of client neural networks. In Fig. 7.7 we present the comparison of performance of various federated learning techniques as a function of the communication cost measured by the number of parameter exchanges between the clients and the server, and by the size (measured in gigabytes) of the model parameters transmitted. FedMA is significantly more efficient when the communication budget constraints stem from the compute infrastructure bandwidth, i.e., communication is measured by the size of the messages being exchanged.

7.7 Theoretical Understanding of Model Fusion

We follow up the preceding exposition of algorithms for model fusion with the statistical properties of model fusion. From an algorithmic perspective, the main challenge of model fusion is establishing a correspondence between separately learned parameters on different clients. This is due to the presence of certain invariances in the model parameterizations. To focus on the statistical issues, we assume the correct correspondence has been established in this section. Equivalently, we assume there are no invariances in the parameterization of the models under consideration.

7.7.1 Preliminaries: Parametric Models

A *statistical model* is a (parameterized) family of probability distributions $\{P_\theta \mid \theta \in \Theta\}$ on a sample space $\mathcal{Z}$. A *parametric model* is a statistical model in which the *parameter* θ is a vector in $\mathbb{R}^d$.[1] We note that this use of the term "model" differs from its typical use in machine learning (ML). In ML, model usually refers to a (parameterized) collection of prediction rules (e.g., all neural networks with a certain architecture). To bridge this difference, associate with each probability distribution in the statistical model a prediction rule, so there is a correspondence between parameters and prediction rules.

In this setup, the main task is estimating θ_* from independent identically distributed (IID) observations $Z_1, \ldots, Z_N \sim P_{\theta_*}$. An *estimator* of θ_* from $Z_{1:N} \triangleq (Z_1, \ldots, Z_N)$ is a random variable $\widehat{\boldsymbol{\theta}}_N \triangleq T_N(Z_{1:N})$. We note that T_N may be complicated and/or implicitly defined: e.g., the maximum likelihood estimator (MLE) is $\widehat{\boldsymbol{\theta}} = T_N(Z_{1:N})$, where

$$T_N(Z_{1:N}) \triangleq \arg\max_{\theta \in \Theta} \frac{1}{N} \sum_{i=1}^{N} \log p(Z; \theta), \tag{7.36}$$

where $p(\cdot, \theta)$ is the density of P_θ. In this chapter, we focus on **asymptotically linear** estimators.

Definition 7.1 An estimator $\widehat{\boldsymbol{\theta}}_N \triangleq T_N(Z_{1:N})$ is *asymptotically linear* iff there is $\varphi : \mathcal{Z} \to \Theta$ (which may depend on θ_*) such that $\mathbb{E}_{\theta_*}[\varphi(Z)] = 0$ and

$$\sqrt{N}(\widehat{\boldsymbol{\theta}}_N - \theta_*) = \frac{1}{\sqrt{N}} \sum_{i=1}^{N} \varphi(Z_i) + o_P(1),$$

where $o_P(1)$ denotes a term that vanishes in probability as n grows. The random vector $\varphi(Z_i)$ is called the *influence function* of Z_i.

---

[1] To keep things simple, we assume the set of all possible parameters Θ, called the *parameter space*, is an open subset of $\mathbb{R}^d$.

Although the definition of asymptotically linear estimators seems restrictive, most estimators we encounter in practice are asymptotically linear. For example, the maximum likelihood estimator is asymptotically linear under certain technical conditions on the underlying statistical model. To see this, we note that the optimality of the MLE $\widehat{\boldsymbol{\theta}}_N$ implies it satisfies the zero-gradient optimality condition

$$0 = \tfrac{1}{N} \sum_{i=1}^{N} \partial_\theta \ell(Z_i; \widehat{\boldsymbol{\theta}}_N),$$

where $\ell(z; \theta) \triangleq \log p(z; \theta)$ is the *log-likelihood* and its gradient with respect to θ is the *score*. We expand the score at θ_* to obtain

$$0 = \tfrac{1}{N} \sum_{i=1}^{N} \partial_\theta \ell(Z_i; \theta_*) + \partial_\theta^2 \ell(Z_i; \theta_*)(\widehat{\boldsymbol{\theta}}_N - \theta_*) + O(\|\widehat{\boldsymbol{\theta}}_N - \theta_*\|_2^2).$$

Rearranging, we have

$$\sqrt{N}(\widehat{\boldsymbol{\theta}}_N - \theta_*) = (\tfrac{1}{N} \sum_{i=1}^{N} \partial_\theta^2 \ell(Z_i; \theta_*))^{-1} (\tfrac{1}{\sqrt{N}} \sum_{i=1}^{N} \partial_\theta \ell(Z_i; \theta_*) + O(\sqrt{N}\|\widehat{\boldsymbol{\theta}}_N - \theta_*\|_2^2).$$
$$(7.37)$$

This is almost the definition of an asymptotically linear estimator. We recognize $\tfrac{1}{N} \sum_{i=1}^{N} \partial_\theta \ell(Z_i; \theta_*)$ as an average of IID random matrices, so it converges to its expected value (under certain tail conditions):

$$\tfrac{1}{N} \sum_{i=1}^{N} \partial_\theta \ell(Z_i; \theta_*) = \mathbb{E}_{\theta_*}[\partial_\theta^2 \ell(Z, \theta_*)] + O_P(\tfrac{1}{\sqrt{N}}).$$

We recognize $\mathbb{E}_{\theta_*}[\partial_\theta^2 \ell(Z, \theta_*)]$ as the *Fisher information*, and we denote it hereafter as $I(\theta_*)$. As long as $I(\theta_*)$ is non-singular, we have

$$(\tfrac{1}{N} \sum_{i=1}^{N} \partial_\theta \ell(Z_i; \theta_*))^{-1} = I(\theta_*)^{-1} + O_P(\tfrac{1}{\sqrt{N}}).$$

It is known that the MLE converges at a $\tfrac{1}{\sqrt{N}}$-rate (under certain technical conditions), so the $O(\sqrt{N}\|\widehat{\boldsymbol{\theta}}_N - \theta_*\|_2^2)$ term is $O_P(\tfrac{1}{\sqrt{N}})$. We combine these facts with (7.37) to obtain

$$\sqrt{N}(\widehat{\boldsymbol{\theta}}_N - \theta_*) = \tfrac{1}{\sqrt{N}} \sum_{i=1}^{N} I(\theta_*)^{-1} \partial_\theta \ell(Z_i; \theta_*)$$
$$+ O_P(\tfrac{1}{\sqrt{N}}) \tfrac{1}{\sqrt{N}} \sum_{i=1}^{N} \partial_\theta \ell(Z_i; \theta_*) + O_P(\tfrac{1}{\sqrt{N}}).$$

Finally, we recognize $\tfrac{1}{\sqrt{N}} \sum_{i=1}^{N} \partial_\theta \ell(Z_i; \theta_*)$ is $O_P(1)^2$ to conclude

$$\sqrt{N}(\widehat{\boldsymbol{\theta}}_N - \theta_*) = \tfrac{1}{\sqrt{N}} \sum_{i=1}^{N} I(\theta_*)^{-1} \partial_\theta \ell(Z_i; \theta_*) + O_P(\tfrac{1}{\sqrt{N}}).$$

---

[2] Its variance is $O(1)$.

Thus the MLE is asymptotically linear with influence function $I(\theta_*)^{-1}\partial_\theta \ell(\cdot;\theta_*)$.

We wrap up with another example from ML. In supervised learning, the observations Z are pairs (X, Y), where X is the feature vector and Y is the target, and the unknown parameter is *identified* by a system of *moment equations*; i.e.,

$$\theta = \theta_* \iff \mathbb{E}_{\theta_*}\big[m(Z;\theta) \mid X\big] = 0, \tag{7.38}$$

where m is a $\mathbb{R}^d$-valued map. In this case, a natural estimator is a root of the empirical version of the moment equations: $\widehat{\theta}_N$ solves

$$0 = \tfrac{1}{N} \sum_{i=1}^{K} m(Z;\widehat{\theta}_N). \tag{7.39}$$

For example, consider linear regression:

$$\widehat{\theta}_N \triangleq \arg\min_{\theta \in \mathbb{R}^d} \tfrac{1}{N} \sum_{i=1}^{n} \tfrac{1}{2}(Y_i - \theta^T X_i)^2.$$

The optimality conditions of the least-squares cost function are the normal equations:

$$0 = \tfrac{1}{N} \sum_{i=1}^{n} X_i(Y_i - \theta^T X_i). \tag{7.40}$$

We note that the normal equations have the form of (7.39). It is not hard to check that the population counterpart of the normal equations

$$0 = \mathbb{E}_{\theta_*}\big[X(Y - \theta^T X) \mid X\big]$$

uniquely identify θ_*. A similar Taylor expansion argument shows that

$$\sqrt{N}(\widehat{\theta}_N - \theta_*) = \tfrac{1}{\sqrt{N}} \sum_{i=1}^{N} V(\theta_*)^{-1} m(Z_i;\theta_*) + O_P(\tfrac{1}{\sqrt{N}}),$$

where $V(\theta) \triangleq \mathbb{E}_{\theta_*}[\partial_\theta m(Z,\theta)]$. Thus this estimator is asymptotically linear with influence function $V(\theta_*)^{-1} m(\cdot;\theta_*)$.

7.7.2   The Benefits and Drawbacks of Model Fusion in Federated Settings

In federated learning, the dataset $Z_{1:N}$ is distributed across J clients. To keep things simple, we assume the samples are distributed evenly across the clients; i.e., each client has $n = \tfrac{N}{J}$ samples. Recall in model fusion, each client independently estimates θ_* from its data to obtain an estimator. The clients send their estimators to a server, and the server averages the client estimators to obtain a global estimator:

$$\tilde{\theta}_N \triangleq \tfrac{1}{J} \sum_{j=1}^{J} \widehat{\boldsymbol{\theta}}_{n,j},$$

where $\widehat{\boldsymbol{\theta}}_{n,j}$ is the j-th client's estimate of θ_* from its n samples. If the $\widehat{\boldsymbol{\theta}}_{n,j}$'s is asymptotically linear, then

$$\tilde{\theta}_N - \theta_* = \tfrac{1}{J} \sum_{j=1}^{J} \widehat{\boldsymbol{\theta}}_{n,j} - \theta_* = \tfrac{1}{N} \sum_{j=1}^{J} \sum_{i=1}^{n} \varphi(Z_{j,i}) + o_P(\tfrac{1}{\sqrt{n}}), \qquad (7.41)$$

where φ is the influence function of the $\widehat{\boldsymbol{\theta}}_{n,j}$'s, and $Z_{j,i}$ is the i-th sample on the j-th client. Ignoring the $o_P(\tfrac{1}{\sqrt{n}})$ term, we see that $\tilde{\theta}_N$ is the average of N IID random variables, so the fluctuation of $\tilde{\theta}_N$ around θ_* is $O_P(\tfrac{1}{\sqrt{N}})$. *This is the same order as the hypothetical centralized estimator.* Further, even the asymptotic distribution of $\tilde{\theta}_N$ matches that of $\widehat{\boldsymbol{\theta}}_N$:

$$\sqrt{n}(\tilde{\theta}_N - \theta_*) - \sqrt{n}(\widehat{\theta}_N - \theta_*) = o_P(1).$$

In other words, the performance of $\tilde{\theta}_N$ and $\widehat{\boldsymbol{\theta}}_N$ in estimating θ_* is indistinguishable.

Unfortunately, there are a few caveats to the rosy picture of model fusion from the preceding section. The first is ignoring the $o_P(\tfrac{1}{\sqrt{n}})$ term in (7.41). This is only a good heuristic when the $o_P(\tfrac{1}{\sqrt{n}})$ is asymptotically negligible compared to the linear term. Recalling the MLE example in Sect. 7.7.1, we see that the $o_P(\tfrac{1}{\sqrt{n}})$ term hides terms that are actually $O_P(\tfrac{1}{n})$. In order for the terms that we neglected to be actually (asymptotically negligible), we must have $J \lesssim \sqrt{N}$. In other words, the number of clients cannot increase arbitrarily without also increasing the number of samples per client. This is less than ideal because it precludes model fusion from scaling to large numbers of clients without upgrading the computational and storage capabilities of the clients.

7.8 Conclusion

We conclude the chapter with recommendations along three directions: (1) summary for practitioners considering model fusion in federated learning applications; (2) promising directions for developing new model fusion and federated learning algorithms; and (3) open problems in theoretical understanding of model fusion.

Practical considerations. The key strength of model fusion is the ability to perform federated learning in a single communication round. This is the only solution in problems where clients have models that have been trained on data that is no longer available. There are several reasons why data may not be available. For example, the data may have been deleted to comply with regulations prohibiting prolonged data storage such as GDPR. The data may also have been lost due to system failures or other unexpected events. Even in applications in which the data is available, setting up the compute and network infrastructures that support frequent communications between the clients and the server, required by optimization based methods such as Federated Averaging, could be expensive or impractical.

We have seen that model fusion is especially effective on simpler models, such as unsupervised models and neural networks with few layers. Unfortunately, the performance of model fusion deteriorates on deep neural networks. For such deep neural networks we recommend FedMA. It does require several communication rounds, precluding the legacy model use case, but it can greatly simplify network load due to its memory efficiency.

As we saw, two key limitations of model fusion and FedMA are

1. They are more challenging to implement than Federated Averaging in practice due to the complexity of the matching algorithms.
2. They do not scale to larger number of clients.

In comparison, Federated Averaging has lower communication efficiency, but is easier to implement and scales readily to thousands of clients due to its simplicity. We note that IBM Federated Learning [33] provides some of the model fusion functionality that alleviates the implementation barriers.

Promising directions for advancing model fusion methodology. Throughout the chapter we have outlined a series of directions for extending and improving model fusion: identifying invariance classes beyond permutations by considering loss landscape of neural networks discussed in Sect. 7.2; developing more sophisticated methods for estimating matching variables for fusion of deep neural networks in Sect. 7.3.3; extending model fusion to clients with time stamps in Sect. 7.4.3; developing posterior fusion techniques supporting sampling methods (e.g., MCMC) and other distribution divergences (e.g., Wasserstein) to broaden KL-fusion applications in Sect. 7.5.2; considering approximate optimization techniques for quadratic assignment to improve the fusion of LSTMs in Sect. 7.6.1. In addition, we note one limitation of FedMA that is worth exploring: current version of FedMA assumes that all clients communicate their corresponding layer weights at every iteration; however, in federated learning applications with large number of clients, it is more practical to consider a stochastic setting, i.e., a random subset of clients communicating at every round. An interesting direction for future work is to study how this stochasticity can be accounted for in the layer-wise FedMA learning algorithm. Finally, we recommend exploring model fusion applications outside of federated learning. For example, experiment presented in Figure 4 in [47] demonstrates that model fusion has potential to correct biases in the data, i.e., reduce the effect of spurious correlations in the client models on the global model.

Open problems in theoretical understanding of model fusion. As we saw, a key statistical limitation of model fusion is the $\sqrt{N}$-barrier. This barrier arises due to the bias in the clients' estimates of the model parameters. Although it is hard to overcome this barrier in general, it may be possible to develop estimators with smaller biases in special cases that permit model fusion to scale to larger number of clients.

References

1. Agueh M, Carlier G (2011) Barycenters in the Wasserstein space. SIAM J Math Anal 43:904–924
2. Banerjee A, Dhillon IS, Ghosh J, Sra S (2005) Clustering on the unit hypersphere using von Mises-Fisher distributions. J Mach Learn Res 6:1345–1382
3. Bardenet R, Doucet A, Holmes C (2017) On Markov chain Monte Carlo methods for tall data. J Mach Learn Res 18(1):1515–1557
4. Betancourt M (2017) A conceptual introduction to Hamiltonian Monte Carlo. arXiv preprint arXiv:170102434
5. Bishop CM (2006) Pattern recognition and machine learning. Springer, New York
6. Blei DM, Jordan MI (2006) Variational inference for Dirichlet process mixtures. Bayesian Anal 1:121–143
7. Blei DM, Kucukelbir A, McAuliffe JD (2017) Variational inference: a review for statisticians. J Am Stat Assoc 112(518):859–877
8. Breiman L (2001) Random forests. Mach Learn 45:5–32
9. Broderick T, Boyd N, Wibisono A, Wilson AC, Jordan MI (2013) Streaming variational Bayes. In: Advances in neural information processing systems
10. Bui TD, Nguyen CV, Swaroop S, Turner RE (2018) Partitioned variational inference: a unified framework encompassing federated and continual learning. arXiv preprint arXiv:181111206
11. Campbell T, How JP (2014) Approximate decentralized Bayesian inference. arXiv:14037471
12. Carli FP, Ning L, Georgiou TT (2013) Convex clustering via optimal mass transport. arXiv:13075459
13. Carpenter B, Gelman A, Hoffman M, Lee D, Goodrich B, Betancourt M, Brubaker MA, Guo J, Li P, Riddell A et al (2017) Stan: a probabilistic programming language. J Stat Softw 76:1–32
14. Claici S, Yurochkin M, Ghosh S, Solomon J (2020) Model fusion with Kullback-Leibler divergence. In: International conference on machine learning
15. Cooper Y (2018) The loss landscape of overparameterized neural networks. arXiv preprint arXiv:180410200
16. Das R, Zaheer M, Dyer C (2015) Gaussian LDA for topic models with word embeddings. In: Proceedings of the 53rd annual meeting of the association for computational linguistics and the 7th international joint conference on natural language processing (Volume 1: Long Papers)
17. Dietterich TG (2000) Ensemble methods in machine learning. In: International workshop on multiple classifier systems
18. Draxler F, Veschgini K, Salmhofer M, Hamprecht F (2018) Essentially no barriers in neural network energy landscape. In: International conference on machine learning
19. EU (2016) Regulation (EU) 2016/679 of the European Parliament and of the Council of 27 April 2016 on the protection of natural persons with regard to the processing of personal data and on the free movement of such data, and repealing Directive 95/46/EC (General Data Protection Regulation). Official Journal of the European Union
20. Ferguson TS (1973) A Bayesian analysis of some nonparametric problems. Ann Stat 1:209–230
21. Garipov T, Izmailov P, Podoprikhin D, Vetrov D, Wilson AG (2018) Loss surfaces, mode connectivity, and fast ensembling of DNNs. arXiv preprint arXiv:180210026
22. Ghahramani Z, Griffiths TL (2005) Infinite latent feature models and the Indian buffet process. In: Advances in neural information processing systems
23. Gromov M, Katz M, Pansu P, Semmes S (1999) Metric structures for Riemannian and non-Riemannian spaces, vol 152. Birkhäuser, Boston

24. Hasenclever L, Webb S, Lienart T, Vollmer S, Lakshminarayanan B, Blundell C, Teh YW (2017) Distributed Bayesian learning with stochastic natural gradient expectation propagation and the posterior server. J Mach Learn Res 18:1–37
25. Hochreiter S, Schmidhuber J (1997) Long short-term memory. Neural Comput 9:1735–1780
26. Hsu TMH, Qi H, Brown M (2019) Measuring the effects of non-identical data distribution for federated visual classification. arXiv preprint arXiv:190906335
27. Jordan MI, Ghahramani Z, Jaakkola TS, Saul LK (1999) An introduction to variational methods for graphical models. Mach Learn 37:183–233
28. Kuhn HW (1955) The Hungarian method for the assignment problem. Nav Res Logist (NRL) 2:83–97
29. LeCun Y, Bottou L, Bengio Y, Haffner P et al (1998) Gradient-based learning applied to document recognition. In: Proceedings of the IEEE
30. Li H, Xu Z, Taylor G, Studer C, Goldstein T (2017) Visualizing the loss landscape of neural nets. arXiv preprint arXiv:171209913
31. Lloyd S (1982) Least squares quantization in PCM. IEEE Trans Inf Theory 28:129–137
32. Loiola EM, de Abreu NMM, Boaventura-Netto PO, Hahn P, Querido T (2007) A survey for the quadratic assignment problem. Eur J Oper Res 176:657–690
33. Ludwig H, Baracaldo N, Thomas G, Zhou Y, Anwar A, Rajamoni S, Ong Y, Radhakrishnan J, Verma A, Sinn M et al (2020) IBM federated learning: an enterprise framework white paper v0. 1. arXiv preprint arXiv:200710987
34. McMahan B, Moore E, Ramage D, Hampson S, y Arcas BA (2017) Communication-efficient learning of deep networks from decentralized data. In: Artificial intelligence and statistics
35. Minka TP (2001) Expectation propagation for approximate Bayesian inference. In: Conference on uncertainty in artificial intelligence
36. Neal RM (2012) Bayesian learning for neural networks. Springer Science & Business Media, Berlin/Heidelberg
37. Neal RM et al (2011) MCMC using Hamiltonian dynamics. Handb Markov Chain Monte Carlo 2(11):2
38. Opper M (1998) A Bayesian approach to on-line learning. On-line Learning in Neural Networks
39. Peyré G, Cuturi M, Solomon J (2016) Gromov-Wasserstein averaging of kernel and distance matrices. In: International conference on machine learning
40. Sahu AK, Li T, Sanjabi M, Zaheer M, Talwalkar A, Smith V (2018) On the convergence of federated optimization in heterogeneous networks. arXiv preprint arXiv:181206127
41. Simonyan K, Zisserman A (2014) Very deep convolutional networks for large-scale image recognition. arXiv preprint arXiv:14091556
42. Singh SP, Jaggi M (2019) Model fusion via optimal transport. arXiv preprint arXiv:191005653
43. Srivastava S, Cevher V, Dinh Q, Dunson D (2015) Wasp: scalable Bayes via barycenters of subset posteriors. In: Artificial intelligence and statistics
44. Teh YW, Grür D, Ghahramani Z (2007) Stick-breaking construction for the Indian buffet process. In: Artificial intelligence and statistics
45. Thibaux R, Jordan MI (2007) Hierarchical Beta processes and the Indian buffet process. In: Artificial intelligence and statistics
46. Wainwright MJ, Jordan MI et al (2008) Graphical models, exponential families, and variational inference. Found Trends® Mach Learn 1:1–305
47. Wang H, Yurochkin M, Sun Y, Papailiopoulos D, Khazaeni Y (2020) Federated learning with matched averaging. In: International conference on learning representations
48. Yurochkin M, Agarwal M, Ghosh S, Greenewald K, Hoang N (2019) Statistical model aggregation via parameter matching. In: Advances in neural information processing systems
49. Yurochkin M, Agarwal M, Ghosh S, Greenewald K, Hoang N, Khazaeni Y (2019) Bayesian nonparametric federated learning of neural networks. In: International conference on machine learning
50. Yurochkin M, Fan Z, Guha A, Koutris P, Nguyen X (2019) Scalable inference of topic evolution via models for latent geometric structures. In: Advances in neural information processing systems

Chapter 8
Federated Learning and Fairness

Annie Abay, Yi Zhou, Nathalie Baracaldo, and Heiko Ludwig

Abstract As federated learning utilization quickly expands to a variety of industries, examining its interactions with, and impact on, machine learning bias becomes increasingly relevant. This chapter is dedicated to the discussion of social fairness in federated learning, as opposed to fairness in equal party-to-party contributions to a global model. Social fairness in machine learning, while multi-faceted, is primarily concerned with techniques to verify that machine learning predictions are fair in spite of dataset features traditionally and historically documented as inducing discriminatory bias, i.e., race, sex, etc. This chapter reviews causes of bias in machine learning that are related to federated learning, the unique challenges that federated learning presents in fairness, and notable work in the field that covers a variety of approaches toward creating and measuring fairer federated models.

8.1 Introduction

As machine learning (ML) becomes increasingly ingrained into the daily lives of the human population, research in the creation of machine learning models that are discrimination-aware has sky-rocketed [11], as well as documentation of the adverse effects of their absence.

In [4, 11], a ML algorithm used by judges in more than 12 states was found twice as likely to incorrectly classify black defendants as high-risk for re-offending, and white defendants as low-risk. Its predictions impacted if defendants "should be let out on bail before trial, [the] type of supervision on inmates... [and] had an impact on the length of [their] sentences." In August 2020, over 300,000 students across the United Kingdom received ML-generated results for their A-level exams, a critical component for college applications. The algorithm took into account multiple features, but included information such as students' school's

A. Abay (✉) · Y. Zhou · N. Baracaldo · H. Ludwig
IBM Research – Almaden, San Jose, CA, USA
e-mail: anniek@ibm.com; yi.zhou@ibm.com; baracald@us.ibm.com; hludwig@us.ibm.com

177

historical performance. Socioeconomic status affects student success (i.e., being able to afford a tutor, access to paid resources, more time for schoolwork v. working a job), and this feature consideration (among others) disadvantaged students from state-funded schools and aided students from private schools. Students from state-funded schools received an average of a full-letter grade below their teacher-predicted scores. Days of protests led to the government's decision to disregard the model's results and award students their teacher-predicted scores.

As mentioned in previous chapters, federated learning (FL) has emerged as a methodology to train machine learning models collaboratively, models that maintain the data privacy of all parties involved. FL's prioritization of data privacy has increased the breadth of its utilization to include various industries with a range of applications. This chapter reviews the discussion of bias and fairness in federated learning, including the similarities and differences in bias mitigation methods, challenges in the federated learning setting, as well as approaches and important gaps affecting the creation and measurement of fairer federated models.

What is Fairness?
Fairness does not have one unified definition and changes based on the context it is provided in. Merriam Webster defines "fairness" as "the quality or state of being fair, especially :fair or impartial treatment: lack of favoritism toward one side or another" [24]. What makes something impartial? How is favoritism enCOMPASsed? How does the perception of these impact the way solutions are designed? We explore these questions in later sections.

Contribution Fairness and Social Fairness: The conversation around analyzing fairness in the context of federated learning is separated into two main tracks: contribution fairness and social fairness.

- Contribution fairness focuses on how different federated learning parties impact the global model (i.e., are parties with a smaller amount of training data being underrepresented in favor of parties with larger amounts of data?). Any type of dataset (i.e., MNIST [29]) can be utilized when assessing contribution fairness, as we are examining the party's performance via accuracy and F1 score, and overall impact on the global model.
- Social fairness focuses on how data attributes that are historically documented as inducing discriminatory bias (i.e., sex, race etc.) impact a data sample's predicted label. Consequently, datasets utilized in ML fairness are about people, such as the Adult dataset [22], which classifies whether individuals have an annual income above or below $50,000.

8.2   Preliminaries and Existing Mitigation Methods

8.2.1   Notation and Terminology

This chapter focuses on social fairness. Two popular perspectives of social fairness within machine learning are individual fairness and group fairness. Individual fairness is based on the idea that similar individuals should receive similar treatment, regardless of their *sensitive attribute* value. Group fairness is based on the idea that underprivileged groups as a whole should receive treatment similar to that of privileged groups.

Sensitive attributes are dataset features that historically have been used to discriminate against a group of people. These include sex, race, age, religion, etc. Bias mitigation methods are usually designed around sensitive attributes, which we elaborate on in Sect. 8.2.2. These *sensitive attributes* will separate a dataset into a *privileged group* and *underprivileged group* based on both the value of the sensitive attribute and what we are analyzing in the dataset.

For example, the Adult and COMPAS datasets are two popular datasets for social fairness analysis. The Adult dataset is composed of data from the 1994 Census database, and as mentioned, classifies whether the individuals in the dataset have an annual income above or below $50,000. If *sex* is the sensitive attribute we are examining, data sample with a sensitive attribute value of *male* would be in the privileged group, and those with a sensitive attribute value of *female* would be in the underprivileged group. This is based on the documented discrepancy in income between males and females [27]. The COMPAS (Correctional Offender Management Profiling for Alternative Sanctions) dataset is composed of data collected from defendants of Broward County, Florida from 2013 and 2014; it contains predictions as to whether a person who has committed a crime will do so again. If *sex* is again the sensitive attribute we are examining, data samples with a sensitive attribute value of *female* would be in the privileged group, and those with a sensitive attribute value of *male* would be in the underprivileged group. This is based on the documented discrepancy in jail sentencing between males and females [28].

Favorable labels are labels that are considered advantageous, based on the dataset and what we are examining. For example, in the Adult dataset, "above 50,000" is the *favorable label*, and "below 50,000" is the *unfavorable label*. Similarly, in the COMPAS dataset, the *favorable label* is "will not commit a crime again" and "will commit a crime again" is the *unfavorable label*.

$D := (X, Y)$ is the training dataset, where X will refer to the feature set and Y will refer to the label set. The set of sensitive attributes is $S \subseteq X$, and s/s_i is a specific sensitive attribute value. In the same vein, x/x_i and y denote a feature vector and label, respectively.

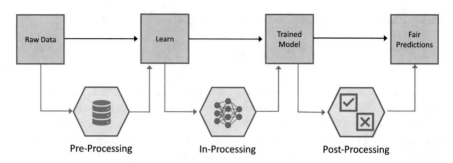

Fig. 8.1 Types of Bias Mitigation Methods

8.2.2 Types of Bias Mitigation Methods

Machine learning algorithms are categorized into three categories of methods to attack bias, organized by which phase of training these methods are applied; these are pre-processing, in-processing, and post-processing algorithms, as illustrated in Fig. 8.1.

Pre-processing methods [9, 10, 17, 32] are designed to be utilized prior to when training begins, and focus on reducing the bias in the dataset. These methods often work in one of two ways. One method is to assign sample weights to different data points; these sample weights can be calculated based on the sensitive attribute value, label, or other attributes in the feature set etc. The protocols to do this vary by method. Pre-processing methods work well for federated learning for two reasons. Firstly, they by nature can be paired with any type of machine learning model, which increases the pool of users. Secondly, each party can perform data pre-processing on their own without affecting the learning process, which helps maintain our protection of data privacy.

In-processing methods [8, 12, 20, 31] are utilized during training, and focus on reducing bias as the model learns. These usually will reduce bias by adjusting the optimization problem, and this can be done in more than one way; for example, by adding a regularizer term to the objective function that reduces a "prejudice index" [21]. Unlike pre-processing methods, in-processing methods have application constrictions, and most methods are linked to specific types of models they can be used with.

Post-processing methods [13, 18, 26] are utilized after training is complete, and focus on reducing bias in the test set's label predictions. These methods treat the model as a black box and use a protocol to change predicted labels to be more fair. One such example is [13], which uses a linear program to find probabilities with which to change data point labels, based on a fairness metric.

8.2.3 Data Privacy and Bias

The primary concern in creating bias mitigation methods that are suitable for federated learning is to design the methods to maintain the privacy of all parties involved. The majority of bias mitigation methods are created with centralized machine learning in mind and require full access to the data to complete the protocol, access like *sensitive attribute* values, for example. While some FL-friendly methods work around this [14, 15], this is still a barrier to broad usage of bias mitigation techniques in federated learning.

In the next section, we will examine sources of bias in machine learning, and how addition sources come into play when we look at bias mitigation in federated learning.

8.3 Sources of Bias

The vast majority of bias mitigation approaches are designed with centralized machine learning in mind. This design approach, more often than not, means the method will require a level of data access that is not compatible with the privacy protocol of federated learning. As mentioned beforehand, evaluating machine learning models for bias oftentimes requires information, such as data points' *sensitive attribute* value. We will examine how machine learning models learn bias, and specifically analyze bias factors unique to federated learning models. Figure 8.2 provides an overview.

8.3.1 Centralized and Federated Causes

Both centralized and federated learning are impacted by what are called "traditional" sources of bias, which include but are not limited to: *prejudice, exclusion bias, negative legacy*, and *underestimation*.

Prejudice: Kamishima et al. [21] defines prejudice as a "statistical dependence between a sensitive variable... and the target variable... or a non-sensitive variable." This is further broken down into three types of prejudice: direct, indirect, and latent. *Direct prejudice* is found when a *sensitive attribute* is utilized in training the machine learning model. The model predictions are then defined by [25] as containing *direct discrimination*. This form of prejudice can be sidestepped by removing the sensitive attribute from the training set. This however leads us to *indirect prejudice*, where in the absence of direct prejudice there is a "statistical dependence between a sensitive variable and a target variable" [21]. In the case where our feature set, outside of the *sensitive attribute*, has little variance, the sensitive attribute is still highly correlated with the label. In another vein, *latent*

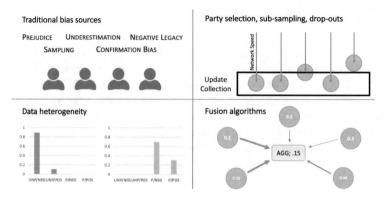

Fig. 8.2 Causes of Bias that Affect Federated Learning Models

prejudice is found when there is high correlation between the sensitive attribute and another feature, or multiple features. While in this case the sensitive attribute does not directly impact the label set Y, its impact still influences the predicted labels, and prejudice is felt.

Exclusion Bias: Exclusion bias [11] is the removing of data features, during data pre-processing or "cleaning," which results in information relevant to the model predictions being excluded from training. Ghoneim [11] uses an example of training a model with data of passengers on the Titanic, predicting whether a passenger would have survived the accident. In this example, randomized passenger ID number is removed during pre-processing as it is deemed inconsequential data. Unknown to the user training the model, passenger ID was correlated with passenger room number, in that passengers with larger ID numbers were passengers with rooms farther away from lifeboats, meaning it was more difficult to escape.

Negative legacy: Also known as sampling bias [16], negative legacy [21] is defined as data sampling or data labeling that is discriminatory. This is a cause of bias in machine learning models that is both quite broad, but potentially difficult to detect in application.

Underestimation: Underestimation occurs when a trained machine learning model is not fully converged, which is the effect of a training dataset that is limited in size. In [21], underestimation is measured by calculating the Hellinger distance between the training set distribution and model-predicted distribution of the training set.

8.3.2 Federated Learning-Specific Causes

In addition to traditional bias sources, federated learning has additional, unique, factors that contribute to a federated model's bias sources. These are defined in [1] as *data heterogeneity*, *fusion algorithms*, and *party selection and subsampling*.

8.3.2.1   Data Heterogeneity

Each party engaged in a federated learning protocol has its own training and testing set, used to train its local model. As federated learning requires data privacy, each party's data composition is unknown to other parties and unknown to the aggregator. Between them, parties may have very different data distributions, which differ largely from the overall composition. Certain federated learning processes allow for dynamic participation of parties, where over the course of the federated learning process, parties may leave and return for later rounds of training. Parties leaving and returning in the FL process may highly impact the overall data composition, both locally and holistically. It is not clear how much this affects the global model.

For example, say a chain of hospitals wants to use federated learning to train an image classifier for detecting heart disease. Each hospital, in a different location, trains their local model with their patients' data. A hospital in a predominantly minority neighborhood of a larger, predominantly non-minority city is likely to have a very different set of patients in its local dataset, relative to the overall composition of the hospital chain's set of patients [1].

8.3.2.2   Fusion Algorithms

The fusion algorithm dictates how party updates are combined, and subsequently incorporated into the global model. As such, they can influence the way that bias is measured in the final model. Some fusion algorithms perform a simple average of parties' model weights, while others perform different weighted averages, one for example is based on party size (i.e., parties with larger datasets influence the global model more than parties with smaller datasets) [23]. Depending on the application of the federated learning task, this could have negative effects on sensitive groups.

Researchers have proposed examining parties' updates against the global model's performance to calculate to what extent model updates will affect the global model's behavior. Many solely examine how model accuracy is affected, which does not inform about the effect on model bias. Some robust aggregation methods will exclude party replies altogether if dissimilar to those of other parties [7, 30], and in a real-world FL task would easily exclude a minority group, for example.

In the same hospital example as above, some hospitals training together have very different dataset sizes. This could be based on some hospitals being located in areas where the population is socioeconomically disadvantaged, thereby those hospitals have less patients that can afford an expensive medical procedure; these hospitals' datasets would be smaller. Involvement in a federated learning process that rewards larger party size with more global model influence would diminish these hospitals' contributions to the global model, and incorrectly give the impression that the model is comprehensively learning from the hospital chain's set of patients. This example can be easily reproduced for situations where other sensitive attributes like age, sex, or race can affect whether or not a user's data is systemically kept out of a federated learning task [1].

8.3.2.3 Party Selection and Subsampling

FL scenarios can usually be categorized into two main groups: (A) parties are small in data and the number of parties is very large, i.e., where parties are cell phones, or (B) parties are large in data and the number of parties is small, i.e., where parties are companies. At each round of training, the aggregator will query parties for their model updates, which are then incorporated into the global model. However, not all parties may be involved in every round of training [5, 30], which can introduce bias. Especially in FL tasks with a large number of parties, the aim may be to satisfy a quota of parties' updates to begin the next round of training. Depending on the federated learning task, different attributes, some bias-correlated, may affect whether a party is included in a training round.

Consider a scenario where a company wants to train a model to improve the user experience in its cell phone app, and engages users in a FL process. In this example, each phone is a party, and the question of which model updates are included is dependent on network speed. Faster devices (i.e., newer and more expensive) are likely to be represented at disproportionately higher rates than slower devices. Likewise, devices in regions with slower networks may be represented at disproportionately lower rates. Inclusion here is correlated with socioeconomic status, and is a systemic source of bias [1].

8.4 Exploring the Literature

8.4.1 Centralized Methods

As mentioned in Sect. 8.2.2, many bias mitigation methods are categorized into 3 categories: pre-processing, in-processing, and post-processing. Examples are included below:

Pre-processing: Reweighing [17] is a bias mitigation method that assigns weights to data points, based on the pairing of their sensitive attribute value and class label. These weights are calculated as the ratio of expected probability of the pairing, over the observed probability of the pairing.

$$W(s, y) := \frac{P_{exp}(s,y)}{P_{obs}(s,y)} = \frac{|(X \in D|S=s)||(X \in D|Y=y)|}{|(X \in D|S=s) \wedge Y=y||D|}, \ \forall s \in S, y \in Y. \tag{8.1}$$

Below is a sample calculation table for the Adult dataset [22]. In this dataset, the sensitive attributes are sex and race. The label is classification of income (above or below 50,000 dollars/year), and as women and people of color are both historically

Table 8.1 Sample reweighing weights calculation for *Adult* dataset

AGE	EDUCATION	SEX	RACE	CLASS	WEIGHT$_{sex}$	WEIGHT$_{race}$
21	MASTERS	FEMALE	WHITE	>50K	1.25	0.83
43	HS-GRAD	FEMALE	BLACK	≤50K	0.75	0.75
38	BACHELORS	MALE	WHITE	>50K	0.83	0.83
45	12TH	FEMALE	WHITE	≤50K	0.75	1.5
43	12TH	MALE	BLACK	≤50K	1.5	0.75
19	MASTERS	FEMALE	BLACK	>50K	1.25	1.25
61	BACHELORS	MALE	WHITE	>50K	0.83	0.83
29	ASSOC-VOC	MALE	BLACK	>50K	0.83	1.25

disadvantaged in relation to pay, the privileged groups are {Male, White} and underprivileged groups are {Female, Black}.

As seen in Table 8.1, reweighing weights change based on the targeted sensitive attribute, as this affects the calculated probabilities, which in turn changes the weights.

In-processing: Prejudice Remover [20] is an in-processing bias mitigation method proposed for centralized ML. This method incorporates a fairness-aware regularizer, $R(D, \Theta)$, to the logistic loss function as follows:

$$L(D; \Theta) + \tfrac{\lambda}{2}||\Theta||_2^2 + \eta R(D, \Theta). \tag{8.2}$$

$\mathcal{L}$ is the regular loss function, $||\Theta||_2^2$ is a ℓ_2 regularizer that protects against over-fitting, R is the fairness regularizer, Θ is the set of model parameters, and λ and η are regularization parameters. R minimizes the amount of bias the model learns by reducing the prejudice index [20], which measures the learned prejudice from the training dataset.

$$R = \sum_{(x_i, s_i) \in D} \sum_{y \in 0,1} M[y|x_i, s_i; \Theta] \ln \frac{\hat{P}r[y|s_i]}{\hat{P}r[y]}, \tag{8.3}$$

$M[y|x_i, s_i; \Theta]$ is the conditional probability of prediction, and $\hat{P}r$ is the sample distribution induced by the training dataset. Evaluating R requires knowledge of local data distribution and may result in data leakage if the evaluation is performed globally.

Post-processing: Reject Option-Based Classification (ROC) [19] is a bias mitigation method that changes classification labels based on a calculated "critical region."

Within this critical region, the method changes labels by assigning positive labels to underprivileged group members and negative labels to privileged group members.

Algorithm 8.1 Reject Option-Based Classification

Input: $\{F_k\}_{k=1}^K$ (K $\geq$ 1 probabilistic classifiers trained on D), X (test set), X^d (deprived group), θ

Output: $\{C_i\}_{i=1}^M$ (labels for instances in X)

Critical Region:

$\forall X_i \in \{Z | Z \in X, \max[p(C^+|Z), 1 - p(C^+|Z)] < \theta\}$:

if $X \in X^d$ **then**

  $C_i = C^+$

end if

if $X \notin X^d$ **then**

  $C_i = C^-$

end if

Standard Decision Rule:

$\forall X_i \in \{Z | Z \in X, \max[p(C^+|Z), 1 - p(C^+|Z)] \geq \theta\}$:

$C_i = \text{argmax}_{\{C^+, C^-\}}[p(C^+|X_i), p(C^-|X_i)]$

8.4.2 Adapting Centralized Methods for FL

In creating bias mitigation methods for federated learning, the primary concern is how to design methods that maintain privacy requirements, when many methods require access to the entire training set. This is not possible in FL. Figure 8.3 illustrates the different types of bias mitigation in a federated learning process.

In the previous section we discussed Reweighing, a pre-processing method used in centralized learning. It is adapted into two FL-friendly methods in [2], called *local reweighing* and *global reweighing with differential privacy*.

In *local reweighing*, each party computes reweighing weights, $W(s, y)$, $\forall s, y$, locally based on its own training dataset. Each party then has a unique set of reweighing weights, which are utilized for local training. Parties do not need to communicate with the aggregator and reveal neither their sensitive attributes nor data sample information; data privacy is maintained. Experiments in [2] demonstrate both high accuracy and effective bias mitigation with *local reweighing* and *partial local reweighing*, where only a subset of the parties employ the bias mitigation method. This makes the method particularly suited for federated learning, where parties may experience dropouts, or only a subset of parties may care to participate in bias mitigation practices.

In *global reweighing with differential privacy* (DP), global reweighing weights $W(s, y)$ are not unique to each party. If parties agree to share sensitive attribute information and noisy data statistics (noisy by adding DP), they can use this method to create a global set of weights that all parties use. Parties can control the amount of noise added to their data statistics via the value of the ϵ parameter, in the range [0,1]. In [2], experiments on the Adult dataset [22] demonstrate this method is both effective in mitigating bias and maintains high accuracy with ϵ values as low as 0.4.

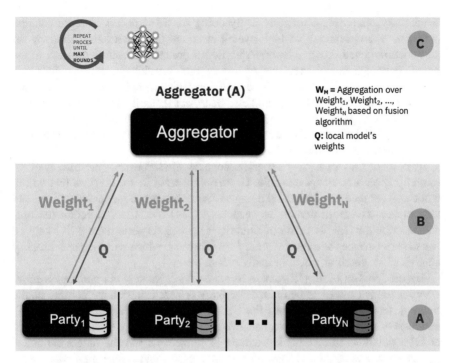

Fig. 8.3 Bias Mitigation in FL | (A) Pre-processing, (B) In-processing, (C) Post-processing

Unlike *local reweighing*, this method is not compatible with dynamic participation, as this would change the overall data composition and therefore the global weight calculations.

Federated prejudice removal, also proposed in [2], is a FL-friendly adaptation of Prejudice Remover [20]. Each party locally uses the Prejudice Remover algorithm to create a less biased local model and shares only the model parameters with the aggregator. The aggregator can then employ existing FL algorithms to update the global model, and bias is iteratively mitigated as training continues.

8.4.3 Bias Mitigation Without Sensitive Attributes

As mentioned in Sect. 8.3, prejudice has to do with a dependence of labels Y or feature set vectors x_i on a sensitive attribute S. Papers working around this have started to explore the creation of bias mitigation techniques without the presence of sensitive attributes in the training set.

Hashimoto et al. [14] works to create fairer models via distributionally robust optimization (DRO). This work utilizes DRO to minimize worst-case risk across all groups in the training set, over using empirical risk minimization (ERM), which only minimizes average loss. By doing so, ERM results in representation disparity, where a model has high overall accuracy, but much lower accuracy for an underprivileged group. The underprivileged group then contributes less to the global federated model; negative effects of this are elaborated on in Sect. 8.3. Representation disparity forms the definition of bias in [14], and by minimizing representation disparity, this method also addresses disparity amplification. This is where bias is magnified by a change in data distribution, in that the representation of the underprivileged group decreases over time because of poor model performance for underprivileged group members.

This work's usecase of a speech recognizer that does not perform well on minority accents is an apt example. Users do not want to use services that do not work well for them, and will stop doing so. With the speech recognizer, this would lead to less data from non-native speakers [3, 14] over time, affecting the user data distribution. Limited data of minority speakers suggests there will likely be worse performance for minority users going forward, which will increase disparity amplification, and is an effect of *underestimation*.

Hebert-Johnson et al. [15] works to create fairer models via *multicalibration*, where a trained model is calibrated for fairness based on defined subsets of the training data. This work frames the question around bias as where "qualified" members of an unprivileged group receive negative labels. To address this, a system of subpopulations of the training dataset is defined, C, and the model calibrates against each subpopulation. The calibration is used to guarantee predictions have high accuracy, by determining if a data sample has membership in a subpopulation defined in C. As C increases, the overall fairness guarantee increases. This method makes two assumptions that should be kept in mind. The first, that membership in a subpopulation is found "efficiently." The model checks for correlation between a data point and subpopulations in C when making predictions, and increasing the size of C complicates model efficiency. The second, that the data distribution demonstrates the underprivileged group is "sufficiently represented" to appear in random sample. This assumption serves as an issue with a common factor of biased machine learning algorithms: not enough data for underprivileged groups. This is what allows things such as *underestimation* to occur, and such an assumption places limits on method application.

These works have not been used in a federated learning setting. Applications with federated learning will require large scaling that may be difficult to incorporate into these methods [16].

8.5 Measuring Bias

It is difficult to answer the question of how something should be quantified when there is more than one way it is defined. In spite of this, multiple ways to quantify fairness and bias have been proposed. Many are based on manipulations of the confusion matrix, a table that classifies different aspects of performance between true and predicted labels. Below we have included calculations of eight popular bias metrics, as collected by [6], including *statistical parity difference, equal opportunity difference, average odds difference, disparate impact, Thiel index, Euclidean distance, Mahalanobis distance*, and *Manhattan distance*.

Statistical parity difference is calculated as the difference between the success rates between the underprivileged group and the privileged group. *Equal opportunity difference* is calculated as the difference between the true positive rates between the underprivileged group and the privileged group. The *average odds difference* is the average of the difference between the true positive rates and false positive rates between the underprivileged group and the privileged group. For these three metrics, the ideal value is 0. Negative values indicate bias against the underprivileged group, and positive values indicate bias against the privileged group. Bellamy et al. [6] defines a fairness region between −0.1 and 0.1, meaning metric values between these bounds are considered "fair." *Disparate impact* is calculated as the ratio of the success rates of the underprivileged group and the privileged group. These four metrics are some of the most in ML fairness literature. For *disparate impact*, the ideal value is 1, and [6] defines the fairness region with the bounds of 0.8 and 1.2.

Additionally, four other metrics are documented in this work. The *Thiel index* is calculated as the measure of entropy between the true and classifier-predicted labels. The *Euclidean distance* is calculated as the average of the Euclidean distance between the underprivileged group and the privileged group. *Mahalanobis distance* is calculated as the average of the Mahalanobis distance between the underprivileged group and the privileged group. And the *Manhattan distance* is calculated as the average of the Manhattan distance between the underprivileged group and the privileged group.

8.6 Open Issues

There are several open issues in the field of fairness and machine learning. One such issue is finding methods that can measure and reduce bias, without directly checking sensitive attributes [16]. While some approaches begin to address this [14, 15], scaling these are difficult, and options are few.

Another, more broadly, is how to design approaches that mitigate bias around multiple sensitive attributes simultaneously. This addresses the point of intersectionality. While a mitigation method reduces bias around one sensitive attribute, and evaluates the effectiveness against that single sensitive attribute, identities are multifaceted and are impacted by multiple factors. It would be naive to assume otherwise, and current bias mitigation methods have yet to address this.

A third open issue is the design of more bias approaches that can account for sensitive attribute value ranges that are not binary. Most methods are designed with one clear privileged group and one clear unprivileged group, which is a framing of bias often inconsistent with the real world. For the COMPAS dataset, for example, one of the first papers utilizing the dataset pre-processes the data so that for the race attribute, all non-white races are grouped together to make two categories, "White" and "Non-white." Not all people of color (POC) experience bias the same way, or to the same degree, and this approach incorrectly amasses and oversimplifies POC experiences, which consequently means bias will not be correctly measured or mitigated. In a parallel vein, the field lacks methods designed for individuals to have more than one sensitive attribute value, i.e., for someone of two or more races. Again, this means bias will not be correctly measured or mitigated.

8.7 Conclusion

In this chapter, we discussed the effects of social bias on machine learning. We explored multiple definitions of fairness, and terminology commonly used in the literature. We conducted a survey of the types of bias mitigation methods, as well as examples of each. Following this, we looked at different fairness metrics how they are calculated. Additionally, we examined sources of bias in machine learning algorithms, both sources that affect centralized and federated learning, and federated learning-specific causes. We discuss approaches that avoid training with sensitive attribute values. Lastly, we discussed open issues in the field, and topics for further research.

References

1. Abay A, Chuba E, Zhou Y, Baracaldo N, Ludwig H (2021) Addressing unique fairness obstacles within federated learning. AAAI RDAI-2021
2. Abay A, Zhou Y, Baracaldo N, Rajamoni S, Chuba E, Ludwig H (2020) Mitigating bias in federated learning. arXiv preprint arXiv:2012.02447
3. Amodei DEA (2012) Deep speech 2 end to end speech recognition in English and Mandarin. In: International conference on machine learning
4. Angwin J, Larson J, Mattu S, Mirchner L. There's software used across the country to predict future criminals. And its biased against blacks. https://github.com/propublica/compas-analysis. Accessed: 20219-10-08
5. Bellamy RK, Dey K, Hind M, Hoffman SC, Houde S, Kannan K, Lohia P, Martino J, Mehta S, Mojsilovic A et al (2018) AI fairness 360: an extensible toolkit for detecting, understanding, and mitigating unwanted algorithmic bias. arXiv preprint arXiv:1810.01943
6. Bellamy RK, Dey K, Hind M, Hoffman SC, Houde S, Kannan K, Lohia P, Martino J, Mehta S, Mojsilovic A et al (2018) AI fairness 360: an extensible toolkit for detecting, understanding, and mitigating unwanted algorithmic bias. arXiv preprint arXiv:1810.01943
7. Blanchard P, Guerraoui R, Stainer J et al (2017) Machine learning with adversaries: byzantine tolerant gradient descent. In: Advances in neural information processing systems, pp 119–129
8. Calders T, Verwer S (2010) Three Naive Bayes approaches for discrimination-free classification. Data Mining Knowl Disc 21(2):277–292
9. Dwork C, Hardt M, Pitassi T, Reingold O, Zemel R (2012) Fairness through awareness. In: Proceedings of the 3rd innovations in theoretical computer science conference, pp 214–226
10. Feldman M, Friedler SA, Moeller J, Scheidegger C, Venkatasubramanian S (2015) Certifying and removing disparate impact. In: Proceedings of the 21th ACM SIGKDD international conference on knowledge discovery and data mining, pp 259–268

11. Ghoneim S (2019) 5 types of bias & how to eliminate them in your machine learning project. Towards Data Science
12. Goh G, Cotter A, Gupta M, Friedlander MP (2016) Satisfying real-world goals with dataset constraints. In: Advances in neural information processing systems, pp 2415–2423
13. Hardt M, Price E, Srebro N (2016) Equality of opportunity in supervised learning. In: Advances in neural information processing systems, pp 3315–3323
14. Hashimoto T, Srivastava M, Namkoong H, Liang P (2018) Fairness without demographics in repeated loss minimization. In: International conference on machine learning
15. Hebert-Johnson U, Kim MP, Reingold O, Rothblum GN (2018) Multicalibration: calibration for the (computationally-identifiable) masses. In: International conference on machine learning
16. Kairouz P, McMahan HB, Avent B, Bellet A, Bennis M, Bhagoji AN, Bonawitz K, Charles Z, Cormode G, Cummings R et al (2019) Advances and open problems in federated learning. arXiv preprint arXiv:1912.04977
17. Kamiran F, Calders T (2011) Data preprocessing techniques for classification without discrimination. Knowl Inf Syst 33:1–33
18. Kamiran F, Karim A, Zhang X (2012) Decision theory for discrimination-aware classification. In: 2012 IEEE 12th international conference on data mining. IEEE, pp 924–929
19. Kamiran F, Karim A, Zhang X (2012) Decision theory for discrimination-aware classification. In: IEEE international conference on data mining
20. Kamishima T, Akaho S, Asoh H, Sakuma J (2012) Fairness-aware classifier with prejudice remover regularizer. In: Proceedings of the European conference on machine learning and principles and practice of knowledge discovery in databases
21. Kamishima T, Akaho S, Asoh H, Sakuma J (2012) Fairness-aware classifier with prejudice remover regularizer. In: Proceedings of the European conference on machine learning and principles and practice of knowledge discovery in databases
22. Kohavi R. Scaling up the accuracy of naive-bayes classifiers: a decision-tree hybrid. http://archive.ics.uci.edu/ml/datasets/Adult. Accessed: 30 Sept 2019
23. McMahan HB, Moore E, Ramage D, Hampson S et al (2016) Communication-efficient learning of deep networks from decentralized data. arXiv preprint arXiv:1602.05629
24. Merriam-Webster (2021) Fairness. https://www.merriam-webster.com/dictionary/fairness. Accessed: 10 Mar 2021
25. Pedreschi D, Ruggieri S, Turini F (2008) Discrimination-aware data mining. In: 14th international conference on knowledge discovery and data mining
26. Pleiss G, Raghavan M, Wu F, Kleinberg J, Weinberger KQ (2017) On fairness and calibration. In: Advances in neural information processing systems, pp 5680–5689
27. Sheth S, Gal S, Hoff M, Ward M (2021) 7 charts that show the glaring gap between men's and women's salaries in the US. Business Insider
28. Starr SB (2012) Estimating gender disparities in federal criminal cases. The social science research network electronic paper collection
29. LeCun Y, Cortes C, Burges CJ (2021) The MNIST database of handwritten digits. http://yann.lecun.com/exdb/mnist/. Accessed: 24 Feb 2021
30. Yin D, Chen Y, Ramchandran K, Bartlett P (2018) Byzantine-robust distributed learning: towards optimal statistical rates. arXiv preprint arXiv:1803.01498
31. Zafar MB, Valera I, Rodriguez MG, Gummadi KP (2015) Fairness constraints: mechanisms for fair classification. arXiv preprint arXiv:1507.05259
32. Zemel R, Wu Y, Swersky K, Pitassi T, Dwork C (2013) Learning fair representations. In: International conference on machine learning, pp 325–333

Part II
Systems and Frameworks

Part II of this book addresses the perspective of federated learning as a distributed system and how the system choices we make have impact on the outcome of the federated learning process.

Chapter gives an overview of federated learning system and specifically addresses issues of enterprise-oriented, cross-silo scenarios, comparing with those of embedded and mobile systems, cross-device scenarios.

Chapter 10 looks at system considerations for local training and aggregator scalability. Parties can run their local training clients on compute platforms that might not typically be used for model training, in particular on devices, phones, or small and local servers where data is present. With large numbers of participants, the aggregator system must also become scalable. The chapter discusses multiple strategies to address this.

Chapter then addresses *straggler management*, of particular importance in enterprise use cases when active management of delayed responses is important. Finally, Chap. 12 focuses on *participation fairness*. Different device performance and connectivity can introduce bias beyond what we find in centralized learning. This chapter presents techniques to ensure all parties are equally included during the model building process.

Chapter 9
Introduction to Federated Learning Systems

Syed Zawad, Feng Yan, and Ali Anwar

Abstract In this chapter, we introduce federated learning from a systems perspective. We go into the details of the different federated learning scenarios that have different system design considerations. We first introduce two most common but quite different federated learning scenarios, namely cross-device federated learning and cross-silo federated learning. Cross-device federated learning typically involves a significant number of parties (e.g., thousands to millions), who are usually less reliable and equipped with mobile or IoT devices that have various computing and communication capabilities. In cross-silo federated learning, the parties are usually a small number of organizations with ample computing power and reliable communications. We first describe the two very different problems that each of them address. We then describe the architectural differences between the two and their corresponding training steps. We also discuss the unique systems challenges that arise due to these properties and give a brief description of current works that have talked about these problems in detail.

9.1 Introduction

Federated Learning system aims at providing system support for training machine learning models collaboratively using distributed data silos such that privacy is maintained, and the model performance is not compromised [20, 23]. The key system design to support training models "in-place," which is quite different from conventional learning systems where the data is collected and managed centrally over a fully controlled distributed cluster. The biggest advantage of such an "in-place" training system is to facilitate privacy and security protections as the

S. Zawad (✉) · F. Yan
University of Nevada, Reno, Reno, NV, USA
e-mail: szawad@nevada.unr.edu; fyan@unr.edu

A. Anwar
IBM Research – Almaden, San Jose, CA, USA
e-mail: ali.anwar2@ibm.com

concerns of which have lead to new legislation such as the General Data Protection Regulation (GDPR) [39] and the Health Insurance Portability and Accountability Act (HIPAA) [32] that prohibits transferring user private data to a centralized location. Such design, however, brings significant new system challenges due to its unique training procedure and privacy and security properties. Data owners typically have intrinsic heterogeneity in both data and computing resources, which makes convectional wisdom of centralized learning system difficult to be adopted here. The "in-place" training method requires more complex coordination of computing and communication resources among data owners.

In general, these system architectures usually consist of either commercial clusters (such as those used by corporations to house their data) or on edge devices (such as sensor arrays and smart devices that store data upon usage). Due to the increasing prevalence of edge devices, the IoT networks are generating a wealth of data each day. Additionally, the growing computational power of these devices coupled with concerns over transmitting private information makes it increasingly attractive to store data locally and push network computation to the edge, which is fundamental to the design of federated systems. This concept of edge computing is not a new one. Indeed, computing simple queries across distributed, low-powered devices is a decades-long area of research that has been explored under the label of query processing in sensor networks, computing at the edge, and fog computing [28]. Recent works have also considered training machine learning models centrally but serving and storing them locally. For example, this is a common approach in mobile user modeling and personalization [7]. One significant challenge of such systems has been the lack of computational capabilities, which have restricted the kind of tasks they are able to perform [35, 36].

However, as pointed out it [2, 4, 31], the growth in the hardware capabilities of IoT devices has made it possible to train models locally. This has led to a growing interest in the feasibility of federated learning as a mainstream method for performing large-scale distributed training [46]. As we discuss in this chapter, learning in such a setting differs significantly from traditional distributed environments requiring fundamental advances in areas such as privacy, large-scale machine learning, and distributed optimization, and raising new questions at the intersection of diverse fields, such as machine learning and systems [4, 24, 46]. The majority of the challenges that were traditionally associated with large-scale distributed training applications are common to federated systems as well. For example, a common challenge for training clusters with a large number of nodes is that some might be slower than others, leading to an overall increase in training times or issues with convergence [22, 38, 47]. This can also occur with federated learning systems [6, 41, 43]. Similar challenges in synchronization schemes [27, 42], device scheduling [5, 6, 12], security [10, 17, 25], and resource efficiency [2, 6]. In addition to these challenges, there are also a full set of unique ones faced by federated systems, which stem from the fundamental difference between it and distributed machine learning—*the lack of control of the training end* [19].

In traditional distributed learning, the system is fully observable and the engineers have a certain level of control over the system such as how many nodes are available and the kind of hardware they have. However, since federated systems contain only the devices from users, the set of hardware available is diverse and dynamic, which adds an extra layer of complexity in terms of systems design. As such, much work has been done to address this problem [2, 6, 21, 37] and yet a lot of issues still need addressing [18].

9.1.1 Chapter Overview

For this chapter, we focus on providing a discussion of all such systems challenges of federated learning and what the state-of-the-art papers do to address them, along with their future works. We start by providing some background on general FL and then by describing the two major categories of federated learning systems— **Horizontal** or **Cross-device** FL and **Vertical** or **Cross-silo** FL. We then break the chapter into two major sections, one for each of the system types. We go into an in-depth discussion of their systems components and how they come together to create the overall architecture. We go into a step-by-step description of the full training process for each of the two types of systems. We then talk about their systems hurdles and the design factors that contribute to them. We then briefly take a look at the challenges faced by the parties during local training such as party selection policies, low computational throughput, etc. and mention the state-of-the-art papers that address them.

9.2 Cross-Device vs. Cross-Silo Federated Learning

We start the chapter by providing a general background on Federated Learning, and explain the two major categories of systems available for it. We discuss their architectures in detail and the step-by-step process for the training. We provide a detailed comparison between the two and discuss their advantages and disadvantages.

The seminal work in federated learning was proposed in [29]. In this paper, the authors give the reasoning for calling their proposed system "Federated Learning"— "...we term our approach Federated Learning, since the learning task is solved by a loose federation of participating devices (which we refer to as parties) which are coordinated by a central aggregator." The motivation for proposing such a system was the need for the capacity to train on data partitioned across a massive number of unreliable devices with limited communication bandwidth. The data cannot be shifted due to privacy constraints resulting in the need to train on them locally. However, the traditional way of using mini-batch SGD is a communication heavy method, which is a huge drawback for training on edge devices due to their limited resources. The authors suggest that instead of transferring the gradients per step, send the fully trained model weights, and perform aggregation on that. While this has the potential to adversely affect the convergence process itself, the significant reduction in communication rounds, bandwidth, and training time is considered a worthwhile trade-off. Additionally, this also enables the addition of differential noise and encryption mechanisms, which allows for a more secure training process. As such, this system has been widely adopted by the community and is considered the mainstream definition of federated learning systems.

The scope of such kind of systems is limited to training on IoT devices containing user-generated data. The other method for secured machine learning involves the shared training of a single model from different data banks in separate data silos. The separate data banks usually contain sensitive information collected by institutions such as financial data, medical records, transaction histories, etc., which have strict privacy restrictions. However, to develop advanced AI models we need vast quantities of data available in such data banks. Additionally, such data banks usually contain largely different data features for the same datapoint. For example, one hospital may contain the medical history of a certain person while another contains recent test results. If we wish to train a model that attempts to predict diseases based on history and test results, we will need to train on both kinds of data with the different features on different hardware. This is also a federated learning challenge but unlike training a model on IoT devices and aggregating on the central aggregator, we have to train different parts of the model in separate silos.

This requires a completely different type of system, and as such even though they are both federated learning, we can broadly categorize them into two separate types. The first type of FL performed on edge devices is called "Cross-device" or "Horizontal" Federated Learning, while the training of data with separate features is called "Cross-silo" or "Vertical" Federated Learning (we will use these terms interchangeably). We next describe each of these system architectures and training steps in details.

9.3   Cross-Device Federated Learning

9.3.1   Problem Formulation

The mainstream federated learning problem means training a single, global machine learning model on data, which is stored separately from each other. The number can range anywhere between single digits to thousands of devices. The model is trained with the limitation that the data must be stored locally and cannot be moved to other parties or even observed. This results in requiring the training of the global model locally on each device individually and then being aggregated in a centralized aggregation server. The paper [24] formalizes the objective function of cross-device federated learning as

$$minimize\ F(w),\ \text{where}\ F(w) = \sum_{k=1}^{n} p_k F_k(w) \tag{9.1}$$

Here, the $F_k(w)$ is the local objective function for device k with model weights w. The p_k is the importance given to the contribution of device k to the global model's objective function $F(w)$. The most common aggregation algorithm for cross-device FL called *FedAvg* [20] directly uses a version of the problem formulation. Specifically, they define the local objective function $F_k(w)$ as the loss minimization function traditionally associated with stochastic gradient descent training for a model with weights w. The term p_k of a device is simply set as

$$p_k = \frac{j_k}{\sum_{l=1}^{n} j_l} \tag{9.2}$$

where j_k is the total number of datapoints the local model w_k was trained on in device k. Therefore, the p_k can be considered as the weighted average of the number of datapoints per-device. For the rest of the chapter, we will use the term aggregation to refer to this *FedAvg* algorithm unless otherwise specified since it is the most commonly used in state of the art.

Using this definition, we can logically describe the objective of a federated learning system as given a set of data owners $k_1, k_2, ...k_n$ containing data $d_1, d_2, ...d_n$, we train the global model W_g using a weight-averaged aggregation of models $w_1, w_2, ..w_k$ trained on $d_1, d_2, ...d_k$. In contrast, the conventional method pulls all the available data together such that $D = d_1 \cup d_2... \cup d_n$ to train the model W_g without giving more weight to any k's model.

There are two additional soft requirements that all federated systems must fulfill. Firstly, it must ensure that no third-parties can access user-end data, and even if data is observed, it must not be able to be associated with a certain device. Secondly, the performance of a model derived using federated learning must be "close" to the performance of the model had it been trained using traditional methods. Fulfillment of these two criteria are additional challenges that complicate the system design aspects, and will be discussed in the following subsections as well.

9.3.2   System Overview

The typical federated learning system has not changed much from the original design proposed in [20]. While many newer works propose novel and technically complex architectures, they are usually variants of the standard party-aggregator system. Figure 9.1 shows the system diagram of a federated learning system.

The architecture usually contains two major components:

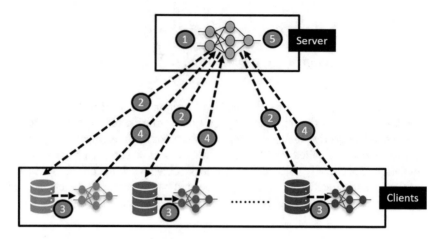

Fig. 9.1 System overview of a cross-device federated learning system

- **Aggregation Server**—This is typically owned by the third-parties who wish to train their models on the user-owned data. The hardware employed may reside in cloud systems or proprietary aggregation servers and does not usually require high-performance hardware to be deployed on. This server is also called the aggregation server or the aggregator due to its major function being the accumulation of the model weights. An aggregation server contains the global model G_n, which it updates periodically using the weights from the devices. As such, it must contain enough memory to store the weights for multiple models. It may also contain a load-balancer in order to manage the large amounts of device connections that may be required for certain systems. The hardware's computational capabilities also must be significant since the security measures can include computationally expensive operations. The aggregation algorithms also vary in complexity depending on the type being used. Certain variations of this system may include different topographies, profiling, and scheduling modules, which we will discuss in the later sections.
- **Parties**—This is the group of user-owned hardware and generally consists of Internet-of-Things devices such as cell phones, tablets, sensor arrays, and other such smart devices. As pointed out in [6], the hardware for each of these devices can vary widely due to the diversity of the hardware used by the data owners. Such devices usually contain very limited computational resources and high frequencies of downtime [4], resulting in the need for high-efficiency training techniques and frameworks to be employed. The party-side devices contain the user-generated data, the model training, and security systems. Since this device-side hardware cannot be controlled, most research work focus on optimizing the local training process.

9.3.3 Training Procedure

The training for a global model in cross-device federated learning occurs in rounds, where one round means one full training step. The steps within each round is explained below.

Step 1 As shown in Fig. 9.1, the first step is taken on the aggregator side. Initially at round 0, we start with an untrained global model G_0 with randomly initialized weights. At the start of each round, we select a subset of all available devices to train on. Which devices and how many significantly impact the training time, model performance, convergence time, and computational cost of the system. As such there have been many works that each provide different policies, which impact the training process of the system in different ways, and we provide a brief discussion on the state of the art on these aspects. The basic federated learning system implementations, however, generally select 10% of all available devices uniform randomly.

Step 2 Once k devices have been selected, the weights of the global model G_n are sent to each of these devices for training. This is a communication-expensive phase of the training process since deep models can be large and transferring them consumes significant portions of the bandwidth. This is especially a concern for IoT devices on metered or fragile connections as pointed out in [4].

Step 3 Once the global model G_n reaches each of the k devices training on their individual datasets D^k is done separately on each device, resulting in g_n^k model being generated on device k. This phase is computationally expensive on the party-side.

Step 4 After the local models have been generated, privacy mechanisms such as differential privacy [40] and secure aggregation [3] are applied in order to anonymize the models such that observation of the models will not reveal any information about the device or the dataset on which the model was trained one. Such mechanisms may be computationally expensive, and so adds another layer of overhead to the training procedure. The encrypted/noisy model weights from each of the devices are then sent back to the aggregator, which incurs more communication cost.

Step 5 The aggregator receives the models g_n^k from each of the devices and uses an aggregation algorithm (the standard being Eq. 9.2) to generate the new global model for the next round G_{n+1}. The process then goes back to **Step 1** and the process repeats from round $n + 1$.

The above steps are repeated until the stop conditions (usually a certain number of rounds or convergence criteria) are met.

9.3.4 Challenges

While the per-round process for cross-device federated learning is relatively simple, the scale and diversity of the underlying infrastructure add layers of complexity to the framework design decisions, which in turn introduces new and interesting systems challenges unique to federated learning. We can break down these challenges into various parts and discuss each of them briefly:

Aggregation Server In the cross-device setting, the aggregation server is usually a powerful central server. It has a lot of responsibilities since it is the main source of management in the system such as communication, aggregation, security, etc. Thus it is required to have a stable and reliable server. The system described above consists of a single server, and so has a single point of failure. As such, one of the challenges of a FL system is to provide a reliable server cluster and design the topography such that the dependence on a single server is removed.

Party Selection As mentioned in the previous section, one of the most impact-ful decisions for a cross-device system is how to select the devices. This is a challenge in the design of the system since it can influence many aspects of the process. One such challenge is that of *resource heterogeneity*. As pointed out in [4, 6, 23, 31], the hardware types can vary widely among parties due to their sheer diversity. This means that the number of devices can be significantly different from each other in terms of training time for the same model. Thus the overall training time of the full system is highly dependent on the latency of the stragglers in a system. Selecting devices consciously such that the training time is controllable (such as selecting faster devices only) is a design decision that needs to be made during party selection.

Each of the devices also have distinct computational limitations such as memory, bandwidth allocation, battery capacity, etc. Since one of the criteria of a FL system could be to make training as non-intrusive [4] for the end-user as possible, a party selection policy that keeps interference to a minimum becomes important. For example, if battery is low training a model may mean the device becoming unusable after a while for the user. As such, criteria must also be fulfilled in order to select a certain party. This also extends to the challenge of availability of devices. Not all devices are available at all times for training. As an example, we can think of solar-powered sensor arrays as devices that can only perform computation-heavy calculations only when enough power is being generated. As such, it is important for the aggregator to be aware of the availability patterns of participating devices and schedule them accordingly. Lastly, the *data heterogeneity* is a defining property of federated learning where device has different quantities and qualities of data, which results in a skewed training process. Some devices tend to have good data, while other devices do not. Deciding on how to choose devices such that biased training is mitigated while getting the most out of all the available data is one of the major challenges in Federated Learning.

Communication Communication can be a critical bottleneck in federated system networks since the actual training nodes (i.e., devices) have diverse communication capabilities. Previously, training on user data would mean transferring the data from the device to the central aggregator once and perform the training on it. However, now the need for privacy requires that the data cannot be transferred. This requires the frequency communication between the data-owning devices and the aggregator. Federated networks can potentially be comprised of a massive number of devices, e.g., millions of smart phones [4], and the network connections among the devices can be significantly (in some cases, orders of magnitudes [26]) slower than that of the local computation. This is mainly due to IoT devices having limited communication capacities. As such, it is often the case that the bottleneck is not due to the computation but the communication, so much so that it can constitute the majority of the training time. Therefore, the communication efficiency of federated systems is an important systems factor in engineering efficient federated systems.

Local Computation While there has been much work done for efficient data center training, papers talking about training models on resource-constrained devices is rare. In addition to the training computation, there are overheads for other parts of the system for encryption/decryption, calculating differential noise, compressing, serializing and sending the model, testing, and so on. Generally speaking, the IoT devices are not meant for such computation heavy workloads and so the local operations are usually un-optimized on both the software and hardware levels. Additional constraints such as limited bandwidth and power capacities, limiting computation and memory usage so as to not interfere with the user experience, managing hardware and OS security, etc. means special considerations must be taken when designing local training frameworks. Such considerations also vary widely depending on the infrastructure. For example, when training on a sensor array, the user experience is not a factor but very limited resources is. Training on a set of cell phones on the other hand would mean being conscious of consuming too much resources at a time, but the total computational capacity may not be a limitation. As such, the design of local frameworks is a very important aspect of federated learning systems.

Aggregation Scheme Lastly, we must consider the impact of the method for aggregation of the model parameters from the diverse set of end-devices. The first challenge for every federated aggregator is to ensure the privacy of the participants. The two most common low cost privacy-preserving techniques—differential privacy and secure aggregation—require a synchronized aggregation. The main idea for differential privacy is that a certain amount of noise is applied to the model weights such that it is indistinguishable from the other device weights. This requires a synchronized aggregation since multiple models must be trained and received at the same time in order to get an estimate of the model weights for each of the devices and their corresponding amount of noise applied. In the case of secure aggregation, it is an encryption method with the key being shared between multiple devices. It is only possible to decrypt the model weights once all the devices have reported back their keys making it a synchronized process as well. However, a synchronized aggregation scheme is much slower than asynchronous ones and so careful design decisions must be made in proposing novel aggregation schemes focusing on training latencies. Another challenge occurs when scaling up training to thousands of devices. Using a single aggregator becomes a bottleneck in such cases, and other novel aggregation schemes such as hierarchical aggregation must be employed to handle scalability. Aggregation schemes also impact the convergence rate and model performance [1, 24] and so must be balanced with mechanisms to reduce training time and computation costs. They can also determine the communication frequency of a system since their policies dictate how devices are trained. As such, developing an aggregation scheme is a delicate balancing act between the various properties of a federated systems and makes it an important yet challenging task.

Based on these factors, we can see that the design decisions that contribute to the development to a fully functioning, efficient cross-device federated learning system is a complex task given the sheer number of challenges, most of which also intersect to some degree. As such, research into better frameworks that mitigate these effects is an important direction. We discuss many such papers in the later sections and show that while there are state-of-the-art approaches to address one of the issues or another, there are yet to be any comprehensive solutions to all of the challenges and much is yet left as future work.

9.4 Cross-Silo Federated Learning

9.4.1 Problem Formulation

The main objective for a cross-silo federated learning system is the same as that of a cross-device system—*a global model that must be trained on disparate data.* The major difference is how the data is structured. Cross-silo (also called vertical federated learning or feature-based federated learning) is applicable when there are two or more datasets that share the ID space but possess different features. For example, we can consider the scenario with two banks who wish to train a model for predicting their user's credit purchase behaviors. One bank has the data for the users' assets and income, while the other bank contains their spending history. Thus the features available for each data silo is different, but the user IDs intersect. In contrast, a cross-device federated learning system would be the scenario where both the banks have user purchase history but different sets of users. We can formally define the objective of vertical federated learning as

$$minimize \ L(W) = \sum_{j=1}^{m} \sum_{i=1}^{n} L(w_j, x_j^i) \qquad (9.3)$$

where $L(W)$ is the global loss function for the global model W, i is the user ID, j is the data provider (i.e., silo) with the local data provider model w_j trained from data feature x_j^i. The equation can be considered as the process of aggregating the training of the different features from different silos with privacy-preserving constraints, and these additional criteria can also be formalized as explained in [46]

$$X_a \neq X_b, \ Y_a \neq Y_b, \ I_a = I_b, \ \forall D_a, D_b \text{ where } a \neq b \qquad (9.4)$$

where X are the feature labels for the data silos a and b, Y are the corresponding labels, I is the user id, and D is the data. Thus, we say that the cross-silo federated learning process is the shared training of a global model between two or more separate silos such that they have different features and labels for the same datapoint across all available data.

There are two additional soft-constraints on the cross-silo federated learning system as well. The first deals with privacy-preservation, stating that none of the parties involved in the training process can associate another party's datapoint ID with its own. As an example using our previous scenario, the first bank trains on its own data and sends the partially trained gradients to the second bank. The second bank would associate its own datapoint IDs with the output gradients (explained in detail in the next sections) but should not be able to interpret the partial gradients to backward-engineer the actual data present in the first bank for that datapoint. This is done in practice using homomorphic encryption—a method of transforming the data such that a mathematical operation on it would result in the same output had it not been transformed but such that the original data is not observable. The second soft-constraint is that the performance of the model trained in this way must be as close to the performance of a model that would have been generated had all the data been present in a single silo as in traditional distributed learning.

9.4.2   System Overview

The cross-device federated learning system is slightly more complicated than the cross-device federated learning system. Unlike the cross-device system, there was no one paper that proposed the definitive architecture for the cross-silo federated learning system. However, over time the industry has developed a standard that is used in a significant majority of the current frameworks. Figure 9.2 shows the system diagram of this federated learning system.

This system also contains two major parts:

Third-Party Aggregator Server This server is separate from the data-owning parties and is owned by a third-party to coordinate the training process and manage the security aspects. They are typically commercial-grade hardware clusters due to the amount of management and computation it has to perform. While this server does not usually perform any operations on the models to be trained themselves in the vast majority of the cases, a significant amount of its workload consists of encryption/decryption of homomorphic encryption, which is extremely expensive. With larger number of participants, this computational cost increases exponentially. This server does not contain any models, data, or training code. However, it may

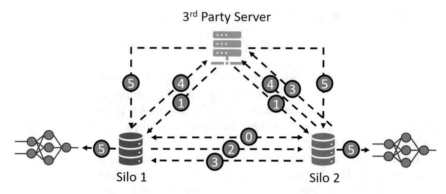

Fig. 9.2 System overview of a cross-silo federated learning system

also require load-balancers depending on the communication load from each of the silos. Some variations of this server also perform other tasks such as evaluation and sanity checks, which further increases the computation cost.

Data Silos Data Silos are the group of user-owned hardware, which contains the proprietary data on which the global model must be sharedly trained. As pointed out in [30], the hardware for each of the silos is usually commercial-grade, less heterogeneous, and have similar training latencies. They also do not have significant computational limitations or long periods of downtime, making them much more suitable for long-term training. The silo-owners also have more control over their architecture. As such, many of the challenges faced by cross-device federated learning such as stragglers, dropouts, and disparate datasets are not relevant to this scenario. However, as will be explained in the later sections, the training process between silos is performed in mini-batches, and the gradients must be first calculated on one silo and then sent to the others in a sequence. This is done until all the data from all the silos have been used to calculate the final loss, after which backpropagation is performed on each silo's models. As such, an increased number of silos can result in significant increase in training latency and so that provides unique system challenges as well. Additionally, the frequent transfer of gradients makes communication overhead a concern, and homomorphic encryption used for every transfer means a computation cost is added on top of the communication cost. As such, a significant portion of the systems research in cross-silo federated learning is focused on reducing these costs.

9.4.3 Training Procedure

The training for cross-silo federated learning also takes place in rounds, but with one key difference being that all data silos participate in every round of training. The process is also more sequential in that partial gradients and losses are exchanged between each of the participants. In the following example, we use two silos to make it simple to understand, but this can be expanded to multiple parties. The training steps (as simplified in [24, 30, 46]) are as follows.

Step 0 Before starting the training loop, the participants must align their datasets using an anonymous data alignment technique. Methods using privacy-preserving protocols [25], secure multi-party communication [34], key-sharing [10], and randomized responses [11] are used for such cases. These methods are usually computationally expensive but generally this step is only run once at the beginning of the training process and so is not too costly in the long run. The output of this step is that the data between the parties are matched such that the datapoint indices are the same for the same ID. Those that could not be matched are usually discarded for training and the privacy for each of the datapoints preserved [44].

Step 1 The third-party aggregator generates and sends encryption key-pairs to each of the participants for secure communication, and the "partial models" that need to be trained on each party is initialized. We use the term "partial model" since, generally speaking, in cross-silo federated learning it is assumed that only the last party in the training chain contains the labels. The other parties contain only the training features and the part of the model that can perform a forward pass using those data features.

Step 2 The first party (Silo 1 in Fig. 9.2) trains on one mini-batch of its local data and generates the output from the forward pass, which is then encrypted via homomorphic encryption and sent to the next party (Silo 2).

Step 3 Silo 2 then uses its own data to do a forward pass. For simplicity, let us assume that this party contains the labels, and so the calculation of the loss function happens here. The intermediate outputs are sent to Silo 1 (after applying homomorphic encryption), which will be later required to update the Silo 1 model weights. The loss is sent to the third-party aggregator.

Step 4 Both silos then calculate their partial gradients or intermediate outputs, add another layer of encryption masks on them, and send them to the third-party aggregator. These partial gradients or intermediate results can be a variety of values depending on the implementation details of the training process. They can include the output vector of the last operations for the partial model [8], the intermediate predictions per silo [26], or even the gradients from estimated loss [45]. The extra masks are added such that one silo cannot gain information about the other party's data from these intermediate outputs.

Step 5 The third-party aggregator decrypts the mask from each of the received intermediate data and uses this with the loss values to determine the exact gradient for each of the partial models in all the participants and sends them to their corresponding silos. The silos then use these received gradients to update their local models individually, generating new models. Then it loops back to Step 1 and begins the process all over again.

The above steps iterate until their convergence criteria are met. Note that many newer systems can have distinctly different frameworks. For example, Chen et al. [8] makes this process asynchronous and does not wait for all the parties to send their intermediate outputs to get the gradients from the third-party aggregators, while [48] does away with the third-party coordinator entirely. However, the overall structure of the systems is still the same—each of the parties perform partial training of the full model and must then be consolidated into calculating the full model's gradients. The gradients are sent to their corresponding parties to update their local partial models, and all of these must be done in a privacy-preserving manner.

9.4.4 Challenges

The standard cross-silo federated learning system shares many of the systems challenges faced in traditional distributed learning systems. For example, the frequent communication overheads due to the transfer of mini-batch gradients, coordination between each of the training nodes, stragglers, etc. are applicable to cross-silo training systems too. In addition to these, there are a few unique challenges to such systems as well.

Resource Heterogeneity Generally speaking, organizations that wish to participate in vertical federated learning tend to have more powerful hardware compared to those in cross-device [8, 24, 46]. As such, the actual training steps are more efficient. However, as described above, the full process is still an inherently synchronous and sequential process, which needs to have each and every silo participate in every step. This means that even one straggler can significantly reduce the overall training latency throughout the whole training period. The training resources for each of the silos must be carefully managed such that performance bottlenecks can be avoided. It must be mentioned, however, that this is a much more manageable problem than for cross-device federated learning. Cross-silo federated learning systems tend to have much more control over their underlying systems, making it easier to handle stragglers.

Single Point of Failure Each of the silos perform a part of the full training, and they are all required to train their portions completely in order to achieve the fully trained model. As pointed out in [8, 13, 49], this inter-dependency and the sequential nature of the per-step training process sets up the system in such a way that failure of any one node will result in the complete failure of the full system. In cross-device federated learning (or even traditional distributed learning to some extent), a device or node failure will not cause the full system to halt since other resources exist for training, but every node or silo is important here in cross-silo federated learning. As such, special systems design considerations such as backup nodes need to be considered for cross-silo frameworks.

Security Overheads There are three mainstream methods for ensuring the privacy of federated learning systems. The most common one being *Differential Privacy* [1, 33]. While it performs sufficiently well for cross-device federated learning with acceptable runtimes [40], recent works have demonstrated it is not effective for cross-silo federated learning due to it having stronger privacy constraints [49]. The other commonly used technique is *Secure Aggregation* [3, 4], but it has two major drawbacks. Firstly, it can allow the third-party aggregators to directly observe the gradients incoming from each of the silos, which in turn can reveal confidential information of their datasets. Secondly, the encryption process involves key-sharing, meaning that each of the silos need to report with their corresponding parts of the keys, which puts a strong limitation on the federated learning systems to become synchronous [49]. The only other option available is *Homomorphic Encryption* [15, 17]. However, it is extremely computationally expensive. In some cases, it can take up as much as 80% of the per-mini-batch computation time [49] and is therefore considered the single largest expensive operation for each silo, making it the most important factor that determines the total training time. While many works have been proposed to reduce this overhead without compromising privacy [9, 16, 49], this overhead still takes up a large majority of the computation time. As such, reducing the security overhead for cross-silo systems is considered the biggest challenge to designing more efficient federated learning systems.

The above challenges make for interesting directions for systems research. However, currently the overwhelming amount of research in cross-silo federated learning focuses on the privacy enhancement and overhead reduction due to it being the largest factor detrimental to the system's efficiency. The other challenges share some similarities to other types of systems such as cloud storage (for mitigating single point of failure issues) and traditional distributed systems (for addressing stragglers). As such, we find that cross-device federated learning provides a wider variety of unique and interesting challenges compared to cross-silo federated learning, and so we focus on the cross-device scenario for the rest of the sections. Next, we break down the complete cross-device FL system into parts and discuss their systems aspects in detail.

9.5 Conclusion

While many more types of Federated Learning systems are being proposed as we speak, they are usually a variation of these two major categories. Completely new Fl techniques such as Federated Neural Architecture Search have recently been proposed, but there are relatively new fields and not enough work has been done for them to become major categories like Vertical and Horizontal FL. Additionally, many of these newer types of systems are usually a variation or an incremental improvement on these two traditional architectures. For example, the recently proposed Federated Transfer Learning [14] can be categorized as a more complex type of Vertical Federated Learning since it uses transfer learning to teach a single model from disparate feature sets. As such, the overview given in this chapter for the two major types of Federated Learning from a systems perspective is a great starting point in understanding the various available Federated Learning frameworks.

References

1. Abadi M, Chu A, Goodfellow I, McMahan HB, Mironov I, Talwar K, Zhang L (2016) Deep learning with differential privacy. In: Proceedings of the 2016 ACM SIGSAC conference on computer and communications security. ACM, pp 308–318
2. Balakrishnan R, Akdeniz M, Dhakal S, Himayat N (2020) Resource management and fairness for federated learning over wireless edge networks. In: 2020 IEEE 21st international workshop on signal processing advances in wireless communications (SPAWC). IEEE, pp 1–5
3. Bonawitz K, Ivanov V, Kreuter B, Marcedone A, McMahan HB, Patel S, Ramage D, Segal A, Seth K (2017) Practical secure aggregation for privacy-preserving machine learning. In: Proceedings of the 2017 ACM SIGSAC conference on computer and communications security. ACM, pp 1175–1191
4. Bonawitz K, Eichner H, Grieskamp W, Huba D, Ingerman A, Ivanov V, Kiddon C, Konecný J, Mazzocchi S, McMahan B, Overveldt TV, Petrou D, Ramage D, Roselander J (2019) Towards federated learning at scale: System design. In: Talwalkar A, Smith V, Zaharia M (eds) Proceedings of machine learning and systems 2019, MLSys 2019, Stanford, CA, USA, March 31–April 2, 2019, mlsys.org. https://proceedings.mlsys.org/book/271.pdf
5. Caldas S, Konečny J, McMahan HB, Talwalkar A (2018) Expanding the reach of federated learning by reducing client resource requirements. Preprint. arXiv:181207210
6. Chai Z, Ali A, Zawad S, Truex S, Anwar A, Baracaldo N, Zhou Y, Ludwig H, Yan F, Cheng Y (2020) Tifl: A tier-based federated learning system. In: Proceedings of the 29th international symposium on high-performance parallel and distributed computing, pp 125–136
7. Chan Z, Li J, Yang X, Chen X, Hu W, Zhao D, Yan R (2019) Modeling personalization in continuous space for response generation via augmented wasserstein autoencoders. In: Proceedings of the 2019 conference on empirical methods in natural language processing and the 9th international joint conference on natural language processing (emnlp-ijcnlp), pp 1931–1940
8. Chen T, Jin X, Sun Y, Yin W (2020) Vafl: a method of vertical asynchronous federated learning. e-prints. arXiv–2007
9. Cheng K, Fan T, Jin Y, Liu Y, Chen T, Papadopoulos D, Yang Q (2019) Secureboost: A lossless federated learning framework. Preprint. arXiv:190108755
10. Du W, Atallah MJ (2001) Secure multi-party computation problems and their applications: a review and open problems. In: Proceedings of the 2001 workshop on new security paradigms, pp 13–22
11. Du W, Zhan Z (2003) Using randomized response techniques for privacy-preserving data mining. In: Proceedings of the ninth ACM SIGKDD international conference on Knowledge discovery and data mining, pp 505–510
12. Dwork C, Hardt M, Pitassi T, Reingold O, Zemel R (2012) Fairness through awareness. In: Proceedings of the 3rd innovations in theoretical computer science conference, pp 214–226
13. Feng S, Yu H (2020) Multi-participant multi-class vertical federated learning. Preprint. arXiv:200111154
14. Gao D, Liu Y, Huang A, Ju C, Yu H, Yang Q (2019) Privacy-preserving heterogeneous federated transfer learning. In: 2019 IEEE international conference on big data (Big Data). IEEE, pp 2552–2559
15. Gentry C et al (2009) A fully homomorphic encryption scheme, vol 20. Stanford University, Stanford

16. Hao M, Li H, Xu G, Liu S, Yang H (2019) Towards efficient and privacy-preserving federated deep learning. In: ICC 2019-2019 IEEE international conference on communications (ICC). IEEE, pp 1–6
17. Hardy S, Henecka W, Ivey-Law H, Nock R, Patrini G, Smith G, Thorne B (2017) Private federated learning on vertically partitioned data via entity resolution and additively homomorphic encryption. Preprint. arXiv:171110677
18. Hosseinalipour S, Brinton CG, Aggarwal V, Dai H, Chiang M (2020) From federated to fog learning: Distributed machine learning over heterogeneous wireless networks. IEEE Commun Mag 58(12):41–47. https://doi.org/10.1109/MCOM.001.2000410
19. Kairouz P, McMahan HB, Avent B, Bellet A, Bennis M, Bhagoji AN, Bonawitz K, Charles Z, Cormode G, Cummings R et al (2019) Advances and open problems in federated learning. Preprint. arXiv:191204977
20. Konecnỳ J, McMahan HB, Felix XY, Richtárik P, Suresh AT, Bacon D (2016) Federated learning: Strategies for improving communication efficiency. CoRR
21. Lalitha A, Shekhar S, Javidi T, Koushanfar F (2018) Fully decentralized federated learning. In: Third workshop on bayesian deep learning (NeurIPS)
22. Li C, Shen H, Huang T (2016) Learning to diagnose stragglers in distributed computing. In: 2016 9th workshop on many-task computing on clouds, grids, and supercomputers (MTAGS). IEEE, pp 1–6
23. Li X, Huang K, Yang W, Wang S, Zhang Z (2019) On the convergence of fedavg on non-iid data. In: International conference on learning representations
24. Li T, Sahu AK, Talwalkar A, Smith V (2020) Federated learning: Challenges, methods, and future directions. IEEE Signal Process Mag 37(3):50–60
25. Liang G, Chawathe SS (2004) Privacy-preserving inter-database operations. In: International conference on intelligence and security informatics. Springer, pp 66–82
26. Liu, Y., Kang, Y., Zhang, X., Li, L., Cheng, Y., Chen, T., . . . & Yang, Q. A Communication efficient vertical federated learning framework. 2019. arXiv preprint arXiv:1912.11187
27. Lo SK, Lu Q, Zhu L, Paik Hy, Xu X, Wang C (2021) Architectural patterns for the design of federated learning systems. Preprint. arXiv:210102373
28. Mao Y, You C, Zhang J, Huang K, Letaief KB (2017) A survey on mobile edge computing: The communication perspective. IEEE Commun Surv Tutorials 19(4):2322–2358
29. McMahan HB, Moore E, Ramage D, Hampson S et al (2016) Communication-efficient learning of deep networks from decentralized data. Preprint. arXiv:160205629
30. McMahan HB, et al (2021) Advances and open problems in federated learning. Found Trends® Mach Learn 14(1):1
31. Nishio T, Yonetani R (2019) Client selection for federated learning with heterogeneous resources in mobile edge. In: ICC 2019-2019 IEEE international conference on communications (ICC). IEEE, pp 1–7
32. O'herrin JK, Fost N, Kudsk KA (2004) Health insurance portability accountability act (hipaa) regulations: effect on medical record research. Ann Surg 239(6):772
33. Pathak MA, Rane S, Raj B (2010) Multiparty differential privacy via aggregation of locally trained classifiers. In: NIPS, Citeseer, pp 1876–1884
34. Scannapieco M, Figotin I, Bertino E, Elmagarmid AK (2007) Privacy preserving schema and data matching. In: Proceedings of the 2007 ACM SIGMOD international conference on Management of data, pp 653–664
35. Shi W, Dustdar S (2016) The promise of edge computing. Computer 49(5):78–81
36. Shi W, Cao J, Zhang Q, Li Y, Xu L (2016) Edge computing: Vision and challenges. IEEE Internet Things J 3(5):637–646
37. Sprague MR, Jalalirad A, Scavuzzo M, Capota C, Neun M, Do L, Kopp M (2018) Asynchronous federated learning for geospatial applications. In: Joint European conference on machine learning and knowledge discovery in databases. Springer, pp 21–28
38. Tandon R, Lei Q, Dimakis AG, Karampatziakis N (2017) Gradient coding: Avoiding stragglers in distributed learning. In: International conference on machine learning, PMLR, pp 3368–3376

39. Tankard C (2016) What the GDPR means for businesses. Netw Secur 2016(6):5–8
40. Wei K, Li J, Ding M, Ma C, Yang HH, Farokhi F, Jin S, Quek TQ, Poor HV (2020) Federated learning with differential privacy: Algorithms and performance analysis. IEEE Trans Inf Forensics Secur 15:3454–3469
41. Wu W, He L, Lin W, Mao R, Maple C, Jarvis SA (2020) Safa: a semi-asynchronous protocol for fast federated learning with low overhead. IEEE Trans Comput 70:655
42. Xie C, Koyejo S, Gupta I (2019) Asynchronous federated optimization. Preprint. arXiv:190303934
43. Xu Z, Yang Z, Xiong J, Yang J, Chen X (2019) Elfish: Resource-aware federated learning on heterogeneous edge devices. Preprint. arXiv:191201684
44. Xu R, Baracaldo N, Zhou Y, Anwar A, Joshi J, Ludwig H (2021) Fedv: Privacy-preserving federated learning over vertically partitioned data. e-prints, pp arXiv–2103
45. Yang K, Fan T, Chen T, Shi Y, Yang Q (2019) A quasi-newton method based vertical federated learning framework for logistic regression. Preprint. arXiv:191200513
46. Yang Q, Liu Y, Chen T, Tong Y (2019) Federated machine learning: Concept and applications. ACM Trans Intell Syst Technol (TIST) 10(2):12
47. Yang R, Ouyang X, Chen Y, Townend P, Xu J (2018) Intelligent resource scheduling at scale: a machine learning perspective. In: 2018 IEEE symposium on service-oriented system engineering (SOSE). IEEE, pp 132–141
48. Yang S, Ren B, Zhou X, Liu L (2019) Parallel distributed logistic regression for vertical federated learning without third-party coordinator. Preprint. arXiv:191109824
49. Zhang C, Li S, Xia J, Wang W, Yan F, Liu Y (2020) Batchcrypt: Efficient homomorphic encryption for cross-silo federated learning. In: 2020 USENIX annual technical conference (USENIX ATC 20), pp 493–506

Chapter 10
Local Training and Scalability
of Federated Learning Systems

Syed Zawad, Feng Yan, and Ali Anwar

Abstract In this chapter, we delve deeper into the systems aspects of Federated Learning. We focus on the two main parts of FL—the participating devices (parties) and the aggregator's scalability. First, we discuss the party-side, where we look into details about various factors that impact local training such as computational resources, memory, network, and so on. We also briefly talk about how there are challenges present in each of these aspects and introduce the state-of-the-art papers that address them. Then we discuss how to develop large-scale Federated Learning aggregation systems. We talk about various aggregation schemes in current literature that aim at reducing the scalability challenges. We discuss each of their advantages and disadvantages and suggest scenarios for which they are most applicable. We also provide a list of state-of-the-art works that use these schemes.

10.1 Party-Side Local Training

In this section, we will discuss the systems complications on the local training side. Here, we break down the resources required for local training into three parts—computation, memory, and network—and then discuss their complications along with the consequent state-of-the-art techniques developed to address them.

We focus specifically on local training, which is the most computationally expensive part of the FL system. It determines the overall time it takes for the full training to complete as well as the amount of resources consumed and therefore is the most important factor to consider when making FL system design decisions.

However, as mentioned earlier, we have very little to no control over the hardware and availability of the parties. In the general case, the cross-device systems are

S. Zawad (✉) · F. Yan
University of Nevada, Reno, Reno, NV, USA
e-mail: szawad@nevada.unr.edu; fyan@unr.edu

A. Anwar
IBM Research – Almaden, San Jose, CA, USA
e-mail: ali.anwar2@ibm.com

designed under the expectation that the resources are significantly less than those available in standard clusters. Additionally, the party-side down times and training interruptions are assumed to be significantly more frequent [1, 2, 20, 29]. Many recent works have been proposed to address these issues, and we discuss a few of the state of the art next. While most of these works were directly introduced in the scope of federated learning, some are taken from other intersecting fields such as edge-computing and can be directly applicable.

10.1.1 Computation

The first most important determinant of training time is computation speed. As such, we look at the state of the art in Federated Learning and Edge-computing that efficiently utilizes them.

The first paper we talk about is *Model Pruning Enables Efficient Federated Learning on Edge Devices* [11]. In this paper, the authors propose a new FL paradigm called PruneFL which aims to reduce the model sizes themselves to reduce overall local training cost. While model pruning is a commonly used strategy in ML, this framework stands out in that it uses the party data to perform the pruning. They do so in two ways—*Distributed* and *Adaptive* pruning. *Distributed* pruning works by including initial pruning on a selected party's model by taking advantage of sparse matrices. Simply put, they throw out the operations whose weights end up training to 0 locally. However, local pruning can be damaging due to data heterogeneity since these weights may vary significantly between parties. In order to mitigate that issue, the authors propose the next step as *Adaptive* pruning. In this stage, the framework continuously ensure that a model's weights are not pruned too much such that the overall performance drops by keeping track of accuracy over time. In doing so, it maintains a fine balance between reducing computation and maintaining performance.

We next talk about *SLIDE: In Defense of Smart Algorithms over Hardware Acceleration for Large-Scale Deep Learning Systems* [3], where the authors here present the framework called SLIDE (Sub-LInear Deep learning Engine). It focuses on efficient deployment of models on the kernel level instead of making models efficiently scale with hardware. This is a highly relevant work in Federated Learning since increasing hardware resources is not an option and it is important to be able to squeeze out every bit of performance algorithmically. The authors use a unique blend of smart randomized algorithms, multi-core parallelism, and workload optimizations. They propose a significant amount of updates to the current machine learning frameworks such as making OpenMP more efficient, applying several novel algorithmic and data-structural choices in designing the LSH based sparsification, utilizing sparse gradient updates to achieve negligible update conflicts, etc. They show that using just a CPU, the number of computations can be drastically reduced to get significant increases in efficiency.

Lastly, we discuss *Accelerating Slide Deep Learning on Modern CPUs: Vectorization, Quantizations, Memory Optimizations, and More* [5]. This work is an improvement on SLIDE. The authors here show how SLIDE's computations allow for a unique possibility of vectorization via AVX (Advanced Vector Extensions)-512, which was not previously utilized. They highlight opportunities for different kinds of memory optimizations such as sparse updates, quantizations, detecting highly active vs. inactive parameters, etc. They too demonstrate that there is much room for improvement on the software side of machine learning and is therefore a very relevant work in making CPUs more efficient.

10.1.2   Memory

Our next challenge in local training is the limited memory size available on parties. The problem here is similar to that of computation in that we have little to no control over it, and that they are usually magnitudes of less powerful than in conventional clusters. As such, here we present a few of the state of the art in addressing this issue.

An important paper here is called *DeepX*: A software accelerator for low-power deep learning inference on mobile devices [14]. It is one of the seminal papers on taking a software-first approach to efficiency; DeepX focuses on the development of deep learning models on mobile devices. While they provide a few improvements from the computation side, their major contribution is in the utilization of memory. They provide two solutions. First, they apply *Runtime Layer Compression* (RLC), which provides runtime control of the memory and computation (along with energy as a side-effect). They simply reduce the operations layer-wise such that only the most important operations use larger bytes. Second, they use *Deep Architecture Decomposition* (DAD), which efficiently identifies unit-blocks of an architecture and allocates them to local or remote memory depending on access frequency. This further utilized memory for efficient model storage. It is unusual to have distributed memory in FL parties, and so DeepX providing algorithms to efficiently use limited memory for deep models is an important work.

Another such paper is *FloatPIM*: In-memory acceleration of deep neural network training with high precision [9]. This work provides a solution at the interface level between software and hardware. Processing in-memory (PIM) is a technique which exploits analog characteristics of non-volatile memory to support matrix multiplication in memory by transferring the digital input data into an analog domain and pass the analog signal through a crossbar ReRAM to compute matrix multiplication. The authors demonstrate that it is possible to reduce the floating-point memory using this technique and propose a framework that can reduce deep model's memory within computational error-bounds. This can be easily integrated in FL parties and provide a trade-off between memory and accuracy depending on the situation.

The last paper we talk about here is *Exploring Processing In-Memory for Different Technologies* [7]; another work that attempts to utilize PiMs but focuses more on making this system applicable to different types of memory. They propose designs which enable PIM in the three major memory technologies—SRAM, DRAM, and non-volatile memories (e.g., NVMs). They exploit the analog properties of different memories to implement logic functions like OR, AND, and majority inside memory. It is then extended to further implement in-memory addition and multiplication. As pointed out in this paper, it is important for machine learning libraries to support such systems since IoT devices can have a variety of memory types.

10.1.3 Energy

Energy efficiency is also one of the major concerns when designing efficient ML training algorithms due to FL parties generally being battery-powered devices. Deep learning is inherently very power-consuming due to the large amounts of computation that needs to be performed. As such, there has been a relatively large amount of work in energy efficiency for IoT devices in the FL scope compared to works for CPU and memory. We present the state of the art in such system here.

In the paper *Energy efficient federated learning over wireless communication networks* [30] the authors argue that the learning process and the communication frequency are the key ingredients that determine the energy efficiency of federated learning systems. This is because they both indirectly impact the number of local training steps, which affects the total energy consumption. They then pose the problem as a joint learning and communication optimization problem with the goal of minimizing energy consumption under latency constraints. Based on this problem definition, they develop an iterative algorithm which derives a closed-form solution for time allocation, bandwidth allocation, power control, computation frequency, and learning accuracy for every local step. However, the iterative algorithm requires an initial feasible solution, and the authors construct the completion time minimization problem and apply a bisection-based algorithm to obtain the optimal solution. The majority of this work is theoretical and the system implementation here is very minimal with good results. This is a key paper in this area since it derives the mathematical relationship between the model's performance, communication overhead, and energy consumption.

Another important paper is *To Talk or to Work: Flexible Communication Compression for Energy Efficient Federated Learning over Heterogeneous Mobile Edge Devices* [15]. This is also an interesting paper since they describe the relationship between resource heterogeneity and power consumption. The paper targets at improving the energy efficiency of FL over mobile edge networks to accommodate the resource heterogeneity among the parties. Based on this, they develop a convergence-guaranteed FL algorithm that enables flexible communication compression. They derive a convergence bound and develop a compression control scheme to balance the energy consumption of local computing against the wireless communication. They target the long-term goal of the system. They apply compression parameters which are specifically chosen for the party participants adapting to their computing and communication environments. This allows for accommodating compression mechanisms dynamically based on device heterogeneity and allows for better control of energy consumption mechanisms.

Meanwhile, the paper *Energy-aware analog aggregation for federated learning with redundant data* [26] takes a different approach to energy efficiency. While the other papers focus on reducing local computations such that they decrease energy consumption, this paper takes a different approach by using scheduling policy to control it. In other words, they design an aggregator selection policy such that energy-aware decisions are made. They define the problem as a budget allocation problem and define an energy consumption budget for the full training process. They also add a redundancy metric which determines if the local data of a party is already present in other more efficient devices. They then propose an energy-aware dynamic party scheduling policy on the aggregator side, which maximizes the average weighted fraction of scheduled parties. This policy requires no future knowledge of energy consumption and is therefore useful for FL systems where the resource usage varies not only across hardware but in temporally as well.

The paper *Federated Learning over Wireless Networks: Optimization Model Design and Analysis* [27] also demonstrates theoretically the relationship between convergence rate and energy consumption. However, instead of showing the relationship, they show that the Federated Learning over wireless networks problem (FEDL) has two main two trade-offs—the learning time versus mobile device energy consumption that falls under the Pareto-optimality curve, and the computation versus communication frequency by finding the optimal number of local optimization steps (i.e., mini-batch size). They prove theoretically that such a problem is non-convex. However, they contain special patterns in CPU cycles and data heterogeneity that allows for breaking them down into smaller sub-problems which can be convex. The first two sub-problems can be solved separately, and their solutions can be used to solve the larger problem scope. The analysis of the closed-form solution provides a Pareto-efficient controlling knob which allows for the tuning between computation (i.e., energy consumed) and communication. This is a notable paper in terms of energy efficiency due to demonstrating that there is a Pareto-optimality involved in the trade-off, which can be utilized by future systems to develop more efficient or tunable FL frameworks.

10.1.4   Network

Lastly, we talk about what some argue to be the largest systems bottleneck in FL, which is networks. A majority of the state-of-the-art machine learning models utilize neural networks which usually contain millions of parameters. In FL systems, the parties have to send these parameters (obfuscated with encryption/differential noise) over the network, which generally translates to up to gigabytes of data depending on the size of the model. However, as mentioned before, many FL scenarios will have low-bandwidth connection devices with high rates of downtime. We now discuss a few papers that attempt to address these challenges.

An interesting approach to network efficiency is provided in *Robust and Communication-Efficient Federated Learning from Non-IID Data* [23]. Methods of reducing communication bandwidth involving compression mechanisms have been used extensively in traditional machine learning and have been applied to Federated learning systems as well. The authors of this paper, however, argue that they are of limited utility since they only compress one direction of communication or are only applicable under certain circumstances which are unrealistic in the general FL scenarios. Based on their observations of the their current works, the authors propose Sparse Ternary Compression, or STC. This is a new compression framework that is specifically designed to meet the requirements of the Federated Learning environment since it extends the existing compression techniques of top-k gradient sparsification which enables downstream compression along with ternarization and optimal Golomb encoding of the weights. They perform experiments on four different learning models and demonstrate that STC can significantly outperform Federated Averaging. They evaluate their method under common Federated Learning scenarios such as with parties that have a high amount of data heterogeneity, parties with small datasets, or a high number of parties, etc. They demonstrate that they can significantly reduce communication bandwidth due to smaller model sizes while reducing the impact of data heterogeneity as well. Since data heterogeneity is such a common problem in FL, the fact that ternarizing gradients can benefit model performance as well is a significant benefit.

FedPAQ: A Communication-Efficient Federated Learning Method with Periodic Averaging and Quantization [21] points out that FL frameworks face multiple systems-oriented challenges. They specifically point out that communication bottlenecks are significant challenges due to many devices trying to interact at the same time. They also say that scalability is a very important aspect of FL too since such systems can contain millions of parties. Due to these systems' challenges as well as data heterogeneity and privacy concerns, Federated Learning can be a very challenging problem to tackle. The authors present FedPAQ as a way of addressing these challenges. FedPAQ is a communication-efficient method that performs periodic aggregation and quantization. In other words, they control the frequency of communication between the parties to reduce the total amount of bandwidth on the party-side. They also devise a mechanism to be robust to partial party participation since not all parties are always available. These features address the communications and scalability challenges in federated learning.

Another unique perspective is given in *CMFL: Mitigating communication overhead for federated learning* [19]. While existing works mainly focus on reducing the total bits transferred in each update via data compression, this paper takes a different view that parties can have updates that are irrelevant to training the global model. The idea is that such updates can be identified before they are transferred to the aggregator and can be precluded before being sent to the aggregator thereby reducing bandwidth consumption. Based on this idea, the authors propose the framework called Communication-Mitigated Federated Learning (CMFL) in this paper. The framework provides parties with the feedback information on the direction of the global model updates. A party's update is considered "irrelevant" if they are too similar to the global model's, which would indicate that the party does not have unique features to aid the training. By avoiding those uploads irrelevant updates to the aggregator, CMFL can substantially reduce the communication overhead while still guaranteeing convergence. This approach indicates an important property of FL systems where certain parties are more important to convergence than others, and we will later talk about systems that can utilize this to make more efficient FL frameworks.

Another work with a unique approach to addressing network overhead management is *FedBoost: Communication-Efficient Algorithms for Federated Learning* [8]. The authors here present a unique method of FL, namely *ensemble training*, to boost model training efficiency. They first prove mathematically that it is possible to train a large model as an ensemble of smaller and more efficient models using Federated Learning. They also show that by offloading only the smaller parts of the ensemble to the parties with predefined intervals can lead to a significant reduction in communication cost compared to training the full models locally (as is done traditionally). They also perform what they call "base predictor training" which is essentially pre-training the ensemble on the aggregator side with controllable and balanced data up to a certain point. The partly trained model is then used as the base global model and then the FL training on the parties actually starts. This results in lower number of FL training rounds, meaning that it leads to an overall reduction in the number of communication rounds too.

10.2   Large-Scale FL Systems

Cross-device FL is considered a large-scale system due to having large number of party participants (can be up to millions [1]). However, the challenges faced are quite distinct as discussed before. The state of the art in large-scale FLs have two large challenges—managing a huge number of connections/aggregations and reducing the impact of stragglers.

Up to this point, we have only discussed a specific type of FL architecture called the *Central Aggregator* architecture since the seminal paper in FL [12] uses this and is the base architecture for all advanced FL systems. However, as discussed before, such a simple architecture poses quite a few challenges (e.g., communication bottlenecks and stragglers), and many newer systems have considerably changed this base architecture into more novel ones that can mitigate some of these challenges. In this section, we broadly group these architectures into 4 groups as pointed out in [17]—Clustered, Hierarchical, Decentralized, and Asynchronous—and discuss them in detail.

10.2.1 Clustered FL

Figure 10.1 shows the base cross-device architecture that we have based our discussions on. As pointed out before in Sect. 10.1.2, such systems face the stragglers' issue where the total training latency per round is bound by the slowest party selected in that round. It also makes no concession to address data heterogeneity challenges.

In order to address these shortcomings, one of the first types of FL is called the Clustered FL systems. The key idea here is to group the parties into clusters such that the party devices within each set have certain similar properties (e.g., data distributions, training latencies, hardware, location, etc.). The system diagram is given in Fig. 10.2. As we can see, the major difference between this system and the basic FL one is not much in terms of infrastructure. Instead, the major difference is in terms of how the parties are organized. The properties which determine the organization of the clusters usually differ in literature depending on what problem is being attempted to be addressed. For example, if we wish to develop an architecture where we want all parties to participate equally but some of them are more prone to dropouts than others, we can group them based on their dropout probabilities. Then we can simply choose to select parties from the higher dropout rates more frequently than those with lower rates, evening out the participation. Similarly, for data heterogeneity challenges, we can group parties based on how balanced their local data are. Creating a balance between the sampling frequencies of unbalanced and balanced datasets can mitigate the data heterogeneity issues.

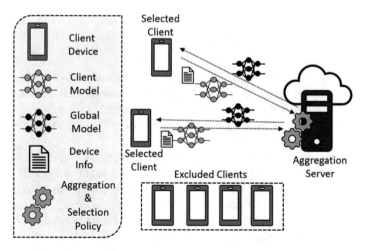

Fig. 10.1 Central Aggregator architecture

Fig. 10.2 Clustered FL
architecture

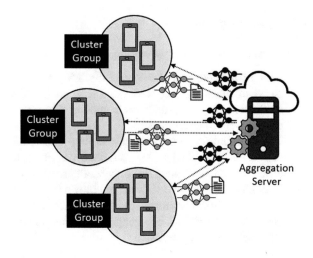

10.2.1.1   Design Challenges

There are a few challenges involved in developing such FL systems. They are—

Clustering Criteria The properties used to group parties are one of the most important design decision that needs to be made in such systems since it will completely define the priorities of the framework. For example, if parties are grouped based on resources and not data, we may develop a system that can control the speed of training but be completely blind to model performance.

Selection Criteria After clustering, the next most important step is how to control the system such that the clustered properties can be taken advantage of. For example, too frequent selection of parties with biased datasets will lead to a biased model, but too few may mean that important features may end up being excluded from training.

Profiling Depending on the clustering properties, the parties need to be profiled for their characteristics. For example, when clustering based on data an accurate method of quantifying and ranking them would be required. Depending on the accuracy of this profiling, devices may end up in the wrong cluster and cause even more problems in defining a good selection policy.

Dynamic Properties Certain properties of parties such as the number of data-points, network connectivity, available training resources, etc. can change over time due to user behavior. As such, the profiling must be done more than once throughout the training process.

Privacy Profiling data is completely against the privacy-centric design principle of FL systems. As such special precautions need to be taken such that even profiling and clustering properties could not be used to identify the specific parties (e.g., Secure Aggregation, Blockchains, Differential Privacy).

10.2.1.2 Pros and Cons

The advantages of such systems are—

- **Ease of implementation**—Since the infrastructure is not much different compared to the basic FL, it is quite easy to implement. The major hurdles of implementation are usually on the policy side when the clustering and selection criteria need to be defined.
- **Complementary to other architectures**—The clustering and selection policies can be easily applied on top of other types of FL architectures with drastically different structures since it is only an algorithmic add-on.
- **Tunable**—Usually, works which employ a form of clustering FL provide control knobs that allow the properties of the system to be tuned. For example, [2] gives parameters which can be changed to compromise between convergence speed and final model performance.

The disadvantages of such a system are—

- **Hard to tune**—As mentioned in the challenges section, the biggest challenge is to define a good clustering and selection policy which usually requires a lengthy tuning process through trial-and-error.
- **Scalability**—The clustering in and of itself does not allow for scalable infrastructures. Clustered FL must be paired with other types of FL architectures such as hierarchical FL to enable the handling of a large number of devices. Additionally, it becomes harder to profile and balance policies with larger number of parties.
- **Overheads**—Profiling, especially on dynamic systems, is an overhead that must be carried by Clustered FL systems. For FL where the party hardware may have limited resources in the first place, this becomes an extra cost over time regardless of how lightweight the profiling process is.

10.2.1.3 Notable Examples in Literature

An important example of a Federated Learning system using such an aggregation scheme is *An Efficient Framework for Clustered Federated Learning* [6]. This paper introduces a clustering system based on the loss values of the party's gradients. The authors provide theoretical analysis on how losses can be used to mitigate the performance loss of a model due to data heterogeneity. The authors analyze the convergence rate of this algorithm with squared loss, generic strongly convex, and smooth loss functions. IFCA is shown as guaranteed to converge, and the authors also discuss the optimality of the statistical error rate. They also propose using IFCA with the weight sharing technique in multi-task learning if clustering is ambiguous. Another such paper is *Clustered Federated Learning* [24], which presents CFL. It is a novel Federated Multi-Task Learning framework, which exploits geometric properties of the loss surface of the FL training systems. CFL groups the parties into clusters based on the cosine similarities of their weights. The idea here is that parties with jointly trainable data distributions help train the system better if selected together. Unlike other existing FMTL approaches, CFL is applicable to general non-convex objectives, requires no modifications to the FL communication protocol, and comes with mathematical guarantees on the clustering quality. However, they focus more on the privacy aspects, proving that it is possible to profile user data and yet keep them anonymous by adding differential noise. *TiFL: A Tier-based Federated Learning System* [2] is another paper which uses this scheme. They propose clustering based on resource as well as data heterogeneity. They first perform extensive experiments in a real distributed system to demonstrate how data heterogeneity impacts model performance and how resource heterogeneity causes straggler issues. They show that the challenges faced due to both these properties can be managed by clustering similarly behaving parties together and making smart scheduling decisions. They propose an automatic and dynamic party selection process using selection probabilities and clustering which significantly reduce the impact of stragglers and data bias.

10.2.2   Hierarchical FL

While Clustered FL is easier to implement and flexible, it lacks scalability. In order to address the challenge of having to communicate with possibly thousands of devices simultaneously, we need a considerable shift in the overall structure. The single most disadvantage of the Central Aggregator system is that it only has one aggregator that handles all the parties. The most obvious way to solve this issue is to add extra aggregation servers, each of which contributes to the development of the global model. Thus, hierarchical FL systems were proposed. As the name implies, it contains levels of aggregators, each in charge of its own set of parties and passes on its aggregated model higher up until they reach a central aggregator which manages the single global model. Figure 10.3 shows the architecture of such systems.

Fig. 10.3 Hierarchical FL architecture

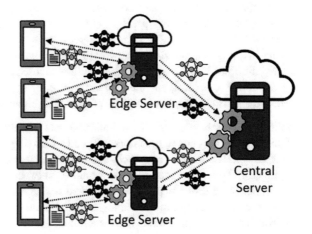

The most notable difference here is the edge aggregators between the parties and the central aggregator. The parties usually communicate with their own edge aggregator, which performs the tasks performed traditionally by the central aggregator such as selecting devices, sending and receiving model weights, encryption/decryption of privacy mechanisms, aggregation, etc. However, these edge aggregators do not contain the final global model. Rather, they have what is called an "intermediate" model which it passes up the hierarchy until it reaches the final central aggregator. There can be any number of hierarchies between the central aggregator and the parties depending on the scale of the infrastructure. The edge aggregators in between other edge aggregators and the central aggregator also usually contain their own aggregation mechanisms. Such hierarchical structures break down the full system into smaller more manageable parts and make it scalable.

10.2.2.1 Design Challenges

Implementation Such systems are usually quite large in scale and require significant modification to the underlying architecture. As such, it requires effort to set up in a distributed system. There are also implementation decisions which will impact the scalability, robustness, ease of management, and efficiency. For example, how many parties to allocate per edge aggregator will decide how many edge aggregators would be required to manage the full system. Too little would mean under-utilization of resources while too much can cause resource contention.

Management The large number of parties is usually magnitudes more than the number of workers in traditional distributed ML systems. The addition of edge aggregators adds even more management problems due to having more nodes to the full system. In the Central Aggregator scenario, we only need to manage one aggregator. With edge aggregators, we need to add extra mechanisms such as fault-tolerance, monitoring, etc. which adds to the complexity of the system.

Synchronization Due to each of the aggregators managing their own set of parties, the synchronization of updates between the edge aggregators becomes a challenge. In FL systems, parties vary greatly between the speeds at which they report back their model weights. With a central aggregator, we only need to wait for all the parties in a single round. However, with multiple edge aggregators at different levels, aggregation needs to be synchronized within each edge aggregator, between edge aggregators in one level, and then between the edge aggregators in each level of the hierarchy. This can amplify straggler problems if not addressed correctly.

Balancing Heterogeneity The imbalanced data distributions between parties also mean that not all edge aggregators will end up having the same quality of "intermediate" models. It will completely depend on the parties allocated under that edge aggregator. This imbalance of model quality between the edge aggregators even within the same levels necessitates the need for advanced aggregation algorithms. Similar to synchronization challenges, the model imbalance can be amplified due to multiple levels of aggregations if we are not careful.

10.2.2.2   Pros and Cons

The advantages of such systems are—

- **Scalability**—By virtue of its architecture, it is a highly scalable system. This can also be a dynamic property of the system since we can easily add and remove edge aggregators depending on the needs of the underlying infrastructure.
- **System efficiency**—Systems can be made more efficient due to the ability to add/remove nodes easily. In Central Aggregator architectures, the node can become overwhelmed with connections and model updates, resulting in resource contention and reduction in system efficiency. Load balancing becomes easier when there is the option to add/remove aggregation nodes.
- **Robustness**—Central Aggregation schemes have a single point of failure. Hierarchical systems have multiple nodes at a time, each containing a relatively newer copy of the "intermediate" models. This means that nodes going offline can easily be handled.

The disadvantages of such a system are—

- **Communication redundancy**—Model weights must be calculated and communicated multiple times as they are moved to the top of the hierarchy. With a central aggregator, the weights are just transferred once per round.
- **Privacy Overheads**—Privacy mechanisms such as encryption are usually resource intensive tasks and are required every time the weights are moved between aggregators. As such, multiple levels of aggregation means the same privacy protocols must be applied repeatedly during every transfer, incurring significant computation cost over time.
- **Security**—Due to the larger number of nodes in a hierarchical system, there are more options for malicious parties to attack. Additionally, it is harder to detect compromised nodes due to the sheer number of edge aggregators. Central aggregation servers, in comparison, are easier to monitor and detect anomalies on.

10.2.2.3   Notable Examples in Literature

One of the seminal papers here is *Towards Federated Learning at Scale: System Design* [1]. This is the first paper to point out the systems challenges in large-scale FL, the authors here propose a template of the hierarchical structure as a means of managing vast number of parties. They focus on ensuring that every party gets to participate without having to worry about resources for the Aggregators and show that hierarchical structures are ideal for such scenarios due to the dynamically changing infrastructure. Another important paper is *Client-Edge-Cloud Hierarchical Federated Learning* [16]. While the previous work focused more on scalability, they did not consider the limitations of using simple aggregation algorithms on their impact on model performance. This paper demonstrated that we may need different aggregation algorithms at different levels, and good aggregators can be designed to mitigate the impact of data heterogeneity, party availability, communication redundancy, and stale weights. *HFEL: Joint Edge Association and Resource Allocation for Cost-Efficient Hierarchical Federated Edge Learning* [18] introduces a novel hierarchical federated learning system called HFEL. They formulate the hierarchical design challenge as a joint computation and communication resource allocation problem. They solve the optimization problem to come up with an effective communication vs. node resources trade-off and propose a scheduler for aggregators which further reduces the resource costs.

10.2.3 Decentralized FL

Another idea with parallels to the Hierarchical FL system is that of Decentralized FL. The main difference between the Hierarchical FL and Decentralized FL is that the aggregation servers in a Decentralized system operate independently from each other as shown in Fig. 10.4. Unlike the former, there is no waiting between aggregators for each other's "intermediate" model weights and passing it up the levels. Instead, the Aggregators rely on blockchains or message passing to coordinate between themselves to train a global model.

The training process is also quite different due to not having a central aggregator. For example, the system BrainTorrent [22] starts by starting each of the parties to perform their local training in parallel. Each of these models is given a version number of 0 initially. A random party among all the devices sends out a ping to the other parties to get each of their local model versions. All parties with a higher version number than the current party's local model are asked to send their model weights, and the asking party takes those weights and aggregates them with its own local model, creating a new version of the global model. Then the training process begins again by randomly selecting another party and continues until all party models converge to the same global model. In such a system, only one party is updated per round instead of many.

The advantages to scalability are obvious in this case as well compared to the Central Aggregation system. Due to using the parties as aggregation and communication nodes, the communication bottleneck issue is completely removed. It also allows for indefinite expansion on the number of parties used for training.

Fig. 10.4 Decentralized FL architecture

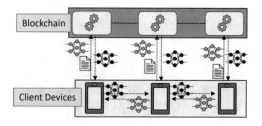

10.2.3.1   Design Challenges

Lightweight Aggregation and Communication Mechanisms One large drawback of such a peer-to-peer based aggregation system is in the reduction of local computational and communication overheads. Since the party hardware are the ones doing the heavy-lifting and usually such hardware have limited resources, it becomes very important to design very lightweight mechanisms. Such systems also become infeasible under certain scenarios such as in low-bandwidth areas.

Unpredictability Since there are usually no aggregators to manage the training process, it becomes impossible to control it. A certain amount of randomness is involved, and properties of the system such as data heterogeneity, stragglers, dropouts, etc. can impact the model performance, convergence speed, and resource usage in unpredictable ways. Without adding control mechanisms such as coordinator servers, it becomes almost impossible to design an efficient Decentralized system.

Storage and Energy Overheads Since parties take turns to aggregate and update the global models, they must incur significant storage costs since they must handle multiple large models at the same time. Given that the devices are already limited in capacity, this becomes another pressing matter; compression techniques or memory utilization methods must be implemented to handle such issues. It may also mean that more energy is expended every time a party is selected.

10.2.3.2   Pros and Cons

The advantages of such systems are—

- **Scalability**—Similar to Hierarchical FL, its architecture is by default a scalable system. Few changes to the structure are needed when adding or removing parties from the system. Dynamic shifts in the party availability and data shifts do not need to be explicitly addressed.
- **Lack of Aggregator Nodes**—A minimal number of nodes are required for training since it is the parties that tend to manage the training process. Aggregators can be added for monitoring, scheduling, etc. but the actual aggregation is not required and so the nodes can have low hardware resources, reducing the cost of implementation.
- **Robustness**—There are as many aggregators as there are parties in the system, meaning that a few node failures will not affect the training process significantly. Since all the parties also tend to keep their own versions of the models, it is easy to retain a certain amount of progress even in case of a large number of party failures.

The disadvantages of such a system are—

- **Local resource usage**—As mentioned before, the fact that parties are the one performing aggregation means that the cost increases per device. This can be a disqualifying reason in many light-on-resources systems such as sensor networks. Additionally, communication bottleneck can become a large concern if too many parties fall behind on training and end up trying to aggregate at the same time.
- **Security**—A managed aggregator is more secure than a decentralized system due to the latter randomly passing around weights across all devices. A single malicious party can completely infiltrate the system if it gets selected for training and is therefore a huge security flaw. Parties must all be trusted completely for such a system to be implemented securely.
- **Lack of control**—The system largely depends on the party devices that get selected to be the aggregator. If the parties are fast and have consistently high bandwidths, the system will converge faster. However, if there are too many dropouts, stragglers, biased datasets, etc. it becomes a much less efficient system. Since there is no central aggregator, there are very few policies that can be implemented to control such systems.

10.2.3.3   Notable Examples in Literature

The first important paper is called *BrainTorrent: A Peer-to-Peer Environment for Decentralized Federated Learning* [22]. This paper introduces BrainTorrent, a FL framework without a centralized aggregation server. They perform peer-to-peer aggregation such that all the participants can train a single global model without needing to have a coordinator. BrainTorrent was designed with medical ML models in mind but can be extended to other types of applications too. *Decentralized Federated Learning with Adaptive Partial Gradient Aggregation* [10] is another paper where the authors present a decentralized aggregation algorithm developed as an extension of FedAvg [12] called FedPGA. They propose that the parties in FedPGA exchange partial gradients instead of the full model weights. Doing so significantly reduces the network load. The partial gradients work as a direction for moving the global weights (similar to gradients in traditional learning). The aggregation algorithm is very similar to FedAvg, but with averaging over partial gradients than weights. The paper *Fully Decentralized Federated Learning* [13] also propose a decentralized federated learning training framework. However, to reduce communication overhead they limit the peer-to-peer connections to only those within one hop away. They theoretically prove that it is possible for decentralized FL mechanisms such as theirs can converge.

Fig. 10.5 Asynchronous FL architecture

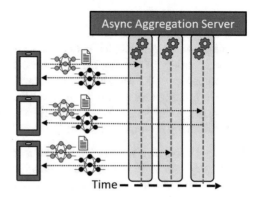

10.2.4   Asynchronous FL

The last unique type of architecture is the Asynchronous FL. So far, we have only talked about synchronized FL algorithms which wait for the weights for all parties to be available before aggregation. However, this is what makes FL systems susceptible to stragglers and bottlenecks in communication, which limits scalability. In asynchronous aggregation mechanisms, the global model does not wait for every party weight but updates the global model as soon as a single party reports back.

Figure 10.5 shows the architecture of a typical asynchronous FL system. The major difference here is the timeline for aggregation. When a party comes in with its latest update, the aggregator performs an aggregation immediately on the received weights and generates a new version of the global model. If another set of weights come in at the same time or immediately after, it is aggregated to the newer iteration of the global model. In other words, the weights are aggregated in a sequence. In this method, no waiting is necessary for all the weights to report back and thus stragglers can come in at any time without holding up the training progress.

10.2.4.1   Design Challenges

Aggregation Techniques Since the process is not synchronized, developers need to come up with novel algorithms that can perform weight aggregation to generate a new global model. This can be a challenging task since asynchronous methods in general tend to perform worse than synchronous averaging, resulting in a worse final model. This can be especially true for FL where the impact of data heterogeneity can be amplified without a good aggregation algorithm to mitigate it. Currently, there are no works that have proposed a theoretically complete asynchronous algorithm that can also be used to mitigate biased training.

Staleness This is a challenge which traditional distributed asynchronous algorithms face as well. If the party weights come in too late, i.e., the global models are already a significant number of steps past when the reporting party had received and updated its local model, the new weights may actually be detrimental to the model training process. As such, it is better to discard the weights that are too "stale." Given the wide variety of training latencies are between the parties, it is natural that the slowest parties will always be stale compared to the faster party weights. This would result in training bias as well as under-utilization of the full dataset. As such, how to design a policy so that slower parties can also contribute without compromising mode accuracy is an open challenge.

Privacy-Preserving Mechanisms Another challenge for Asynchronous FL is from the privacy aspect. Currently, the two major privacy-preserving techniques (secure aggregation and differential privacy) require the participation of multiple parties per round. In differential privacy, a certain amount of noise is added to each of the party model weights at such a level that they cannot be distinguished from each other. It can only be done if there are multiple party models, otherwise it is not possible to know how much noise to apply to a single party's model weights. Similarly, for secure aggregation, the keys are shared among multiple parties such that the model weights can only be decrypted when all the models report back. In asynchronous aggregation, parties come in one at a time, making it hard to apply any of these techniques.

10.2.4.2 Pros and Cons

The advantages of such systems are—

- **Nullifies the straggler effect**—Due to not being synchronous, the FL system can tolerate a significant amount of training latency in the parties. Without having to wait for all the parties to come in, the per-round time is not bound by the slowest arriving party anymore. The training process becomes free running.
- **Used with other architectures**—Like clustered FL, this too can be used with other FL architecture types since it requires changes on the policy level instead of the infrastructure. For example, it can be applied in hierarchical FL where every edge aggregator asynchronously trains each of the devices it is in charge of. As such, it can gain the benefits of scalability as well as completely side-step the impact of stragglers.
- **Robust to dropouts**—Usually, asynchronous FL will have multiple parties running at the same time and does not usually care about when these parties report back their models. As such, if a model times out or gets interrupted and does not end up sending its weights, the training still proceeds as normal since the other parties can keep on contributing and updating the global model.

The disadvantages of such a system are—

- **Prone to staleness**—The reason why it can avoid stragglers is also the same reason it is more prone to stale weights. As mentioned before, the variety of training latencies mean that some slower parties are always bound to contribute with older weight updates which can be detrimental to the training process.

- **Privacy**—Also discussed above, current privacy systems tend to be reliant on multiple parties synchronized in reporting back their weights. Asynchronous design is directly opposite of these mechanisms, meaning that new types of privacy-preserving methods need to be developed, or a hybrid asynchronous approach (such as controlling the number of asynchronous parties per round) needs to be made.

10.2.4.3   Notable Examples in Literature

One of the first papers that introduced the asynchronous FL as a viable solution is *Asynchronous Online Federated Learning for Edge Devices* [25]. They propose the ASO-fed framework, where the edge devices perform online learning. They propose a system under the premise that the party data is constantly changing, and the parties are constantly training locally on the ever-changing data while providing periodic updates to the aggregator. They use a simple running average as their asynchronous system and do not discuss the privacy implications. However, being a seminal work in this field, they do show that stragglers' problems are completely mitigated by asynchronous aggregation. *Asynchronous Federated Optimization* [28] is another paper where the authors propose a system with an aggregator which is also a coordinator. It manages a queue with all the parties which are set to run in parallel. Every few updates, the coordinator puts a party in the waiting queue and swaps it with another idle party from the queue. In this way, they manage the system by providing control over the training procedure and can mitigate the errors accrued due to stale weights. They also prove the convergence of the proposed approach by posing it as a non-convex problem. In the paper *Communication-Efficient Federated Deep Learning with Asynchronous Model Update and Temporally Weighted Aggregation* [4]. The authors propose an asynchronous learning strategy where different layers of the deep neural networks are categorized into shallow and deeps layers. The parameters of the deep layers are updated less frequently than those of the shallow layers. A temporally weighted aggregation strategy is introduced on the aggregator to make use of the previously trained local models, thereby enhancing the accuracy and convergence of the central model, while keeping a reduced communication overhead.

10.3   Conclusion

For this chapter, we talked about various types of aggregation schemes that exist to address specific problems for Federated Learning systems. Most of the current state-of-the-art frameworks use a version of one or more of these architectures. We provide a general overview of their pros and cons and provide some example work to demonstrate how they are used in real frameworks. We also pointed out the seminal and most unique approaches that are available in current literature. While other works exist, these papers provide the distinct techniques which lay the groundwork for most of the other state-of-the-art works. As such, this chapter provides a good coverage of the different solutions available to address the certain Federated Learning systems challenges.

References

1. Bonawitz K, Eichner H, Grieskamp W, Huba D, Ingerman A, Ivanov V, Kiddon C, Konecný J, Mazzocchi S, McMahan B, Van Overveldt T, Petrou D, Ramage D, Roselander J (2019) Towards federated learning at scale: System design. In Talwalkar A, Smith V, and Zaharia M (eds) Proceedings of machine learning and systems 2019, MLSys 2019, Stanford, CA, USA, March 31–April 2, 2019. mlsys.org
2. Zheng Chai, Ahsan Ali, Syed Zawad, Stacey Truex, Ali Anwar, Nathalie Baracaldo, Yi Zhou, Heiko Ludwig, Feng Yan, and Yue Cheng (2020) Tifl: A tier-based federated learning system. In: Proceedings of the 29th international symposium on high-performance parallel and distributed computing, pp 125–136
3. Chen B, Medini T, Farwell J, Tai C, Shrivastava A (2020) Slide: in defense of smart algorithms over hardware acceleration for large-scale deep learning systems. Proceedings of Machine Learning and Systems 2:291–306
4. Chen Y, Xiaoyan Sun X, Yaochu Jin Y (2019) Communication-efficient federated deep learning with asynchronous model update and temporally weighted aggregation. Preprint. arXiv:1903.07424
5. Daghaghi S, Meisburger N, Zhao M, Shrivastava A (2021) Accelerating slide deep learning on modern cpus: Vectorization, quantizations, memory optimizations, and more. Proc Mach Learn Syst 3:156
6. Ghosh A, Chung J, Yin D, Ramchandran K (2020) An efficient framework for clustered federated learning. Preprint. arXiv:2006.04088
7. Gupta S, Imani M, Rosing T (2019) Exploring processing in-memory for different technologies. In: Proceedings of the 2019 on great lakes symposium on VLSI, pp 201–206
8. Hamer J, Mohri M, Suresh AT (2020) Fedboost: A communication-efficient algorithm for federated learning. In: International conference on machine learning. PMLR, pp 3973–3983
9. Imani M, Gupta S, Kim Y, Rosing T (2019) Floatpim: In-memory acceleration of deep neural network training with high precision. In 2019 ACM/IEEE 46th annual international symposium on computer architecture (ISCA). IEEE, pp 802–815
10. Jiang J, Hu L (2020) Decentralised federated learning with adaptive partial gradient aggregation. CAAI Trans Intell Technol 5(3):230–236
11. Jiang Y, Wang S, Valls V, Ko BJ, Lee WH, Leung KK, Tassiulas L (2019) Model pruning enables efficient federated learning on edge devices. Preprint. arXiv:1909.12326
12. Konecnỳ J, McMahan HB, Yu FX, Richtárik P, Suresh AT, Bacon D (2016) Federated learning: Strategies for improving communication efficiency. CoRR
13. Lalitha A, Shekhar S, Javidi T, Koushanfar F (2018) Fully decentralized federated learning. In: Third workshop on bayesian deep learning (NeurIPS)
14. Lane ND, Bhattacharya S, Georgiev P, Forlivesi C, Jiao L, Qendro L, Kawsar F (2016) Deepx: A software accelerator for low-power deep learning inference on mobile devices. In: 2016 15th ACM/IEEE international conference on information processing in sensor networks (IPSN). IEEE, pp 1–12
15. Li L, Shi D, Hou R, Li H, Pan M, Han Z (2020) To talk or to work: Flexible communication compression for energy efficient federated learning over heterogeneous mobile edge devices. Preprint. arXiv:2012.11804

16. Liu L, Zhang J, Song SH, Letaief KB (2020) Client-edge-cloud hierarchical federated learning. In: ICC 2020-2020 IEEE international conference on communications (ICC), pp 1–6. IEEE
17. Lo SK, Lu Q, Zhu L, Paik HY, Xu X, Wang C Architectural patterns for the design of federated learning systems. Preprint. arXiv:2101.02373, 2021.
18. Luo S, Chen X, Wu Q, Zhou Z, Yu S (2020) Hfel: Joint edge association and resource allocation for cost-efficient hierarchical federated edge learning. IEEE Trans Wirel Commun 19(10):6535–6548
19. Luping W, Wei W, Bo L (2019) Cmfl: Mitigating communication overhead for federated learning. In: 2019 IEEE 39th international conference on distributed computing systems (ICDCS). IEEE, pp 954–964
20. Kairouz P, McMahan HB, Avent B, Bellet A, Bennis M, Bhagoji AN et al (2021) Advances and open problems in federated learning. Foundations and Trends® in Machine Learning 14(1-2):1–210
21. Reisizadeh A, Mokhtari A, Hassani H, Jadbabaie A, Pedarsani R (2020) Fedpaq: A communication-efficient federated learning method with periodic averaging and quantization. In: International conference on artificial intelligence and statistics. PMLR, pp 2021–2031
22. Roy AG, Siddiqui S, Pölsterl S, Navab N, Wachinger C (2019) Braintorrent: A peer-to-peer environment for decentralized federated learning. Preprint. arXiv:1905.06731
23. Sattler F, Wiedemann S, Müller KR, Samek W (2019) Robust and communication-efficient federated learning from non-iid data. IEEE Trans Neural Netw Learn Syst 31(9):3400–3413
24. Sattler F, Müller KR, Samek W (2020) Clustered federated learning: Model-agnostic distributed multitask optimization under privacy constraints. IEEE Trans Neural Netw Learn Syst 32:3710
25. Sprague MR, Jalalirad A, Scavuzzo M, Capota C, Neun M, Do L, Kopp M (2018) Asynchronous federated learning for geospatial applications. In: Joint European conference on machine learning and knowledge discovery in databases. Springer, pp 21–28
26. Sun Y, Zhou S, Gündüz D (2020) Energy-aware analog aggregation for federated learning with redundant data. In: ICC 2020-2020 ieee international conference on communications (ICC). IEEE, pp 1–7
27. Tran NH, Bao W, Zomaya A, Nguyen MN, Hong CS (2019) Federated learning over wireless networks: Optimization model design and analysis. In: IEEE INFOCOM 2019-IEEE conference on computer communications. IEEE, pp 1387–1395
28. Xie C, Koyejo S, Gupta I (2019) Asynchronous federated optimization. Preprint. arXiv:1903.03934
29. Xu Z, Yang Z, Xiong J, Yang J, Chen X (2019) Elfish: Resource-aware federated learning on heterogeneous edge devices. Preprint. arXiv:1912.01684
30. Yang Z, Chen M, Saad W, Hong CS, Shikh-Bahaei M (2020) Energy efficient federated learning over wireless communication networks. IEEE Trans Wirel Commun 20:1935

Chapter 11
Straggler Management

Syed Zawad, Feng Yan, and Ali Anwar

Abstract For this chapter, we elaborate on one of the most common challenge in Federated Learning—*stragglers*. The chapters "Local Training and Scalability of Federated Learning Systems" and "Introduction to Federated Learning Systems" have talked briefly about it, and we delve even deeper here. We first provide an introduction on what the problem is and why it is important. We talk about a study to show the effect of stragglers in a practical setting. As an example, we then talk about TiFL, a framework that proposes to solve such a problem using grouping. Empirical results are presented to show how such systems may help mitigate the effect of stragglers.

11.1 Introduction

As discussed before, Federated Learning shines light on a new emerging high performance computing paradigm by addressing the security and privacy challenges through utilizing decentralized data that is training local models on the local data of each party (or data party) and using a central aggregator to accumulate the learned gradients of local models to train a global model. Though the computing resource of individual party may be far less powerful than the computing nodes in conventional supercomputers, the computing power from the massive number of parties can accumulate to form a very powerful "decentralized virtual supercomputer." Depending on the usage scenarios, FL is usually categorized into *cross-silo* FL and *cross-device* FL [9]. In cross-device FL, the parties are usually a massive number of mobile or IoT devices with various computing and communication capacities [9, 10, 14] while in cross-silo FL, the parties are a small number of organizations with ample

S. Zawad (✉) · F. Yan
University of Nevada, Reno, Reno, NV, USA
e-mail: szawad@nevada.unr.edu; fyan@unr.edu

A. Anwar
IBM Research – Almaden, San Jose, CA, USA
e-mail: ali.anwar2@ibm.com

computing power and reliable communications [9, 16]. In this section, we focus on the cross-device FL which intrinsically pushes the heterogeneity of computing and communication resources to a level that is rarely found in datacenter distributed learning and cross-silo FL. More importantly, the data in FL is also owned by parties where the quantity and content can be quite different from each other, causing severe heterogeneity in data that usually does not appear in datacenter distributed learning, where data distribution is well controlled.

We first demonstrate a case study from TiFL [8] to quantify how data and resource heterogeneity in parties impacts the performance of FL with FedAvg in terms of training performance, and model accuracy and summarize the key findings: (1) training throughput is usually bounded by slow parties (a.k.a. stragglers) with less computational capacity and/or slower communication, which is named as the *resource heterogeneity*. (2) Different parties may train on different quantity of samples per training round and results in different round time that is similar to the straggler effect, which impacts the training time and potentially also the accuracy. This observation is called the *data quantity heterogeneity*. (3) In FL, the distribution of data classes and features depends on the data owners, thus resulting in a non-uniform data distribution, known as non-Identical Independent Distribution (*non-IID data heterogeneity*). The experiments show that such heterogeneity can significantly impact the training time and accuracy.

While *resource heterogeneity* and *data quantity heterogeneity* information can be reflected in the measured training time, the *non-IID data heterogeneity* information is difficult to capture. This is because any attempt to measure the class and feature distribution violates the privacy-preserving requirements. To solve this challenge, TiFL offers an *adaptive* party selection algorithm that uses the accuracy as indirect measure to infer the *non-IID data heterogeneity* information and adjust the tiering algorithm on-the-fly to minimize the training time and accuracy impact. Such approach also serves as an online version to be used in an environment where the characteristics of heterogeneity change over time.

11.2 Heterogeneity Impact Study

Compared with datacenter distributed learning and cross-silo FL, one of the key features of cross-device FL is the significant resource and data heterogeneity among parties, which can potentially impact both the training throughput and the model accuracy. Resource heterogeneity arises as a result of vast number of computational devices with varying computational and communication capabilities involved in the training process. The data heterogeneity arises as a result of two main reasons— (1) the varying number of training data samples available at each party and (2) the non-uniform distribution of classes and features among the parties.

11.2.1 Formulating Standard Federated Learning

Cross-device FL is performed as an iterative process whereby the model is trained over a series of global training rounds, and the trained model is shared by all the involved parties. We define K as the total pool of parties available to select from for each global training round, and C as the set of parties selected per round. In every global training round, the aggregator selects a random fraction of parties C_r from K. The aggregator first randomly initializes weights of the global model denoted by ω_0. At the beginning of each round, the aggregator sends the current model weights to a subset of randomly selected parties. Each selected party then trains its local model with its local data and sends back the updated weights to the aggregator after local training. At each round, the aggregator waits until all selected parties respond with their corresponding trained weights. This iterative process keeps on updating the global model until a certain number of rounds are completed or a desired accuracy is reached.

The state-of-the-art cross-device FL system proposed in [3] adopts a party selection policy where parties are selected randomly. A coordinator is responsible for creating and deploying a master aggregator and multiple child aggregators for achieving scalability as the real-world cross-device FL system can involve up to tens of thousands of parties [3, 9, 12]. At each round, the master aggregator collects the weights from all the child aggregators to update the global model.

11.2.2 Heterogeneity Impact Analysis

The resource and data heterogeneity among involved parties may lead to varying response latencies (i.e., the time between a party receives the training task and returns the results) in the cross-device FL process, which is usually referred as the straggler problem. We denote the response latency of a party c_i as L_i, and the latency of a global training round is defined as

$$L_r = Max\left(L_1, L_2, L_3, L_4 \ldots L_{|C|}\right), \tag{11.1}$$

where L_r is the latency of round r. From Eq. (11.1), we can see the latency of a global training round is bounded by the maximum training latency of parties in C, i.e., the slowest party.

Let us define τ levels of parties, where within the same level, the parties have almost similar response latencies. Assume that the total number of levels is m and τ_m is the slowest level with $|\tau_m|$ parties inside. In the baseline case, the aggregator selects the parties randomly, resulting in a group of selected parties with composition spanning multiple party levels.

We formulate the probability of selecting $|C|$ parties from all party levels except the slowest level τ_m as follows:

$$Pr = \frac{\binom{|K|-|\tau_m|}{|C|}}{\binom{|K|}{|C|}}. \tag{11.2}$$

Accordingly, the probability of at least one party in C comes from τ_m can be formulated as:

$$Pr_s = 1 - Pr. \tag{11.3}$$

Theorem 11.1 $\frac{a-1}{b-1} < \frac{a}{b}$, *while* $1 < a < b$.

Proof Since $1 < a < b$, we could get $ab - b < ab - a$, that is $(a - 1)b < (b - 1)a$ *and* $\frac{a-1}{b-1} < \frac{a}{b}$.   □

$$\begin{aligned} Pr_s &= 1 - \frac{\binom{|K|-|\tau_m|}{|C|}}{\binom{|K|}{|C|}} \\ &= 1 - \frac{(|K| - |\tau_m|)\dots(|K| - |\tau_m| - |C| + 1)}{|K|\dots(|K| - |C| + 1)} \\ &= 1 - \frac{|K| - |\tau_m|}{|K|}\dots\frac{|K| - |\tau_m| - |C| + 1}{|K| - |C| + 1}. \end{aligned}$$

$$\tag{11.4}$$

By applying Theorem 11.1, we get:

$$\begin{aligned} Pr_s &> 1 - \frac{|K| - |\tau_m|}{|K|}\dots\frac{|K| - |\tau_m|}{|K|} \\ &= 1 - (\frac{|K| - |\tau_m|}{|K|})^{|C|}. \end{aligned}$$

$$\tag{11.5}$$

In real-world scenarios, large number of parties can be selected at each round, which makes $|K|$ extremely large. As a subset of K, the size of C can also be sufficiently large. Since $\frac{|K|-|\tau_m|}{|K|} < 1$, we get $(\frac{|K|-|\tau_m|}{|K|})^{|C|} \approx 0$, which makes $Pr_s \approx 1$, meaning in a standard cross-device FL training process, the probability of selecting at least one party from the slowest level is reasonably high for each round. According to Eq. (11.1), the random selection strategy adopted by state-of-the-art cross-device FL system may suffer from a slow training performance.

Algorithm 11.1 Federated averaging training algorithm

1: **Aggregator:** initialize weight w_0
2: **for** each round $r = 0$ **to** $N - 1$ **do**
3:     $C_r =$ (random set of $|C|$ parties)
4:     **for** each party $c \in C_r$ **in parallel do**
5:         $w_{r+1}^c = TrainParty(c)$
6:         $s_c =$ (training size of c)
7:     **end for**
8:     $w_{r+1} = \sum_{c=1}^{|C|} w_{r+1}^c * \frac{s_c}{\sum_{c=1}^{|C|} s_c}$
9: **end for**

11.2.3   Experimental Study

To experimentally verify the above analysis and demonstrate the impact of resource heterogeneity and data quantity heterogeneity, we show a study with a setup similar to the paper [7]. The testbed is briefly summarized as follows—

- A total of 20 parties are used and each party is further divided into 5 groups with 4 party per group.
- 4 CPUs, 2 CPUs, 1 CPU, 1/3 CPU, 1/5 CPU are allocated for every party from group 1 through 5, respectively, to emulate the resource heterogeneity.
- The model is trained on the image classification dataset CIFAR10 [11] using the standard cross-device FL process in Sect. 11.2.1 (model and learning parameters are detailed in Sect. 11.4).
- Experiments with different data size for every party are conducted to produce data heterogeneity results.

As shown in Fig. 11.1a, with the same amount of CPU resource, increasing the data size from 500 to 5000 results in a near-linear increase in training time per round. As the amount of CPU resources allocated to each party increases, the training time gets shorter. Additionally, the training time increases as the number of data points increase with the same number of CPUs. These preliminary results imply that the straggler issues can be severe under a complicated and heterogeneous FL environment.

To evaluate the impact of data distribution heterogeneity, the same CPU resources for every party (i.e., 2 CPUs) are kept and generate a biased class and feature distribution following [17]. Specifically, the dataset is distributed in such a way that every party has equal number of images from 2 (non-IID(2)), 5 (non-IID(5)) and 10 (non-IID(10)) classes, respectively. We train the model on Cifar10 dataset using the standard FL system as described in Sect. 11.2.1 with the model and training parameters detailed in Sect. 11.4. As seen in Fig. 11.1b, there is a clear difference in the accuracy with different non-IID distributions. The best accuracy is given by

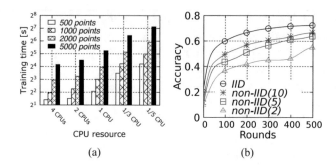

Fig. 11.1 (**a**) Training time per round (logscale) for one party with varying amount of resource and training data quantity (number of training points); (**b**) accuracy under varying number of classes per party (non-IID) with fixed amount of computational resources

the IID since it represents a uniform class and feature distribution. As the number of classes per party is reduced, we observe a corresponding decrease in accuracy. Using 10 classes per party reduces the final accuracy by around 6% compared to IID (it is worth noting that non-IID(10) is not the same as IID as the feature distribution in non-IID(10) is skewed compared to IID). In the case of 5 classes per party, the accuracy is further reduced by 8%. The lowest accuracy is observed in the 2 classes per party case, which has a significant 18% drop in accuracy.

These studies demonstrate that the data and resource heterogeneity can cause significant impact on training time and training accuracy in cross-device FL. One framework proposed to tackle this problem is called *TiFL—A Tier-based Federated Learning Framework* [8]. It is a heterogeneity-aware party selection methodology that selects the most profitable parties during each round of the training to minimize the heterogeneity impact while preserving the FL privacy proprieties, thus improving the overall training performance of cross-device FL.

11.3 Design of TiFL

In this section, we describe the design of TiFL. The key idea of a tier-based system is that given the global training time of a round is bounded by the slowest party selected in that round (see Eq. 11.1), selecting parties with similar response latency in each round can significantly reduce the training time. We first give an overview of the architecture and the main flow of TiFL system. Then we introduce the profiling and tiering approach. We explain how a tier selection algorithm can potentially mitigate the heterogeneity impact through a straw-man proposal as well as the limitations of such static selection approach. We then discuss the proposed adaptive tier selection algorithm to address the limitations of this straw-man system.

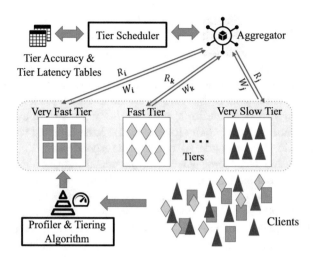

Fig. 11.2 Overview of TiFL

11.3.1   System Overview

The overall system architecture of TiFL is presented in Fig. 11.2. TiFL follows the system design to the state-of-the-art FL system [3] and adds two new components: a *tiering module* (a profiler and tiering algorithms) and a *tier scheduler*. These newly added components can be incorporated into the coordinator of the existing FL system [4]. TiFL supports master-child aggregator design for scalability and fault tolerance.

In TiFL, the first step is to collect the latency metrics of all the available parties through a lightweight profiling as detailed in Sect. 11.3.2. The profiled data is further utilized by the tiering algorithm. This groups the parties into separate logical pools called *tiers*. Once the scheduler has the tiering information (i.e., tiers that the parties belong to and the tiers' average response latencies), the training process begins. Different from standard FL that employs a random party selection policy, in TiFL the scheduler selects a tier and then randomly selects targeted number of parties from that tier. After the selection of parties, the training proceeds as state-of-the-art FL system does. By design, TiFL is non-intrusive and can be easily plugged into any existing FL system in that the tiering and scheduler module simply regulate party selection without intervening the underlying training process.

11.3.2   Profiling and Tiering

Given the global training time of a round is bounded by the slowest party selected in that round (see Eq. 11.1), if the system can select parties with similar response latency in each round, the training time can be improved. However, in FL, the response latency is unknown a priori, which makes it challenging to carry out the above idea. To solve this challenge, a process through which the parties are *tiered* (grouped) by the *Profiling and Tiering* module is proposed as shown in Fig. 11.2. As the first step, all available parties are initialized with a response latency L_i of 0. The profiling and tiering module then assigns all the available parties the profiling tasks. The profiling tasks execute for *sync_rounds* rounds and in each profiling round, the aggregator asks every party to train on the local data and waits for their acknowledgement for T_{max} seconds. All parties that respond within T_{max} have their response latency value RT_i incremented with the actual training time, while the ones that have timed out are incremented by T_{max}. After *sync_rounds* rounds are completed, the parties with $L_i >= sync\_rounds * T_{max}$ are considered *dropouts* and excluded from the rest of the calculation. The total overhead incurred by the offline profiling would be $sync\_rounds * T_{max}$. As TiFL would run the profiling for *sync_rounds* rounds and each round would take time T_{max} to complete the training. The collected training latencies through profiling of parties create a histogram, which is split into m groups and the parties that fall into the same group form a tier. The response latency of each party is then stored by the scheduler and recorded persistently which is used later for scheduling and selecting tiers. The profiling and tiering can be conducted periodically for parties with changing computation and communication performance over the time so that parties can be adaptively grouped into the right tiers.

11.3.3 Straw-Man Proposal: Static Tier Selection Algorithm

In this section, we discuss a naive static tier-based party selection policy and discuss its limitations, which motivates the development of an advanced adaptive tier selection algorithm in the next section. While the profiling and tiering module introduced in Sect. 11.3.2 groups parties into m tiers based on response latencies, the tier selection algorithm focuses on how to select parties from the proper tiers in the FL process to improve the training performance. The natural way to improve training time is to prioritize toward faster tiers, rather than selecting parties randomly from all tiers (i.e., the full K pool). However, such selection approach reduces the training time without taking into consideration of the model accuracy and privacy properties. To make the selection more general, one can specify each tier n_j is selected based on a predefined probability, which sums to 1 across all tiers. Within each tier, $| C |$ parties are uniform randomly selected.

In a real-world FL scenarios, there can be a large number of parties involved in the FL process (e.g., up to 10^{10}) [3, 9, 12]. Thus in the tiering-based approach, the number of tiers is set such that $m << |K|$ and number of parties per tier n_j is always greater than $|C|$. The selection probability of a tier is controllable, which results in different trade-offs. If the users' objective is to reduce the overall training time, they may increase the chances of selecting the faster tiers. However, drawing parties only from the fastest tier may inevitably introduce training bias due to the fact that different parties may own a diverse set of heterogeneous training data spread across different tiers; as a result, such bias may end up affecting the accuracy of the global model. To avoid such undesired behavior, it is preferable to involve parties from different tiers so as to cover a diverse set of training datasets. We show an empirical analysis on the latency-accuracy trade-off in Sect. 11.4.

11.3.4 Adaptive Tier Selection Algorithm

While the above naive static selection method is intuitive, it does not provide a method to automatically tune the trade-off to optimize the training performance nor adjust the selection based on changes in the system. In this section, we describe TiFL's adaptive tier selection algorithm that can automatically strike a balance between training time and accuracy and adapt the selection probabilities adaptively over training rounds based on the changing system conditions.

The observation here is that heavily selecting certain tiers (e.g., faster tiers) may eventually lead to a biased model; TiFL needs to balance the party selection from other tiers (e.g., slower tiers). The question being which metric should be used to balance the selection. Given the goal here is to minimize the bias of the trained model, TiFL can monitor the accuracy of each tier throughout the training process. A lower accuracy value of a tier t typically indicates that the model has been trained with less involvement of this tier, therefore tier t should contribute more in the next training rounds. To achieve this, TiFL can increase the selection probabilities for tiers with lower accuracy. To achieve good training time, it also needs to limit the selection of slower tiers across training rounds. Therefore, the idea of $Credits_t$ is introduced, which is a constraint that defines how many times a certain tier can be selected.

Specifically, a tier is initialized randomly with equal selection probability. After the weights are received and the global model is updated, the global model is evaluated on every party for every tier on their respective $TestData$ and their resulting accuracies are stored as the corresponding tier t's accuracy for that round r. This is stored in A_t^r, which is the mean accuracy for all the parties in tier t in training round r. In the subsequent training rounds, the adaptive algorithm updates the probability of each tier based on that tier's test accuracy at every I rounds. This is done in the function $ChangeProbs$, which adjusts the probabilities such that the lower accuracy tiers get higher probabilities to be selected for training; then with the new tier-wise selection probabilities ($NewProbs$), a tier which has remaining $Credits_t$ is selected from all available tiers τ. The selected tier will have its $Credits_t$ decremented. As parties from a particular tier get selected over and over throughout the training rounds, the $Credits_t$ for that tier ultimately reduces down to zero, meaning that it will not be selected again in the future. This is a way of limiting the number of times a tier can be selected so as to control the training time by controlling the maximum number of times the slower tiers are selected. This

---

Algorithm 11.2 Adaptive Tier Selection Algorithm. $Credits_t$: the credits of Tier t, I: the interval of changing probabilities, $TestData_t$: evaluation dataset specific to that tier t, A_t^r: test accuracy of tier t at round r, τ: set of Tiers

---

1: **Aggregator:** initialize weight w_0, $currentTier = 1$, $TestData_t$, $Credits_t$, equal probability with $\frac{1}{T}$, for each tier t.
2: **for** each round $r = 0$ **to** $N - 1$ **do**
3:     **if** $r\%I == 0$ and $r \geq I$ **then**
4:         **if** $A_{currentTier}^r \leq A_{currentTier}^{r-I}$ **then**
5:             $NewProbs = ChangeProbs(A_1^r, A_2^r \ldots A_T^r)$
6:         **end if**
7:     **end if**
8:     **while** $True$ **do**
9:         $currentTier = $ (select one tier from T tiers with $NewProbs$)
10:         **if** $Credits_{currentTier} > 0$ **then**
11:             $Credits_{currentTier} = Credits_{currentTier} - 1$
12:             **break**
13:         **end if**
14:     **end while**
15:     $C_r = $ (random set of $|C|$ parties from $currentTier$)
16:     **for** each party $c \in C_r$ **in parallel do**
17:         $w_r^c = TrainParty(c)$
18:         $s_c = $ (training size of c)
19:     **end for**
20:     $w_r = \sum_{c=1}^{|C|} w_{r+1}^c * \frac{s_c}{\sum_{c=1}^{|C|} s_c}$
21:     **for** each t in τ **do**
22:         $A_t^r = Eval(w_r, TestData_t)$
23:     **end for**
24: **end for**
25:
26: **function** ChangeProbs$AccuraciesByTier$
27:     $A = SortAscending(AccuraciesByTier)$
28:     $D = n * (n - 1)/2$ where $n = $ # of tiers with $Credits_t > 0$
29:     $NewProbs = []$
30:     **for** each Index i, Tier t in A **do**
31:         $NewProbs[t] = i/D$
32:     **end for**
33:     **return** $NewProbs$
34: **end function**

---

serves as a control knob for the number of times a tier is selected and by setting this upper-bound, TiFL can limit the amount of times a slower tier contributes to the training, thereby effectively gaining some control over setting a soft upper-bound on the total training time. For the straw-man implementation, the paper used a skewed probability of selection to manipulate training time. Since TiFL wishes to adaptively change the probabilities, the $Credits_t$ is added to gain control over limiting training time.

On the one hand, the tier-wise accuracy A_r^t essentially makes TiFL's adaptive tier selection algorithm *data heterogeneity aware*; as such, TiFL makes the tier selection decision by taking into account the underlying dataset selection biasness and automatically adapts the tier selection probabilities over time. On the other hand, $Credits_t$ is introduced to intervene the training time by enforcing a constraint over the selection of the relatively slower tiers. While $Credits_t$ and A_t^r mechanisms optimize toward two different and sometimes contradictory objectives—training time and accuracy, TiFL cohesively synergizes the two mechanisms to strike a balance for the training time-accuracy trade-off. More importantly, with TiFL, the decision making process is automated, thus relieving the users from intensive manual effort. A: Accuracy—One concern is that the uneven probability selection might impact the overall accuracy of the model due to overfitting on data from a particular tier or training too long on tiers with "bad" data. However, it has been found that if "bad" tiers get selected too often, the impact is reflected on other tiers as well, i.e., other tiers' accuracies get decreased. As such, in the next rounds other tiers are given a higher selection probability and are chosen more often, thereby mitigating the effect of the "bad" tiers, thanks to the adaptive capability of the dynamic algorithm. The adaptive algorithm is summarized in Algorithm 11.2.

11.3.5 Training Time Estimation Model

In real-life scenarios, the training time and resource budget are typically finite. As a result, FL users may need to compromise between training time and accuracy. A training time estimation model would facilitate users to navigate the training time-accuracy trade-off curve to effectively achieve desired training goals. Therefore, they build a training time estimation model that can estimate the overall training time based on the given latency values and the selection probability of each tier:

$$L_{all} = \sum_{i=1}^{n} (max(L_{tier\_i}) * P_i) * R, \tag{11.6}$$

where L_{all} is the total training time, L_{tier_i} is the response latency of all the parties in tier i, P_i is the probability of tier i, and R is the total number of training rounds. The model is a sum of products of the tier and maximum latency of each tier, which gives the latency expectation per round. This is multiplied by the total number of training rounds to get the total training time.

Assume that for party c_i, one round of local training using a differentially private algorithm is (ϵ, δ)-differentially private, where ϵ bounds the impact any individual may have on the algorithm's output and δ defines the probability that this bound is violated. Smaller ϵ values therefore signify tighter bounds and a stronger privacy guarantee. Enforcing smaller values of ϵ requires more noise to be added to the model updates sent by parties to the FL aggregator which leads to less accurate models. Selecting parties at each round of FL has distinct privacy and accuracy implications for party-level privacy-preserving FL approaches. For simplicity it is assumed that all parties are adhering to the same privacy budget and therefore same (ϵ, δ) values. Let us first consider the scenario wherein C is chosen uniformly at random each round. Compared with each party participating in each round, the overall privacy guarantee, using random sampling amplification [2], improves from (ϵ, δ) to $(O(q\epsilon), q\delta)$ where $q = \frac{|C|}{|K|}$. This means that there is a stronger privacy guarantee *with the same noise scale*. Parties may therefore add less noise per round or more rounds may be conducted without sacrificing privacy. For the tiered approach the guarantee also improves. Compared to (ϵ, δ) in the all party scenario, the tiered approach improves to an $(O(q_{max}\epsilon), q_{max}\delta)$ privacy guarantee where the probability of selecting tier with weight θ_j is given by $\frac{1}{n_{tiers}} * \theta_j$,

$q_{max} = \max_{j=1\ldots|n_{tiers}|} q_j$ and $q_j = (\frac{1}{n_{tiers}} * \theta_j)\frac{|C|}{|n_j|}$.

11.4 Experimental Evaluation

We discuss the paper's prototyped TiFL results on both the naive and the adaptive selection approach and perform extensive testbed experiments under three scenarios: resource heterogeneity, data heterogeneity, and resource plus data heterogeneity.

11.4.1 Experimental Setup

Testbed As a proof of concept case study, a FL testbed is built for the synthetic datasets by deploying 50 parties on a CPU cluster where each party has its own exclusive CPU(s) using TensorFlow [1]. In each training round, 5 parties are selected to train on their own data and send the trained weights to the server which aggregates them and updates the global model similar to [4, 14]. Bonawitz et al. [4] introduces multiple levels of server aggregators in order to achieve scalability and fault tolerance in extreme scale situations, i.e., with millions of parties. In the prototype, the authors simplify the system to use a powerful single aggregator as it is sufficient for the purpose here, i.e., the system does not suffer from scalability and fault tolerance issues, though multiple layers of aggregator can be easily integrated into TiFL.

TiFL is developed by extending the widely adopted large scale distributed FL framework LEAF [6] in the same way. LEAF provides inherently non-IID with data quantity and class distributions heterogeneity. LEAF framework does not provide the resource heterogeneity among the parties, which is one of the key properties of any real-world FL system. The current implementation of the LEAF framework is a simulation of a FL system where the parties and aggregator are running on the same machine. To incorporate the resource heterogeneity they first extend LEAF to support the distributed FL where every party and the aggregator can run on separate machines, making it a real distributed system. Next, they deploy the aggregator and parties on their own dedicated hardware. This resource assignment for every party is done through uniform random distribution resulting in equal number of parties per hardware type. By adding the resource heterogeneity and deploying them to separate hardware, each party mimics a real-world edge device. Given that LEAF already provides non-IIDness, with the newly added resource heterogeneity feature the new framework provides a real-world FL system which supports data quantity, quality, and resource heterogeneity. For the setup, the authors use exactly the same sampling size used by the LEAF [6] paper (0.05) resulting in a total of 182 parties, each with a variety of image quantities. Accuracy—The test sets for all the datasets are generated through sampling 10% of the total data per party. As such, the test distribution is representative of the distribution of the training set.

11.4.1.1 Experimental Results

Models and Datasets TiFL uses four image classification applications for evaluation. They use *MNIST*[1] *and Fashion-MNIST* [15], where each contains 60,000 training images and 10,000 test images, where each image is 28×28 pixels. TiFL uses a CNN model for both datasets, which starts with a 3×3 convolution layer with 32 channels and ReLu activation, followed by a 3×3 convolution layer with 64 channels and ReLu activation, a MaxPooling layer of size 2×2, a fully connected layer with 128 units and ReLu activation, and a fully connected layer with 10 units and ReLu activation. Dropout 0.25 is added after the MaxPooling layer, dropout 0.5 is added before the last fully connected layer. They use *Cifar10* [11], which contains richer features compared to MNIST and Fashion-MNIST. There is a total of 60,000 color images, where each image has 32×32 pixels. The full dataset is split evenly between 10 classes and partitioned into 50,000 training and 10,000 test images. The model is a four-layer convolution network ending with two fully connected layers before the softmax layer. It was trained with a dropout of 0.25. Lastly they also use the FEMNIST data set from LEAF framework [6]. This is an image classification dataset which consists of 62 classes and the dataset is inherently non-IID with data

---

[1] http://yann.lecun.com/exdb/mnist/.

quantity and class distributions heterogeneity. The standard model architecture as provided in LEAF [5] is used in this case.

Training Hyperparameters TiFL uses RMSprop as the optimizer in local training and set the initial learning rate (η) as 0.01 and decay as 0.995. Local batch size of each party is 10, and local epochs is 1. For CIFAR10 the total number of parties ($|K|$) is 50 and the number of participated parties ($|C|$) at each round is 5. For FEMNIST they use the same number of total parties and parties per round as CIFAR10 and default training parameters provided by the LEAF Framework (SGD with lr 0.004, batch size 10). TiFL trains for a total of 2000 rounds for FEMNIST and 500 rounds for the synthetic datasets. Every experiment runs 5 times and uses the average values.

Heterogeneous Resource Setup All the parties are split into 5 groups with equal parties per group. For MNIST and Fashion-MNIST, each group is assigned with 2 CPUs, 1 CPU, 0.75 CPU, 0.5 CPU, and 0.25 CPU per part, respectively. For the larger Cifar10 and FEMINIST model, each group is assigned with 4 CPUs, 2 CPUs, 1 CPU, 0.5 CPU, and 0.1 CPU per part, respectively. This leads to varying training time for parties belong to different groups. By using the tiering algorithm of TiFL, there are 5 tiers.

Heterogeneous Data Distribution FL differs from the datacenter distributed learning in that the parties involved in the training process may have non-uniform data distribution in terms of amount of data per party and the non-IID data distribution. For *data quantity heterogeneity*, the training data sample distribution is 10%, 15%, 20%, 25%, 30% of total dataset for difference groups, respectively, unless otherwise specifically defined. For *non-IID heterogeneity*, the TiFL paper uses different non-IID strategies for different datasets. For MNIST and Fashion-MNIST, they adopt the setting in [14], where they sort the labels by value first, divide into 100 shards evenly, and then assign each party two shards so that each party holds data samples from at most two classes. For Cifar10, they shared the dataset unevenly in a similar way and limit the number of classes to 5 per party (non-IID(5)) following [13, 17] unless explicitly mentioned otherwise. In the case of FEMINIST they use its default non-IID-ness.

Scheduling Policies They evaluate several different naive scheduling policies of the proposed tier-based selection approach, defined by the selection probability from each tier, and compare it with the state-of-the-practice policy (or no policy) that existing FL works adopt, i.e., randomly select 5 parties from all parties in each round [4, 14], agnostic to any heterogeneity in the system named $\boxed{standard}$ (called *vanilla* in the plots and by the original paper [8]). $\boxed{fast}$ is a policy that TiFL only selects the fastest parties in each round. $\boxed{random}$ demonstrates the case where the selection of the fastest tier is prioritized over slower ones. $\boxed{uniform}$ is a base case for the tier-based naive selection policy where every tier has an equal probability

Table 11.1 Scheduling policy configurations

DataSet	Policy	Selection probabilities				
Cifar10/FEMNIST		Tier 1	Tier 2	Tier 3	Tier 4	Tier 5
	standard	N/A	N/A	N/A	N/A	N/A
	slow	0.0	0.0	0.0	0.0	1.0
	uniform	0.2	0.2	0.2	0.2	0.2
	random	0.7	0.1	0.1	0.05	0.05
	fast	1.0	0.0	0.0	0.0	0.0
MNIST/FMNIST	standard	N/A	N/A	N/A	N/A	N/A
	uniform	0.2	0.2	0.2	0.2	0.2
	fast1	0.225	F0.225	0.225	0.225	0.1
	fast2	0.2375	0.2375	0.2375	0.2375	0.05
	fast3	0.25	0.25	0.25	0.25	0.0

of being selected. $\boxed{slow}$ is the worst policy that TiFL only selects parties from the slowest tiers and is included to demonstrate the worst-case scenario. The paper also demonstrates the sensitivity analysis when the policy prioritizes more aggressively toward the fast tier, i.e., from $\boxed{fast1}$ to $\boxed{fast3}$, the slowest tier's selection probability has reduced from 0.1 to 0 while all other tiers got equal probability. They also include the $\boxed{uniform}$ policy for comparison, which is the same as in CIFAR-10. Table 11.1 summarizes all these scheduling policies by showing their selection probabilities.

11.4.1.2 Training Time Estimation via Analytical Model

In this section, we discuss the accuracy of the training time estimation model on different naive tier selection policies by comparing the estimation results of the model with the measurements obtained from testbed experiments. The estimation model takes as input of the profiled average latency of each tier, the selection probabilities, and total number of training rounds to estimate the training time. The authors use mean average prediction error (MAPE) as the evaluation metric, which is defined as follows:

$$\text{MAPE} = \frac{|L_{all}^{est} - L_{all}^{act}|}{L_{all}^{act}} * 100, \tag{11.7}$$

where L_{all}^{est} is the estimated training time calculated by the estimation model and L_{all}^{act} is the actual training time measured during the training process. Table 11.1

Table 11.2 Estimated VS
actual training time

Policy	Estimated (s)	Actual (s)	MAPE (%)
Slow	46,242	44,977	2.76
Uniform	12,693	12,643	0.4
Random	5143	5053	1.8
Fast	1837	1750	5.01

demonstrates the comparison results. The results suggest the analytical model is very accurate as the estimation error never exceeds more than 6% (Table 11.2).

11.4.2 Resource Heterogeneity

In this section, we show the performance of TiFL with static selection policies in terms of training time and model accuracy in a resource heterogeneous environment as depicted in Sect. 11.4.1 and assume there is no data heterogeneity. TiFL is evaluated with adaptive selection policy in Sect. 11.4.5. In practice, data heterogeneity is a norm in FL, and this scenario demonstrates how TiFL can tame resource heterogeneity alone. The scenario with both resource and data heterogeneity is shown in Sect. 11.4.4.

The results are organized in Fig. 11.3 (column 1), which clearly indicate that when prioritized toward the fast tiers, the training time reduces significantly. Compared with *standard*, *fast* achieves almost 11 times improvement in training time, see Fig. 11.3a. One interesting observation is that even *uniform* has an improvement of over 6 times over the *standard*. This is because the training time is always bounded by the slowest party selected in each training round. In TiFL, selecting parties from the same tier minimizes the straggler issue in each round and thus greatly improves the training time. For accuracy comparison, Fig. 11.3c shows that the difference between polices is very small, i.e., less than 3.71% after 500 rounds. However, if we look at the accuracy over wall-clock time, TiFL achieves much better accuracy compared to *standard*, i.e., up to 6.19% better if training time is constraint, thanks to the much faster per round training time brought by TiFL, see Fig. 11.3e. Note here that different policies may take very different amount of wall-clock time to finish 500 rounds.

11.4.3 Data Heterogeneity

In this section, the performance of TiFL under data heterogeneity due to both *data quantity heterogeneity* and *non-IID heterogeneity* is shown in Sect. 11.4.1. To demonstrate only the impact from data heterogeneity, they allocate homogeneous resource to each party, i.e., 2 CPUs per party.

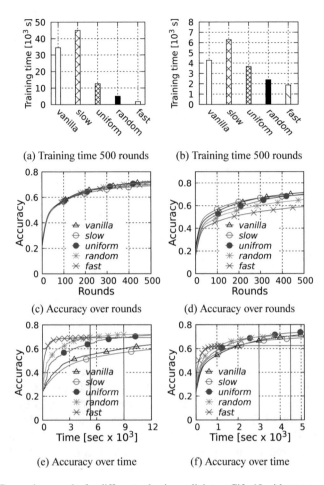

Fig. 11.3 Comparison results for different selection policies on Cifar10 with resource heterogeneity (0.5–4 CPUs) and homogenous data quantity (Column 1), and data quantity heterogeneity with homogenous resources (2 CPUs per party) (Column 2). (**a**) Training time 500 rounds. (**b**) Training time 500 rounds. (**c**) Accuracy over rounds. (**d**) Accuracy over rounds. (**e**) Accuracy over time. (**f**) Accuracy over time

Data Quantity Heterogeneity The training time and accuracy results are shown in Fig. 11.3 (column 2). From the training time comparison in Fig. 11.3b, it is interesting that TiFL also helps in data heterogeneity only case and achieves up to 3 times speedup. The reason is that *data quantity heterogeneity* may also result in different round time, which shares the similar effect as resource heterogeneity. Figure 11.3d and f shows the accuracy comparison, where we can see *fast* has relatively obvious drop compared to others because Tier 1 only contains 10% of the data, which is a significant reduction in volume of the training data. *slow* is also a heavily biased policy toward only one tier, but Tier 5 contains 30% of the

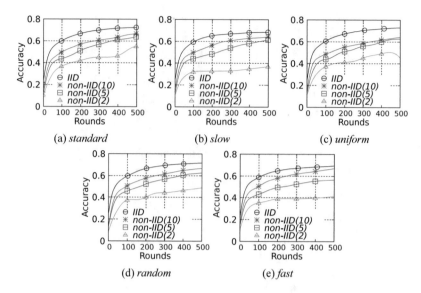

Fig. 11.4 Comparison results for different selection policies on Cifar10 with different levels of non-IID heterogeneity (Class) and fixed resources. (**a**) *standard*. (**b**) *slow*. (**c**) *uniform*. (**d**) *random*. (**e**) *fast*

data thus *slow* maintains good accuracy while worst training time. These results imply that like resource heterogeneity only, data heterogeneity only can also benefit from TiFL. However, policies that are too aggressive toward faster tier needs to be used very carefully as parties in fast tier achieve faster round time due to using less samples. It is also worth to point out that in the experiments the total amount of data is relatively limited. In a practical case where data is significantly more, the accuracy drop of *fast* is expected to be less pronounced.

Non-IID Heterogeneity Non-IID heterogeneity effects the accuracy. Figure 11.4 shows the accuracy over rounds given 2, 5, and 10 classes per party in a non-IID setting. The IID results in plots are also shown for comparison. These results show that as the heterogeneity level in non-IID heterogeneity increases, the accuracy impact also increases for all policies due to the strongly biased training data. Another important observation is that *standard* case and *uniform* have a better resilience than other policies, thanks to the unbiased selection behavior, which helps minimize further bias introduced during the party selection process.

11.4.4   Resource and Data Heterogeneity

This section presents the most practical case study with static selection policies, since here they evaluate with both resource and data heterogeneity combined.

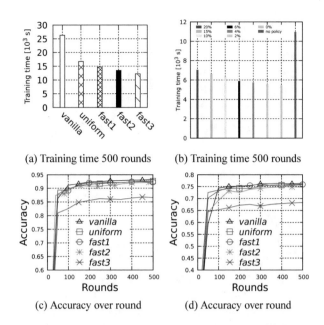

Fig. 11.5 Comparison results for different selection policies on MNIST (Column 1) and FMNIST (Column 2) with resource plus data heterogeneity. (**a**) Training time 500 rounds. (**b**) Training time 500 rounds. (**c**) Accuracy over round. (**d**) Accuracy over round

TiFL is evaluated for both resource and data heterogeneity combined with adaptive selection policy in Sect. 11.4.5.

MNIST and **Fashion-MNIST (FMNIST)** results are shown in Fig. 11.5 columns 1 and 2, respectively. Overall, policies that are more aggressive toward the fast tiers bring more speedup in training time. For accuracy, all polices of TiFL are close to *standard*, except *fast3* falls short as it completely ignores the data in Tier 5.

Cifar10 results are shown in Fig. 11.6 column 1. It presents the case of resource heterogeneity plus non-IID data heterogeneity with equal data quantities per party and the results are similar to resource heterogeneity only since non-IID data with the same amount of data quantity per party results in a similar effect of resource heterogeneity in terms of training time. However, the accuracy degrades slightly more here as because of the non-IID-ness the features are skewed, which results in more training bias among different classes.

Figure 11.6 column 2 shows the case of resource heterogeneity plus both the data quantity heterogeneity and non-IID heterogeneity. As expected, the training time shown in Fig. 11.6b is similar to Fig. 11.6a since the training time impact from different data amounts can be corrected by TiFL. However, the behaviors of round accuracy are quite different here as shown in Fig. 11.6d. The accuracy of *fast* has degraded a lot more due to the data quantity heterogeneity as it further amplifies the training class bias (i.e., the data of some classes become very little to none) in the already very biased data distribution caused by the non-IID heterogeneity. Similar

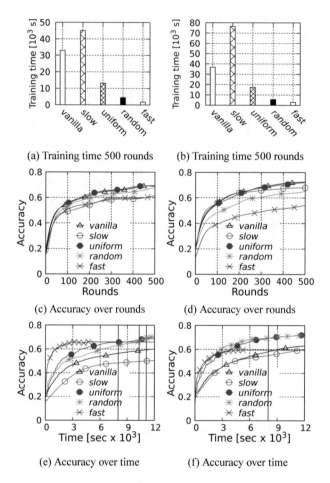

Fig. 11.6 Comparison results for different selection policies on Cifar10 with resource plus non-IID heterogeneity (Column 1) and resource, data quantity, and non-IID heterogeneity (Column 2). (**a**) Training time 500 rounds. (**b**) Training time 500 rounds. (**c**) Accuracy over rounds. (**d**) Accuracy over rounds. (**e**) Accuracy over time. (**f**) Accuracy over time

reasons can explain for other policies The best performing policy in accuracy here is the *uniform* case and is almost the same as *standard*, thanks to the even selection nature which results in little increase in training class bias. Figure 11.6f shows the wall-clock time accuracy. As expected, the significantly improved per round time in TiFL shows its advantage here as within the same time budget, more iterations can be done with shorter round time and thus remedies the accuracy disadvantage per round. *fast* still falls short than *standard* in the long run as the limited and biased data limits the benefits of more iterations. *fast* also perform worse than *standard* as it has no training advantage.

11.4.5 Adaptive Selection Policy

The above evaluation demonstrates the naive selection approach in TiFL that can significantly improve the training time but sometimes can fall short in accuracy, especially when strong data heterogeneity presents as such approach is data heterogeneity agnostic. In this section, we show the evaluation results of the proposed *adaptive* tier selection approach of TiFL, which takes into consideration of both resource and data heterogeneity when making scheduling decisions without privacy violation. *Adaptive, standard* (i.e., *vanilla* in the plots) and *uniform* are compared, and the latter is the best accuracy performing static policy.

Figure 11.7 shows that the *adaptive* policy outperforms both *standard* and *uniform* policies in both training time and accuracy for resource heterogeneity with data quantity heterogeneity (Amount) and non-IID heterogeneity (Class), thanks to the data heterogeneity-aware schemes. In the combined resource and data heterogeneity case (Combine), *adaptive* achieves comparable accuracy with *standard* with almost half of the training time and a slightly higher training time compared to *uniform*. The time difference arises when the adaptive policy tries to balance training time and accuracy, i.e., the 10% difference in training time is for the trade-off of achieving around 5% better accuracy. The other policy which achieves this accuracy is *standard*, which has almost 2× more training time. Considering this, we note that the training time difference is not significant and performs similar as *uniform* in training time while improves significantly in accuracy.

The above robust performance of *adaptive* is credited to both the resource and data heterogeneity-aware schemes. The accuracy over rounds for different policies under different non-IID heterogeneity are shown in Fig. 11.8. It is clear that *adaptive* consistently outperforms *standard* and *uniform* in different level of non-IID heterogeneity.

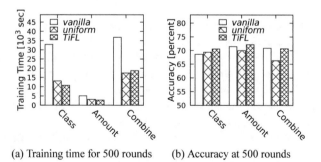

(a) Training time for 500 rounds   (b) Accuracy at 500 rounds

Fig. 11.7 Comparison results for different selection policies on Cifar10 with data quantity heterogeneity (Amount), non-IID heterogeneity (Class), and resource plus data heterogeneity (Combine). (**a**) Training time for 500 rounds. (**b**) Accuracy at 500 rounds

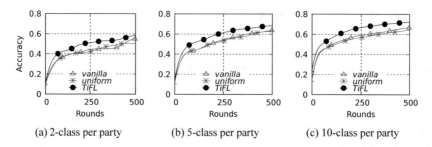

Fig. 11.8 Comparison results of Cifar10 under non-IID heterogeneity (Class) for different party selection policies with fixed resources (2 CPUs) per party. (**a**) 2-class per party. (**b**) 5-class per party. (**c**) 10-class per party

11.4.6 Adaptive Selection Policy

This section provides the evaluation of TiFL using a widely adopted large scale distributed FL dataset FEMINIST from the LEAF framework [6]. This uses exactly the same configurations (data distribution, total number of parties, model and training hyperparameters) as mentioned in [6] resulting in total number of 182 parties, i.e., deploy-able edge devices. Since LEAF provides its own data distribution among devices the addition of resource heterogeneity results in a range of training times thus generating a scenario where every edge device has a different training latency. The system here incorporates TiFL's tiering module and selection policy to the extended LEAF framework. The profiling modules collect the training latency of each parties and creates a logical pool of tiers which is further utilized by the scheduler. The scheduler selects a tier and then the edge parties within the tier in each training round. For the experiments with LEAF, the paper limits the total number of tiers to 5 and during each round we select 10 parties, with 1 local epoch per round.

Figure 11.9 shows the training time and accuracy over rounds for LEAF with different party selection policies. Figure 11.9a shows the training time for different selection policies. The least training time is achieved by using the *fast* selection policy; however, it impact the final model accuracy by almost 10% compared to *standard* selection policy. The reason for the least accuracy for *fast* is the result of less training point among the parties in tier 1. One interesting observation is that the *slow* policy out performs the *fast* policy in terms of accuracy even though each of these selection policies rely on data from only one tier. It must be noted that the slow tier is not only the reason of less computing resources but also the higher quantity of training data points. These results are consistent with the observations from the results presented in Sect. 11.4.3.

Figure 11.9b shows the accuracy over rounds for different selection policies. The proposed *adaptive* selection policy achieves 82.1% accuracy and outperforms the *slow* and *fast* selection policies by 7% and 10%, respectively. The *adaptive* policy is on par with the *standard* and *uniform* (82.4% and 82.6%, respectively), when

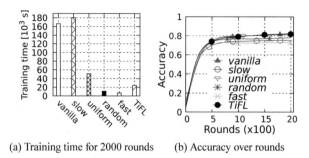

(a) Training time for 2000 rounds   (b) Accuracy over rounds

Fig. 11.9 Comparison results for different selection policies on LEAF with default data hetero-geneity (quantity, non-IID heterogeneity), and resource heterogeneity. (**a**) Training time for 2000 rounds. (**b**) Accuracy over rounds

comparing the total training time for 2000 rounds *adaptive* achieves $7\times$ and $2\times$ improvement compared to *standard* and *uniform*, respectively. *fast* and *random* both outperformed the *adaptive* in terms of training time; however, even after convergence the accuracy for both of these selection policies shows a noticeable impact on the final model accuracy. The results for FEMNIST using the extended LEAF framework for both accuracy as well as training time are also consistent with the results reported in Sect. 11.4.5.

11.5   Conclusion

In this section, we have investigated and quantified the heterogeneity impact on "decentralized virtual supercomputer"—FL systems. We observe that stragglers are indeed an issue, and it can be further complicated by data heterogeneity. We presented a framework called TiFL and discussed it in detail as a means of demonstrating the impact of stragglers as well as an example of potential mitigation solutions.

References

1. Abadi M, Barham P, Chen J, Chen Z, Davis A, Dean J, Devin M, Ghemawat S, Irving G, Isard M et al (2016) Tensorflow: A system for large-scale machine learning. In: 12th USENIX symposium on operating systems design and implementation (OSDI 16), pp 265–283
2. Beimel A, Kasiviswanathan SP, Nissim K (2010) Bounds on the sample complexity for private learning and private data release. In: Theory of cryptography conference. Springer, pp 437–454
3. Bonawitz K, Ivanov V, Kreuter B, Marcedone A, McMahan HB, Patel S, Ramage D, Segal A, Seth K (2017) Practical secure aggregation for privacy-preserving machine learning. In: Proceedings of the 2017 ACM SIGSAC conference on computer and communications security. ACM, pp 1175–1191

4. Bonawitz K, Eichner H, Grieskamp W, Huba D, Ingerman A, Ivanov V, Kiddon C, Konecny J, Mazzocchi S, McMahan HB et al (2019) Towards federated learning at scale: System design. Preprint. arXiv:1902.01046
5. Caldas S, Konečny J, McMahan HB, Talwalkar A (2018) Expanding the reach of federated learning by reducing client resource requirements. Preprint. arXiv:1812.07210
6. Caldas S, Wu P, Li T, Konečný J, McMahan HB, Smith V, Talwalkar A (2018) Leaf: A benchmark for federated settings. Preprint. arXiv:1812.01097
7. Chai Z, Fayyaz H, Fayyaz Z, Anwar A, Zhou Y, Baracaldo N, Ludwig H, Cheng Y (2019) Towards taming the resource and data heterogeneity in federated learning. In: 2019 USENIX conference on operational machine learning (OpML 19), pp 19–21
8. Chai Z, Ali A, Zawad S, Truex S, Anwar A, Baracaldo N, Zhou Y, Ludwig H, Yan F, Cheng Y (2020) Tifl: A tier-based federated learning system. In: Proceedings of the 29th international symposium on high-performance parallel and distributed computing, pp 125–136
9. Kairouz P, McMahan HB, Avent B, Bellet A, Bennis M, Bhagoji AN, Bonawitz K, Charles Z, Cormode G, Cummings R et al (2019) Advances and open problems in federated learning. Preprint. arXiv:1912.04977
10. Konečný J, McMahan HB, Yu FX, Richtárik P, Suresh AT, Bacon D (2016) Federated learning: Strategies for improving communication efficiency. Preprint. arXiv:1610.05492
11. Krizhevsky A, Nair V, Hinton G (2014) The cifar-10 dataset. Online: http://www.cs.toronto.edu/kriz/cifar.html, 55
12. Li T, Sahu AK, Talwalkar A, Smith V (2019) Federated learning: Challenges, methods, and future directions. Preprint. arXiv:1908.07873
13. Liu L, Zhang J, Song SH, Letaief KB (2019) Edge-assisted hierarchical federated learning with non-iid data. Preprint. arXiv:1905.06641
14. McMahan HB, Moore E, Ramage D, Hampson S et al (2016) Communication-efficient learning of deep networks from decentralized data. Preprint. arXiv:1602.05629
15. Xiao H, Rasul K, Vollgraf R (2017) Fashion-mnist: a novel image dataset for benchmarking machine learning algorithms. Preprint. arXiv:1708.07747
16. Yang Q, Liu Y, Chen T, Tong Y (2019) Federated machine learning: Concept and applications. ACM Trans Intell Syst Technol (TIST) 10(2):12
17. Zhao Y, Li M, Lai L, Suda N, Civin D, Chandra V (2018) Federated learning with non-iid data. Preprint. arXiv:1806.00582

Chapter 12
Systems Bias in Federated Learning

Syed Zawad, Feng Yan, and Ali Anwar

Abstract Data parties typically vary significantly in data quality, hardware resources, and stability, which results in challenges such as increased training times, higher resource costs, sub-par model performance and biased training. They result in hard systems challenges, and many existing works tend to address each of these challenges in isolation. Specifically, the bias in hardware and data consequently causes biased models. Additional challenges are introduced when party dropouts are considered. While the "Stragglers Management" chapter focuses mostly on the impact of stragglers, this chapter focuses on the impact of biasness. We take a look at how factors such as device dropouts, biased device data, biased participation, etc. affect the FL process from a systems perspective. We present a characterization study that empirically demonstrates how these challenges together impact important performance metrics such as model error, fairness, cost, and training time, and why it is important to consider them together instead of in isolation. We then talk about a method called *DCFair* which is a framework that comprehensively considers the multiple aforementioned important challenges of practical FL systems. Discussions on the characterization study and possible solutions are useful to gain a much deeper understanding of the inter-dependency of systems properties in Federated Learning.

12.1 Introduction

The prevalence of mobile and internet-of-things (IoT) devices in recent years has led to massive amount of data that can be potentially used to train state-of-the-art machine learning models. However, regulations such as HIPAA [3, 36] and GDPR [16, 43] limit the access and transmission of personal data in consideration

S. Zawad (✉) · F. Yan
University of Nevada, Reno, Reno, NV, USA
e-mail: szawad@nevada.unr.edu; fyan@unr.edu

A. Anwar
IBM Research – Almaden, San Jose, CA, USA
e-mail: ali.anwar2@ibm.com

of security and privacy. To enable training ML models using personal data with privacy protection, Federated Learning (FL) has been proposed [23, 31], where the main idea is that the data owners/parties train ML models locally and only sent trained weights to the third-party for aggregation. For example, [30] uses manufacturing robot data to predict their failure across different factories without revealing their actual manufacturing process. Leroy et al. [25] utilizes app-generated data for training keyword spotting models. In conventional ML training, data is typically owned by a single party and maintained in a centralized location. Both data and computation can be controllably distributed over a cluster of computing nodes. However, data in FL is generated and owned by parties and the privacy requirements prevent accessing or moving the personal data during training. This leads to a few challenges unique to FL. One such problem is that data can vary significantly in quality and quantity among parties (termed *data heterogeneity*). This can result in biased model training and an overall sub-par model [26, 40, 45]. Bias in FL models is generally measured by the *variance of accuracies* of the global model after being evaluated on the data of individual party's test datasets (termed *good-intent fairness* [33]). The quantity of datapoints per device also varies widely, resulting in different amounts of *training cost* incurred by each device.

Another challenge is that the local training times vary greatly across parties depending on their hardware resources (known as *resource heterogeneity*) which can result in stragglers and thus longer overall training time. Furthermore, mobile and IoT devices are not dedicated to the training tasks. Only when parties meet certain criteria for device properties such as battery status, idle time, training time, and network status [6], they can participate in training. This can change even in the middle of training, resulting in the phenomenon of *party drops*, which further leads to imbalanced training.

Data heterogeneity, resource heterogeneity, and party drops are important characteristics of FL and the study (Sect. 12.3) demonstrates that they can heavily impact model error, fairness, cost, and training time of a model. However, previous works consider them in isolation, resulting in sub-optimal models and long training time. For example, [6, 8, 11, 17, 35] partially addressed the straggler problem but did not consider model performance, cost nor fairness issues. Systems such as [6, 44] pointed out that the party drop phenomenon is common and important in federated learning but did not provide detailed analysis nor a solution to addressing its negative impact. Other works like [9, 39, 47] considered local training costs but ignored other factors. Mohri et al. [33] and Yu et al. [48] discussed the training fairness without considering the impact of party drop and problems associated with resource heterogeneity. The line of work on novel aggregation schemes [37, 40] mainly focuses on data heterogeneity challenges.

In this chapter, we discuss a framework called *DCFair*, a holistic approach that considers the impact of data heterogeneity, resource heterogeneity, and party drop on model error, fairness, cost, and training time when selecting parties to query. Based on the quantified impact of these factors on model convergence and training time, *DCFair* makes judicious party selection decisions to achieve the most profitable training results. *DCFair* formulates the problem as a multi-objective optimization and considers two key properties when making a scheduling decision: *selection probability* and *selection mutualism*. Both properties are derived by taking into consideration data heterogeneity, resource heterogeneity, and party drops. Selection probability describes how often a party should be selected and its quantification is empowered by a training efficiency assessment approach that employs Underestimation Index (UEI) [21] as a unified measure to represent model error, fairness, and cost while preserving the privacy requirements. Selection mutualism captures the mutualism among parties in terms of training time in a specific training round and aims at minimizing the straggler and party drop effects to improve the overall training time.

DCFair is implemented on a real distributed cluster where parties and aggregation server are deployed on their own hardware. It is then evaluated on the system using three benchmarks (FEMNIST, Cifar10, and Shakespeare). *DCFair* is also compared with the state-of-the-art large-scale FL systems [6] and the widely used bare-bone FL system [31]. We see that *DCFair* achieves better Pareto frontier and outperforms the state-of-the-art and state-of-the-practice systems with between 4% and 10% better model accuracy, improved good-intent fairness, lower cost, and faster training time [46]. In this chapter, we provide these discussions and evaluations as they can be a very useful way of gaining a better understanding how the many properties in FL systems come together to impact each other.

12.2 Background

For this section we discuss about the important FL properties to provide some context and background for the rest of the chapter. The current state of the art in systems which address the various corresponding properties are also discussed.

12.2.1 Fairness in Machine Learning

Fairness of models is an extensively explored concept in traditional ML [7, 12–14, 18, 38] and many works have defined their own notions of "fairness". For example, [24] introduces *counterfactual fairness* where a decision is considered fair towards an individual if the decision taken by a model would be the same if that individual belonged to a different sample group. This topic was recently explored in FL by Abay et al. [2]. Dwork et al. [15] talks about *classification fairness* which measures how much a model is biased during inference towards or against a particular target class. Hardt et al. [18] proposes a criterion for discrimination against sensitive attributes for protected classes in general supervised learning. Mehrabi et al. [32] extensively discusses current fairness issues in ML. While these approaches focus on mitigating bias for unprivileged groups, e.g., race or gender, the fairness definition does not consider such protected attributes.

12.2.2 Fairness in Federated Learning

Good-intent fairness was defined as the variance of party test accuracies of a model in [33]. If a model performs well on one party's dataset and bad on another, it indicates that the model is biased against the features of the worse-performing party and therefore is not fair. In this section we use this fairness definition. Mohri et al. [33] also propose a minimax optimization framework called Agnostic Federated Learning (AFL) to reduce overfitting on local party data by optimizing with learning bounds on the parties with the highest losses. However, AFL does not consider resource usage or the biased participation of parties which are important practical concerns in FL.

Balakrishnan et al. [4] talks about utilizing resources fairly but does not take resource heterogeneity or data heterogeneity into account. Li et al. [27] proposes q-FFL, which is a method to reduce biasness in the global model by making the party accuracies more uniform (i.e., increasing *good-intent fairness*). They do this by assigning more weights to the party updates with higher empirical loss values, thereby ensuring that the worst party updates can still contribute enough to the global model and get a more uniform testing accuracy across parties. For this work, we use the same definition of fairness and the objective is the same. However, instead of focusing on the aggregation algorithm, we focus on the party drops phenomenon of FL (i.e., how to be fair to parties if they do not consistently contribute to the FL training process). This work works under the same assumptions as [33] in that they assume equal participation of all parties. Yu et al. [48] and Khan et al. [22] talk about fairness not in terms of *good-intent fairness* but how much value a party gets from participation. Costs are considered in terms of monetary compensation, which is orthogonal to this work. In this chapter, the focus is on the cost in terms of resource efficiency (total samples used in training) instead.

12.2.3 Resource Usage in Federated Learning

Most works in FL focus on communication and energy efficiency [39, 42, 44, 47], but few have explored policy-driven schedulers. Wang et al. [44] theoretically analyzes the tradeoff between local update and global parameter aggregation to minimize the loss function under a given resource budget. Mulya Saputra et al. [39] uses reinforcement learning for optimizing caching, local computation and communication efficiency. Nishio and Yonetani [35] selects parties every round such that they can complete training within a given time limit, thereby controlling the amount of resources consumed per round. Caldas et al. [9] focuses on reducing model size using compression methods and update frequencies resulting in less resources used overall. Muhammad et al. [34] proposed a novel aggregation and global model distribution scheme that reduces time to converge and reduces communication cost early in the training process. Reisizadeh et al. [37] introduces *FedPAQ* with the aim to reduce communication overhead of too many devices trying to communicate with the central aggregator at the same time. Bonawitz et al. [6] proposes a comprehensive system to enable large-scale distributed FL frameworks. Hosseinalipour et al. [19] focuses on scaling up wireless communication systems for edge devices. Some works also focus on resource and data heterogeneity [8, 29, 41]. Li et al. [26] introduces *FedProx*, an aggregation algorithm that takes into account data heterogeneity to get a better model. Chai et al. [11] proposes a novel system to mitigate the effect of stragglers without compromising model performance. For this chapter, we take a different perspective on resource usage focusing less on optimizing resource usage at the local and global levels and more on optimizing *resource efficiency*, which to our knowledge is the first work addressing this issue in the scope of FL.

12.3   Characterization Study

Since systematically characterizing multiple variability points of FL has not been studied in the past, in this section we start by formally defining multiple important metrics in FL for better understanding the problem. We then discuss a characterization study that demonstrates the tradeoffs involved between them.

12.3.1   Performance Metrics

The incentive for the *global model owners* in FL is to train a highly accurate and generalizable ML model using other parties' private data that would otherwise be unavailable. Model performance and training time are two relevant metrics that impact final applications for model owners. On the other hand, the incentive for the *local data owners* in FL is to get better services from global model owners by contributing their data to training under privacy protection. Thus data owners usually prefer good user experience (i.e., with as less cost as possible) and fair reward (i.e., the trained model performs well on their data). We identify four important performance metrics when evaluating FL: *model error*, *training time*, *cost*, and *fairness*. They are defined as follows:

- *Model Error* is defined as the test accuracy error on all datasets, i.e., *mean* error of the global model on each of the party's sampled test data, i.e., $1 - \frac{\sum_{i=0}^{n} A^i}{n}$ where A^i is the accuracy of global model on test data of i and n is the total number of parties.

- *Training time* is defined as the wall-clock time of training. Wall-clock time is chosen instead of training rounds, as the round time can differ significantly due to data and resource heterogeneity.
- *Cost* is defined as the samples that have been used for training. Note that even if a party drops out during training, the used data samples also count into the cost. Training samples is used instead of resource hours as the resource in FL is highly heterogenous across parties. It is worth noting that more sophisticated cost metrics can also be used such as carbon footprint, executed floating point operations, but we use this for the sake of simplicity.
- *Fairness* is defined as *good-intent fairness* [33] that measures the accuracy variance when the global model is evaluated using test datasets of individual parties, i.e., the *variance* of the accuracies of the global model on each of the party's test data represented as $\sqrt{\frac{\sum_{i=0}^{n}(A^i-\bar{A})^2}{n-1}}$ where A^i is the accuracy of global model on test data of i, n is the total parties and $\bar{A}$ is the mean accuracy. Good-intent fairness was chosen as it effectively reflects the bias issue among parties—the main contributors in FL. The lower the *Fairness* value, the more fair the model is.

These metrics quantify the different performance aspects of federated learning systems. In order to understand how these metrics influence each other, next we discuss a set of characterization studies.

12.3.2 Tradeoff Between Fairness and Training Time

One of the focuses of state-of-the-art large-scale FL systems such as [6] is on the reduction of overall training time. Due to the highly heterogeneous nature of the local parties, the training latencies (defined as party's local training time) vary greatly. Given the round training time is bounded by the slowest party (i.e., straggler), the straggler effects significantly impact the overall training time. To address this, Bonawitz et al. [6] suggests selecting 130% of parties but only use the weights from the first 100% for training the global model and discard the weights of the slowest 30%. While this does handle the straggler problem, it also results in biased training since this approach always drops out the slower parties.

Figure 12.1a demonstrates the tradeoff between fairness and training time (see the Evaluation section for the experiment setup). The *GLOBAL* curve represents the mean of all the parties' error. The *FAST* curve presents the mean error of the global model on the fastest 70% of the parties. *SLOW* is the mean error of the slowest 30% of parties. Finally, *DEFAULT* shows the mean *global* error for default FL systems where no party update is dropped. We observe a difference in test error of around 15% between the fastest 70% and slowest 30% of the parties, showing a significant difference in model performance between faster and slower parties, leading to poor *fairness*. However, Fig. 12.1b shows a significant reduction in total training time if the [6] policy (*LS-FL*) is implemented, proving that it can indeed

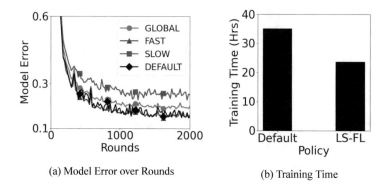

(a) Model Error over Rounds   (b) Training Time

Fig. 12.1 Tradeoff of fairness and training time. Figure (**a**) shows the test accuracies of parties with different training speeds over time. Using the LS-FL policy results in higher model error for the slower parties (SLOW), resulting in overall sub-par performance compared to that of the DEFAULT FL policy. (**b**) shows that LS-FL does reduce training time due to throwing away straggler updates

reduce training time. We observe from these results that the choice of party selection policy greatly affects model's fairness and training time, which demonstrates a tradeoff relationship.

12.3.3   Impact of Dropout on Fairness and Model Error

Apart from policy, party drops can also be caused on the party-side. As pointed out in [6, 9, 20, 28], one major issue of training on IoT devices is the availability as parties can dropout even in the middle of training. As such, FL training may suffer from dropped weights due to party-side downtime (*party-side dropouts*) even with a selection policy which is all-inclusive. This *party-side dropout* is non-deterministic and thus can be modeled as a probability. To study the effects of *party-side dropouts*, a randomly assigned probability called the *Dropout Ratio (DR)* is given to every party with an exponential distribution with a scale of 0.4, resulting a skewed *DR* distribution across parties. During training, whenever a party is selected it has the probability of dropping out equal to its assigned *DR*. The same experiment is run again as in the previous section using LEAF FEMNIST with the default policy instead of [6]. The CCDF of each of the party's test error distribution at convergence is presented in Fig. 12.2a for party *DROPOUTS* and compare it to the test error distribution derived if none of the parties had any *DR* (*DEFAULT*).

We observe that for the party *DROPOUTS* error distribution has a significantly longer tail than the *DEFAULT* distribution. The parties on that end are those with higher dropout ratios (*DR > 0.7*) and they tend to perform much worse than other parties. This demonstrates that dropping out parties from the training process results in the global model from being unable to train well on them and thus performing

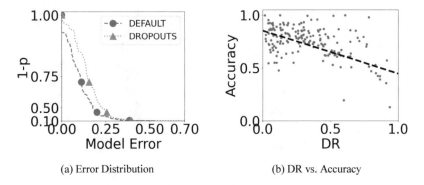

(a) Error Distribution   (b) DR vs. Accuracy

Fig. 12.2 (**a**) CCDF of the global model error on all the parties' local test datasets. The right tail of the error distribution is worse for parties that dropout, indicating that participation is vital for a more fairer training process. (**b**) Party test accuracy vs. Dropout ratio (DR). The higher the probability a party to drop out of training (i.e., increased DR), the higher the chances of the global model performing worse on its test dataset

very badly on their test dataset. We also observe that the mean model error in the case of party *DROPOUTS* is also significantly higher than party *DEFAULT*, indicating that the loss of training data due to party drops adversely effects the model's performance. In Fig. 12.2b we see the correlation between the *DR* assigned to a set of parties and their corresponding test accuracies at convergence. We observe a clear trend where it shows that with lower *DR*, the parties tend to participate more in the training process and so achieves higher accuracies and vice versa. From these results, we conclude that party drops reduce participation of parties throughout the whole training process resulting in reduced global accuracy as well as unfairness. In order to make up for the skewed participation due to party drops, we shall discuss a selection policy that can increase participation of high-dropout parties without a significant bump in cost.

12.3.4   Tradeoff Between Cost and Model Error

One simple method of increasing overall participation is increasing the number of parties selected per round. This causes all the parties to have more opportunity to be selected in a round, thereby increasing total participation. The same experiment as in Fig. 12.2a is performed again with party drops, but by increasing total number of parties selected per round from 10 to 20 and observe its impact on *Cost* and *Model Error* in Fig. 12.3. In Fig. 12.3a we show the total number of samples trained over the same number of rounds (2000) for different number of parties selected per round for comparison. As expected, increasing the total participants by a factor of two also yields a twofold increase in resources consumed. While this does cause a decrease in the mean error (Fig. 12.3b) of the global model due to the increased

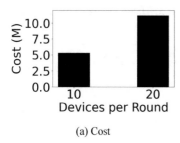
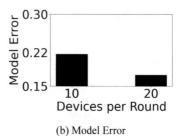

(a) Cost   (b) Model Error

Fig. 12.3 (**a**) Cost (total number of datapoints trained in 2000 rounds in Millions) vs. number of parties selected in each round. Increasing the selection amount significantly impacts overall cost for the system. (**b**) Mean model error for different numbers of parties selected per round. Even though more parties selected per round increases cost, it also results in better model performance thus demonstrating the tradeoff between these two factors

participation, a twofold increase is a significant burden on the local parties which are already resource constrained. From this experiment, we conclude that increasing participation can benefit in the training of the global model, but with a significant increase in local resource usage. In the next section, we detail the *DCFair*'s proposed party selection approach based on these findings.

12.4   Methodology

For this section, we define the problem formally and then use the observations from the characterization study to discuss how they contributed to the development of the *DCFair* framework.

12.4.1   Problem Formulation

The goal is to design an effective party selection scheduler that optimizes the performance metrics in FL. The scheduling parameter is defined as the selection probability of a party in each training round. Given there are four performance metrics (model error, fairness, cost, and training time) to consider, we can formulate the problem as multi-objective optimization. Assume that the FL system trains a global model G on a set of parties $D = [d_1, d_2, d_3, \ldots d_n, \ldots d_N]$ according to a party selection scheduler S defined as the selection probability of each party in training round i: $S_i = [s_1^i, s_2^i, s_3^i, \ldots s_n^i, \ldots s_N^i]$. Let the evaluation error of G on the data of individual party in D as $A = [a_1, a_2, a_3, \ldots a_n, \ldots a_N]$. The goal is to optimize the model's mean test error defined as $a(S) = 1 - mean(A)$, good-intent fairness defined as $f(S) = var(A)$, total training cost $c(S)$ defined as the total number of data points processed (including dropped out data points), and the training time $t(S)$:

$$minimize\ (a(S),\ f(S),\ c(S),\ t(S)).\qquad(12.1)$$

12.4.2 DCFair Overview

Simultaneously optimizing model error, fairness, cost, and training time in FL is challenging as the data distribution is not accessible due to privacy requirements. In addition, the scheduling probability is difficult to be directly connected with these performance metrics. To solve the above challenges, the key idea is to find a measurable metric that have the following properties: (1) preserve the privacy requirements; (2) easy to be modeled with scheduling probability; (3) can represent and unify some or all the optimization metrics.

The Underestimation Index (UEI) proposed in [21] has potential to meet the above requirements. UEI is a metrics for measuring the distance between a model's prediction results and the actual labels, which is a good indicator of how well a model has learned the features of that dataset. It is defined as:

$$UEI_n\ =\ \frac{1}{\sqrt{2}}||\sqrt{P_n^{pr}}-\sqrt{P_n^{act}}||_2,\qquad(12.2)$$

where n is the party index number, P_n^{act} is the class distribution of the training dataset, and P_n^{pr} is the predicted class distribution of global model. UEI values range from 0.0 to 1.0, where higher UEI means more bias against the training dataset. A party with a high UEI value indicates that the features in the data of this party are not well captured in the global model, thus the party is "disenfranchised" so far and more training involvement of this party helps fairness. In addition, reducing UEI across all parties means the features of global data has been well captured and thus improve model error. Furthermore, the participation of parties with low UEI benefits less the training progress, thus such participation may reduce resource efficiency and incur high cost. For parties with the same UEI, their resource efficiency can be different, e.g., to reduce UEI by 10%, some party needs to train 10,000 samples while other party may only need to train 500 samples. To reflect the resource efficiency difference, *DCFair* introduces cost normalized UEI, defined as

$$CUEI_n\ =\ \frac{UEI_n}{c_n}.\qquad(12.3)$$

For optimizing training time, the main idea is to minimize the straggler and party drop effects. Here *DCFair* proposes the idea of selection mutualism, which captures the mutualism among training time of parties in a specific training round.

Specifically, parties with similar round training latency are given higher probability to be selected in the same round to reduce the straggler effects and the average dropout ratio of all parties in a round needs to be smaller than a user defined threshold. The selection mutualism is inspired by the tiered FL approach proposed in [11]. The proposed selection mutualism approach is more general as it removes the fixed tiers in [11] and adds support to mitigate party drop effects to optimize training time. *DCFair* employs the above methods to make optimal party selection scheduling decisions. Next, we introduce in detail provided in *DCFair* on how to quantify selection probability and selection mutualism and combine them to solve the multi-objective optimization problem defined in Eq. 12.1.

12.4.3 Selection Probability

Due to the party drop effects in FL, the eventual participation rate of a party, termed PR_n, depends on both the selection probability of a party S_n and its dropout ratio DR_n:

$$PR_n = S_n \times (1 - DR_n). \tag{12.4}$$

To design a party selection scheduler that can minimize model error, fairness, and cost, the selection probability can be set so that $CUEI$ is minimized. In other words, party with higher $CUEI$ needs higher selection probability. In addition, parties with high-dropout ratio also need to be compensated with higher selection probability so that their eventual participation rate can be consistent with their selection probability. Therefore, first the participation rate of party n as a function of $CUEI$ and then add the party drop ratio to compute selection probability. The function used by *DCFair* is a standard exponential function as it produces a proper skew from $CUEI$ to participation rate. Specifically:

$$PR_n^i = f(CUEI_n^i) = \sigma * \frac{1}{e^{-CUEI_n^i}}, \tag{12.5}$$

where i is the round index and n is the party index. σ is a normalization term that converts $CUEI$ based metrics into a probability based metrics. By adding the party drop ratio, the selection probability of a party n at round i would be:

$$S_n^i = \begin{cases} \frac{PR_n^i}{1-DR_n^i} = \sigma * \frac{1}{e^{-CUEI_n^i} \times (1-DR_n^i)} & if \ \ DR_n^i < 1.0 \\ PR_n^i = \sigma * \frac{1}{e^{-CUEI_n^i}} & if \ \ DR_n^i = 1.0. \end{cases} \tag{12.6}$$

Algorithm 12.1 DCFair Algorithm. w_i: the global model for round i, D: List of all participating parties, R: Total # of training rounds, I: Metric update frequency, UEI, c, DR, L: List of UEI, C, DR and *training time* metrics for each party, DR_{max}: minimum average DR in a round

---

 1: **Aggregator:** initialize weight w_0.
 2: **for** each round $i = 1$ **to** R **do**
 3:     **if** $i\%I == 0$ **then**
 4:         $Send Global Model(w_i, D)$
 5:         $UEI, c, DR, L = Get Party Metrics(D)$
 6:     **end if**
 7:     $S = $ (Calculate using Eq. 6 and 7 with UEI, c, DR)
 8:     $d = $ (randomly select one party from all parties using S)
 9:     $S' = $ (Calculate using Eq. 8 and 9 S, L)
10:     $s = $ (randomly select n parties using S' such that DR_{max} is met)
11:     $w_{i+1} = Train(s + d)$
12: **end for**

Because the party selection probability sums to 1 ($\sum_{n=1}^{N} S_n^i = 1$). We can compute σ as:

$$\sigma = \begin{cases} \dfrac{1}{\sum_{n=1}^{N} \frac{1}{e^{-CUEI_n^i \times (1-DR_n^i)}}} & if \ DR_n^i < 1.0 \\[4mm] \dfrac{1}{\sum_{n=1}^{N} \frac{1}{e^{-CUEI_n^i}}} & if \ DR_n^i = 1.0. \end{cases} \tag{12.7}$$

12.4.4 Selection Mutualism

As the round training time is bonded by the slowest party (i.e., straggler), the key idea to minimize the straggler effect is to adjust the selection probability so that parties with similar training latency can be selected in the same round. Specifically, in a training round, after selecting the first party, *DCFair* uses its training latency as the standard of this round, denoted as L. *DCFair* adjusts parties' selection probability based on the training latency difference between theirs and L. They formulate the mutualism adjusted selection probability as:

$$S_n'^i = f(S_n^i, L_n, L) = \theta * S_n^i * e^{|L_n - L|}, \tag{12.8}$$

where L_n is the training latency of party n. They select exponential function as an example to reflect the training latency difference's impact on selection probability and such function can be changed to adjust the impact. θ is a normalization coefficient such that $\sum_{i=1}^{N} S_n'^i = 1$ and can be computed as

$$\theta = \frac{1}{\sum_{n=1}^{N} S_n^i * e^{|L_n - L|}}. \tag{12.9}$$

To minimize the party drop effects, the average dropout ratio of selected parties should be below a threshold DR_{max}. This is to avoid the situation where too many parties dropped out and the remaining number of parties could not meet the requirement of minimum participants (e.g., too few participants may result in failed privacy protection such as differential privacy [1] and secure aggregation [5]). DR_{max} can be configured based on the specific scenarios.

The total training time $t(S)$ can be computed as:

$$t(S) = \sum_{i=1}^{I} LS^i, \tag{12.10}$$

where i is the training round index and I is the total number of rounds. LS^i is the training latency of the slowest party selected in round i, which is impacted by the mutualism based selection probability adjustment above. The detailed algorithm of DCFair is presented in Algorithm 12.1.

12.5 Evaluation

In this section, *DCFair* is compared against [6] (named *LS-FL* for convenience) and the *DEFAULT* FL system [31] by comparing the four performance metrics (model error, fairness, cost, and training time) using different applications.

Benchmarks *DCFair* is evaluated on a real distributed cluster with three benchmarks. They use **Cifar10**, which has been widely used in FL literature [2, 11, 49]. They also use **FEMNIST** (image classification) and **Shakespeare** (character prediction) that are from the federated learning framework *LEAF* [10], which provides a realistic data heterogeneous distribution between devices and has been considered as the new standard for recent state-of-the-art FL works [9, 11].

Testbed Setup The cluster is set up by deploying the aggregation server exclusively on a 32-CPU node and every party is deployed on separate 2-CPU nodes. The parties are launched on separate individual hardware (details in Table 12.1). The system is implemented using TensorFlow Keras and communication is handled via the socket protocol. Training latencies are injected on each party by instrumenting the training system via a *sleep* function approach. It is worth noting that the testbed is among the most practical and largest scale in FL research (to the best of knowledge, only [6, 11] used similar testbed) and most existing FL works, even the latest ones, use simulation based testbeds [5, 26, 49].

Table 12.1 Training setup

Dataset	Model	Train/Test split	Total parties/selected per round	Learning rate/Batch size
FEMNIST	2 conv 2 dense	53,839/5383	179/10	0.004/10
CIFAR10	4 conv 2 dense	50,000/10,000	100/10	0.0005/32
Shakespeare	256 cell lstm 1 dense	115,135/11,513	30/3	0.0003/2

Resource Heterogeneity It is generated by randomly assigning training latencies per party using a Gaussian distribution sampling with a mean of 5 s and a standard deviation of 1.5 s following [11]. This generates a set of parties with variable training latencies to reflect resource heterogeneity and help the training time analysis. Party drop ratios (probability of dropping out during a round) are also assigned to each party using an exponential distribution of 0.4, which provides enough high-dropout and low-dropout parties to have a noticeable impact on training.

12.5.1 Cost Analysis

Data Heterogeneity For Cifar10, data heterogeneity is generated using the class-wise distribution as defined in [11, 49]. For FEMNIST and Shakespeare, the default data heterogeneity and data quantity provided by *LEAF* are used. Further details of the datasets and training setup are given in Table 12.1.

12.5.2 Model Error and Fairness Analysis

We first see how *DCFair* performs under the fairness metric without any constraints. Figure 12.4 shows the *CCDF* of the test errors on the parties at convergence. For all cases, we note that the worst-performing system is *LS-FL* due to its large tail, as well as distinctly higher median and mean values than *DEFAULT* and *DCFair*. The reasons are twofold: (1) *LS-FL* actively discards slower parties, and (2) it has no mechanism to handle party drops, thereby falling victim to skewed participation. Since *DEFAULT* does not have an active party drop policy, the participation of parties is not as low as *LS-FL* and therefore achieves an overall lower mean and median error distributions. *DCFair* performs best overall in both mean and variance in distribution. This can be attributed to the policy taking into account the parties' dropout probability as well as *CUEI* when making participation decisions. Since *DCFair* promotes the participation of high-dropout parties as well as parties with data on which the model is under-fitting (parties with high *UEI*), for every round it

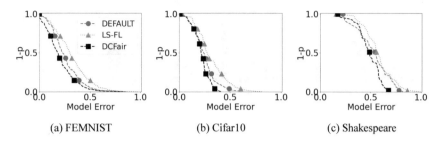

Fig. 12.4 Model error CCDF distribution of global model. DCFair can consistently have a lower distribution variance of accuracies (fairness) between devices compared to the *LS-FL* and *DEFAULT*. (**a**) FEMNIST. (**b**) Cifar10. (**c**) Shakespeare

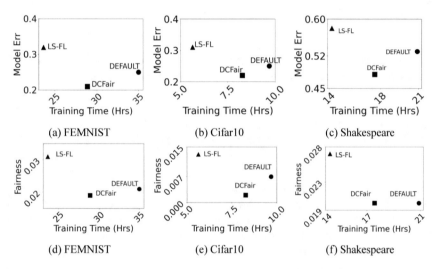

Fig. 12.5 Comparison of the training time, model error, and fairness (lower is better) for the frameworks *LS-FL, DEFAULT*, and DCFair. While *LS-FL* can achieve lower training time, DCFair performs better in terms of model error and fairness. The results are consistent across the three benchmarks. Please note that the lower the *fairness* value, the more fair the model is. (**a**) FEMNIST. (**b**) Cifar10. (**c**) Shakespeare. (**d**) FEMNIST. (**e**) Cifar10. (**f**) Shakespeare

chooses the parties that are most important to the model performance and fairness which actively benefits the overall training process.

12.5.3 Training Time Analysis

Next we evaluate how *DCFair* compares against other frameworks in terms of training time. In Fig. 12.5, we see the total training time at convergence against the mean and variance (i.e., fairness) for error distributions. We observe that *DCFair*

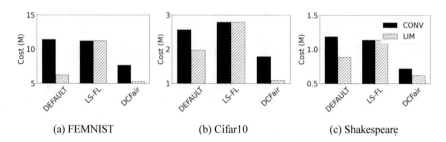

(a) FEMNIST   (b) Cifar10   (c) Shakespeare

Fig. 12.6 Cost comparison at the convergence time (CONV) and by limiting training time (LIM) to 24, 5, and 14 h for FEMNIST, Cifar10, and Shakespeare datasets, respectively. M stands for Millions. Time constraints are selected based on the fastest framework to converge, i.e., *LS-FL*. DCFair achieves consistently lower cost, thanks to its cost-aware design. (**a**) FEMNIST. (**b**) Cifar10. (**c**) Shakespeare

consistently has lower mean and variance of the test error distributions. *DEFAULT* has the highest training time across the board, with *LS-FL* performing the best since the former has no mechanism of handling stragglers and the latter simply selects the faster 75% of devices only. For example, in FEMNIST comparison, *DEFAULT* has a training time of 35 h, while *LS-FL* is around 24 h (around 0.7), while *DCFair* takes around 29 h. *LS-FL* is expected to perform the best in training time here since the slowest 25% of the parties never had a chance to participate in the training due to being biasedly dropped. However, *LS-FL* has the highest variance and test error across the board, showing that the training time reduction comes with a compromise of model error and fairness. *DCFair* performs better than the *DEFAULT* but is slower than *LS-FL* since it does not discriminate against slower parties. However, DCFair does perform better than *DEFAULT* thanks to its *selection mutualism*. It allows for more consistency of training time within rounds by grouping together parties such that within any round it only selects faster or slower parties, but not both. This reduces the probability of selecting slower parties in each round to reduce the overall training time.

For this section, we observe how DCFair performs in terms of cost compared to the other frameworks. Figure 12.6 shows the total cost incurred at convergence and within a time constraint. We observe that DCFair has lower cost than the other two systems across all datasets. Using $CUEI$ for party selection enables DCFair to be cost-aware, and as a result, DCFair tends to prioritize the selection of lower cost parties. Since both *DEFAULT* and *LS-FL* have no mechanisms to handle cost, they train significantly more datapoints and so have higher amount of cost.

12.5.4 Pareto Optimality Analysis

Lastly, we analyze all the performance metrics together in a full end-to-end manner by looking at the Pareto frontier. Figure 12.7 shows the model error and

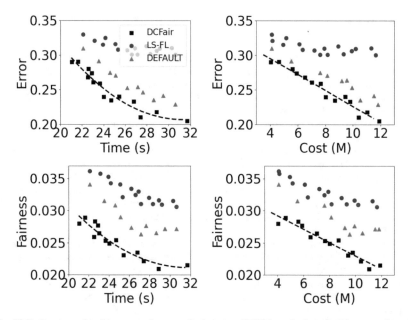

Fig. 12.7 Pareto optimality comparison results between DCFair and others in fairness and model error against training time and cost (datapoints trained). DCFair demonstrates the best tradeoffs

fairness against cost and training time for the FEMNIST dataset. *DCFair*, *LS-FL*, and *DEFAULT* are tuned such that they train for varying amounts of time and cost and plot their corresponding error and fairness values. Across each of the metric combinations, *DCFair* outperforms the other two frameworks by achieving a better Pareto frontier. *LS-FL* consistently performs worse with model error due to dropping out parties intentionally, which is also detrimental to fairness. While *DEFAULT* performs better, it is still unable to handle party drops and results in poor training time and cost. Therefore, DCFair achieves the best tradeoff among all the performance metrics.

12.6 Conclusion

In this chapter, we talk about the different properties of FL such as device drops, hardware heterogeneity, and fairness. We analyze how they are affected by each other, how their biasness can be detrimental to the model quality, and how they impact the overall training process and outcome. We talk about *DCFair*, the first system that comprehensively takes into consideration resource heterogeneity, data heterogeneity, and party drops to optimize a set of important performance metrics simultaneously in order to mitigate the impact of biasness. We use this framework under different tunings to demonstrate how the various metrics interact with each other and in this process gain a better understanding of the overall FL process.

References

1. Abadi M, Chu A, Goodfellow I, McMahan HB, Mironov I, Talwar K, Zhang L (2016) Deep learning with differential privacy. In: Proceedings of the 2016 ACM SIGSAC Conference on computer and communications security. ACM, pp 308–318
2. Abay A, Zhou Y, Baracaldo N, Rajamoni S, Chuba E, Ludwig H (2020) Mitigating bias in federated learning. Preprint. arXiv:2012.02447
3. Accountability Act (1996) Health insurance portability and accountability act of 1996. Public Law 104:191
4. Balakrishnan R, Akdeniz M, Dhakal S, Himayat N (2020) Resource management and fairness for federated learning over wireless edge networks. In: 2020 IEEE 21st international workshop on signal processing advances in wireless communications (SPAWC). IEEE, pp 1–5
5. Bonawitz K, Ivanov V, Kreuter B, Marcedone A, McMahan HB, Patel S, Ramage D, Segal A, Seth K (2017) Practical secure aggregation for privacy-preserving machine learning. In: Proceedings of the 2017 ACM SIGSAC conference on computer and communications security. ACM, pp 1175–1191
6. Bonawitz K, Eichner H, Grieskamp W, Huba D, Ingerman A, Ivanov V, Kiddon C, Konecny J, Mazzocchi S, McMahan HB et al (2019) Towards federated learning at scale: System design. Preprint. arXiv:1902.01046
7. Buolamwini J, Gebru T (2018) Gender shades: Intersectional accuracy disparities in commercial gender classification. In: Conference on fairness, accountability and transparency, pp 77–91
8. Cai L, Lin D, Zhang J, Yu S (2020) Dynamic sample selection for federated learning with heterogeneous data in fog computing. In: 2020 IEEE international conference on communications, ICC 2020, Dublin, Ireland, June 7–11, 2020. IEEE, pp 1–6
9. Caldas S, Konečny J, McMahan HB, Talwalkar A (2018) Expanding the reach of federated learning by reducing client resource requirements. Preprint. arXiv:1812.07210
10. Caldas S, Wu P, Li T, Konečný J, McMahan HB, Smith V, Talwalkar A (2018) Leaf: A benchmark for federated settings. Preprint. arXiv:1812.01097
11. Chai Z, Ali A, Zawad S, Truex S, Anwar A, Baracaldo N, Zhou Y, Ludwig H, Yan F, Cheng Y (2020) Tifl: A tier-based federated learning system. In: Proceedings of the 29th international symposium on high-performance parallel and distributed computing, pp 125–136
12. Chen IY, Szolovits P, Ghassemi M (2019) Can AI help reduce disparities in general medical and mental health care? AMA J Ethics 21(2):167–179
13. Cortes C, Mohri M, Medina AM (2015) Adaptation algorithm and theory based on generalized discrepancy. In: 21st ACM SIGKDD conference on knowledge discovery and data mining, KDD 2015. Association for Computing Machinery, pp 169–178
14. DiCiccio C, Vasudevan S, Basu K, Kenthapadi K, Agarwal D (2020) Evaluating fairness using permutation tests. In: Gupta R, Liu Y, Tang J, Prakash BA (eds) KDD '20: The 26th ACM SIGKDD conference on knowledge discovery and data mining, Virtual Event, CA, USA, August 23–27, 2020. ACM, pp 1467–1477
15. Dwork C, Hardt M, Pitassi T, Reingold O, Zemel R (2012) Fairness through awareness. In: Proceedings of the 3rd innovations in theoretical computer science conference, pp 214–226
16. General Data Protection Regulation (2016) Regulation (EU) 2016/679 of the European parliament and of the council of 27 April 2016 on the protection of natural persons with regard to the processing of personal data and on the free movement of such data, and repealing directive 95/46. Off J Eur Union (OJ) 59(1–88):294
17. Gudur GK, Balaji BS, Perepu SK (2020) Resource-constrained federated learning with heterogeneous labels and models. In: 3rd International workshop on artificial intelligence of things (AIoT'20), KDD 2020
18. Hardt M, Price E, Srebro N (2016) Equality of opportunity in supervised learning. In: Advances in neural information processing systems, pp 3315–3323

19. Hosseinalipour S, Brinton CG, Aggarwal V, Dai H, Chiang M (2020) From federated to fog learning: Distributed machine learning over heterogeneous wireless networks. IEEE Commun Mag 58(12):41–47

20. Kairouz P, McMahan HB, Avent B, Bellet A, Bennis M, Bhagoji AN, Bonawitz K, Charles Z, Cormode G, Cummings R et al (2019) Advances and open problems in federated learning. Preprint. arXiv:1912.04977

21. Kamishima T, Akaho S, Sakuma J (2011) Fairness-aware learning through regularization approach. In: 2011 IEEE 11th international conference on data mining workshops. IEEE, pp 643–650

22. Khan LU, Pandey SR, Tran NH, Saad W, Han Z, Nguyen MNH, Hong CS (2020) Federated learning for edge networks: Resource optimization and incentive mechanism. IEEE Commun Mag 58(10):88–93

23. Konečný J, McMahan HB, Yu FX, Richtárik P, Suresh AT, Bacon D (2016) Federated learning: Strategies for improving communication efficiency. Preprint. arXiv:1610.05492

24. Kusner MJ, Loftus J, Russell C, Silva R (2017) Counterfactual fairness. In: Advances in neural information processing systems, pp 4066–4076

25. Leroy D, Coucke A, Lavril T, Gisselbrecht T, Dureau J (2019) Federated learning for keyword spotting. In: ICASSP 2019–2019 IEEE international conference on acoustics, speech and signal processing (ICASSP). IEEE, pp 6341–6345

26. Li T, Sahu AK, Zaheer M, Sanjabi M, Talwalkar A, Smith V (2018) Federated optimization in heterogeneous networks. Preprint. arXiv:1812.06127

27. Li T, Sanjabi M, Beirami A, Smith V (2019) Fair resource allocation in federated learning. In: International conference on learning representations

28. Li T, Sahu AK, Talwalkar A, Smith V (2020) Federated learning: Challenges, methods, and future directions. IEEE Signal Process Mag 37(3):50–60

29. Lu S, Zhang Y, Wang Y (2020) Decentralized federated learning for electronic health records. In: 2020 54th Annual conference on information sciences and systems (CISS). IEEE, pp 1–5

30. Machine learning to augment shared knowledge in federated privacy-preserving scenarios (musketeer). https://musketeer.eu/wp-content/uploads/2019/10/MUSKETEER_D2.1-v1.1.pdf. Accessed 03 Jan 2021

31. McMahan HB, Moore E, Ramage D, Hampson S et al (2016) Communication-efficient learning of deep networks from decentralized data. Preprint. arXiv:1602.05629

32. Mehrabi N, Morstatter F, Saxena N, Lerman K, Galstyan A (2019) A survey on bias and fairness in machine learning. Preprint. arXiv:1908.09635

33. Mohri M, Sivek G, Suresh AT (2019) Agnostic federated learning. In: international conference on machine learning. PMLR, pp 4615–4625

34. Muhammad K, Wang Q, O'Reilly-Morgan D, Tragos E, Smyth B, Hurley N, Geraci J, Lawlor A (2020) Fedfast: Going beyond average for faster training of federated recommender systems. In: Proceedings of the 26th ACM SIGKDD international conference on knowledge discovery & data mining, pp 1234–1242

35. Nishio T, Yonetani R (2019) Client selection for federated learning with heterogeneous resources in mobile edge. In: ICC 2019–2019 IEEE international conference on communications (ICC). IEEE, pp 1–7

36. O'herrin JK, Fost N, Kudsk KA (2004) Health insurance portability accountability act (HIPAA) regulations: effect on medical record research. Ann Surg 239(6):772

37. Reisizadeh A, Mokhtari A, Hassani H, Jadbabaie A, Pedarsani R (2020) Fedpaq: A communication-efficient federated learning method with periodic averaging and quantization. In: International conference on artificial intelligence and statistics. PMLR, pp 2021–2031

38. Saleiro P, Rodolfa KT, Ghani R (2020) Dealing with bias and fairness in data science systems: A practical hands-on tutorial. In Gupta R, Liu Y, Tang J, Prakash BA (eds) KDD '20: The 26th ACM SIGKDD conference on knowledge discovery and data mining, Virtual Event, CA, USA, August 23–27, 2020. ACM, pp 3513–3514

39. Saputra YM, Hoang DT, Nguyen DN, Dutkiewicz E, Mueck MD, Srikanteswara S (2019) Energy demand prediction with federated learning for electric vehicle networks. In: 2019 IEEE global communications conference (GLOBECOM). IEEE, pp 1–6

40. Sattler F, Wiedemann S, Müller KR, Samek W (2019) Robust and communication-efficient federated learning from non-iid data. IEEE Trans Neural Netw Learn Syst 31(9):3400–3413

41. Savazzi S, Nicoli M, Rampa V (2020) Federated learning with cooperating devices: A consensus approach for massive IoT networks. IEEE Internet Things J 7(5):4641–4654

42. Sun Y, Zhou S, Gündüz D (2020) Energy-aware analog aggregation for federated learning with redundant data. In: ICC 2020–2020 IEEE international conference on communications (ICC). IEEE, pp 1–7

43. Voigt P, Von dem Bussche A (2017) The EU general data protection regulation (GDPR). A Practical Guide, 1st Ed., Springer International Publishing, Cham 10:3152676

44. Wang S, Tuor T, Salonidis T, Leung KK, Makaya C, He T, Chan K (2019) Adaptive federated learning in resource constrained edge computing systems. IEEE J Sel Areas Commun 37(6):1205–1221

45. Wang X, Han Y, Leung VCM, Niyato D, Yan X, Chen X (2020) Convergence of edge computing and deep learning: A comprehensive survey. IEEE Commun Surv Tutorials 22(2):869–904

46. Xu Z, Yang Z, Xiong J, Yang J, Chen X (2019) Elfish: Resource-aware federated learning on heterogeneous edge devices. Preprint. arXiv:1912.01684

47. Yang Z, Chen M, Saad W, Hong CS, Shikh-Bahaei M (2020) Energy efficient federated learning over wireless communication networks. IEEE Trans Wirel Commun 20:1935

48. Yu H, Liu Z, Liu Y, Chen T, Cong M, Weng X, Niyato D, Yang Q (2020) A fairness-aware incentive scheme for federated learning. In: Proceedings of the AAAI/ACM conference on AI, ethics, and society, pp 393–399

49. Zhao Y, Li M, Lai L, Suda N, Civin D, Chandra V (2018) Federated learning with non-iid data. Preprint. arXiv:1806.00582

Part III
Privacy and Security

Privacy and security are two important aspects that need to be considered for the applicability of federated learning in the enterprise and consumer space. In this part of the book, we cover in detail these aspects.

Chapter provides a comprehensive overview of existing privacy threats to federated learning systems and provides a detailed presentation of popular defenses and their advantages and disadvantages. The threats described include *model inversion*, *training data extraction*, *membership inference*, and *property attacks*. The chapter discusses multiple state-of-the-art defenses including multiple crypto systems, pair-wise masking, differential privacy among others. The chapter also describes which defenses can be applied to a diverse set of scenarios. Chapter 14 covers two of the existing defenses in more detail, while Chap. 15 presents in great detail popular gradient-based data extraction attacks.

Federated learning opens up new ways for malicious parties to compromise the learning process and the outcome. Security concerns including byzantine, poisoning, and evasion threats are addressed in Chap. 16. Several defenses against these manipulation attacks are also reviewed. Chapter 17 takes a deep dive into the byzantine threats and defenses when training neural networks.

Chapter 13
Protecting Against Data Leakage in Federated Learning: What Approach Should You Choose?

Nathalie Baracaldo and Runhua Xu

Abstract Federated learning (FL) is an example of privacy by design where the primary benefit and inherent constraint is to ensure data is never transmitted. In this paradigm, data remains with its owner. Unfortunately, multiple attacks capable of extracting private training data by inspecting the resulting machine learning models or the information exchanged during the FL training process have been demonstrated. As a result, a plethora of defenses have surfaced. In this chapter, we overview existing inference attacks to assess their associated risks and take a close look at the significant corpus of popular defenses designed to mitigate them. Additionally, we analyze common scenarios to help provide clarity on what defenses are most suitable for different use cases. We demonstrate that one size does not fit all when selecting the right defense.

13.1 Introduction

Privacy by design has been one of the main drivers of federated learning (FL), where each training party maintains their data locally while collaboratively training a machine learning (ML) model. Compared to existing ML techniques that require the collection of training data into a central place, this new paradigm represents a big improvement that balances utility of data and privacy. Given this significant change in data collection, FL is becoming a popular approach to protect data privacy.

In its basic form, Fig. 13.1, the FL training process requires an aggregator and parties to exchange model updates. Although no training data is shared during this process, this basic setup has been shown to be vulnerable to *inference* of private training data. Risks of inference of private information are prevalent in multiple stages of the FL process. Inference attacks can be classified based on the attack surface used to obtain private data into (1) those carried out on the final model

N. Baracaldo (✉) · R. Xu
IBM Research – Almaden, San Jose, CA, USA
e-mail: baracald@us.ibm.com; runhua@ibm.com

© The Author(s), under exclusive license to Springer Nature Switzerland AG 2022
H. Ludwig, N. Baracaldo (eds.), *Federated Learning*,
https://doi.org/10.1007/978-3-030-96896-0_13

281

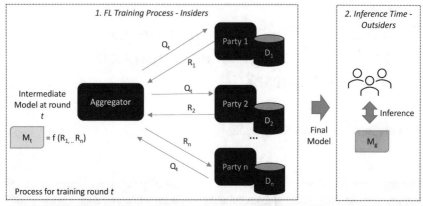

Fig. 13.1 FL system with n parties $P_1, \ldots P_n$ without any privacy protection techniques. All entities participating in the FL training process are considered insiders while entities getting access to the final predictive model are consider outsiders. The table shows the attack surface for each of the entities; this information is available for multiple rounds

produced by the federation and (2) those executed using information exchanged during the FL training process.

Attacks executed exploiting the final model are applicable to *all* ML models and are not exclusive to FL. In other words, threats of inference based on a model itself are inherent to ML *regardless* of how the model was trained. Attacks in this category include data extraction, membership inference, model inversion, and property inference attacks, e.g., [12, 57, 70]. All of them infer information about the training data and in some cases the training data itself.

The second category of attacks are *specific* to FL and represent a new threat. These attacks take advantage of the information transmitted during the FL training process. For example, attack techniques such as [28, 39, 89, 91] use the newly exchanged information, exercise the potential for manipulating the training process or both. In some cases, this new information creates more successful attacks, while in some other cases the threats are not as worrisome as attacks are only applicable to some artificial settings. In this chapter, we present and contrast the differences and similarities of carrying out attacks based on the model and the FL process in Sect. 13.2.

The existence of inference risks have led to the proposal of a plethora of defenses that aim to provide different privacy guarantees. These include the introduction of differential privacy at the party side [30, 80, 84] and at the aggregator side, the integration of different cryptosystems to enable secure aggregation including Pailler [60], Threshold Pailler [19], different flavors of functional encryption [2–4, 8, 11], pairwise masking [10], homomorphic encryption (HE) [29], or the combination of multiple of those. The number of options available is substantial, and some of them have been shown to be vulnerable to inference attacks on themselves.

With this plethora of attacks and defenses, important questions remain to be answered: *How critical are inference attacks? How much should we worry about them? If we need to mitigate the risk, what defense should we use?*

In this chapter, we set out to help answer these questions. Selecting a defense technique for privacy preservation is not an easy task, and in the current landscape, it is challenging to compare the assumptions, disadvantages, and advantages of different solutions.

We show that *one size does not fit all* and that it is imperative to consider what initially may be seen as nuances of different defenses. In fact, as it is often the case in the security and privacy disciplines, a thorough risk assessment is needed to identify a suitable threat model for a particular application[1]. In some cases, more than one defense needs to be applied to mitigate stringent privacy requirements.

Unfortunately, there is *no free lunch* when incorporating defenses into the FL process. Incorporating them may result in higher training times, lower model performance, or more expensive deployments. For this reason, it is imperative to decide the *right* level of protection for a particular federation. This sweet spot highly depends on the scenario where FL has been applied to. For example, consider a use case where a single company has stored its data in multiple clouds and wants to run FL to obtain a ML model. Here, potential inference attacks may not be relevant. However, if multiple competitors decide to train a model together, there may exist mistrust and there may be an incentive for them to try to infer private information of some other participating parties. In other cases, regulation and compliance requirements may require additional protections, as it is the case for healthcare data and personal information [14, 63].

In this chapter, our aim is to provide clarity in the area of inference attacks to FL and to help provide some understanding of which attacks are relevant and which defenses are suitable. In Sect. 13.2, we begin by presenting in detail the system entities, assumptions and attack surfaces of the FL system as well as the potential adversaries. We then provide a thorough overview of relevant state-of-the-art inference attacks. Then, in Sect. 13.3 we characterize existing defenses in terms of functionality offer, disadvantages, and existing vulnerabilities. To this end, we carefully define various privacy goals and then compare defenses based on the

---

[1] A threat model defines the trust assumptions place in each of the entities in a system. For example, it determines if parties fully trust the aggregator or whether they only trust it to a certain degree.

different privacy guarantees offered and the attacks they can prevent. In some cases, we also discuss potential vulnerabilities of some defenses. In Sect. 13.4 we discuss some guidelines that demonstrate how to map threat models to some of the defenses and subsequently conclude our chapter.

13.2   System Entities, Attack Surfaces, and Inference Attacks

Figure 13.1 presents an overview of the entities involved during the training and inference process. We categorize relevant entities as *insiders* and *outsiders*, where insiders are entities involved during the training process while outsiders are entities that uniquely have access to the final model produced by the federation.

There are two main insider entities involved in the FL training process: the *parties* who own the data and the *aggregator* who orchestrates the learning process. Figure 13.1 presents a *plain FL* system; we refer to it as plain because no privacy-preserving provisions have been added to it. The federation consists of n parties P_1, P_2, ... P_n, where each party P_i owns its private dataset D_i. The training process starts with the aggregator sending *queries* Q_t requesting aggregated information needed to build a ML model to the parties, and the parties use their training data to answer that query. We refer to the reply sent back to the aggregator as a *model update* and denote the model update of party P_i as R_i. When the aggregator receives all the replies R_1, R_2... R_n., it proceeds to fuse or aggregate them creating a new *intermediate global model*. At the end of round t, the intermediate global model is denoted as M_t. The global model is then used as input to create the next query for the parties and the process repeats until the target model performance is reached.

Most privacy attacks are designed for gradient descent and its variants, which can be used to train neural networks, linear models, support vector machines (SVMs), among other model types. Hence, we briefly revisit how this algorithm is applied in FL. For a training round t, the aggregator sends the initial weights of the model to each party. Then, each party P_i runs gradient descent or stochastic gradient decent (SGD) using its own dataset D_i for a pre-specified number of epochs and hyperparameters. The party then sends back the resulting model weights, which are referred to as *model updates*. Alternatively, the party may send gradients as model updates.[2] The aggregator in its simplest form takes all the model updates and fuses them. The most popular way to aggregate model updates is to use FedAvg [53] where the aggregator uses the number of samples of each party to compute a weighted average of the model updates. The new model weights are sent back to the parties, and the process repeats until the target accuracy is reached or the maximum

---

[2] Experimentally, we have found that exchanging model weights leads to faster convergence than exchanging gradients.

rounds elapses. Notice that the fusion model updates obtained by the aggregator can be used to acquire the ML model.

13.2.1 System Setup, Assumptions, and Attack Surfaces

To fully understand the attack surface in FL and the existing risks, we begin by clearly defining the assumptions commonly made about the FL distributed system setup and the information that is typically assumed to be known by insiders before the training process begins.

From the machine learning perspective, there are multiple common assumptions. Before FL starts, it is often assumed that data at each of the parties is already annotated, that the set of labels are known, and that the model definition is available [24, 50]. Thus, prior to the federation starting, the aggregator and the parties know that there are, for example, five classes, and that each of the parties has pre-processed its dataset in the same fashion. If structural data is used, all parties and aggregator also know the number, order of the feature set, categorization, and the pre-processing technique to be used. Accordingly, in this chapter, we assume this information is known by all insiders participating in the FL process. Additionally, the aggregator has access to the *model updates* sent by all the parties and the intermediate models, while parties have access to their training data, aggregated models, and model updates.

From the system perspective, it is important to set up the distributed system in a secure way. In particular, throughout the FL process, messages exchanged between parties and the aggregator need to be properly authenticated and secure channels need to be established. In this way, external entities are prevented from snooping messages exchanged between the aggregator and parties. Man-in-the middle attacks where an adversary may try to impersonate a targeted party or the aggregator are also prevented. Hence, in the following discussion we assume external entities cannot snoop private information by inspecting exchanged messages or impersonate insiders.

13.2.2 Potential Adversaries

In FL, the training data is not explicitly shared. Hence, we focus on inference threats. The *goal* of an adversary is to infer private data based on the information they have access to (we will elaborate on the attackers' goals in more detail later as they differ for each attack).

The table in Fig. 13.1 summarizes the information that is available for each potential adversary and the information they can manipulate to their advantage. As we can see, a malicious aggregator has the opportunity to manipulate the aggregation process producing incorrect intermediate models and queries. It may also manipulate the final model. Similarly, a malicious party may manipulate the model updates sent to the aggregator. Finally, outsiders have the power to manipulate the inference queries to the final model to try to infer private information.

The adversaries can be further characterized according to their behavior as follows:

- A *Curious and passive aggregator* may try to infer private information based on all model updates exchanged during the training process. If the final or intermediate models are known, this adversary may also try to use the model to infer private data from participants.
- A *curious and active aggregator* may utilize the same information to perform inferences but may try to manipulate the queries sent to each party, the intermediate and final model updates.
- A *curious party* may try to infer private information based on the queries received and its knowledge about the learning objective.
- A *curious and active party* may try to infer private information based on the queries received and may also carefully craft the attacks to try to further determine information about other parties' private data.
- *Colluding parties* are conspiring parties who may try to infer information by orchestrating attacks together. These parties may share information among them and may actively or passively attack the system to infer private information about other parties.
- *Curious colluding aggregator and parties* it is possible for a curious aggregator and a few malicious parties to try to passively or actively launch privacy attacks to infer information about the training data and properties of a targeted party(ies).
- *Curious outsiders* are impersonated by external adversaries who do not participate in the training process but have access to the final ML model. They may try to use the final model and its predictions to infer private training data or related information.

Any such adversaries may carry out attacks to try to infer private information. In the following we overview multiple inference attacks.

13.2.3   *Inference Attacks to Federated Learning*

Multiple inference attacks have been demonstrated in FL systems. They can be classified according to their inference objective in the following four categories:

1. *Training data extraction:* These attacks aim to recover the *exact* individual training samples used during the training process. In other words, for an image based task, the output of the attack is a pixel-by-pixel output of the training data, while for a text-classification task, the output of the attack is the word-by-word text of the training corpus.

2. *Membership Inference:* In this type of attacks, the objective of the adversary is to determine if a particular sample was part of the training data used to create a model. This clearly constitutes a privacy violation if being part of a training set reveals, for example, a medical condition, social or political association.
3. *Model Inversion:* Here, the goal of the adversary is to construct a representative of each of the classes. This type of attack leads to great privacy violations in cases where each class contains samples that are similar among them. Consider the case of a face recognition model where a class contains information about a single individual. In this case, the results of the model inversion are visually similar to images of the person that were used during the training process. However, if the class members are *not* all similar, the results do not look like the training data [70], and hence this attack may produce innocuous results.
4. *Property Inference:* This type of attack focuses on revealing properties of the training data that are not relevant for the training task at hand. Attacks in this settings can extract global properties of the training dataset such as the ratio of the samples included or can focus on extracting properties of sub-populations of the training data.

We summarize the goals and outputs of these four types of attacks in Fig. 13.2. A variety of attacks have been demonstrated. In Table 13.1 we characterize representative attacks based on the adversaries that execute them and the information that they use to infer private training data. Again, we emphasize that attacks which can be carried out by outsiders are not specific to FL.

We now briefly describe in some detail how the attacks are carried out. This will help understand the vulnerabilities of the system and determine what defenses are more suitable to mitigate different attacks in Sect. 13.3.

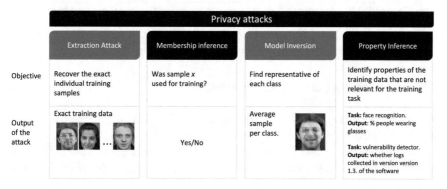

Fig. 13.2 Inference attacks to machine learning. The figure contrasts the different objectives of each attack and presents sample outputs for different attacks. We use the faces ORL dataset of the AT&T Laboratories Cambridge for illustration purposes. The output for the model inversion attack was generated using the attack presented in [26]

Table 13.1 Inference attacks to undefended FL systems

Attacks	Attack surface		Adversary
	Model	Model update	
Training data extraction attacks			
Henderson et al. [33], Carlini et al. [12], Zanella et al. [87], Carlini et al. [13]	Black-box		Outsider, insiders
DLG [91], iDLG [89], Geping et al. [28], Cafe [39]		Gradients	Curious aggregator
Wang et al. [81], Wei et al. [83]	White-box	Gradients	Curious aggregator
Song et al. [75]		Gradients	Aggregator active and passive modes
Membership attacks			
Shokri et al. [70], Salem et al. [68], Hayes et al. [32], Choquette et al. [16]	Black-box		Outsider, insiders
Nasr et al. [57] (multiple attacks)	Black-box		Curious aggregator or curious parties
		Gradients	Active curious party
		Gradients	Active curious aggregator
		Gradients	Active isolating aggregator
Model inversion			
Fredrikson et al. [25, 26]	Black-box		Outsider, insider
Hitaj et al. [34]		Gradients	Curious party
Property attacks			
Ateniese et al. [7], Ganju et al. [27]	White-box		Outsider, insider
Melis et al. [54]		Gradients	Curious party

13.2.3.1 Training Data Extraction Attacks

Multiple attacks have been proposed to extract data samples used during the training process by querying a published model [12, 13, 33, 73, 87]. Most of them are devoted to extracting training data from neural networks [12], text-based models [13, 87], and embeddings [73], but there are some approaches for other types of models such as [33].

Data memorization of generative text models has been studied for traditional and FL settings in [12] and [78], respectively. Generative text models help users to auto-complete their phrases to speed texting and email writing. The adversary's goal in this setting is to extract secret sequences of text to learn private information; for example, inputting to the system the sentence *my social security number is* may lead to the exposure of social security numbers used during the training process. In [12]

Carlini et al. proposed metrics to measure memorization in these types of models when the adversary has black-box[3] access to a published model that can query at will. Later, Thakkar et al. [78] compare differences between the memorization exhibited by models trained in FL versus those trained in traditional fashion. Their experiments suggest that training in FL may help reduce memorization and they hypothesize the diverse distribution of data among users is responsible the situation. However, both studies are based on black-box access. That is, they only consider the role of an outsider. Further analysis is required to identify how much information can be leaked when the adversary has more information, as it is the case for FL insiders.

A substantial number of data extraction attacks in FL for general models have been highlighted and the great majority make use of model updates that contain gradients [5, 28, 39, 81, 89, 91]. To understand how these attacks are possible, consider a simple text-based classifier that uses a bag-of-words as a way to encode the training data. Recall that once a party has been queried with an initial model, it uses its local training data to train for a few epochs and then returns the gradients to the aggregator. A curious aggregator may observe the gradients received from a party, which may be sparse, and can trivially know whether a word was part of the training data because a non-zero gradient is only possible if an original word was present in the party's training data [53]. Gradient-based attacks have also been demonstrated for images including number digits [5], faces [91], and other datasets [28].

An efficient attack known as *Deep leakage from gradients* (DLG) was proposed in [91], where a curious aggregator can obtain the training samples and the labels of a victim party. DLG was demonstrated to have high attack success rates in a few optimization interactions for some network topologies; however, it does not work unless the batch size utilized is small, for example, eight. A subsequent attack [81] increases the recovery accuracy for more general model initialization by using a distance metric based on a Gaussian kernel based on gradients. However, the approach also has limited attack success rates for larger batch sizes. This is clear a limitation that has raised the speculation on whether using higher batch sizes could deter the problem.

Unfortunately, increasing the batch size to prevent these attacks has been shown to be a naive defense in [39], where a new gradient-based attack called CAFE was presented. CAFE's attack success was demonstrated even under larger batch sizes of 40. The attack uses a surrogate or synthetic dataset created offline. By assuming the model is known, which is clearly the case for the aggregator, the aggregator can pass the surrogate data through the model and compute fake gradients. After that, the aggregator compares the resulting surrogate gradients with the real gradients coming from the victim party and tries to minimize the difference by updating its surrogate

---

[3] Black-box access in ML refers to a scenario where the adversary cannot access model parameters and can only query the model. White-box access, conversely, refers to settings where the adversary has access to the inner-works of the model.

dataset. This process can continue over the course of training to recreate the original dataset. An improvement in label guessing was presented in [89]. Additionally, studies and approaches such as the one presented by Geiping et al. [28] have shedded light into the vulnerabilities of commonly used vision network architectures beyond the topologies originally considered by DLG. Other attack optimizations inspired by the original DLG continue to emerge in the literature.

While most attacks are based on gradient-based inferences, gradient-based algorithms in FL are frequently trained exchanging model weights. As mentioned before, we have experimentally observed that exchanging model weights leads to faster convergence times. In some circumstances, a curious aggregator could still carry out the same inference procedure by computing the gradient between two iterations with the same party. This, however, would require for the victim party to be queried in subsequent training rounds that are not too far apart. Hence, further experimental analysis in this direction is needed to evaluate the attacks under those system setups. Finally, as our overview of these attacks shows, there is a trend to continuously improve limitations of existing data extraction attacks.

13.2.3.2   Membership Inference Attacks

Multiple membership inference attacks [57, 68, 70] have demonstrated that it is possible to know if a sample, called a *target sample*, was part of a training dataset, violating the privacy when the inclusion in a training set is itself sensitive. For example, consider an ML model trained based on photographs of people who suffer from a taboo disease or a political or group affiliation. In this case, getting to know if a person is part of the training set will reveal its health condition or that it is part of a particular group.

Attacks in this category take advantage of the notoriously higher prediction confidence that models exhibit when they are queried on their training data. This is mainly caused by overfitting of the model to training samples. Traditional attacks in this category assume the adversary has black-box access to the model (does not know the model parameters) and can only query the model for prediction. The adversary queries the model multiple times to understand how it behaves with respect to a set of engineered inputs to reconstruct the loss surface of the model. After multiple such queries, the adversary can determine if a sample was part of the training dataset. Most attacks in this classification make use of the confidence scores associated with a prediction query to guide the creation of informative queries.

Because a variety of attacks make use of the confidence score to achieve their objective, some attempts to hide confidence scores or reduce the number of queries a classifier can answer have been proposed as potential mitigation techniques [38, 68, 85]. However, those solutions do not effectively prevent membership inference. It has been demonstrated that even when a model does not expose the confidence scores associated with its classification, it is possible for adversaries to successfully execute a membership attack [16]. In the label-only attack presented in [16], for a target sample, the adversary generates multiple perturbed samples and queries the classifier to determine the robustness of the model to modifications. Data augmentation and adversarial samples are used to generate the perturbed samples. Based on the classifier's replies, it is possible to determine if there is a strong or weak membership signal.

Membership inference attacks can also be carried out in FL based on messages exchanged. In particular, Nasr et al. [57] present a variety of attacks that take advantage of the model updates exchanged during the training process. All the attacks proposed in [57] inspect the gradients of each of the neural network layers separately to take advantage of the fact that higher layers of the neural network are fine-tuned to better extract high level features that may reveal private information at higher rates than lower level layers of the network. To make decisions on whether the network was trained using the target sample, the attack may use an unsupervised inference model based on auto-encoders or a shallow model for scenarios where the adversary has some background knowledge on the victim's data.

Nasr et al. propose attacks that can be launched by a curious aggregator or a curious party in *active* or *passive* mode. A *curious aggregator* can carry three types of attacks (1) a passive attack where it observes each of the model updates of the parties and tries to determine membership of a target sample for that party, (2) an active attack where the aggregator manipulates the aggregated model according to the previous discussion, and (3) an active and isolating attack where the curious aggregator does not aggregate the model updates of other participants to increase the attack success rate.

The attack proposed for malicious parties is limited with respect to the one launch by an aggregator because parties can only see intermediate models. Hence, the object of a curious party is limited to determine if *any* of the other parties has the target sample. For malicious parties, the attack can be passive (no manipulation of the model update) or active.

Active membership attacks are exclusive to FL; an active attack refers to whether the adversary manipulates the model updates or model to induce observable gradient behaviors on member samples. In the case of an active curious party trying to infer the membership of a target sample x, the adversary will run gradient *ascent* on sample x by the model and update its local model in the direction of the increasing loss on the sample. This modified model weights are shared with the aggregator, who then sends the new model updates with the parties. Interestingly, when a party has sample x, its local SGD will abruptly reduce the gradient of the loss on x. This signal can be captured to infer the party has the target sample by a supervised or unsupervised inference model. Because multiple rounds are required to train a model, the adversary has several opportunities to manipulate the model updates, which leads to higher attack success rates.

In conclusion, FL does offer new opportunities for adversaries to carry out membership inferences attacks.

13.2.3.3 Model Inversion Attacks

The goal of model inversion attacks is to construct representative of each of the classes. Model inversion attacks to traditional training processes include [25, 26]. Typically these attacks use the confidence score output by the model to guide the reconstruction of data samples of a known and targeted label. For example, Fredrikson et al. showed that an adversary with access to the model and some demographic information about a patient can predict the patient's genetic markers [25].

Model inversion attacks have been tailored to FL in [34], where Hitaj et al. proposed a Generative Adversarial Network (GAN)-based procedure that operates on model updates and forces victims to reveal more information by carefully crafting gradients. The attack can be carried out by any participating party, where such curious party aims to gain information about a target label. The adversary trains a GAN based on observed model updates. The GAN subsequently generates prototypical samples of the private training set. The curious party can create inputs to the training process that force the victim to release more accurate private information. Concretely, the adversary will generate errors for recovered samples, ensuring the victim party tunes the model and reveals further information about the target sample. Here, the adversary actively manipulates the training process.

Finally, model inversion attacks are particularly dangerous if there is a meaningful *average* representation of a class. Otherwise, the output may not provide useful information to the attacker.

13.2.3.4 Property Inference Attacks

An adversary can infer information about the properties of the inputs, such as the environment where the data was produced, even when the model task is completely independent of that extracted information. Some approaches have focus on extracting *global properties* of the training set [7, 27] while more recent approaches have focused on extracting *sub-population properties* [54]. Global properties include the distribution of different classes in the training dataset, for example, a neural network trained to classify smiling faces may leak relative attractiveness of the individuals in the training set. Another less innocuous example of inference of global properties was presented by Ganju et al. [27] who demonstrated models may help an adversary determine whether a machine where the training logs were collected contained two important vulnerabilities that could lead the adversary to illicitly gain bitcoin by exploiting such vulnerabilities. Sub-population properties refer to properties of a particular sample in the training data, for example, in a model classifying medical reviews, an adversary may be capable of inferring the medical speciality from which they came [54].

Global property inference was first proposed by Ateniese et al. in [7], who demonstrated that SVMs and hidden Markov models (HMM) easily leak global properties. Their approach uses multiple meta-classifiers trained in surrogate datasets that exhibit the tested properties and requires white-box access to the model. After that, the approach was extended to work into fully connected neural networks in [27].

With the introduction of FL, attacks that can further isolate *sub-population properties* have been proposed. Melis et al. recently proposed a novel and more fine-grained attack that combines membership and property inference in FL settings in [54]. Here, a curious party can first identify the presence of a particular record in the training dataset of a victim party by running a membership inference attack. Similarly to global property attacks, this attack also trains meta-classifiers using auxiliary labeled data and uses them to determine if a property exist or not in one sub-populations. Experimental results show that the attack is successful for image and text-classifiers, and that are possible with two to 30 parties. In this case, FL opens attack surfaces for malicious insiders to leak more private information.

13.3 Mitigating Inference Threats in Federated Learning

While the amount of attacks to FL system keeps increasing, so do the number of solutions that is proposed to address these threats. Defenses in this area are diverse in nature and protect different aspects of the FL training process or deployment of the model. Broadly speaking, defenses in this area include:

1. *Modification of the training procedure and restrictive interfaces to query a model:* Examples in this category include pruning [37] or compressing the model updates sent to the aggregator to deter gradient-based attacks, adding regularizers to reduce overfitting [70] and prevent extraction attacks, minimize the number of queries or avoid reporting confidence of models to prevent membership attacks [85]. All these defenses have been demonstrated to fail under adaptive attacks [16, 83, 83]. Because these solutions have been demonstrated to lead to a false sense of privacy, we omit them in the following analysis.
2. *Syntactic and perturbation techniques:* Techniques in this category include adaptations of k-anonymity [76] and differential privacy (DP) [22], which have been incorporated to FL training process by multiple defenses including [80] and [17], respectively. We will see these approaches may help deter *some* of the attacks previously presented.
3. *Secure aggregation and secure hardware techniques:* This category includes approaches that use different cryptosystems to ensure the aggregator cannot access individual model updates, as well as approaches that require specialized hardware execution environments to ensure the execution flow is followed and the computation is maintained private.

How Do All These Techniques Compare? In the following, we characterize these techniques based on the information that is maintained private according to the following definitions which are illustrated in Fig. 13.3.

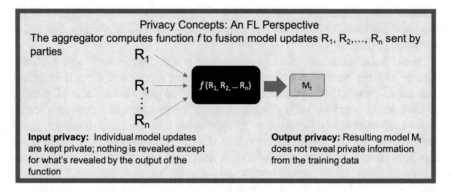

Fig. 13.3 Illustration of the *input privacy* and *output privacy* concepts in the FL context. Function f computed by the aggregator fusions model updates $R_1, R_2, \ldots, R_n$ sent by parties

1. *Input Privacy*: Solutions that fall in this category preserve the privacy of the model updates shared by each of the parties. In other words, a solution provides privacy of the input if the FL training process does not reveal anything other than what can be inferred by the resulting final ML model.

   Techniques in this category can be used to protect against threats coming from malicious aggregators trying to isolate or infer private information from parties. Multi-party computation techniques are often used to maintain this type of privacy. However, we will see that these techniques come in many flavors and provide different guarantees and *some* are vulnerable to inference attacks.
2. *Output Privacy*: Techniques in this category ensure that the final model or intermediate models do not leak private information of the training data. Defenses that fall in this category are designed to prevent attacks coming from outsiders and insiders that make inferences base on the models. Differential privacy is a defense in this category, but as we will see, it has its own limitations.
3. *Privacy of the Input and Output*: Approaches in this category preserve both the privacy of the input and the output.

Based on these three definitions, we categorize some representative defenses in Table 13.2. In the table, we also added a column to highlight that existing techniques also differ on whether the intermediate or final models are exposed in plaintext to the aggregator. This reflects a different trust model that is important to select the right solution for different use cases. Note that even if the model is not revealed in plaintext to the aggregator, a solution may not protect the privacy of the output because the final model, when decrypted by the parties, is still vulnerable to output inference. In the following, we briefly present existing defenses and provide information about the attacks that they address and the ones they cannot thwart.

Table 13.2 Privacy-preserving defenses for Federated Learning. The second column refers to the intended privacy goal; we will show that some of the approaches may lead to privacy leakage if not deployed correctly. In particular, plaintext results may suffer from disaggregation inferences if not deployed in conjunction with additional provisions

	Privacy of		Aggregator's access to
Existing proposals	Input	Output	fusion model updates
Secure aggregation:			
– Partial HE [58, 88]	✓	✗	Encrypted
– Fully HE [67]	✓	✗	Encrypted
– Threshold Paillier (TP) [80]	✓	✗	Plaintext
– Garbled Circuit (GC)	✓	✗	Plaintext
– Pairwise Mask (PM) [10, 40, 72]	✓	✗	Plaintext
– TPA-based Functional encryption (FE) [84]	✓	✗	Plaintext
Differential privacy:			
– Global DP (Aggregator)	✗	✓	Plaintext
– Local DP (Party side) [6, 30, 80, 84]	✓	✓	Plaintext with DP Noise
Hybrid approaches			
– DP with Threshold Paillier [80]	✓	✓	Plaintext with DP Noise
– DP with Functional Encryption [84]	✓	✓	Plaintext with DP Noise
– DP with pairwise masking [42]	✓	✓	Plaintext with DP Noise
Special hardware support			
– Truda (three aggregators)[15]	✓	✗	Partial model in plaintext
Approaches specific for XGBoost			
– Pairwise mask [48]	✓	✗	Plaintext
– Oblivious TEE-based XGBoost [46]	✓	✗	Plaintext

13.3.1   Secure Aggregation Approaches

To achieve privacy of the input, secure multi-party computation has been proposed to achieve *secure aggregation (SA)*. SA allows an entity to compute a function f that takes as input $R_1, R_2, \ldots, R_n$ without getting to know any of the inputs. There are multiple ways to preform SA including pairwise masking and modern cryptographic schemes such as fully homomorphic encryption, functional encryption, partial homomorphic encryption, threshold Pailler, among others. These techniques differ in multiple aspects as shown in Tables 13.2 and 13.3. The key differences are:

- *Supported threat model:* The threat model they cover differs on the trust each of the SA techniques puts on the aggregator. Some approaches expose the intermediate models to the aggregator while others only expose encrypted data (see Table 13.2). Additionally, some techniques have verification provisions to

Table 13.3 Behavior of the solutions under dynamic party participation

Approach	Verified aggregation	New parties	Dropout	Special hardware
Secure aggregation				
– Partial HE [58, 88]	✓	No rekeying	✗	✗
– Fully HE [67]	✓	No rekeying	✓	✗
– Threshold Paillier [80]	✗	Rekeying	✓	✗
– Pairwise Mask [10, 40, 72]	✗	Rekeying	✓	✗
– HybridAlpha (TPA-based FE) [84]	✓	No rekeying	✓	✗
Special hardware support				
– Truda (three aggregators)[15]	✓	No rekeying	✓	✓
Federated boosted model				
– Secure Federated GBM [48]	✗	Rekeying	✓	✗
– Oblivious TEE-based XGBoost [46]	✓	No rekeying	✓	✓

ensure that the aggregator can only fusion a minimum number of model updates (see first column of Table 13.3).

- *Dynamic participation:* Another aspect where SA techniques differ is their adaptability to parties joining the federation in the middle of the training process and their tolerance to some parties leaving the federation intentionally or accidentally. Some techniques require fully halting the training process to re-key the system when a party drops or joins, while others are more resilient against these changes and can continue training without modifications (see third column Table 13.3). Clearly, techniques that do not require the entire system to be rekeyed are preferred.

- *Infrastructure:* Each approach may require a different infrastructure setup. Some of them require the use of additional fully trusted authorities or special hardware, while others need multiple non-colluding servers that may increase the cost of deployment a solution or special hardware.

In the following, we present in more detail each of SA techniques and discuss their advantages and shortcomings.

13.3.1.1   Homomorphic Encryption-Based Secure Aggregation

Homomorphic encryption (HE) can be classified into partial HE and fully HE depending on the types of operations that can be computed over encrypted data.

Partially Homomorphic Encryption schemes enable the computation of additive operations over encrypted data by an untrusted entity. Paillier cryptosystem [60] and its variants [56, 59] are some of the most used cryptosystems in this category. Partial

HE cryptosystems satisfy the following property:

$$\text{Enc}(m_1) \circ \text{Enc}(m_2) = \text{Enc}(m_1 + m_2).$$

Where $\text{Enc}(m_1)$ and $\text{Enc}(m_2)$ are encrypted and $\circ$ represents a predefined function. In other words, an untrusted entity receiving $\text{Enc}(m_1)$ and $\text{Enc}(m_2)$ can compute their addition without decrypting the m_1 and $m2$. Notice that the untrusted entity only obtains the result $\text{Enc}(m_1 + m_2)$ in encrypted form.

Because most aggregation functions in FL require uniquely additive operations, these cryptosystems are a popular option for FL [58, 88]. We now examine how this cryptosystem is applied to FL in more detail. During setup all parties agree upon a public/private key pair(pk, sk), and the aggregator receives a public encryption key pk. The parties encrypt their model updates using pk before sending them to the aggregator. Once the aggregator receives all encrypted model updates, it computes the addition of model updates using its public key pk obtaining the encrypted result. The encrypted result is then forwarded to the parties, who in turn decrypt them using sk and continue the training in plaintext.

Compared to other SA approaches, the final result of the aggregation is never revealed to the aggregator, who always obtains the fusion result encrypted. Additionally, no rekeying is needed in case of new parties join or drop the federation. However, according to [84], partial HE may result in high computation and communication costs compared to competing options. To overcome this downside, *BatchCrypt*, an approach where each party *quantizes* its gradient values into low-bit integer representations and then encodes a batch of quantized values to a long integer for encryption, was recently proposed in [88]. These pre-processing operations allow faster encryption/decryption times.

Approaches in this category are limited to additive operations. For this reason, fusion algorithms that require more than a simple average of the model updates cannot be implemented with partially HE schemes.

Fully Homomorphic Encryption schemes support all operations over encrypted data. Thus, they can be used to implement many more fusion algorithms. However, they are more expensive computationally speaking. One such approach was presented in [67].

All HE approaches by themselves are capable of preventing a curious aggregator from inferring data from the intermediate and final models. This is the case because the aggregator can only see this information encrypted. However, parties can decrypt the models and, hence, are still able to infer information from the model itself. Therefore, in Table 13.2, we have marked HE approaches as not providing privacy of the output.

13.3.1.2 Threshold Paillier-Based Secure Aggregation

The Threshold Paillier cryptosystem [19] is a variant of the Paillier encryption system [60] and, thus, supports the additive HE property. The main difference is that in this variant, a predefined number of trusted parties t are required to collaboratively decrypt the ciphertext.

The use of this cryptosystem in FL settings was proposed in [80] to prevent collusion attacks by making sure the final fusion result cannot be obtained unless t trusted parties collaboratively decrypt it. During setup, each participant obtains a public key pk for encryption and a private *share secret key* sk_i for partial decryption, and the aggregator receives a *combing decryption key* dk. Each participant uses pk to encrypt its model update and sends it to the aggregator. The aggregator fuses the received encrypted model updates and sends the result back to t parties. Each party then uses its sk_i to perform its partial decryption and sends it back to the aggregator. Finally, after the aggregator receives at least t partially decrypted results, it makes use of dk to acquire the aggregated model updates in plaintext.

Threshold Pailler approaches for secure aggregation are suitable for federations where the aggregator is trusted to obtain the aggregated model updates in plaintext. Additionally, they provide an interesting property that ensures that at least t trusted parties need to agree to decrypt the model update. However, this functionality comes at the cost of more communication rounds between parties and aggregator resulting in longer training times.

13.3.1.3 Pairwise Mask-Based Secure Aggregation

Each party conceals their model update using pairwise random masks between users to hide the individual input. Following that, the aggregator simply adds up those masked model updates to obtain the global model. The protocol is built in such a way that the pairwise random masks are cancelled out after all model updates are aggregated. The initial pairwise mask design [10] relies on *t-of-n secret sharing* [69] and requires four rounds of communication among parties and the aggregator. The overhead of [10] grows quadratically with the number of parties. To further improve the efficiency and scalability of [10], Turbo-Aggregate [72] employs additive secret sharing and a multi-group circular strategy for model aggregation, while FastSecAgg [40] proposes a novel multi-secret sharing scheme based on a finite-field version of the Fast Fourier Transform.

Although pairwise mask-based approaches support parties dropping out, applying these techniques increases the number of messages exchanged between parties and the aggregator increasing the training time. Additionally, the approach requires rekeying when new parties arrive to the federation.

13.3.1.4 Functional Encryption-Based Secure Aggregation

Functional encryption (FE) [11] is an emerging family of cryptosystems that allow the computation of a function f over a set of encrypted inputs, where the decrypter obtains the final result of f in plaintext. To compute the function, the decrypter entity needs to use a *functional key* that depends on the function evaluated f and the encrypted data. This functional key is provided by a trusted third party (TPA). However, recently, new cryptosystems are removing the need for this entity.

The use of functional encryption for FL was first proposed by Xu et al. in [84], where the concrete cryptosystem of choice is the multi-input inner-point functional encryption [31]. This cryptosystem enables the aggregator to obtain in plaintext the result of the function $f(x, y) = \sum_i x_i y_i$, where x_i corresponds to the private encrypted model update coming from party i and y_i corresponds to the weight used to fuse model update x_i. That is, when y is a vector with ones, all model updates have the same weight. When $y_i = 0$, it means that x_i will be ignored during the aggregation.

During setup, given a maximum number of supported parties, the TPA initializes the functional cryptosystem. Each party then receives its party-specific secret key sk_i from the TPA. During training, each party encrypts its model update using sk_i and sends the resulting ciphertext to the aggregator. After a predefined amount of time elapses, the aggregator fusions all encrypted model updates received and prepares vector y. Vector y is used to request a functional decryption key dk_y from the TPA to fusion the received model updates and obtain the result in plaintext. If some parties did not reply, the aggregator does not include a position for them in vector y. This enables for a clean manage of dropouts without rekeying or further communication rounds. Vector y is then sent to the TPA, who inspects y to ensure providing the functional key for the received vector would not allow for inference attacks where a malicious aggregator may try to isolate the model updates of one or very few parties. If y complies with a pre-specified number of replies, the functional encryption key is sent to the aggregator. Otherwise, the aggregator cannot compute the aggregated result.

A distinctive advantage of the approach proposed in [84] is that it provides support for verification of the number of aggregated model updates preventing attacks from a curious aggregator trying to isolate a few model replies. It also supports dynamic party dropout and new parties joining, without requiring more than one message exchanged between the parties and the aggregator compared to other SA approaches such as Pairwise Masking or threshold Pailler. On the downside, the approach proposed in [84] requires the use of a TPA. This limitation could be solved with new advances in the functional encryption field.

13.3.1.5 Summary Secure Aggregation

We overviewed representative SA approaches and highlighted their differences and shortcomings. Some of the approaches are limited to additive fusion algorithms and do not react fast to dynamic settings when parties join and drop requiring expensive rekeying operations. Another salient difference is the number of messages that need to be exchanged between the aggregator and parties to obtain the SA results, with functional encryption solutions requiring a single message, while pairwise mask solutions require four messages, and the threshold Pailler requiring three messages.

Another main difference between SA solutions is their trust assumption in the aggregator. Most approaches trust the aggregator with the SA results in plaintext (Table 13.2), with the exception of Partial HE schemes where resulting SA model updates are encrypted and remains unknown to the aggregator. In scenarios where the aggregator deploys the final model and has test data to score the model performance, having the model in plaintext is useful. For use cases in which the aggregator is fully untrusted, Partial HE-based techniques may be more suitable.

SA techniques where the resulting SA fusion results are in *plaintext* have been recently shown to be vulnerable to disaggregation attacks [45, 71]. In disaggregation attacks, a curious aggregator may use SA results from multiple rounds to infer the model of a single party. When the party's model is isolated, the curious aggregator can carry any attack presented in Table 13.1 that takes as input the model itself making SA pointless. To prevent this type of attack, careful sub-sampling of parties has been proposed in [71]. This approach adds an additional layer of protection and requires the integration of TEE to bootstrap trust (Sect. 13.3.3). We note, however, that the solution only works for federations with large number of parties. Therefore, given the current state of the art, to prevent disaggregation attacks in small federations, it is best to use SA techniques where the result is encrypted.

Finally, adequately employing SA defenses[4] can only prevent inference based on individual model updates. SA on itself does not prevent any of the attacks based on the final or intermediate models. For that purpose, syntactic and perturbation approaches have been proposed and we overview them in the following.

13.3.2   Syntactic and Perturbation Approaches

Defenses in this category include solutions that aim to prevent inference attacks that use the final or model updates. K-anonymity and differential privacy are among those inference prevention techniques.

13.3.2.1   K-Anonymity-Based Approaches

K-anonymity is a technique to anonymize data that relies in hiding records in groups of k [76], these groups are defined by quasi-identifiers which are information that may serve to identify a potential individual records. For example, zip code, age, gender and race are quasi-identifiers that may serve to identify an individual. Defining the quasi-identifiers requires identifying in advance what background information an adversary may use to identify records in the training data.

---

[4] By *adequate* we mean applying additional sub-sampling techniques required for SA approaches vulnerable to disaggregation attacks as previously explained.

An adaptation of k-anonymity for FL training was presented in [17], where Choudhury et al. claim the approach is legally compliant with privacy legislations in the US, Canada and Spain (2020). The approach works for tabular data and applies k-anonymity in each of the parties independently prior to training resulting in multiple anonymization schemas. When the model is going to be used at inference time, the input is pre-processed according to the closest k-anonymity schema of each party.

One positive aspect of this approach is the interpretability of the parameters used: hiding a sample in a group of k is a very intuitive concept. One of the mayor drawbacks of k-anonymity-based approaches is the fact that the construct depends on the defender's ability to effectively anticipate the background information an adversary will have. Hence, if the adversary has more background information than anticipated, privacy can be compromised as demonstrated in [51]; albeit these attacks have not been demonstrated in the FL setting. *Differential privacy* is an alternative approach that addresses this shortcoming.

13.3.2.2 Differential Privacy-Based Approaches

Differential privacy (DP) [21] is a mathematical framework design to provide a rigorous measure of information disclosure about individual records used in the computation of a function. A training algorithm is described as deferentially private if and only if the inclusion of a single sample in the training data causes only statistically insignificant changes to the algorithm's output.

The formal definition of DP is the following [21]: A randomized mechanism[5] $\mathcal{M}$ provides (ϵ, δ)-*differential privacy* if for any two neighboring database D_1, D_2 that differ only in a single entry, $\forall S \subseteq Range(\mathcal{M})$,

$$\Pr[\mathcal{M}(D_1) \in S] \leq e^\epsilon \cdot \Pr[\mathcal{M}(D_2) \in S] + \delta.$$

When $\delta = 0$, $\mathcal{M}$ is said to satisfy ϵ-*differential privacy*.

The smaller the ϵ, the higher the protection. The additive term δ allows for a relaxation of the definition and enables mechanisms to provide higher utility, in the case of FL, this means increasing the model performance. To create a differential private mechanism, noise proportional to the sensitivity of the output is added to the algorithm's output. The sensitivity measures the maximum change of the output caused by the inclusion of a single data sample. Laplacian and Gaussian mechanisms are popular functions to achieve DP and to reduce the sensitivity, clipping values is a common practice.

An extensive set of mechanisms to optimally add DP noise while training ML models when all data is in a central place are available in the literature, e.g., [1, 36, 74, 79]. In most cases, these mechanisms can be adapted to FL.

---

[5] A mechanism can be understood as an algorithm designed to inject DP noise.

There are three ways in which DP can be applied in FL settings: *local, centralized* and using a *Hybrid* approach that encompasses DP with SA. Applying one or the other addresses a different threat models.

- **Local DP** [41, 65] is applicable in settings where parties do not trust the aggregator. For this reason, each party independently adds noise to the model updates before sending them to the untrusted aggregator. The downside of this approach is that the amount of noise typically causes the model to perform poorly.
- **Global DP** is applicable when the aggregator can be trusted to add DP noise to the model. Global DP ensures that less noise is added resulting in higher model performance compared to local DP.
- **SA-and-Local DP Hybrid** [80] This approach was created to overcome the limitations of local DP to ensure faster convergence times and better model performance. It is applicable to settings where parties do not trust the aggregator. The approach consists of using a SA technique in combination to local DP. This combination ensures that the model updates are not visible in isolation to the aggregator. Hence, each party can add less noise to its model update while maintaining the same privacy guarantee that would result by simply applying local differential privacy. In other words, this hybrid approach ensures the same DP guarantee can be obtained while the amount of total noise injected is reduced by a factor of n, where n is the number of parties. A mathematical formulation of why this is the case can be found in [80]. The SA approach proposed in [80] was Threshold Pailler, but the technique also works for other types of SA as demonstrated in [84].

A very appealing feature of DP is the mathematical guarantee it provides without making assumptions on the background information an adversary may have. At the same time, challenges of applying DP include defining the right ϵ value required for a use case, and the fact that adding noise may result in lower performing models. While ideally ϵ values should not be higher than 0.1 and the recommendation is not higher than one [23], in reality simple queries require epsilon up to one and for classical ML models epsilons up to 10. For neural networks, ϵ values greater than 100 are common [16], raising concerns of the actual protection provided by adding DP. The difficulty in interpretability of ϵ and the large values have motivated the approach based on k-anonymity previously described in this section. However, to ease this difficulty, Holohan et al. [35] proposed a methodology to map the *Randomize response* survey technique [82] to epsilon values, which provides some interpretability to ϵ.

Contrary to popular believe, DP is not a silver bullet against *all* privacy attacks presented in Sect. 13.2. Membership inference, inversion attacks and extraction attacks can be prevented by DP [12]. Let us revisit the DP definition to understand why other attacks may not be prevented by adding noise. According to its privacy definition, DP provides a quantitative privacy guarantee for indistinguishably of individual samples. However, the general information on the population is still available. In fact, one may argue that this is the main reason one would apply DP in the first place: obtaining a good generalizable model without compromising the privacy of individual records. Another assumption of DP is that records in the same dataset are independent [18, 44], which may not be the case in FL or general ML settings. Based on its definition and assumptions, it has been shown that using DP does not deter property based attacks [7, 27, 54]. The reason for which DP does not work against property attacks is the fact that they rely on aggregated properties of the population, while DP focuses on protecting individual samples. It is important to understand this difference to avoid falling into a false sense of privacy.

Currently, DP is one of the best ways to protect the privacy of individual samples and is still an area of ongoing research. An interesting discussion of open challenges and pitfalls of applying DP in real situations has been presented by Domingo et al. in [20]. Additional research focuses on reducing the amount of noise under the interactive nature of FL, where multiple rounds of communication between the aggregator and parties are often required.

13.3.3 Trusted Execution Environments (TEE)

Another set of defenses rely on *trusted execution environments* (TEE) [46]. A TEE is a secure area of computer's main processor that is designed to protect the confidentiality and integrity of the code and data loaded inside. Examples TEEs include IBM Hyper Protect™ [62], Intel SGX™ [52], and AME Memory Encryption™ [43]. One of the core features of TEEs is their ability to perform *remote attestation*. Remote attestation allows a remote client to verify that a specific software version has been securely loaded into the enclave. Hence, TEEs are suitable to run code in untrusted environments while ensuring the code run is the expected.

Thanks to the secure attestation feature, it is possible to ensure an aggregator running on a TEE uses the expected code preventing active attacks where a malicious aggregator may deviate from the expected behavior. Hence, we can ensure that the aggregator combines a minimum number of model updates preventing active attacks that isolate one or a few model updates. It is also possible to ensure it performs Global DP adequately and that, in general, it follows the right protocol. If deploying special hardware is possible, using TEE at the parties side can also improve the security of the system ensuring all participants run the pre-specified source code. This, however, may not be easy to achieve in settings where parties are consumer devices, run in legacy hardware or there is limited budget.

Potential vulnerabilities of the TEE include side-channel attacks and *cukoo attacks* [64] that enable an adversary to compromise the privacy of the data loaded by the TEE. To prevent potential side-channel attacks Law et al. [46] redesigned an adaptation of XGBoost for FL to be data-oblivious. Finally, using a TEE requires special hardware and setting up correctly the keys in the system to prevent cukoo attacks that compromise the cryptographic keys of the system [64].

In [15], data extraction attacks where a curious aggregator takes advantage of the gradients exchanged are deterred by decentralizing the aggregation process. The solution, called *Truda*, changes the FL architecture by introducing three TEE-aggregators that receive a partial view of the model. Parties agree on what pieces of the model sent to each aggregator, and no aggregator obtains all the model. Truda works for fusion algorithms that only require average of model updates. More advanced algorithms such as the ones required to train tree-boosting models based or PFMN [86] cannot be adapted to this architecture.

13.3.4 Other Techniques for Distributed Machine Learning and Vertical FL

In addition to the privacy-preserving approaches discussed above, there exist other techniques designed for distributed ML that have slightly different architectures to FL. These include Helen [90], Private Aggregation of Teacher Ensembles (PATE) [61] and its variant [49], as well as SecureML [55], where data is distributed among two non-colluding servers who jointly train a model using two-party computation. Because their architectures are different to FL, we do not expand on them in this chapter.

In this chapter, we focus our attention to the horizontal FL case, where all parties have the same input data and, thus, can train locally their own models. Vertical FL (Chap. 18) and split learning (Chap. 19) work for different setups. In vertical FL, each party holds only a partial set of the features and only one party typically holds the label while in split learning a different piece of the model may be trained in different parties. Thus, a single party cannot train a model on their own. Inference attacks and defenses have also been presented for these different setups [39, 47]. The threats in these settings may differ, for example, label inference is a potential attack [47]. Determining their vulnerabilities and designing defenses is still an open question.

13.4 Selecting the Right Defense

We have reviewed existing attacks and defenses and are now ready to define what defenses are applicable in different cases. Unfortunately, there is no free lunch when applying a defense as it was highlighted during their detail presentation in the previous section. Incorporating different defenses may lead to longer training times, lower performing models or more expensive deployment. Thus, it is necessary to find a sweet spot to prevent *relevant* attacks for each federation. We now analyze different scenarios.

13.4.1 Fully Trusted Federations

Consider a scenario where all the parties engage in the federation are owned by the same company. This case is embodied, for example, by a company who has stored data in different clouds, data warehouses, countries or has acquired other companies resulting in fragmented datasets. Another example of this type of scenario is a federation where the training task does not involve the use of sensitive data; yet, participants do not want to transfer the data to a single place due to its large volume.

These are low risk scenarios where there is no reason to mistrust each of the parties or the aggregator. Therefore, it is possible to employ plain FL without other protections, other than our assumed secure-and-authenticated channels. With respect to potential outsiders, it is possible to use Global DP to minimize attacks based on the final model. Otherwise, no DP needs to be added.

13.4.2 Ensuring that the Aggregator Can Be Trusted

There are multiple ways to ensure a federation can trust an aggregator. A common way to ensure this is the case is to run the aggregator by a trusted party. Alternatively, the aggregator may be run as a service where through contractual clauses trust may be achieved. The aggregator can also be required to be run using TEE so that parties can verify the aggregator is running the correct code through attestation. Such an aggregator as a service is a practical way to ensure fast deployment in consortium cases where all participants can find trusted company to host the aggregator.

In some cases, adding proper *accountability* to the process may also help boost trust in the aggregator and parties. Recently, an accountability framework for FL was proposed in [9], where all entities engage can subsequently audited if needed. It is in the best interest of a company running aggregators as a service to comply with its contracts, and it is even more critical to do so if it can be audit. Accountability services help ensure there is a way to verify different entities behaved as expected while offering a way for potentially mistrusting parties to verifying during the training process the system is behaving properly. Although accountability cannot prevent inference of information by inspecting the results of the well-executed process, it can help ensure the aggregator is a honest-but-curious adversary, meaning that it adheres to the protocol but may try to infer information based on the information it obtains in the process.

Whenever possible it is beneficial to trust the aggregator. One big advantage of this type of deployment is that the aggregator can offer additional features that require the inspection of individual replies. Among them are running robust learning algorithms that are resilient against noisy model updates or failures in the setup, as well as algorithms to detect and mitigate potential active attacks performed by misbehaving parties. In other words, a risk assessment to see what is more important for the federation is needed. In some cases, mitigating the aggregator's capabilities to perform inference by using the above listed techniques is deemed as enough mitigation to have it as an ally to prevent attacks from parties. In some other scenarios, the risk exposure may be unbearable.

13.4.3 Federations with an Untrusted Aggregator

In some cases, the aforementioned provisions may not be enough to trust the aggregator to obtain individual parties' model updates. For instance, the consumer space, where users may fear their private information is obtained by big companies. In these cases, SA techniques and DP can be applied. In fact, Google, Apple, and other consumer companies are already using local DP to provide privacy of the input and create trust among their users while enabling service improvement [66, 77].

Not all SA techniques offer the same protection. Different SA mechanisms are more suitable than others depending on whether the aggregator and other parties are mistrusted simultaneously. Let us analyze these cases.

The Aggregator Is Trusted to Offer Additional Functionality that Requires Access to the Model Consider the case where the federation wants to make use of an extended set of services offered by the aggregator that require this entity to access the model in plaintext. For example, the federation may want the aggregator to evaluate the performance of a model based on public data or to deploy the resulting model as a service. Solutions where the SA enables the aggregator to see the model in plaintext are adequate for these use cases.

We also highlight that SA mechanisms that enable the aggregator to obtain the model in plaintext need to be complemented with additional provisions to prevent disaggregation attacks. To mitigate disaggregation inference risks, it is necessary to ensure that *all* parties are selected and their model updates aggregated, or that sub-sample parties selections of multiple rounds do not lead to inference. Clearly, the first solution only works for small federations, while the second one can only be applied to large federations. For small federations, the modification is not particularly taxing, as it is typical to include all parties in all rounds to fully leverage their data.

To mitigate the risk of a malicious aggregator isolating replies of a few parties, HybridAlpha [84], the functional encryption-based SA presented in the previous section provides an inference module that verifies a minimum number of replies that have been aggregated before providing a functional key to obtain the model. This module prevents this type of attack. Another potential solution is to run in a TEE to verify the specified number of parties is indeed aggregated.

Now let us consider a federation where parties do not fully trust each other and are concerned other parties may try to obtain private information. Example scenarios in this category include multiple competitors collaborating to detect fraud, where each competing party may benefit from learning data about other parties. In the case the federation fears inference from different parties, for example, fearing attacks where a few curious parties may collude. In those cases, the solution presented in [80] which encompasses SA and DP is particularly useful. The cryptosystem of choice, threshold Paillier, allows for verification that ensures a subset of t trusted out of n total parties need to contribute to obtain the model in plaintext. As a bonus, the solution prevents inference by reducing the noise compared to local DP as outlined in the previous section.

The Aggregator Is Not Trusted with the Model In certain cases, a federation may find too risky to provide the model to the aggregator. In these cases, HE techniques are recommended. Notice that the final model is going to be accessible by parties owning the cryptographic keys. If inferences over the model are relevant, then DP may be added.

In the above analysis, we have avoided discussing particular regulations, as regulations keep evolving and, to date, there is no clear mapping between regulation requirements and technical solutions. This is a relevant open question that we expect will be solved as FL is increasingly applied in regulated settings.

13.5   Conclusions

FL is a privacy by design system that has substantially improved the state-of-the-art techniques that require transmitting private data to a central place. From the privacy perspective, there is a clear benefit over other approaches that move data, as the data can always remain with its rightful owner. Although, some inference attacks have been demonstrated, current defenses and research efforts can be incorporated to mitigate them. Inference of private data in FL systems is a relevant risk for *some* federations where exposing private data is an important consideration. In this chapter, we have overviewed the attack surfaces, the threats, and the defenses to help provide a holistic view of the risk inherent to participating in FL. We also presented multiple attacks characterizing them based on the attack surface, the objective the adversary had and also providing some details on how they may be carried out. As highlighted by our literature review, some attacks may not be realistic as simply changing hyperparameters of the training process can easily deter them, while in some other cases FL creates new relevant threats.

We also presented multiple defenses and highlighted their benefits and drawbacks showing that one size does not fit all. The details of the design of each defense imply various trust assumptions and made them suitable to different applications as they have inherently diverse computational and transmission costs. Matching the right level of protection to the use case is imperative to ensure only necessary overheads are incurred by adding defenses, while deterring relevant risks. Without a doubt, new attacks will emerge creating an arm's race between defenders and adversaries. As future work, enhancing different techniques to reduce overheads and understanding how different legislation and regulation can be mapped to concrete technologies is required. We hope this chapter has helped clarify the state of the art of attacks and available defenses to improve and facilitate the decision making process.

References

1. Abadi M, Chu A, Goodfellow I, McMahan HB, Mironov I, Talwar K, Zhang L (2016) Deep learning with differential privacy. In: Proceedings of the 2016 ACM SIGSAC conference on computer and communications security, pp 308–318
2. Abdalla M, Bourse F, De Caro A, Pointcheval D (2015) Simple functional encryption schemes for inner products. In: IACR international workshop on public key cryptography. Springer, pp 733–751
3. Abdalla M, Benhamouda F, Gay R (2019) From single-input to multi-client inner-product functional encryption. In: International conference on the theory and application of cryptology and information security. Springer, pp 552–582
4. Ananth P, Vaikuntanathan V (2019) Optimal bounded-collusion secure functional encryption. In: Theory of cryptography conference. Springer, pp 174–198
5. Aono Y, Hayashi T, Wang L, Moriai S et al (2017) Privacy-preserving deep learning via additively homomorphic encryption. IEEE Trans Inf Forens Secur 13(5):1333–1345
6. Asoodeh S, Calmon F (2020) Differentially private federated learning: An information-theoretic perspective. In: Proc. ICML-FL
7. Ateniese G, Mancini LV, Spognardi A, Villani A, Vitali D, Felici G (2015) Hacking smart machines with smarter ones: How to extract meaningful data from machine learning classifiers. Int J Secur Netw 10(3):137–150
8. Attrapadung N, Libert B (2010) Functional encryption for inner product: Achieving constant-size ciphertexts with adaptive security or support for negation. In: International workshop on public key cryptography. Springer, pp 384–402
9. Balta D, Sellami M, Kuhn P, Schöpp U, Buchinger M, Baracaldo N, Anwar A, Sinn M, Purcell M, Altakrouri B IFIP EGOV (2021) Accountable Federated Machine Learning in Government: Engineering and Management Insights (Best paper award), IFIP EGOV 2021
10. Bonawitz K, Ivanov V, Kreuter B, Marcedone A, McMahan HB, Patel S, Ramage D, Segal A, Seth K (2017) Practical secure aggregation for privacy-preserving machine learning. In: Proceedings of the 2017 ACM SIGSAC conference on computer and communications security, pp 1175–1191
11. Boneh D, Sahai A, Waters B (2011) Functional encryption: Definitions and challenges. In: Theory of cryptography conference. Springer, pp 253–273
12. Carlini N, Liu C, Erlingsson Ú, Kos J, Song D (2019) The secret sharer: Evaluating and testing unintended memorization in neural networks. In: 28th USENIX security symposium (USENIX Sec 19), pp 267–284
13. Carlini N, Tramer F, Wallace E, Jagielski M, Herbert-Voss A, Lee K, Roberts A, Brown T, Song D, Erlingsson U et al (2020) Extracting training data from large language models. Preprint. arXiv:2012.07805
14. Centers for Medicare & Medicaid Services: The Health Insurance Portability and Accountability Act of 1996 (HIPAA) (1996) Online at http://www.cms.hhs.gov/hipaa/
15. Cheng PC, Eykholt K, Gu Z, Jamjoom H, Jayaram K, Valdez E, Verma A (2021) Separation of powers in federated learning. Preprint. arXiv:2105.09400
16. Choquette-Choo CA, Tramer F, Carlini N, Papernot N (2021) Label-only membership inference attacks. In: Meila M, Zhang T (eds) Proceedings of the 38th international conference on machine learning, Proceedings of machine learning research. PMLR, vol 139, pp 1964–1974. http://proceedings.mlr.press/v139/choquette-choo21a.html
17. Choudhury O, Gkoulalas-Divanis A, Salonidis T, Sylla I, Park Y, Hsu G, Das A (2020) A syntactic approach for privacy-preserving federated learning. In: ECAI 2020. IOS Press, pp 1762–1769

18. Clifton C, Tassa T (2013) On syntactic anonymity and differential privacy. In: 2013 IEEE 29th international conference on data engineering workshops (ICDEW). IEEE, pp 88–93
19. Damgård I, Jurik M (2001) A generalisation, a simpli. cation and some applications of paillier's probabilistic public-key system. In: International workshop on public key cryptography. Springer, pp 119–136
20. Domingo-Ferrer J, Sánchez D, Blanco-Justicia A (2021) The limits of differential privacy (and its misuse in data release and machine learning). Commun ACM 64(7):33–35
21. Dwork C (2008) Differential privacy: A survey of results. In: International conference on theory and applications of models of computation. Springer, pp 1–19
22. Dwork C, Lei J (2009) Differential privacy and robust statistics. In: STOC, vol 9. ACM, pp 371–380
23. Dwork C, Roth A et al (2014) The algorithmic foundations of differential privacy. Found Trends Theor Comput Sci 9(3–4):211–407
24. FederatedAI: Fate (federated AI technology enabler). Online at https://fate.fedai.org/
25. Fredrikson M, Lantz E, Jha S, Lin S, Page D, Ristenpart T (2014) Privacy in pharmacogenetics: An end-to-end case study of personalized warfarin dosing. In: 23rd {USENIX} Security Symposium ({USENIX} Security 14), pp 17–32
26. Fredrikson M, Jha S, Ristenpart T (2015) Model inversion attacks that exploit confidence information and basic countermeasures. In: Proceedings of the 22nd ACM SIGSAC conference on computer and communications security, pp 1322–1333
27. Ganju K, Wang Q, Yang W, Gunter C.A, Borisov N (2018) Property inference attacks on fully connected neural networks using permutation invariant representations. In: Proceedings of the 2018 ACM SIGSAC conference on computer and communications security, pp 619–633
28. Geiping J, Bauermeister H, Dröge H, Moeller M (2020) Inverting gradients–how easy is it to break privacy in federated learning? Preprint. arXiv:2003.14053
29. Gentry C (2009) A fully homomorphic encryption scheme. Ph.D. thesis, Stanford University
30. Geyer RC, Klein T, Nabi M (2017) Differentially private federated learning: A client level perspective. Preprint. arXiv:1712.07557
31. Goldwasser S, Gordon S.D, Goyal V, Jain A, Katz J, Liu FH, Sahai A, Shi E, Zhou HS (2014) Multi-input functional encryption. In: Annual international conference on the theory and applications of cryptographic techniques. Springer, pp 578–602
32. Hayes J, Melis L, Danezis G, De Cristofaro E (2017) Logan: Membership inference attacks against generative models. Preprint. arXiv:1705.07663
33. Henderson P, Sinha K, Angelard-Gontier N, Ke NR, Fried G, Lowe R, Pineau J (2018) Ethical challenges in data-driven dialogue systems. In: Proceedings of the 2018 AAAI/ACM conference on AI, ethics, and society, pp 123–129
34. Hitaj B, Ateniese G, Perez-Cruz F (2017) Deep models under the GAN: information leakage from collaborative deep learning. In: Proceedings of the 2017 ACM SIGSAC conference on computer and communications security, pp 603–618
35. Holohan N, Leith DJ, Mason O (2017) Optimal differentially private mechanisms for randomised response. IEEE Trans Inf Forens Secur 12(11):2726–2735
36. Holohan N, Braghin S, Mac Aonghusa P, Levacher K (2019) Diffprivlib: the IBM^{TM} differential privacy library. Preprint. arXiv:1907.02444
37. Huang Y, Su Y, Ravi S, Song Z, Arora S, Li K (2020) Privacy-preserving learning via deep net pruning. Preprint. arXiv:2003.01876
38. Jia J, Salem A, Backes M, Zhang Y, Gong NZ (2019) Memguard: Defending against black-box membership inference attacks via adversarial examples. In: Proceedings of the 2019 ACM SIGSAC conference on computer and communications security, pp 259–274
39. Jin X, Du R, Chen PY, Chen T (2020) Cafe: Catastrophic data leakage in federated learning. OpenReview - Preprint
40. Kadhe S, Rajaraman N, Koyluoglu OO, Ramchandran K (2020) Fastsecagg: Scalable secure aggregation for privacy-preserving federated learning. Preprint. arXiv:2009.11248
41. Kairouz P, Oh S, Viswanath P (2014) Extremal mechanisms for local differential privacy. Preprint. arXiv:1407.1338

42. Kairouz P, Liu Z, Steinke T (2021) The distributed discrete gaussian mechanism for federated learning with secure aggregation. Preprint. arXiv:2102.06387
43. Kaplan D, Powell J, Woller T (2016) Amd memory encryption. White paper
44. Kifer D, Machanavajjhala A (2011) No free lunch in data privacy. In: Proceedings of the 2011 ACM SIGMOD international conference on management of data, pp 193–204
45. Lam M, Wei GY, Brooks D, Reddi VJ, Mitzenmacher M (2021) Gradient disaggregation: Breaking privacy in federated learning by reconstructing the user participant matrix. Preprint. arXiv:2106.06089
46. Law A, Leung C, Poddar R, Popa R.A, Shi C, Sima O, Yu C, Zhang X, Zheng W (2020) Secure collaborative training and inference for xgboost. In: Proceedings of the 2020 workshop on privacy-preserving machine learning in practice, pp 21–26
47. Li O, Sun J, Yang X, Gao W, Zhang H, Xie J, Smith V, Wang C (2021) Label leakage and protection in two-party split learning. Preprint. arXiv:2102.08504
48. Liu Y, Ma Z, Liu X, Ma S, Nepal S, Deng R (2019) Boosting privately: Privacy-preserving federated extreme boosting for mobile crowdsensing. Preprint. arXiv:1907.10218
49. Liu C, Zhu Y, Chaudhuri K, Wang YX (2020) Revisiting model-agnostic private learning: Faster rates and active learning. Preprint. arXiv:2011.03186
50. Ludwig H, Baracaldo N, Thomas G, Zhou Y, Anwar A, Rajamoni S, Ong Y, Radhakrishnan J, Verma A, Sinn M et al (2020) Ibm^{TM} federated learning: an enterprise framework white paper v0. 1. Preprint. arXiv:2007.10987
51. Machanavajjhala A, Kifer D, Gehrke J, Venkitasubramaniam M (2007) l-diversity: Privacy beyond k-anonymity. ACM Trans Knowl Discov Data (TKDD) 1(1):3–es
52. McKeen F, Alexandrovich I, Berenzon A, Rozas C.V, Shafi H, Shanbhogue V, Savagaonkar UR (2013) Innovative instructions and software model for isolated execution. In: Proceedings of the 2nd international workshop on hardware and architectural support for security and privacy, HASP '13. Association for Computing Machinery, New York
53. McMahan B, Moore E, Ramage D, Hampson, S, y Arcas BA (2017) Communication-efficient learning of deep networks from decentralized data. In: Artificial intelligence and statistics. PMLR, pp 1273–1282
54. Melis L, Song C, De Cristofaro E, Shmatikov V (2019) Exploiting unintended feature leakage in collaborative learning. In: 2019 IEEE symposium on security and privacy (SP). IEEE, pp 691–706
55. Mohassel P, Zhang Y (2017) Secureml: A system for scalable privacy-preserving machine learning. In: 2017 IEEE symposium on security and privacy (SP). IEEE, pp 19–38
56. Naccache D, Stern J (1997) A new public-key cryptosystem. In: International conference on the theory and applications of cryptographic techniques. Springer, pp 27–36
57. Nasr M, Shokri R, Houmansadr A (2019) Comprehensive privacy analysis of deep learning: Stand-alone and federated learning under passive and active white-box inference attacks. 2019 IEEE symposium on security and privacy (SP)
58. Nikolaenko V, Weinsberg U, Ioannidis S, Joye M, Boneh D, Taft N (2013) Privacy-preserving ridge regression on hundreds of millions of records. In: 2013 IEEE symposium on security and privacy. IEEE, pp 334–348
59. Okamoto T, Uchiyama S (1998) A new public-key cryptosystem as secure as factoring. In: International conference on the theory and applications of cryptographic techniques. Springer, pp 308–318
60. Paillier P (1999) Public-key cryptosystems based on composite degree residuosity classes. In: International conference on the theory and applications of cryptographic techniques. Springer, pp 223–238
61. Papernot N, Abadi M, Erlingsson U, Goodfellow I, Talwar K (2016) Semi-supervised knowledge transfer for deep learning from private training data. Preprint. arXiv:1610.05755
62. Park S, McMullen A (2021) Announcing secure build for ibm^{TM} cloud hyper protect virtual servers. IBM^{TM} Cloud Blog. https://www.ibm.com/cloud/blog/announcements/secure-build-for-ibm-cloud-hyper-protect-virtual-servers
63. Parliament E of the European Union C (2016) General data protection regulation (GDPR) – official legal text. https://gdpr-info.eu/

64. Parno B (2008) Bootstrapping trust in a "trusted" platform. In: HotSec
65. Qin Z, Yang Y, Yu T, Khalil I, Xiao X, Ren K (2016) Heavy hitter estimation over set-valued data with local differential privacy. In: Proceedings of the 2016 ACM SIGSAC conference on computer and communications security, pp 192–203
66. Radebaugh C, Erlingsson U.: Introducing TensorFlow privacy: Learning with differential privacy for training data. https://blog.tensorflow.org/2019/03/introducing-tensorflow-privacy-learning.html
67. Roth H, Zephyr M, Harouni A (2021) Federated learning with homomorphic encryption. NVIDIA^{TM} Developer Blog. https://developer.nvidia.com/blog/federated-learning-with-homomorphic-encryption/
68. Salem A, Zhang Y, Humbert M, Berrang P, Fritz M, Backes M (2018) Ml-leaks: Model and data independent membership inference attacks and defenses on machine learning models. Preprint. arXiv:1806.01246
69. Shamir A (1979) How to share a secret. Commun ACM 22(11):612–613
70. Shokri R, Stronati M, Song C, Shmatikov V (2017) Membership inference attacks against machine learning models. In: 2017 IEEE symposium on security and privacy (SP). IEEE, pp 3–18
71. So J, Ali RE, Guler B, Jiao J, Avestimehr S (2021) Securing secure aggregation: Mitigating multi-round privacy leakage in federated learning. Preprint. arXiv:2106.03328
72. So J, Güler B, Avestimehr AS (2021) Turbo-aggregate: Breaking the quadratic aggregation barrier in secure federated learning. IEEE J Sel Areas Inf Theory 2:479
73. Song C, Raghunathan A (2020) Information leakage in embedding models. In: Proceedings of the 2020 ACM SIGSAC conference on computer and communications security, pp 377–390
74. Song S, Chaudhuri K, Sarwate AD (2013) Stochastic gradient descent with differentially private updates. In: 2013 IEEE global conference on signal and information processing. IEEE, pp 245–248
75. Song M, Wang Z, Zhang Z, Song Y, Wang Q, Ren J, Qi H (2020) Analyzing user-level privacy attack against federated learning. IEEE J Sel Areas Commun 38(10):2430–2444
76. Sweeney L (2002) k-anonymity: A model for protecting privacy. Int J Uncertainty Fuzziness Knowl Based Syst 10(05):557–570
77. Team DP (2017) Learning with privacy at scale. Machine Learning Research at Apple^{TM}
78. Thakkar O, Ramaswamy S, Mathews R, Beaufays F (2020) Understanding unintended memorization in federated learning. Preprint. arXiv:2006.07490
79. Tramèr F, Boneh D (2020) Differentially private learning needs better features (or much more data). Preprint. arXiv:2011.11660
80. Truex S, Baracaldo N, Anwar A, Steinke T, Ludwig H, Zhang R, Zhou Y (2019) A hybrid approach to privacy-preserving federated learning. In: Proceedings of the 12th ACM workshop on artificial intelligence and security, pp 1–11
81. Wang Y, Deng J, Guo D, Wang C, Meng X, Liu H, Ding C, Rajasekaran S (2020) Sapag: a self-adaptive privacy attack from gradients. Preprint. arXiv:2009.06228
82. Warner SL (1965) Randomized response: A survey technique for eliminating evasive answer bias. J Am Stat Assoc 60(309):63–69
83. Wei W, Liu L, Loper M, Chow KH, Gursoy ME, Truex S, Wu Y (2020) A framework for evaluating gradient leakage attacks in federated learning. Preprint. arXiv:2004.10397
84. Xu R, Baracaldo N, Zhou Y, Anwar A, Ludwig H (2019) Hybridalpha: An efficient approach for privacy-preserving federated learning. In: Proceedings of the 12th ACM workshop on artificial intelligence and security, pp 13–23
85. Yang Z, Shao B, Xuan B, Chang EC, Zhang F (2020) Defending model inversion and membership inference attacks via prediction purification. Preprint. arXiv:2005.03915
86. Yurochkin M, Agarwal M, Ghosh S, Greenewald K, Hoang N, Khazaeni Y (2019) Bayesian nonparametric federated learning of neural networks. In: International conference on machine learning. PMLR, pp 7252–7261
87. Zanella-Béguelin S, Wutschitz L, Tople S, Rühle V, Paverd A, Ohrimenko O, Köpf B, Brockschmidt M (2020) Analyzing information leakage of updates to natural language models.

In: Proceedings of the 2020 ACM SIGSAC conference on computer and communications security, pp 363–375
88. Zhang C, Li S, Xia J, Wang W, Yan F, Liu Y (2020) Batchcrypt: Efficient homomorphic encryption for cross-silo federated learning. In: 2020 USENIX annual technical conference (USENIX ATC 20), pp 493–506
89. Zhao B, Mopuri KR, Bilen H (2020) IDLG: Improved deep leakage from gradients. Preprint. arXiv:2001.02610
90. Zheng, W Popa RA, Gonzalez JE, Stoica I (2019) Helen: Maliciously secure coopetitive learning for linear models. In: 2019 IEEE symposium on security and privacy (SP). IEEE, pp 724–738
91. Zhu L, Han S (2020) Deep leakage from gradients. In: Federated learning. Springer, pp 17–31

Chapter 14
Private Parameter Aggregation for Federated Learning

K. R. Jayaram and Ashish Verma

Abstract Federated learning enables multiple distributed participants (potentially on different datacenters or clouds) to collaborate and train machine/deep learning models by sharing parameters or gradients. However, sharing gradients, instead of centralizing data, may not be as private as one would expect. Reverse engineering attacks on plain text gradients have been demonstrated to be practically feasible. This problem has been made more insidious by the fact that participants or aggregators may reverse engineer model parameters while participating honestly in the protocol (the so-called *honest, but curious* trust model). Existing solutions for differentially private federated learning, while promising, lead to less accurate models and require nontrivial hyperparameter tuning. In this chapter, we (1) describe various trust models in federated learning and their challenges, (2) explore the use of secure multi-party computation techniques in federated learning, (3) explore how additive homomorphic encryption can be used efficiently for federated learning, (4) compare these techniques with others like the addition of differentially private noise and the use of specialized hardware, and (5) illustrate these techniques through real-world examples.

14.1 Introduction

Some of the early success of distributed machine and deep learning (ML/DL) in several application domains [29, 31] has been in the context of massive centralized data collection, either at a single datacenter or at a cloud service. However, centralized data collection at a (third-party) cloud service can be incredibly privacy-invasive and can expose organizations (customers of the cloud service) to large legal liability when there is a data breach. This is especially true in the case of healthcare data, voice transcripts, home cameras, financial transactions, etc.

K. R. Jayaram (✉) · A. Verma
IBM Research, Yorktown Heights, NY, USA
e-mail: jayaramkr@us.ibm.com; Ashish.Verma1@ibm.com

Centralized data collection often results in "loss of control" over data once it is uploaded. A frequently asked question to which users often do not get a satisfactory answer is "Is the cloud service using my data as promised? Is it actually deleting my data when it claims to do so?". Organizations that have not been convinced by privacy violations and loss of control have been forced by governmental regulations (like HIPAA and GDPR [34]) to restrict data sharing with third-party services.

Federated learning (FL) aims to mitigate these aforementioned issues while maintaining accuracy of ML/DL models. An entity in an FL job can be as small as a smart phone/watch or as large as an organization with multiple data centers. An FL algorithm aims to train an ML/DL model, e.g., a specific neural network model or an XGBoost model, on multiple entities, each with its own "local" dataset, without exchanging any data. This results in multiple "local models," which are then combined (aggregated) by exchanging only parameters (e.g., the weights of a neural network model). An FL algorithm may use a central coordinator to collect parameters of all local models for aggregation, or it may be a peer-to-peer algorithm (broadcast, overlay multicast, etc.)

Initially, it was believed that the exchanged model updates in Federated Learning (FL) communications would contain far less, if any, information about the raw training data. Thus, sharing model updates was considered to be "privacy-preserving." However, even if not discernible immediately, training data information is still embedded in the model updates. Recent research [14, 15, 17, 33, 42, 48, 50, 51] has demonstrated the feasibility and ease of inferring private attributes and reconstructing large fractions of training data by exploiting model updates, thereby challenging the privacy promises of FL in the presence of honest-but-curious aggregation servers.

14.2 Focus, Trust Model, and Assumptions

FL is *typically* deployed in two scenarios: *cross-device* and *cross-silo* [23]. The *cross-device* scenario involves a large number of parties (>1000), but each party has a small number of data items, constrained compute capability, and limited energy reserve (e.g., mobile phones or IoT devices). They are highly unreliable and are expected to drop and join frequently. Examples include a large organization learning from data stored on employees' devices and a device manufacturer training a model from private data located on millions of its devices (e.g., Google Gboard [4]). A *trusted authority*, which performs aggregation and orchestrates training, is typically present in a *cross-device* scenario. Contrarily, in the *cross-silo* scenario, the number of parties is small, but each party has extensive compute capabilities (with stable access to electric power or equipped with hardware machine learning (ML) accelerators) and large amounts of data. The parties have reliable participation throughout the entire FL training lifecycle but are more susceptible to sensitive data leakage. Examples include multiple hospitals collaborating to train a tumor detection model on radiographs, multiple banks collaborating to train a credit card fraud detection model, etc. In *cross-silo* scenarios, there exists *no presumed central trusted authority*. All parties involved in the training are *equal* collaborators. The deployments often involve hosting aggregation in public clouds, or alternatively one of the parties acting as, and providing infrastructure for aggregation. In this chapter, we focus on private parameter aggregation in *cross-silo* scenarios.

We assume that each participant is convinced of the benefits (improvements in accuracy, robustness, etc.) of federated learning. We note that convincing participants to collaborate by projecting potential gains in accuracy due to federated learning is an open research problem. We focus on the so-called *honest, but curious* trust model. Here, each participant is convinced enough that it follows the steps of the federation protocol and does not collude with the coordinator to break the protocol. But the participant may be curious about the data of others, and it may be in their interest to reverse engineer the model parameters to try and discover other participants' data. We also assume that the coordinator is honest but curious with respect to individual participants' data. The participants want to reduce the required amount of trust in the coordinator as much as possible. We also assume that each participant does not attempt to poison or skew the global model by maliciously generating weights. This trust model also includes the simpler case where participants are forbidden from sharing data or the model parameters derived from the data due to regulatory reasons (e.g., FedRAMP, EU data protection guidelines [34]).

Examples and Use Cases This trust model is predominantly found in enterprise federated learning; for example, a multinational bank having branches in multiple countries and so regulated locally (BankA, BankA US, BankA UK, BankA India, etc.), where the bank wants to learn a fraud detection model across data of all its subsidiaries, but the data cannot be transferred to a central location due to governmental data sovereignty and jurisdiction laws [34]. Each participant here is a subsidiary with its national data center(s) and the coordinator might be located in a cloud platform or a global datacenter. Another example is a set of hospitals that want to collaborate to train a tumor detection model; each hospital is unable to trust the others and unwilling to trust and transfer data to a central service. Another example is a cloud-hosted machine learning service (e.g., Azure ML) that has multiple (competing) corporate clients which do not trust each other but have some level of trust in the cloud service to facilitate and secure federated learning.

14.3 Differentially Private Federated Learning

Differential privacy [12, 30] is a framework which deals with publishing the results of computations or queries made on a dataset in such a way that limits the disclosure of private information. In simple words, a computation on a dataset is called differentially private if an observer seeing its output cannot tell if a specific individual's information was used in the computation. ϵ-differential privacy, the typically used notion of differential privacy, is a mathematical definition for the privacy loss associated with any release of derivative information from a dataset.

Differential privacy has been recently used by a number of researchers during model training in a federated setting [1, 2, 32, 43]. A typical way to incorporate differential privacy into a federated learning setup is to add random noise in each of the model/data derivatives shared externally by a participant during the training process. The amount of random noise to be added depends on the level of privacy required by the participant. This noise addition adversely impacts the accuracy of the trained model. While there have been some studies that demonstrate that impact on accuracy can be minimized by adding a small amount of noise and carefully choosing the hyperparameters of the training [1, 43], there is no systematic study on how exactly the noise level impacts model convergence. This may require a long empirical process to determine achievable level of privacy without sacrificing too much accuracy. In this section, we describe this problem in more detail, with empirical evidence.

14.3.1 Background: Differential Privacy (DP)

Differential privacy literature has gained significant attention in the computer science field in the recent past. Here we briefly cover the differential privacy (DP) framework in the interest of keeping this chapter self-contained. For a detailed description of the differential privacy, we refer the reader to [12]. Differential privacy literature states that a (randomized) function $f(\cdot)$ is ϵ-differentially private if for all datasets X, X' that differ by only a single data item and all values of t

$$\left| \ln \frac{P(f(X) = t)}{P(f(X') = t)} \right| \le \epsilon \tag{14.1}$$

The parameter ϵ quantifies the privacy risk; lower ϵ means higher privacy. We work with a practical variant of the original differential privacy definition described in [12], called (ϵ, δ) differential privacy defined for the function $f(\cdot)$:

$$P(f(X) = t) \le e^{\epsilon} P(f(X') = t) + \delta \tag{14.2}$$

This definition is interpreted as saying that $f(\cdot)$ is ϵ-differentially private with probability $1 - \delta$. In order to achieve differential privacy, a noise term is added to the output of $f(\cdot)$ whose variance is dependent on the parameters ϵ and δ. Furthermore, it has been demonstrated that using an additive Gaussian noise term with 0 mean and variance

$$\sigma^2 = \Delta f \cdot \frac{2 \ln \frac{1.25}{\delta}}{\epsilon} \tag{14.3}$$

ensures (ϵ, δ)-differential privacy [12], where Δf is the sensitivity of $f(\cdot)$. Sensitivity is a measure of how much the computation can change when a single element of the underlying dataset changes. More formally, the sensitivity Δf is given by

$$\Delta f = \sup_{(X,X')} (\| f(X) - f(X') \|_l) \tag{14.4}$$

The above definitions deal with differential privacy of a single query, and however training a neural network requires many iterations, that is, many queries of the data. This means we cannot restrict ourselves to merely choosing a single value of ϵ but must consider the total privacy loss over the course of training. According to the composition theorem 3.16 in [12], using multiple (ϵ, δ)-DP will still be differentially private, but with a total privacy loss, referred to as the privacy budget (β), equal to the sum over all used ϵ. This means if we train for T iterations, using a constant value ϵ_0 at each iteration will take a total privacy budget of $\beta = \sum_{i=1}^{T} \epsilon_0 = T\epsilon_0$.

Of course, one's privacy budget does not have to be spent in this naive manner. Several authors have considered different strategies for spending privacy budgets with various improvements in mind. Shokri and Shmatikov [41] propose a way to limit the total number of queries, thus reducing overall privacy loss, and the work [35] uses a generalization of differential privacy to achieve tighter bounds on privacy loss per query.

14.3.2   Incorporating DP into SGD

Stochastic gradient descent (SGD) is one of the most popular optimization techniques used to train deep learning and various other machine learning models [5]. In distributed SGD, a mini-batch of samples is distributed over multiple learners who compute gradients of the model on their local share of the mini-batch and then share the gradients with a centralized parameter server. The parameter server computes the average gradient across the learners, updates the model parameters, and distributes the updated model parameters back to the learners.

In a typical federated learning setting, gradients of the model parameters are computed by the participants on their private dataset and shared with the centralized aggregator. Furthermore, concepts from differential privacy literature are utilized to modify SGD algorithm in order to prevent any leakage of information about the data through the shared gradients. This modification results in the so-called differentially private SGD [1], which is described below.

Let us now discuss how to use (ϵ, δ)-differential privacy in the context of federated learning using SGD. Here, $f(\cdot)$ corresponds to computing the gradient of the model parameters on the local dataset.

The update rule for mini-batch SGD with batch size S and learning rate η is

$$\theta_{k+1} = \theta_k - \eta \frac{1}{S} \sum_{i=1}^{S} g(x_i) \tag{14.5}$$

where $g(x_i)$ is the gradient of the loss function evaluated on data point $x_i \in X$. The quantity that is shared during gradient exchange is $f(x) = \eta \frac{1}{S} \sum_{i=1}^{S} g(x_i)$. Note, for the above formulation, the sensitivity would depend upon the gradient of only one data point which is different in the two datasets X and X', i.e.,

$$\Delta f = \|f(X) - f(X')\|_l$$

$$= \frac{\eta}{S} \left\| \sum_{i=1}^{S} (g(x_i) - g(x_i')) \right\|_l$$

$$= \frac{\eta}{S} \|g(x_j) - g(x_j')\|_l$$

$$\leq \frac{\eta}{S} \cdot 2C$$

where the constant C is an upper bound on the gradient. Hence, following the result from [12] above, the variance of the Gaussian noise term to be added to the gradients to make them (ϵ, δ)-differentially private is $2C \frac{\eta}{S} \frac{\ln \frac{1.25}{\delta}}{\epsilon}$. We describe the high-level algorithm for differentially private SGD [1] below.

Algorithm 14.1 (ϵ, δ)-differentially private SGD

Inputs: learning rate: η, batch size: S, clipping length: C, privacy budget: $T\epsilon_0$, initial weights: θ_0

    for t = 0, 1, 2, 3, ..., T **do**

    Sample S data points uniformly at random

    compute $f(X) = \sum_{i=1}^{S} g(x_i)$

    clip gradient: $f(X) \leftarrow f(X) / \max(1, \frac{\|f(X)\|}{C})$

    sample Z_k from the distribution $N(0, 2C \frac{\eta}{S} \frac{\ln \frac{1.25}{\delta}}{\epsilon})$

    $\theta_{t+1} \leftarrow \theta_t - \eta f(X) + Z_k$

    end for

14.3.3   Experiments and Discussion

Let us now discuss some experiments to understand the challenges in applying DP to federated learning in detail. Consider training two models on two different datasets—Resnet-18 model [19] on the CIFAR-10 [27] dataset and the Resnet-50 [19] model on the SVHN [37] dataset. All models are trained for 200 epochs using plain and DP versions of SGD. The learning rate schedule was to use 0.1 for the first 80 epochs, 0.01 for the next 40, and 0.001 for the remaining 80 epochs.

14.3.3.1   Accuracy vs ϵ

Figures 14.1 and 14.2 illustrate the convergence plots of different training runs for various values of batch size and ϵ. The very first observation is that the models converge to a lower accuracy as we decrease the value of ϵ. From Fig. 14.1, for a given batch size, say 1024, we observe that training schemes with lower noise added (corresponding to higher values of ϵ like 10, 1, and 0.1) have accuracy closer to the accuracy of non-private training. Once enough noise is added (corresponding to ϵ of 0.05 and lower), accuracy drops and does so precipitously for 0.01 and lower values

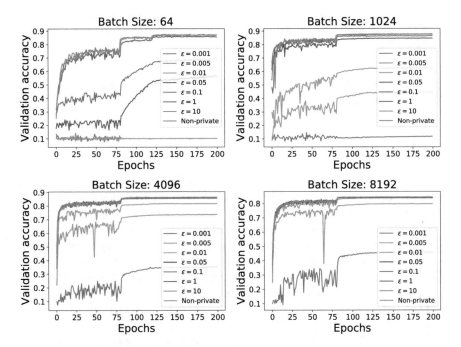

Fig. 14.1 Validation accuracy of Resnet18 on CIFAR10 per epoch for different batch sizes (64, 1K, 4K, and 8K) and ϵ (from 10 down to 0.001). Lower ϵ implies more noise added and hence more privacy

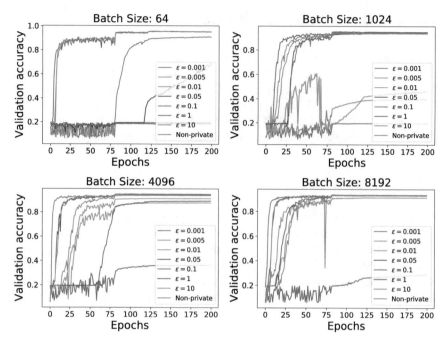

Fig. 14.2 Validation accuracy of Resnet18 on CIFAR10 per epoch for different batch sizes (64, 1K, 4K, and 8K) and ϵ (from 10 down to 0.001). Lower ϵ implies more noise added and hence more privacy

of ϵ. We see this phenomenon throughout Figs. 14.1 and 14.2 for all the models for a given batch size.

14.3.3.2 Accuracy vs Batch Size (Fixed ϵ)

An important observation from Figs. 14.1 and 14.2 is that even though the model accuracy levels for the non-private versions are pretty much unchanged for various values of batch sizes, the private version of training is highly sensitive to the batch size S. We have been able to achieve similar accuracy levels for private version of the models as those of the non-private versions even at very low values of ϵ by simply increasing the batch size. This can be prominently observed in the case of CIFAR10+Resnet18 in Fig. 14.1 for $\epsilon = 0.005$ and 0.01 (green and orange plots in Fig. 14.1, viewed left to right) but is true for all values of ϵ.

Quantitatively, for greater privacy corresponding to $\epsilon = 0.01$ and 0.005, final accuracy increases by approx. **37%** and **78.5%** by increasing batch size from 1024 to 8192, respectively, for CIFAR10+Resnet18. Similar trends can be observed from Fig. 14.2. Overall, while differential privacy is a promising approach, it does seem to require careful hyperparameter tuning to minimize impact on accuracy. This may involve spawning multiple FL jobs corresponding to different hyperparameters.

14.4 Additive Homomorphic Encryption

Research on secure and private federated learning and gradient descent is predominantly based either on (1) clever use of cryptography—homomorphic encryption and secure multi-party computation [3, 8, 10, 13, 16, 25, 26, 38] or on (2) modifying model parameters or gradients through the addition of statistical noise to get differential privacy [1, 2, 43]. Some techniques [45, 47] combine both.

Homomorphic encryption allows computation on ciphertexts, generating an encrypted result which, when decrypted, matches the result of the operations as if they had been performed on the plaintext [18]. Fully homomorphic encryption is expensive, in terms of both encryption/decryption time and the size of the ciphertext. However, averaging gradient vectors in federated gradient descent requires only addition (division by the total number of participants can be done before or after encrypted aggregation). Hence, we can easily employ additive homomorphic encryption like the Paillier cryptosystem [22] to ensure privacy of gradients during federated training. The Paillier cryptosystem is an asymmetric algorithm for public key cryptography. Given only the public key and the encryption of m_1 and m_2, one can compute the encryption of $m_1 + m_2$. The definition of homomorphic encryption beautifully illustrates the challenge in applying it to private gradient aggregation. All participants have to encrypt their gradients/parameters with the *same* public key. This needs a trusted key generator/distributor. But, if all the participants send their Paillier encrypted gradients to this key generator/distributor, it can decrypt them and potentially reverse engineer the data. So, ideally, aggregation has to happen outside the key generator/distributor. Furthermore, the key generator/distributor has to be completely trusted to not leak the private key to any participant. Several designs are possible to satisfy these constraints. In this section, we describe one system— Mystiko [20] as an illustration—and compare it with differential privacy and secure multi-party computation (SMC).

We emphasize that Mystiko is one of many systems that use Paillier cryptography to secure aggregation. In [3, 38], the participants jointly generate a Paillier key pair and send the encrypted gradient vectors to the coordinator who is completely untrusted, except to add Paillier encrypted weights. The participants can then decrypt the aggregated gradient vectors. This, however, requires each participant to collaborate with the others to generate the Paillier keys and a high level of trust that participants do not collude with the untrusted coordinator to decrypt individual gradient vectors. One untrusted participant can leak the Paillier keys and potentially lead to privacy loss.

Secure multi-party computation (SMC) is a subfield of cryptography with the goal of creating methods for parties to jointly compute a function over their inputs while keeping those inputs private. Unlike traditional cryptographic tasks, where cryptography assures security and integrity of communication or storage and the adversary is outside the system of participants (an eavesdropper on the sender and receiver), the cryptography in this model protects participants' privacy from each other. SPDZ [10], and its variants (Overdrive [16, 25, 26]), optimizes classic SMC protocols. The advantage of such protocols is that they work with any number of honest+curious peers, do not change final accuracy of the trained model, and require a large number of colluding peers to break. The drawback, however, is efficiency—SMC protocols are computationally expensive (Sect. 14.6).

14.4.1 Participants, Learners, and Administrative Domains

Logically speaking, it is helpful to define federated learning algorithms in terms of *administrative domains*. An administrative domain is a *set of computing entities* (servers, VMs, desktops, laptops, etc.). Each entity inside an administrative domain trusts the other, malice is not a concern, and there are no legal/regulatory hurdles to sharing data and information derived from data. Note that an administrative domain does not necessarily mean a company or a non-corporate organization. It may be a project within a company handling confidential data; it may be not be located within a corporate datacenter, and instead be associated with an account on the public cloud. An organization can have multiple administrative domains. Each *participant* in a federated learning algorithm corresponds to an administrative domain. Computationally, the actual learning process (typically running on a GPU) performing the neural network training is called a *Learner*. Each learner works on (a batch of) data within the participant to compute the gradient vector.

14.4.2 Architecture

The main characteristics of any privacy preserving federated learning scheme revolve around (a) what methodology is used to encrypt the data (or noise addition) of the participants and (b) the communication protocol being used among the participants and the coordinator (if any) for aggregation of the model parameters or gradients. In MYSTIKO all participants encrypt individual data using a single Paillier public encryption key, adding encrypted gradient vectors and decrypting only the sum. Thus, only the participants are able to view their individual data, ensuring privacy. The question now is (1) how to distribute a common Paillier public key to all participants while keeping the corresponding private key secret? and (2) how to prevent anyone from decrypting individual weights? These are explained in the following sections.

For simplicity, we assume that there is exactly one learner per participant. We will relax this assumption in Sect. 14.4.4. MYSTIKO is typically deployed as a cloud service that mediates several participants. It involves a Job Manager,

Membership Manager, a Key Generator, and a Decryptor. The Membership Manager is responsible for establishing the relationship between each participant and MYSTIKO and also keeping track of participants that belong to each federated learning job. The Job Manager manages an FL job through its lifecycle—it keeps track of participants, helps participants agree on hyperparameters, detects failures, and updates to memberships. While the focus of this chapter is attacks on the privacy of data from within a federation, traditional communication security is nevertheless essential to prevent outside attacks on the federation. For this, MYSTIKO and the participants (learners) agree to use a common public key infrastructure (PKI) [44]. The PKI helps ensure confidentiality of communications between the MYSTIKO components and the learners and also helps bootstrap the Paillier infrastructure. The PKI provides certification authorities (CAs), along with corresponding intermediate and Root CAs, creating a web of trust between the learners and MYSTIKO. MYSTIKO creates a bidirectional TLS channel [39] using the PKI for the security of control messages. The TLS channel is created using strong but ordinary (non-homomorphic) cryptographic algorithms (e.g., RSA for key exchange/agreement and authentication, AES for message confidentiality, and SHA for message authentication [39, 44]). The TLS channel is not used for gradient aggregation, but rather for all other communications, like registration of learners with the MYSTIKO topology formation, rank assignment, transmitting the Paillier public key to each learner, and transmitting the decrypted aggregated gradient vector to each learner.

14.4.3 MYSTIKO *Algorithms*

In this section, we describe MYSTIKO algorithms, starting with the simple ring-based algorithm, before adding parallelism and resiliency through broadcast and All-Reduce based communication.

14.4.3.1 Basic Ring-Based Algorithm

The basic ring-based aggregation algorithm is illustrated in Figs. 14.3 and 14.4. This algorithm operates across P participants, each in its own administrative domain and represented by a learner (L). The algorithm starts with each participant registering with MYSTIKO. MYSTIKO acts as the coordinator. The learners need not fully trust MYSTIKO; they only need to trust it to generate good encryption key pairs, keep the private key secret and follow the protocol.

MYSTIKO's Membership Manager starts the federated learning protocol once all expected learners are registered. The first step is to arrange the learners along a ring topology (Fig. 14.3). This can be done in several straightforward ways: (1) by location—minimizing geographic distance between participants, (2) by following a hierarchy based on the name of the participants (ascending or descending order), or

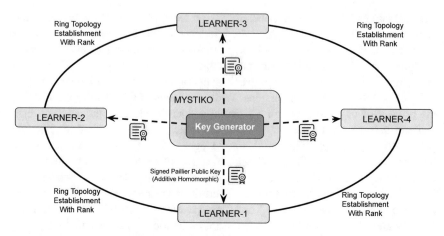

Fig. 14.3 Topology establishment and key distribution

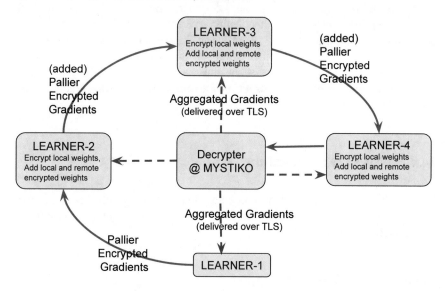

Fig. 14.4 Basic aggregation protocol over a ring topology

(3) by using consistent hashing [24] on the name/identity of the participants. Once the learners have been arranged in a ring, each learner gets a rank (from 1 to P) based on its position in the ring.

MYSTIKO's Paillier Key Generator, which generates a Paillier public and private key pair for each federated learning job. Typically, a unique Paillier key pair is generated for each federated learning job and the Paillier public key securely distributed (over TLS) to all the learners. For long jobs, a separate key pair may be generated either once every epoch or once every h minutes (this is configurable). MYSTIKO's Decryptor is responsible for decrypting the Paillier encrypted aggregated gradient vector for distribution to the learners. Each learner receives a Paillier encrypted gradient vector from the previous learners on the ring, encrypts its own gradient vector with the Paillier public key, and adds (aggregates) the two Paillier encrypted vectors. This aggregated, Paillier encrypted gradient vector is then transmitted to the next learner on the ring. The last learner on the ring transmits the fully aggregated, encrypted gradient vector to MYSTIKO for decryption. MYSTIKO's Decryptor decrypts the aggregated vector and transmits the same securely over TLS to each of the learners.

Security Analysis Data never leaves a learner and by extension any administrative domain. This ensures privacy of data, provided each server inside the administrative domain has adequate defenses against intrusion. Unencrypted gradient vectors do not leave the learner. If there are P participants, in $P - 1$ cases, only aggregated Paillier encrypted gradient vectors leave the learner. Only MYSTIKO has the private key to decrypt these. For the first learner in the ring, the non-aggregated gradient vector is transmitted to the second learner, but it is Paillier encrypted and cannot be decrypted by the same. In fact, none of the participants are able to view even partially aggregated gradient vectors. After decryption, the aggregated gradient vector is distributed securely to the participants over TLS. Reverse engineering attacks, like the ones in [15] and [14], are intended to find the existence of a specific data record in a participant or to find data items that lead to specific changes in gradient vectors; both of which are extremely difficult when several gradient vectors computed from large datasets are averaged [21]. Decryption after averaging ensures the privacy of gradients. MYSTIKO only sees aggregated gradients and cannot get access to individual learner's data or gradients.

Colluding Participants The basic ring-based algorithm is resistant to collusion among $P - 2$ participants. That is, assuming an honest uncorrupted MYSTIKO deployment, it can be broken only if $P - 1$ learners collude. For learner L_i's gradients to be exposed, learners $L_1, L_2, \ldots, L_{i-1}$ and $L_{i+1}, \ldots, L_P$ should collude, i.e., all of them should simply pass on incoming encrypted gradient vector to the next learner, without adding any gradient of their own.

Fault Tolerance The disadvantage of a ring-based aggregation algorithm is that rings can break; for good performance, it is essential that the connectivity between each learner and MYSTIKO remains strong. Traditional failure detection techniques, based on heartbeats and estimation of typical round-trip times, may be used. Distributed synchronous gradient descent consists of a number of iterations, with gradients being averaged at the end of each iteration. If the failure of a learner is detected, the averaging of gradient vectors is paused until the learner is eliminated from the ring by the MYSTIKO's Membership Manager, or connection to the learner is established again. Pausing gradient averaging can also be done when connection to the MYSTIKO is temporarily lost.

14.4.3.2   Broadcast Algorithm

One of the main drawbacks of the ring-based algorithm is the establishment and maintenance of the ring topology. To mitigate this, an alternative is to use group membership and broadcast. Except for the establishment of the topology, the setting remains the same. Learners register with the MYSTIKO's Membership Manager, agree on a common PKI, and know the identity and number of participants. MYSTIKO generates a Paillier public–private key pair for each federated job and distributes the public key securely to each learner.

Each learner Paillier encrypts its gradient vector and broadcasts the encrypted vector to all other learners. Each learner, upon receipt of encrypted vectors from $P - 1$ learners, adds them and sends the Paillier encrypted sum to the MYSTIKO for decryption. After decryption, the aggregated gradient vector is transmitted securely to all learners over TLS. The broadcast algorithm is redundant and wasteful, as every learner computes the aggregate. But, with redundancy comes increased failure resiliency. With the ring, the failure of one participant can lead to partial loss of aggregated gradients, which is not the case for broadcast.

Colluding Participants The objective of breaking this algorithm is to determine the plaintext gradient vector of a specific LA. This algorithm is resistant to collusion and can be broken only if $P - 1$ participants collude, which is highly unlikely. Also, in the event that $P - 1$ participants collude to Paillier encrypt zero vectors instead of their actual gradient vectors, the broadcasted Paillier ciphertexts from all the $P - 1$ learners will be the same, which serves as a red flag enabling collusion detection. In fact, given that data is likely to be different at each participant, getting exactly the same Paillier encrypted gradient vector from even two learners is red flag.

14.4.3.3   All-Reduce

Ring-based All-Reduce [28] is essentially a parallel version of the ring-based aggregation protocol described in Sect. 14.4.3.1. It is illustrated in Fig. 14.5. The problem with the basic ring protocol in Sect. 14.4.3.1 is that each learner has to wait for its predecessor. However, in All-Reduce, the Paillier encrypted gradient vector is divided into P chunks where P is the number of participants. All learners then aggregate Paillier encrypted chunks in parallel. For example, in Fig. 14.5, there are three learners, and the gradient vectors are divided into three chunks each. Learner-2 does not wait for the entire vector of Learner-1 to be received. Instead, while it is receiving the first chunk of Learner-1, it transmits its own second chunk to Learner-3, which in parallel transmits its third chunk to Learner-3. In Step 2, Learner-2 transmits the partially aggregated chunk-1 to Learner-3, which transmits partially aggregated chunk-2 to Learner-1, which transmits the partially aggregated chunk-3 to Learner-2. At the end of Step-2, each learner has Paillier encrypted, aggregated chunks, which are transmitted to MYSTIKO's Decryptor for concatenation and decryption.

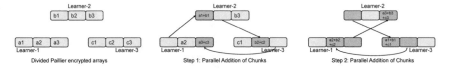

Fig. 14.5 MYSTIKO ring All-Reduce over Paillier encrypted arrays

Security Analysis We note that All-Reduce is the most efficient MYSTIKO proto-col. With P learners, All-Reduce is essentially an instantiation of $P - 1$ rings (of Sect. 14.4.3.1), all operating in parallel. In Fig. 14.5, the first ring starts at the first chunk of Learner-1, the second ring starts at the second chunk of Learner-2, and the third starts at the third chunk of Learner-3. This implies that the security guarantees of All-Reduce are the same as that of the basic ring protocol.

14.4.4 Multiple Learners Per Administrative Domain

For presentation simplicity, we have assumed that there is exactly one learning pro-cess (learner) per participant. More realistically, within an administrative domain, data is partitioned among servers and multiple training processes (learners), which periodically synchronize their gradient vectors using an aggregator process local to the administrative domain. This is done for various reasons, including datasets being large, compute resources being cheap and available, and the desire to reduce training time. MYSTIKO's protocols can be applied in a straightforward manner to this case, with MYSTIKO's protocols running between local aggregators (LAs) instead of between learners. Local aggregation is not Paillier encrypted and non-private because compute resources within an administrative domain are trusted and can share even raw data. But aggregation between LAs follows MYSTIKO's protocols. This is illustrated in Fig. 14.6.

14.5 Trusted Execution Environments

A trusted execution environment (TEE) [40, 46] is a secure area of a main processor. TEEs are isolated execution environment that provide key security features such as isolated execution, integrity of applications executing with the TEE, along with confidentiality of their data assets. TEEs establish an isolated execution environment that runs in parallel with the standard operating system, such as Linux and Microsoft Windows; its aim is to defend sensitive code and data against privileged software attacks from a potentially compromised native OS. ARM TrustZone, IBM Hyperprotect, and Intel SGX are examples of TEE technologies, which use a combination of hardware and software mechanisms to protect sensitive assets. TEEs

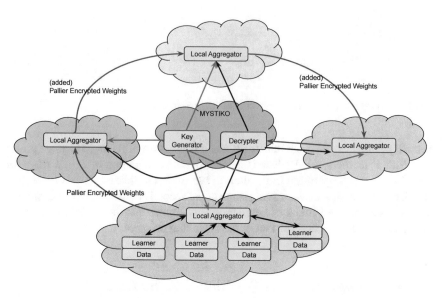

Fig. 14.6 Federating gradient descent

are often designed so that only trusted applications, whose integrity is verified at load time, can access the full power of the server's processors, peripherals, and memory. Hardware isolation provided by the TEE protects the applications inside it from other installed applications (including malware and viruses) on the host operating system. If multiple applications are contained within a TEE, software and cryptographic isolation often protect them from each other.

To prevent the simulation of TEEs with attacker- or user-controlled software on the server, TEEs involve a "hardware root of trust." This involves embedding a set of private keys directly into the TEE at the time of chip manufacturing, using one-time programmable memory. These cannot be changed, even after device resets or restarts. The public counterparts of these keys reside in a manufacturer database, together with a non-secret hash of a public key belonging to the trusted party (usually the chip vendor) which is used to sign trusted firmware alongside the circuits doing cryptographic operations and controlling access. The TEE hardware is designed in a way which prevents all software not signed by the trusted party's key from accessing the privileged features. The public key of the vendor is provided at runtime and hashed; this hash is then compared to the one embedded in the chip. If the hash matches, the public key is used to verify a digital signature of trusted vendor-controlled firmware (such as a chain of bootloaders on Android devices or 'architectural enclaves' in SGX). The trusted firmware is then used to implement remote attestation.

In this chapter, we describe how one system—TRUDA [7]—leverages Trusted Execution Environments (TEEs) to protect the model fusion process. Other examples include [6, 49] and [36]; each of which also has several optimizations. We restrict our treatment here to the basic aggregation process with TEEs using TRUDA as an example. TRUDA runs every aggregator within an encrypted virtual machine (EVM) via AMD Secure Encrypted Virtualization (SEV). All in-memory data are kept encrypted at runtime during model aggregation. To bootstrap trust between parties and aggregators, it uses a two-phase attestation protocol and develops a series of tools for integrating/automating confidential computing in FL. Each party can authenticate trustworthy aggregators before participating in FL training. End-to-end secure channels, from the parties to the EVMs, are established after attestation to protect model updates in transit.

14.5.1 Trustworthy Aggregation

TRUDA enforces cryptographic isolation for FL aggregation via SEV. The aggregators execute within EVMs. Each EVM's memory is protected with a distinct ephemeral VM Encryption Key (VEK). Therefore, TRUDA can protect the confidentiality of model aggregation from unauthorized users, e.g., system administrators, and privileged software running on the hosting servers. AMD provides attestation primitives for verifying the authenticity of individual SEV hardware/firmware. TRUDA employs a new attestation protocol upon the primitives to bootstrap trust between parties and aggregators in the distributed FL setting. This FL attestation protocol consists of two phases:

Phase 1: Launching Trustworthy Aggregators First, TRUDA securely launches SEV EVMs with aggregators running within. To establish the trust of EVMs, attestation must prove that (1) the platform is an authentic AMD SEV-enabled hardware providing the required security properties and (2) the Open Virtual Machine Firmware (OVMF) image to launch the EVM is not tampered. Once the remote attestation is completed, TRUDA provisions a secret, as a unique identifier of a trustworthy aggregator, to the EVM. The secret is injected into EVM's encrypted physical memory and used for aggregator authentication in Phase 2.

The EVM owner instructs the AMD Secure Processor (SP) to export the certificate chain from the Platform Diffie-Hellman (PDH) Public Key down to the AMD Root Key (ARK). This certificate chain can be verified by the AMD root certificates. The digest of OVMF image is also included in the attestation report along with the certificate chain.

The attestation report is sent to the attestation server, which is provisioned with the AMD root certificates. The attestation server verifies the certificate chain to authenticate the hardware platform and check the integrity of OVMF firmware. Thereafter, the attestation server generates a launch blob and a Guest Owner Diffie–Hellman Public Key (GODH) certificate. They are sent back to the SP on the aggregator's machine for negotiating a Transport Encryption Key (TEK) and a Transport Integrity Key (TIK) through Diffie–Hellman Key Exchange (DHKE) and launching the EVMs.

TRUDA retrieves the OVMF runtime measurement through the SP by pausing the EVM at launch time. It sends this measurement (along with the SEV API version and the EVM deployment policy) to the attestation server to prove the integrity of UEFI booting process. Only after verifying the measurement, the attestation server generates a packaged secret, which includes an ECDSA prime251v1 key. The hypervisor injects this secret into the EVM's physical memory space as a unique identifier of a trusted aggregator and continues the launching process.

Phase 2: Aggregator Authentication Parties participating in FL must ensure that they are interacting with trustworthy aggregators with runtime memory encryption protection. To enable aggregator authentication, in Phase 1, the attestation server provisions an ECDSA key as a secret during EVM deployment. This key is used for signing challenge requests and thus serves to identify a legitimate aggregator. Before participating in FL, a party first attests an aggregator by engaging in a challenge response protocol. The party sends a randomly generated nonce to the aggregator. The aggregator digitally signs the nonce using its corresponding ECDSA key and then returns the signed nonce to the requesting party. The party verifies that the nonce is signed with the corresponding ECDSA key. If the verification is successful, the party then proceeds to register with the aggregator to participate in FL. This process is repeated for all aggregators.

After registration, end-to-end secure channels can be established to protect communications between aggregators and parties for exchanging model updates. TRUDA enables TLS to support mutual authentication between a party and an aggregator. Thus, all model updates are protected both within EVMs and in transit.

14.6   Comparing HE- and TEE-Based Aggregation with SMC

In this section, we compare all the three Mystiko algorithms with a state-of-the-art protocol for secure multi-party computation (SPDZ [10, 11, 25]), and schemes for differential privacy (DP) through the addition of statistical noise. We employ a variety of image processing neural network models and datasets of various sizes: (1) 5-Layer CNN (small, 1MB) trained on MNIST dataset (60K handwritten digit images) (2) Resnet-18 (small–medium, 50MB) trained on the SVHN dataset (600K street digit images), (3) Resnet-50 (medium-sized model, 110 MB) trained on CIFAR-100 dataset (60K color images of 100 classes), and (4) VGG-16 (large model, 600MB) trained on Imagenet-1K dataset (14.2 million images of 1000 classes).

Experiments were conducted on a 40-machine cluster to evaluate all the algorithms on a varying number of participants from 2 to 40. No more than one participant was ever run on any machine, each of which was equipped with 8 Intel Xeon E5-4110 (2.10 GHz) cores, 64GB RAM, 1 NVIDIA V100 GPU, and a 10GbE network link. The machines were spread over four datacenters, and in every experiment, participants were uniformly distributed across datacenters. In every experiment, the dataset was *uniformly and randomly* partitioned across participants. Mystiko was executed on a dedicated machine in one datacenter. All data points henceforth are computed by averaging 10 experiment runs.

14.6.1   Comparing MYSTIKO *and SPDZ*

In federated learning, learners (or local aggregators) learn on local data for a specific number of iterations before federated gradient aggregation and model update. Privacy loss happens during gradient aggregation, which is where MYSTIKO and other systems like SPDZ intervene. Hence, we use the following two metrics to evaluate MYSTIKO and SPDZ: (1) total synchronization time, which measures the total time needed for privacy preserving gradient transformations (Paillier encryption in MYSTIKO, share generation in SPDZ, etc.) and the time required to communicate the transformed gradients to participants for federation and (2) communication time, which only measures communication time.

Figure 14.7 plots total synchronization time and communication time against the number of parties involved in federation, for all of our model/dataset combinations. From Fig. 14.7, we observe that All-Reduce is the most scalable of all the protocols, as the number of participants increases. This is mainly because it is a parallel protocol, where each learner/LA is constantly transmitting a small portion of the gradient array. The basic ring protocol is the least scalable because it is sequential.

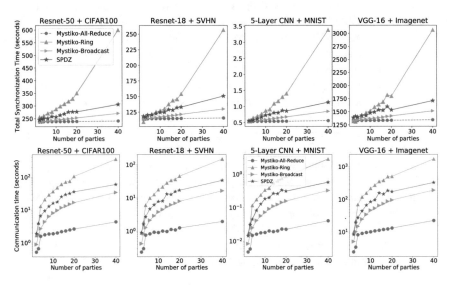

Fig. 14.7 MYSTIKO: total synchronization time (seconds) vs. number of parties (top plots) and total communication time (seconds) vs. number of parties (bottom). Recall that total synchronization time is the sum of the communication time and the gradient transformation time

Table 14.1 MYSTIKO: performance slowdown of Broadcast, Ring, and SPDZ relative to All-Reduce. Full trend available at Fig. 14.7

Communication time				
	MYSTIKO			
# Parties	All-reduce	Broadcast	Ring	SPDZ
20	1	6.4–7.2×	38.2–39.9×	13.1–14.2×
40	1	7.9–8.6×	70.5–80.8×	13.8–14.7×
Total synchronization time				
	MYSTIKO			
# Parties	All-reduce	Broadcast	Ring	SPDZ
20	1	12–51%	27–100%	2.2–6.1×
40	1	6–25%	15–59%	1.3–2.6×

Broadcast performs and scales better than the basic ring protocol because each participant is broadcasting without waiting for the others. SPDZ performs and scales worse than broadcast because its communication pattern is close to (but not exactly) a dual broadcast—each item of the gradient vector at each participant is split into secret shares and broadcast to the other participants; after secure aggregation, the results are broadcast back. MYSTIKO obviates the need for dual broadcast through the use of Paillier encryption and centralized decryption of aggregated gradients. Table 14.1 illustrates the performance impact of using other protocols for two cases (20 and 40 parties from Fig. 14.7).

However, the "enormous" speedups of using All-Reduce do not materialize when total synchronization time is considered. The scalability trends among the four protocols remain the same; the speedups in total synchronization time *remain significant* (as elucidated for two cases in Table 14.1). But the speedups are lower than the speedups due to communication. This demonstrates that the predominant overhead of private gradient descent in MYSTIKO and SPDZ vs. non-private gradient descent is gradient transformation prior to communication. From Fig. 14.7 and Table 14.1, we also observe that for small models (5-Layer CNN and Resnet-18), communication time plays a larger role. But for large models (Resnet-50 and VGG-16), gradient transformation plays a larger role.

Lastly, we observe that when compared to training time (illustrated using epoch time), synchronization time for private gradient descent is significantly larger than non-private gradient descent. This is primarily because training happens on V100 GPUs (with thousands of cores), while gradient transformation happens on CPUs. While there is a GPU accelerated version of fully homomorphic encryption ([9], which has worse performance than Paillier on CPUs), we are not aware of any GPU accelerated version of the Paillier algorithm.

14.6.2   Overheads of Using TEEs: AMD SEV

Compared to MYSTIKO and SPDZ, the biggest advantage is that the overheads are very low. The overhead of using TEEs comes from performing aggregation inside the EVMs. Overhead, measured as an increase in end-to-end latency, was between 2% and 4% per federated synchronization round [7]. And there was no difference in accuracy or convergence rate [7].

14.7   Concluding Remarks

In this chapter, we have examined various options for private parameter aggregation for federated learning. It is clear that each method has a unique set of advantages and disadvantages, and the search for a perfect solution is an active area of research. Differential privacy is very promising from an information leakage standpoint, but it (1) decreases the accuracy of the model and (2) involves nontrivial hyperparameter tuning (batch size, learning rate schedule) to obtain optimal results (or even near-optimal results). Hyperparameter tuning by trying different parameters may not be possible in federated settings due to the fact that all participants are not guaranteed to be available for extended time periods and also because running multiple experiments increases overall latency. Homomorphic encryption and secure multi-party computation do not alter accuracy or convergence rate and do not require hyperparameter tuning. But they incur high runtime overhead—using additive homomorphic encryption can minimize this to a great extent as illustrated by MYSTIKO, if the aggregation protocol is carefully designed. Finally, using TEEs has the potential to make aggregation overhead negligible without incurring any accuracy or convergence penalty, but it does require the use of specialized hardware.

References

1. Abadi M, Chu A, Goodfellow I, McMahan HB, Mironov I, Talwar K, Zhang L (2016) Deep learning with differential privacy. In: Proceedings of the 2016 ACM SIGSAC conference on computer and communications security, association for computing machinery, New York, NY, CCS '16, pp 308–318
2. Agarwal N, Suresh AT, Yu FXX, Kumar S, McMahan B (2018) cpSGD: Communication-efficient and differentially-private distributed SGD. In: NeurIPS 2018
3. Aono Y, Hayashi T, Trieu Phong L, Wang L (2016) Scalable and secure logistic regression via homomorphic encryption. In: Proceedings of the Sixth ACM conference on data and application security and privacy. Association for Computing Machinery, New York, NY, CODASPY '16, pp 142–144
4. Bonawitz K, Eichner H, Grieskamp W, Huba D, Ingerman A, Ivanov V, Kiddon C, Konečný J, Mazzocchi S, McMahan HB et al (2019) Towards federated learning at scale: System design. Preprint. arXiv:190201046

5. Bottou L (1998) On-line learning and stochastic approximations. In: On-line learning in neural networks. Cambridge University Press, New York
6. Chen Y, Luo F, Li T, Xiang T, Liu Z, Li J (2020) A training-integrity privacy-preserving federated learning scheme with trusted execution environment. Inf Sci 522:69–79
7. Cheng P, Eykholt K, Gu Z, Jamjoom H, Jayaram KR, Valdez E, Verma A (2021) Separation of powers in federated learning. CoRR abs/2105.09400. https://arxiv.org/abs/2105.09400, http://2105.09400
8. Cramer R, Damgrd IB, Nielsen JB (2015) Secure multiparty computation and secret sharing, 1st edn. Cambridge University Press, Cambridge
9. Dai W, Sunar B (2016) cuHE: A homomorphic encryption accelerator library. In: Cryptography and information security in the Balkans. Springer International Publishing
10. Damgård I, Pastro V, Smart N, Zakarias S (2012) Multiparty computation from somewhat homomorphic encryption. In: Proceedings of the 32nd annual cryptology conference on advances in cryptology — CRYPTO 2012 - Volume 7417. Springer-Verlag, Berlin, Heidelberg, pp 643–662
11. Data61 C (2020) Multi-Protocol SPDZ. https://github.com/data61/MP-SPDZ
12. Dwork C, Roth A (2014) The algorithmic foundations of differential privacy. Found Trends Theor Comput Sci 9(3–4):211–407
13. Evans D, Kolesnikov V, Rosulek M (2018) A pragmatic introduction to secure multi-party computation. Found Trends Privacy Secur 2:70
14. Fredrikson M, Lantz E, Jha S, Lin S, Page D, Ristenpart T (2014) Privacy in pharmacogenetics: An end-to-end case study of personalized warfarin dosing. In: 23rd USENIX security symposium (USENIX Security 14). USENIX Association, San Diego, CA, pp 17–32. https://www.usenix.org/conference/usenixsecurity14/technical-sessions/presentation/fredrikson_matthew
15. Fredrikson M, Jha S, Ristenpart T (2015) Model inversion attacks that exploit confidence information and basic countermeasures. In: Proceedings of the 22nd ACM SIGSAC conference on computer and communications security. Association for Computing Machinery, New York, NY, CCS '15, pp 1322–1333
16. Garg S, Sahai A (2012) Adaptively secure multi-party computation with dishonest majority. In: Safavi-Naini R, Canetti R (eds) Advances in Cryptology – CRYPTO 2012. Springer Berlin Heidelberg, Berlin, Heidelberg, pp 105–123
17. Geiping J, Bauermeister H, Dröge H, Moeller M (2020) Inverting gradients–how easy is it to break privacy in federated learning? Preprint. arXiv:200314053
18. Gentry C (2010) Computing arbitrary functions of encrypted data. Commun ACM 53(3):97–105
19. He K, Zhang X, Ren S, Sun J (2016) Deep residual learning for image recognition. In: 2016 IEEE conference on computer vision and pattern recognition (CVPR), pp 770–778
20. Jayaram KR, Verma A, Verma A, Thomas G, Sutcher-Shepard C (2020) Mystiko: Cloud-mediated, private, federated gradient descent. In: 2020 IEEE 13th international conference on cloud computing (CLOUD). IEEE Computer Society, Los Alamitos, CA, pp 201–210
21. Jayaraman B, Evans D (2019) Evaluating differentially private machine learning in practice. In: 28th USENIX Security Symposium (USENIX Security 19), USENIX Association, Santa Clara, CA, pp 1895–1912. https://www.usenix.org/conference/usenixsecurity19/presentation/jayaraman
22. Jost C, Lam H, Maximov A, Smeets BJM (2015) Encryption performance improvements of the paillier cryptosystem. IACR Cryptology ePrint Archive. https://eprint.iacr.org/2015/864
23. Kairouz P, McMahan HB, Avent B, Bellet A, Bennis M, Bhagoji AN, Bonawitz K, Charles Z, Cormode G, Cummings R et al (2019) Advances and open problems in federated learning. Preprint. arXiv:191204977
24. Karger D, Lehman E, Leighton T, Panigrahy R, Levine M, Lewin D (1997) Consistent hashing and random trees: Distributed caching protocols for relieving hot spots on the world wide web. In: Proceedings of the twenty-ninth annual ACM symposium on theory of computing. Association for Computing Machinery, New York, NY, STOC '97, pp 654–663

25. Keller M, Yanai A (2018) Efficient maliciously secure multiparty computation for ram. In: EUROCRYPT (3). Springer, pp 91–124. https://doi.org/10.1007/978-3-319-78372-7_4
26. Keller M, Pastro V, Rotaru D (2018) Overdrive: Making SPDZ great again. In: Nielsen JB, Rijmen V (eds) Advances in cryptology – EUROCRYPT 2018. Springer International Publishing, Cham, pp 158–189
27. Krizhevsky A (2009) Learning multiple layers of features from tiny images. https://www.cs.toronto.edu/~kriz/learning-features-2009-TR.pdf, https://www.cs.toronto.edu/~kriz/cifar.html
28. Kumar V, Grama A, Gupta A, Karypis G (1994) Introduction to parallel computing: design and analysis of algorithms. Benjamin-Cummings Publishing, California
29. Lecun Y, Bengio Y, Hinton G (2015) Deep learning. Nat Cell Biol 521(7553):436–444
30. Lee J, Clifton C (2011) How much is enough? choosing ϵ for differential privacy. In: Lai X, Zhou J, Li H (eds) Information security. Springer Berlin Heidelberg, Berlin, Heidelberg, pp 325–340
31. Marr B (2018) 27 Incredible examples of AI and machine learning in practice. Forbes Magazine
32. McMahan HB, Andrew G (2018) A general approach to adding differential privacy to iterative training procedures. CoRR abs/1812.06210. http://arxiv.org/abs/1812.06210
33. Melis L, Song C, De Cristofaro E, Shmatikov V (2019) Exploiting unintended feature leakage in collaborative learning. In: 2019 IEEE symposium on security and privacy. IEEE, pp 691–706
34. Millard C (2013) Cloud computing law. Oxford University Press
35. Mironov I (2017) Rényi differential privacy. In: 2017 IEEE 30th computer security foundations symposium (CSF), pp 263–275
36. Mo F, Haddadi H, Katevas K, Marin E, Perino D, Kourtellis N (2021) PPFL: Privacy-preserving federated learning with trusted execution environments. In: Proceedings of the 19th annual international conference on mobile systems, applications, and services. Association for Computing Machinery, New York, NY, MobiSys '21, pp 94–108
37. Yuval Netzer, Tao Wang, Adam Coates, Alessandro Bissacco, Bo Wu, Andrew Y. Ng Reading digits in natural images with unsupervised feature learning NIPS workshop on deep learning and unsupervised feature learning 2011
38. Phong LT, Aono Y, Hayashi T, Wang L, Moriai S (2018) Privacy-preserving deep learning via additively homomorphic encryption. Trans Inf Forens Secur 13(5):1333–1345
39. Rescorla E (2018) The transport layer security (TLS) protocol version 1.3. RFC 8446
40. Sabt M, Achemlal M, Bouabdallah A (2015) Trusted execution environment: What it is, and what it is not. In: 2015 IEEE Trustcom/BigDataSE/ISPA, vol 1, pp 57–64. https://doi.org/10.1109/Trustcom.2015.357
41. Shokri R, Shmatikov V (2015) Privacy-preserving deep learning. In: ACM CCS '15
42. Shokri R, Stronati M, Song C, Shmatikov V (2017) Membership inference attacks against machine learning models. In: 2017 IEEE symposium on security and privacy (SP), pp 3–18
43. Song S, Chaudhuri K, Sarwate A (2013) Stochastic gradient descent with differentially private updates. In: 2013 IEEE global conference on signal and information processing, GlobalSIP 2013 - Proceedings, 2013 IEEE global conference on signal and information processing, GlobalSIP 2013 - Proceedings, pp 245–248
44. Stallings W (2013) Cryptography and network security: principles and practice, 6th edn. Prentice Hall Press, Upper Saddle River
45. Truex S, Baracaldo N, Anwar A, Steinke T, Ludwig H, Zhang R, Zhou Y (2019) A hybrid approach to privacy-preserving federated learning. In: Proceedings of the 12th ACM workshop on artificial intelligence and security. Association for Computing Machinery, New York, NY, AISec'19, pp 1–11
46. Volos S, Vaswani K, Bruno R (2018) Graviton: Trusted execution environments on GPUs. In: 13th USENIX symposium on operating systems design and implementation (OSDI 18). USENIX Association, Carlsbad, CA, pp 681–696. https://www.usenix.org/conference/osdi18/presentation/volos
47. Xu R, Baracaldo N, Zhou Y, Anwar A, Ludwig H (2019) Hybridalpha: An efficient approach for privacy-preserving federated learning. In: Proceedings of the 12th ACM workshop on

artificial intelligence and security. Association for Computing Machinery, New York, NY, AISec'19, pp 13–23. https://doi.org/10.1145/3338501.3357371

48. Yin H, Mallya A, Vahdat A, Alvarez JM, Kautz J, Molchanov P (2021) See through gradients: Image batch recovery via GradInversion. Preprint. arXiv:210407586

49. Zhang X, Li F, Zhang Z, Li Q, Wang C, Wu J (2020) Enabling execution assurance of federated learning at untrusted participants. In: IEEE INFOCOM 2020 - IEEE conference on computer communications, pp 1877–1886

50. Zhao B, Mopuri KR, Bilen H (2020) iDLG: Improved deep leakage from gradients. Preprint. arXiv:200102610

51. Zhu L, Liu Z, Han S (2019) Deep leakage from gradients. In: Advances in neural information processing systems, pp 14774–14784

Chapter 15
Data Leakage in Federated Learning

Xiao Jin, Pin-Yu Chen, and Tianyi Chen

Abstract Federated learning (FL) is a recent distributed machine learning paradigm, which allows the data owners to participate in the training process while keeping the data privacy. However, recent studies have shown that data can be still leaked through the gradient sharing mechanism in FL. Increasing batch size is often viewed as a promising defense strategy against data leakage. In this chapter, we provide an overview of data leakage problems in FL, revisit this attack premise, and propose an advanced data leakage attack to efficiently recover batch data from the aggregated gradients. We name our proposed method as *catastrophic data leakage in federated learning (CAFE)*. Comparing to existing data leakage attacks, CAFE demonstrates the ability to perform large-batch data leakage attack with high recovery quality. Our experimental results suggest that data participated in FL, especially the vertical case, have a high risk of being leaked from the training gradients.

15.1 Introduction

In this section, we introduce the motivation and then present the necessary background of federated learning.

15.1.1 Motivation

Deep neural networks are currently one of the most frequently used models on AI tasks such as classifying and recognizing [27]. Since the performances of deep learning techniques are closely related to the amount of training data, large

X. Jin (✉) · T. Chen
Rensselaer Polytechnic Institute, Troy, NY, USA
e-mail: xj2285@columbia.edu; chent18@rpi.edu

P.-Y. Chen
IBM Thomas J. Watson Research Center, Yorktown Heights, NY, USA
e-mail: pin-yu.chen@ibm.com

© The Author(s), under exclusive license to Springer Nature Switzerland AG 2022   337
H. Ludwig, N. Baracaldo (eds.), *Federated Learning*,
https://doi.org/10.1007/978-3-030-96896-0_15

companies such as Google and Facebook may achieve massive amounts of training data collection from their users and vast computational power to train their model on a large scale. However, the centralized collection of data has some privacy leakage risks. The users who provide personal sensitive data may contain sensitive information neither have the chance to delete then nor have the access to the training process. Furthermore, in some domains such as medicine and finance, it is illegal to share or gather individual data.

Under such a circumstance, the concept of FL is proposed as a new machine learning system in which multiple parties jointly train machine learning models such as neural networks without sharing their own local datasets. This new protocol has the advantage that companies can spend less time and money gathering data. Meanwhile, more potential training data can be mined to participate in the model training once privacy protection is granted.

Consequently, it is essential to develop privacy-preserving algorithms and defense strategies against information leakage during the FL process.

15.1.2 Background and Related Work

We will provide necessary background for both FL and data leakage from gradients.

15.1.2.1 Federated Learning

The core part of FL is to build machine learning models or deep learning models on data distributed across multiple devices while data privacy is preserved. The data owner set (N owners) can be denoted by $\{\mathcal{F}_1, \mathcal{F}_2, \ldots, \mathcal{F}_N\}$.

We also define their corresponding datasets as [36]

$$\mathcal{D}_{\text{fed}} = \{\mathcal{D}_1, \mathcal{D}_2, \ldots, \mathcal{D}_N\}. \tag{15.1}$$

A machine learning model trained by conventional integrated data $\mathcal{D}$ can be denoted by $\mathcal{M}_{\text{sum}}$, where [36]

$$\mathcal{D} = \mathcal{D}_1 \cup \mathcal{D}_2 \cdots \cup \mathcal{D}_N. \tag{15.2}$$

We denote the feature space of the data $\mathcal{D}$ participating in the FL as X, the label space as $\mathcal{Y}$, and the sample identity (ID) space as $\mathcal{I}$. As a result, federated dataset $\mathcal{D}_{\text{fed}}$ can be rewritten as [36]

$$\mathcal{D}_{\text{fed}} = \{\mathcal{I}, X, \mathcal{Y}\}. \tag{15.3}$$

According to the data distribution in the feature and sample ID space, FL can be briefly classified into horizontal federated learning, vertical federated learning, and federated transfer learning.

Horizontal federated learning (HFL) represents the scenarios in which data distributed in different parties share the same feature space but different sample IDs. We summarize HFL mathematically as [36]

$$\mathcal{X}_i = \mathcal{X}_j, \mathcal{Y}_i = \mathcal{Y}_j, \mathcal{I}_i \neq \mathcal{I}_j, \forall \mathcal{D}_i, \mathcal{D}_j, i \neq j. \tag{15.4}$$

Figure 15.1 illustrates an example of general architecture for a HFL system. In the system, there are k parties participating the model training under the control from a parameter aggregator.

The training process can be summarized into several steps below:

1. The aggregator broadcasts the model to all parties.
2. Parties use the current model to compute the gradients, encrypt them, and upload them to the aggregator.
3. The aggregator aggregates the uploaded encrypted gradients securely according to some gradients aggregation algorithms such as FedAvg [21].
4. The aggregator uses the aggregated gradients to update the model.

Each round will go through all the steps above until the federated model is fully converged. For an enterprise user, the case is probably different. We will introduce the case for enterprise uses in the next part.

In some cases, some local datasets may share the same ID space but differ in feature space. As a result, vertical federated learning (VFL) is applicable to those cases. In a VFL system, we have [36]

$$\mathcal{X}_i \neq \mathcal{X}_j, \mathcal{Y}_i \neq \mathcal{Y}_j, \mathcal{I}_i = \mathcal{I}_j, \forall \mathcal{D}_i, \mathcal{D}_j, i \neq j. \tag{15.5}$$

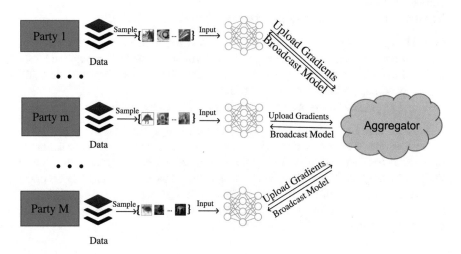

Fig. 15.1 Architecture for a HFL system [36]

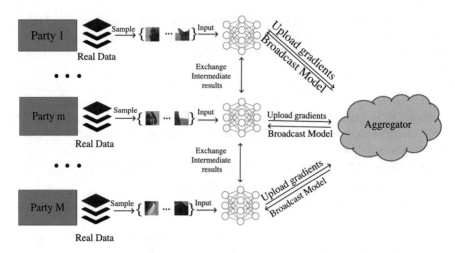

Fig. 15.2 Architecture for a VFL system [36]

Figure 15.2 illustrates an example of a general architecture for a VFL system. In the system, we suppose several cooperators participate in the VFL training process under the authority of a third trusted party (the aggregator). The training process consists of two parts.

Entity Alignment
Firstly, the systems need to align the data by the encryption-based data ID techniques [16, 26] without exposing their own data.

Model Training
The training process can be summarized into several steps below:

(s1) The aggregator creates encryption pairs and sends public key to parties.
(s2) All parties encrypt and exchange their intermediate results compute the loss.
(s3) All parties compute their respective gradients. All parties encrypt their gradients and upload them to the aggregator.
(s4) The aggregator decrypts the gradients and updates the model.

It is generally regarded safe to let parties exchange intermediate results. The dimensionality of intermediate results is so small compared with the dimensionality of original training data that privacy protection can be guaranteed. Since both of the parties only have data with incomplete feature space, they can only compute incomplete gradients of the model. The trusted third party, namely the aggregator, to some extent, needs to aggregate the incomplete gradients together to get the complete gradients of the whole model.

15.1.3   Privacy Protection

Different from the traditional centralized machine learning scenarios, FL requires collaboration between distrustful parties. Therefore, each honest part has to contribute to the FL based on its own local data while defending various attacks from other attackers [20, 35]. An attack can be launched from any device in the whole FL system including both the local and the aggregator part [10].

Adversarial Inference   In [6], the authors proposed a model inversion method to infer features characterizing each class and thus construct representatives of these classes. Hitaj et al. [11], in another way, trained a GAN to infer class representatives. Membership inference is one of the simplest inference attacks in FL. Publications such as [4, 19, 28] show the black-box membership inference techniques on a machine learning model.

Deep Leakage from Gradients (DLG)   Although attackers can launch inference attacks in many ways, it was widely believed that sharing gradients of the federated model between parties and the aggregator is safe and will not leak training data. However, Zhu et al. [39] proposed an effective way to reconstruct training data from shared gradients. Despite the fact that some previous studies [6, 11, 22] have already revealed some data information from the gradients, DLG, however, does not need any generative models and prior knowledge. It can directly infer the data and labels only from the shared public gradient through a few optimization rounds.

Figure 15.3 gives the overview of the DLG algorithm. All variables to be changed are marked with bold borders. Suppose we define the real data as $\mathbf{X}$, the real labels as $\mathbf{y}$, the fake data as $\hat{\mathbf{X}}$, and the fake labels as $\hat{\mathbf{y}}$. We also have federated model $\mathcal{L}(\mathbf{X}; \mathbf{y}, \theta)$ in which θ is the model parameters and the output is the loss. The real gradients $\nabla_\theta \mathcal{L}(\mathbf{X}^t; \mathbf{y}^t, \theta^t)$ and the fake gradients $\nabla_\theta \mathcal{L}(\hat{\mathbf{X}}^t; \hat{\mathbf{y}}^t, \theta^t)$ at tth iteration can be denoted by

$$\nabla_\theta \mathcal{L}(\mathbf{X}^t; \mathbf{y}^t, \theta^t) = \frac{1}{N} \sum_i^N \nabla_\theta \mathcal{L}(\mathbf{X}_i^t; \mathbf{y}_i^t, \theta^t) \tag{15.6}$$

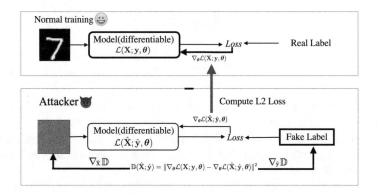

Fig. 15.3 The overview of DLG [39]

$$\nabla_\theta \mathcal{L}(\mathbf{X}_i^t; \mathbf{y}_i^t, \boldsymbol{\theta}^t) = \frac{\partial \mathcal{L}(\mathbf{X}_i^t; \mathbf{y}_i^t, \boldsymbol{\theta}^t)}{\partial \boldsymbol{\theta}^t} \tag{15.7}$$

$$\nabla_\theta \mathcal{L}(\hat{\mathbf{X}}^t; \hat{\mathbf{y}}^t, \boldsymbol{\theta}^t) = \frac{1}{N} \sum_i^N \nabla_\theta \mathcal{L}(\hat{\mathbf{X}}_i^t; \hat{\mathbf{y}}_i^t, \boldsymbol{\theta}^t) \tag{15.8}$$

$$\nabla_\theta \mathcal{L}(\hat{\mathbf{X}}_i^t; \hat{\mathbf{y}}_i^t, \boldsymbol{\theta}^t) = \frac{\partial \mathcal{L}(\hat{\mathbf{X}}_i^t; \hat{\mathbf{y}}_i^t, \boldsymbol{\theta}^t)}{\partial \boldsymbol{\theta}^t} \tag{15.9}$$

$$\mathbb{D}(\hat{\mathbf{X}}^t; \hat{\mathbf{y}}^t) = \|\nabla_\theta \mathcal{L}(\mathbf{X}^t; \mathbf{y}^t, \boldsymbol{\theta}^t) - \nabla_\theta \mathcal{L}(\hat{\mathbf{X}}^t; \hat{\mathbf{y}}^t, \boldsymbol{\theta}^t)\|^2, \tag{15.10}$$

where $\nabla_\theta \mathcal{L}(\mathbf{X}_i^t; \mathbf{y}_i^t, \boldsymbol{\theta}^t)$ and $\nabla_\theta \mathcal{L}(\hat{\mathbf{X}}_i^t; \hat{\mathbf{y}}_i^t, \boldsymbol{\theta}^t)$ represent the local gradients uploaded by the local part i and its corresponding fake gradients, respectively. $\mathbb{D}(\hat{\mathbf{X}}^t; \hat{\mathbf{y}}^t)$ in Eq. (15.10) measures the distance between real gradients and fake gradients. The terms $\mathbf{X}_i^t/\hat{\mathbf{X}}_i^t$ and $\mathbf{y}_i^t/\hat{\mathbf{y}}_i^t$ indicate the real/fake training data and the real/fake training labels at tth round from local part i. The real/fake gradients of the whole training data can be denoted by Zhu et al. [39]

$$\nabla_\theta \mathcal{L}(\mathbf{X}; \mathbf{y}, \boldsymbol{\theta}) = \frac{\partial \mathcal{L}(\mathbf{X}; \mathbf{y}, \boldsymbol{\theta})}{\partial \boldsymbol{\theta}} \tag{15.11}$$

$$\nabla_\theta \mathcal{L}(\hat{\mathbf{X}}; \hat{\mathbf{y}}, \boldsymbol{\theta}) = \frac{\partial \mathcal{L}(\hat{\mathbf{X}}; \hat{\mathbf{y}}, \boldsymbol{\theta})}{\partial \boldsymbol{\theta}} \tag{15.12}$$

$$\mathbb{D}(\hat{\mathbf{X}}; \hat{\mathbf{y}}) = \|\nabla_\theta \mathcal{L}(\mathbf{X}; \mathbf{y}, \boldsymbol{\theta}) - \nabla_\theta \mathcal{L}(\hat{\mathbf{X}}; \hat{\mathbf{y}}, \boldsymbol{\theta})\|^2. \tag{15.13}$$

The objective function of DLG can be denoted by Zhu et al. [39]

$$\hat{\mathbf{X}}^*; \hat{\mathbf{y}}^* = arg \min_{\hat{\mathbf{X}}; \hat{\mathbf{y}}} \mathbb{D}(\hat{\mathbf{X}}; \hat{\mathbf{y}}). \tag{15.14}$$

Since newly defined objective function is differentiable with respect to both $\hat{\mathbf{X}}$ and $\hat{\mathbf{y}}$. We can optimize $\hat{\mathbf{X}}$ and $\hat{\mathbf{y}}$ through their gradients of the objective function [39]

$$\nabla_{\hat{\mathbf{X}}} \mathbb{D}(\hat{\mathbf{X}}; \hat{\mathbf{y}}) = \frac{\partial \mathbb{D}(\hat{\mathbf{X}}; \hat{\mathbf{y}})}{\partial \hat{\mathbf{X}}} \tag{15.15}$$

$$\nabla_{\hat{\mathbf{y}}} \mathbb{D}(\hat{\mathbf{X}}; \hat{\mathbf{y}}) = \frac{\partial \mathbb{D}(\hat{\mathbf{X}}; \hat{\mathbf{y}})}{\partial \hat{\mathbf{y}}}. \tag{15.16}$$

Algorithm 15.1 gives the pseudo-code of DLG algorithm. DLG performs well on many datasets and fully recovers training data in just a few rounds.

Figure 15.4 demonstrates the visualization of DLG on images from several datasets. The algorithm outperforms all previous data leakage methods both in pixel accuracy and in speed.

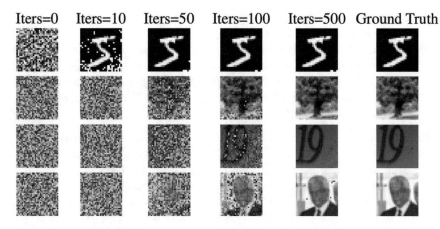

Fig. 15.4 DLG visualization on images from MNIST, CIFAR-100, SVHM, and LFW [39]

Algorithm 15.1 Deep leakage from gradients (DLG) [39]

1: Input differentiable model $\mathcal{L}(.)$, θ parameter weights, $\nabla\theta$ real gradients computed by real training data, learning rate η
2: **procedure** DLG($\mathcal{L}(.), \theta, \nabla_\theta\mathcal{L}(\mathbf{X}; \mathbf{y}, \theta)$)
3: Initialize dummy data $\hat{\mathbf{X}}^1$ and dummy labels $\hat{\mathbf{y}}^1$
4: **for** $t \leftarrow 1, \ldots, n$ **do**
5:    compute fake loss $\mathcal{L}(\hat{\mathbf{X}}^t, \hat{\mathbf{y}}^t, \theta)$
6:    compute fake gradient $\nabla_\theta\mathcal{L}(\hat{\mathbf{X}}^t; \hat{\mathbf{y}}^t, \theta) \leftarrow \frac{\partial\mathcal{L}(\hat{\mathbf{X}}^t, \hat{\mathbf{y}}^t, \theta)}{\partial\theta}$
7:    $\mathbb{D}(\hat{\mathbf{X}}^t; \hat{\mathbf{y}}^t) \leftarrow \|\nabla_\theta\mathcal{L}(\mathbf{X}; \mathbf{y}, \theta) - \nabla_\theta\mathcal{L}(\hat{\mathbf{X}}^t; \hat{\mathbf{y}}^t, \theta)\|^2$
8:    Compute $\nabla_{\hat{\mathbf{X}}}\mathbb{D}(\hat{\mathbf{X}}^t; \hat{\mathbf{y}}^t), \nabla_{\hat{\mathbf{y}}}\mathbb{D}(\hat{\mathbf{X}}^t; \hat{\mathbf{y}}^t)$
9:    $\hat{\mathbf{X}}^{t+1} \leftarrow \hat{\mathbf{X}}^t - \eta\nabla_{\hat{\mathbf{X}}}\mathbb{D}(\hat{\mathbf{X}}^t; \hat{\mathbf{y}}^t)$
10:    $\hat{\mathbf{y}}^{t+1} \leftarrow \hat{\mathbf{y}}^t - \eta\nabla_{\hat{\mathbf{y}}}\mathbb{D}(\hat{\mathbf{X}}^t; \hat{\mathbf{y}}^t)$
11: **end for**
12: **return** $\hat{\mathbf{X}}^{t+1}, \hat{\mathbf{y}}^{t+1}$
13: **end procedure**

Although DLG raises a big threat to FL, there are still some defense strategies which can successfully prevent data leakage.

(s1) **Noisy gradients and random masks [9, 29].** Simulations indicate that when the variance of noise add on real gradients is larger than 10^{-2}, it can successfully prevent data leakage [12]. However, the model accuracy will also be affected.

(s2) **Pruning and compression.** Pruning [13] and compression [1] are good ways to defend against data leakage. However, it is still important to find a balance between privacy protection and model accuracy.

(s3) **Perturbation.** Another defending strategy is to perturb data representation so that the recovered data is degrade [32].

(s4) **Large batch.** DLG only works for batch size less than 8 and image resolution less than 64×64. As a result, increasing large batch size is regarded as a good way to defend against data leakage. However, we proposed a new algorithm based on DLG to enable large-batch data leakage. We will discuss the new algorithm in detail in the next subsection.

Other Related Works Some other works based on [34, 39] have improved the algorithm from many aspects. Zhao et al. [37] proposed an effective way to extract the ground-truth labels analytically with 100% accuracy when the model is trained with cross-entropy loss on single-label classification tasks and [15] raised several strategies to prevent labels leakage in VFL. Li et al. [14] discuss the information leakage on federated approximated logistic regression models. Geiping et al. [7] discussed the relationship between network structure and DLG performances. They found that, generally, an untrained network is easier to leak data than a trained one. Convolutional layers with more channels have a larger potential of being attacked than ones with fewer channels. In [7], a newly designed cost function that compares the cosine similarity between the real and fake gradients has been proposed to replace the cost function in DLG (15.14)

$$\hat{\mathbf{X}}^*; \hat{\mathbf{y}}^* = arg \min_{\hat{\mathbf{X}} \in [0,1]^n; \hat{\mathbf{y}}} 1 - \frac{< \nabla_\theta \mathcal{L}(\mathbf{X}; \mathbf{y}, \boldsymbol{\theta}), \nabla_\theta \mathcal{L}(\hat{\mathbf{X}}; \hat{\mathbf{y}}, \boldsymbol{\theta}) >}{\|\nabla_\theta \mathcal{L}(\mathbf{X}; \mathbf{y}, \boldsymbol{\theta}) - \nabla_\theta \mathcal{L}(\hat{\mathbf{X}}; \hat{\mathbf{y}}, \boldsymbol{\theta})\|} + \alpha \mathrm{TV}(\hat{\mathbf{X}}). \tag{15.17}$$

The loss function in (15.17) consists of the cosine similarity between real and fake gradients and the total variation norm of $\hat{\mathbf{X}}$. Furthermore, it also proposed in [7] that the input to a fully connected layer with bias can be derived analytically which indicates that most fully connected layers are vulnerable to the attack. It is an essential point which we made fully use of it when we designed our algorithm CAFE. Additionally, [38] extends the attack in [7] to more general CNNs and FCNs with or without bias terms.

15.2 Data Leakage Attack in FL

The contributions of this chapter are summarized in the following:

1. We develop an advanced data leakage attack that we term CAFE to overcome the limitation of current data leakage attacks on FL. CAFE is able to recover large-scale data both in VFL and in HFL.
2. Our large-batch data recovery is based on the novel use of data index alignment and internal representation alignment in FL, which can significantly improve the recovery performance.

3. The effectiveness and practical risk induced from our data leakage algorithm is justified in the dynamic FL training setting when the model is updated every round.

15.2.1 Catastrophic Data Leakage from Batch Gradients

To realize large-scale data recovery from aggregated gradients, we propose our algorithm named as *CAFE: Catastrophic dAta leakage in Federated lEarning*. While CAFE can be applied to any type of data, without loss of generality, we use image datasets throughout the chapter.

15.2.1.1 Why Large-Batch Data Leakage Attack Is Difficult?

We start by providing some intuition on the difficulty of performing large-batch data leakage from aggregated gradients based on the formulation of DLG [39]. Assume that N images are selected as the input for a certain learning round. We define the data batch as $\mathcal{X} = \{\mathbf{x}_n, y_n | \mathbf{x}_n \in \mathbb{R}^{H \times W \times C}, n = 1, 2, \ldots, N\}$, where H, W, and C represent the height, the width, and the channel number of each image. Likewise, the batched "recovered data" is denoted by $\hat{\mathcal{X}} = \{\hat{\mathbf{x}}_n, \hat{y}_n | \hat{\mathbf{x}}_n \in \mathbb{R}^{H \times W \times C}, n = 1, 2, \ldots, N\}$, which have the same dimension as $\mathcal{X}$. Then. the objective function is

$$\hat{\mathcal{X}}^* = \arg\min_{\hat{\mathcal{X}}} \left\| \frac{1}{N} \sum_{n=1}^{N} \nabla_{\boldsymbol{\theta}} \mathcal{L}(\boldsymbol{\theta}, \mathbf{x}_n, y_n) - \frac{1}{N} \sum_{n=1}^{N} \nabla_{\boldsymbol{\theta}} \mathcal{L}(\boldsymbol{\theta}, \hat{\mathbf{x}}_n, \hat{y}_n) \right\|^2. \tag{15.18}$$

Note that in (15.18), the dimension of the aggregated gradients is fixed. However, as the N increases, the dimension of $\hat{\mathcal{X}}$ and $\mathcal{X}$ grows. When N is sufficiently large, it will be more challenging to find the "right" solution $\hat{\mathcal{X}}$ of (15.18) corresponding to the ground-truth dataset $\mathcal{X}$. On the other hand, CAFE addresses this large-batch issue by data index alignment, which can effectively exclude undesired solutions.

CAFE vs DLG Suppose that $N = 3$ and (15.18) can be rewritten as

$$\hat{\mathcal{X}}^* = \arg\min_{\hat{\mathcal{X}}} \left\| \frac{1}{3} \sum_{n=1}^{3} \nabla_{\boldsymbol{\theta}} \mathcal{L}(\boldsymbol{\theta}, \mathbf{x}_n, y_n) - \frac{1}{3} \sum_{n=1}^{3} \nabla_{\boldsymbol{\theta}} \mathcal{L}(\boldsymbol{\theta}, \hat{\mathbf{x}}_n, \hat{y}_n) \right\|^2. \tag{15.19}$$

We assume that there is a global optimal solution for (15.19) as

$$\hat{\mathcal{X}}^* = [\{\mathbf{x}_1, y_1\}; \{\mathbf{x}_2, y_2\}; \{\mathbf{x}_3, y_3\}]. \tag{15.20}$$

However, besides the optimal solution, there might be other undesired solutions, such as $\mathcal{X}^*$ shown in (15.21), whose gradients satisfy (15.22).

$$\hat{\mathcal{X}}^* = [\{\hat{\mathbf{x}}_1^*, \hat{y}_1^*\}; \{\hat{\mathbf{x}}_2^*, \hat{y}_2^*\}; \{\mathbf{x}_3, y_3\}] \tag{15.21}$$

$$\sum_{n=1}^{2} \nabla_\theta \mathcal{L}(\theta, \mathbf{x}_n, y_n) = \sum_{n=1}^{2} \nabla_\theta \mathcal{L}(\theta, \hat{\mathbf{x}}_n^*, \hat{y}_n^*)$$

$$\nabla_\theta \mathcal{L}(\theta, \mathbf{x}_n, y_n) \neq \nabla_\theta \mathcal{L}(\theta, \hat{\mathbf{x}}_n^*, \hat{y}_n^*). \tag{15.22}$$

Although the solutions (15.20) and (15.21) have the same loss value in (15.19), solution (15.21) is not an ideal solution for data recovery, which needs to be eliminated by introducing more constraints. When the number N increases, the number of both optimal and undesired solutions explodes. It is hard to find an approach which can converge to a certain solution through only one objective function. However, in CAFE, the number of objective functions can be as many as $\binom{N}{N_b}$. As the case above, suppose $N_b = 2$. Then, we can list all the objective functions

$$\begin{cases} \hat{\mathcal{X}}^{0*} = \arg\min_{\hat{\mathcal{X}}^0} \left\| \frac{1}{2} \sum_{n=1}^{2} \nabla_\theta \mathcal{L}(\theta, \mathbf{x}_n, y_n) - \frac{1}{2} \sum_{n=1}^{2} \nabla_\theta \mathcal{L}(\theta, \hat{\mathbf{x}}_n, \hat{y}_n) \right\|^2 \\ \hat{\mathcal{X}}^{1*} = \arg\min_{\hat{\mathcal{X}}^1} \left\| \frac{1}{2} \sum_{n=2}^{3} \nabla_\theta \mathcal{L}(\theta, \mathbf{x}_n, y_n) - \frac{1}{2} \sum_{n=2}^{3} \nabla_\theta \mathcal{L}(\theta, \hat{\mathbf{x}}_n, \hat{y}_n) \right\|^2 \\ \hat{\mathcal{X}}^{2*} = \arg\min_{\hat{\mathcal{X}}^2} \left\| \frac{1}{2} \sum_{n=1, n\neq 2}^{3} \nabla_\theta \mathcal{L}(\theta, \mathbf{x}_n, y_n) - \frac{1}{2} \sum_{n=1, n\neq 2}^{3} \nabla_\theta \mathcal{L}(\theta, \hat{\mathbf{x}}_n, \hat{y}_n) \right\|^2. \end{cases} \tag{15.23}$$

Comparing with (15.19), (15.23) has more constraint functions which restrict $\hat{\mathcal{X}}$ and dramatically reduces the number of undesired solutions. Solution (15.21) thus can be eliminated by the second and the third equations in (15.23). It suggests that CAFE helps the fake data converge to the optimal solution.

As a motivating example, Fig. 15.5 compares our proposed attack with DLG on a batch of 40 images. The recovery quality of DLG is far from satisfactory, while CAFE can successfully recover all images in the batch. It is worth noting that because DLG is not effective on large-batch recovery, it is suggested in [39] that increasing batch size could be a promising defense. However, the successful recovery of CAFE shows that such defense premise gives a false sense of security in data leakage and the current FL is at risk, as large-batch data recovery can be accomplished.

CAFE in VFL In VFL, the aggregator sends public keys to parties and decides the data index in each round of training and evaluation [2, 3, 36]. During the training process, parties exchange their intermediate results with others to compute gradients and upload them. Therefore, the aggregator has the access to both the

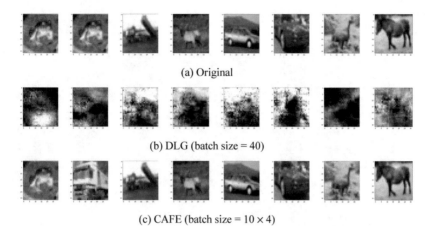

(a) Original

(b) DLG (batch size = 40)

(c) CAFE (batch size = 10 × 4)

Fig. 15.5 Illustration of large-batch data leakage on CIFAR-10 from shared gradients in FL. (**a**) Original. (**b**) DLG (batch size = 40). (**c**) CAFE (batch size = 10 × 4)

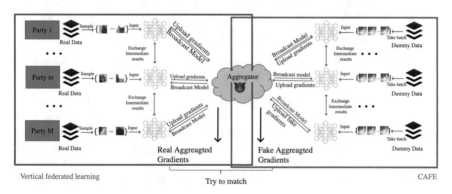

Fig. 15.6 Overview of CAFE in VFL

model parameters and their gradients. Notably, CAFE can be readily applied to existing VFL protocols where the batch data index is assigned.

Figure 15.6 gives an overview of CAFE in the VFL setting. The blue part represents a normal VFL paradigm and the red part represents the CAFE attack. Since data are vertically partitioned among different parties, data index alignment turns out to be an inevitable step in the vertical training process, which provides the aggregator (the attacker) an opportunity to control the selected batch data index. Suppose that there are M parties participating FL and the batch size is N. The aggregated gradients can be denoted by

$$\nabla_\theta \mathcal{L}(\theta, \mathcal{X}^t) = \frac{1}{N_b} \sum_{n=1}^{N_b} \nabla_\theta \mathcal{L}(\theta, \mathcal{X}_n^t) \quad \text{with} \quad \mathcal{X}_n^t = [\mathbf{x}_{n1}^t, \mathbf{x}_{n2}^t, \dots, \mathbf{x}_{nM}^t, y_n^t].$$

$$(15.24)$$

Algorithm 15.2 CAFE in VFL (regular VFL protocol and CAFE protocol)

1: **for** $t = 1, 2, \ldots, T$ **do**
2:     The aggregator broadcasts the model to all M parties
3:         **for** $m = 1, 2, \ldots, M$ **do**
4:            Party m takes real batch data
5:            Party m computes the intermediate results and exchanges them with other parties
6:            Party m uses the exchanged intermediate results to compute local aggregated gradients
7:            Party m uploads real local aggregated gradients to the aggregator.
8:         **end for**
9:         The aggregator computes real global aggregated gradients $\nabla_\theta \mathcal{L}(\theta, \mathcal{X}^t)$
10:        The aggregator computes the fake global aggregated gradients $\nabla_\theta \mathcal{L}(\theta, \hat{\mathcal{X}}^t)$
11:        The aggregator computes CAFE loss: $\mathbb{D}(\mathcal{X}^t; \hat{\mathcal{X}}^t)$ and $\nabla_{\hat{\mathcal{X}}^t} \mathbb{D}(\mathcal{X}^t; \hat{\mathcal{X}}^t)$
12:        The aggregator updates the batch data $\hat{\mathcal{X}}^t$ with $\nabla_{\hat{\mathcal{X}}^t} \mathbb{D}(\mathcal{X}^t; \hat{\mathcal{X}}^t)$
13:     The aggregator updates the model parameters θ with $\nabla_\theta \mathcal{L}(\theta, \mathcal{X}^t)$
14: **end for**

A benign aggregator will perform legitimate computations designed by FL protocol. However, as shown in Fig. 15.6, a curious aggregator can provide the same legitimate computation as a benign aggregator while simultaneously perform data recovery in a stealthy manner. The aggregator symmetrically generates fake images corresponding to the real ones. Once a batch of original data is selected, the aggregator takes the corresponding fake batch and obtains the fake gradients as

$$\nabla_\theta \mathcal{L}(\theta, \hat{\mathcal{X}}^t) = \frac{1}{N_b} \sum_{n=1}^{N_b} \nabla_\theta \mathcal{L}(\theta, \hat{\mathcal{X}}_n^t) \quad \text{with} \quad \hat{\mathcal{X}}_n^t = [\hat{\mathbf{x}}_{n1}^t, \hat{\mathbf{x}}_{n2}^t, \ldots, \hat{\mathbf{x}}_{nM}^t, \hat{y}_n^t].$$

$$(15.25)$$

Algorithm 15.2 gives a pseudo-code that implements our CAFE attack in VFL cases. The key part in our algorithm is aligning the real data batch indices with the fake ones. We define the squared ℓ_2 norm of the difference between the real and fake aggregated gradients in (15.26). Since the aggregator has the access to the model parameters, the attacker is able to compute the gradient of fake data from the loss in (15.26) and optimize the fake data for the purpose of recovering real data.

$$\mathbb{D}(\mathcal{X}^t; \hat{\mathcal{X}}^t) = \left\| \nabla_\theta \mathcal{L}(\theta, \mathcal{X}^t) - \nabla_\theta \mathcal{L}(\theta, \hat{\mathcal{X}}^t) \right\|^2.$$

$$(15.26)$$

Auxiliary Regularizers In addition to the gradient matching loss in (15.26), we further introduce two regularization terms—*internal representation regularization* and *total variance (TV) norm*. Motivated by Geiping et al. [7], the input vectors of the first fully connected layer can be directly derived from the gradients, we define the real/fake inputs of the first fully connected layer at the tth round as $\mathcal{Z}^t/\hat{\mathcal{Z}}^t \in \mathbb{R}^{N \times P}$, and we use ℓ_2 norm of their difference as what we call internal representation regularization.

To promote the smoothness of the fake images, we assume the TV norm of the real images as a constant, ξ, and compare it with the TV norm of the fake ones, $\mathrm{TV}(\hat{X})$. For each image $\mathbf{x} \in \mathbb{R}^{H \times W \times C}$ in data batch X^t, its TV norm is denoted by $\mathrm{TV}(\mathbf{x}) = \sum_{c} \sum_{h,w} \left[|\mathbf{x}_{h+1,w,c} - \mathbf{x}_{h,w,c}| + |\mathbf{x}_{h,w+1,c} - \mathbf{x}_{h,w,c}| \right]$.

As a result, the loss at the tth round $\mathbb{D}(X^t, \hat{X}^t)$ can be rewritten as

$$\mathbb{D}(X^t; \hat{X}^t) = \left\| \nabla_{\boldsymbol{\theta}} \mathcal{L}(\boldsymbol{\theta}, X^t) - \nabla_{\boldsymbol{\theta}} \mathcal{L}(\boldsymbol{\theta}, \hat{X}^t) \right\|^2 + \beta T V(\hat{X}^t) \cdot \mathbb{1}_{\{TV(\hat{X}^t) - \xi \geq 0\}}$$
$$+ \gamma \left\| \mathcal{Z}^t - \hat{\mathcal{Z}}^t \right\|_F^2, \tag{15.27}$$

where β and γ are coefficients and $\mathbb{1}_{\{TV(\hat{X}^t) - \xi \geq 0\}}$ is the indicator function. We will provide an ablation study in Sect. 15.3.5 to demonstrate the utility of these regularizers.

CAFE in HFL Similarly, we can apply our CAFE algorithm to HFL as well. Let X_m^t denote the original batch data and labels taken by local part m at the tth round. The gradients of the parameters at the tth round is

$$\nabla_{\boldsymbol{\theta}} \mathcal{L}(\boldsymbol{\theta}, X^t) = \frac{1}{M} \sum_{m=1}^{M} \nabla_{\boldsymbol{\theta}} \mathcal{L}(\boldsymbol{\theta}, X_m^t), \, X^t = \{X_1^t, X_2^t, \ldots, X_m^t, \ldots, X_M^t\}. \tag{15.28}$$

Similarly, we define the batch fake data and fake aggregated gradients as

$$\nabla_{\boldsymbol{\theta}} \mathcal{L}(\boldsymbol{\theta}, \hat{X}^t) = \frac{1}{M} \sum_{m=1}^{M} \nabla_{\boldsymbol{\theta}} \mathcal{L}(\boldsymbol{\theta}, \hat{X}_m^t), \, \hat{X}^t = \{\hat{X}_1^t, \hat{X}_2^t, \ldots, \hat{X}_m^t, \ldots, \hat{X}_M^t\}. \tag{15.29}$$

Algorithm 15.3 gives a pseudo-code that implements our CAFE attack in VFL cases and Fig. 15.7 shows the overview of CAFE in HFL settings. The left blue part indicates a normal HFL process and the right red part represents the attack. According to our simulation, more than 2000 private images from 4 parties can be leaked by CAFE.

Algorithm 15.3 CAFE in HFL (regular HFL protocol and CAFE protocol)

1: **for** $t = 1, 2, \ldots, T$ **do**

2:    The aggregator broadcasts the model to all M parties

3:        **for** $m = 1, 2, \ldots, M$ **do**

4:            Party m takes batched data $\mathcal{X}_m^t$

5:            Party m uses the received model to compute $\mathcal{L}(\theta, \mathcal{X}_m^t)$ and gradients $\nabla_\theta \mathcal{L}(\theta, \mathcal{X}_m^t)$

6:            Party m uploads real local aggregated gradients to the aggregator

7:        **end for**

8:        The aggregator computes real global aggregated gradients $\nabla_\theta \mathcal{L}(\theta, \mathcal{X}^t)$

9:        **for** $m = 1, 2, \ldots, M$ **do**

10:            The aggregator takes corresponding batched data $\hat{\mathcal{X}}_m^t$

11:            The aggregator uses the received model to compute $\mathcal{L}(\theta, \hat{\mathcal{X}}_m^t)$ and $\nabla_\theta \mathcal{L}(\theta, \hat{\mathcal{X}}_m^t)$

12:        **end for**

13:        The aggregator computes fake global aggregated gradients $\nabla_\theta \mathcal{L}(\theta, \hat{\mathcal{X}}^t)$

14:        The aggregator computes CAFE loss: $\mathbb{D}(\mathcal{X}^t; \hat{\mathcal{X}}^t)$ and $\nabla_{\hat{\mathcal{X}}_m^t} \mathbb{D}(\mathcal{X}^t; \hat{\mathcal{X}}^t)$

15:        **for** $m = 1, 2, \ldots, M$ **do**

16:            The aggregator updates the batch data $\hat{\mathcal{X}}_m^t$

17:        **end for**

    The aggregator updates the model parameters θ with $\nabla_\theta \mathcal{L}(\theta, \mathcal{X}^t)$

18:    **end for**

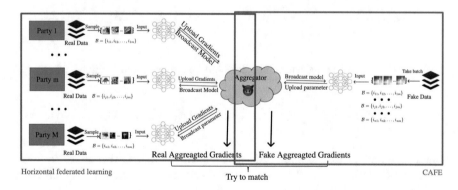

Fig. 15.7 Overview of CAFE in HFL

15.3 Performance Evaluation

In this chapter, we introduce our algorithm named CAFE. We provide simulation results in both plot and table form to support our algorithm.

15.3.1 Experiment Setups and Datasets

We conduct experiments on CIFAR-10, CIFAR-100, and Linnaeus 5 datasets in both HFL and VFL settings. All the fake data are initialized uniformly and optimized by the normalized gradient descent method. Our algorithm can recover all the data participating in FL with a relatively large batch size (more than 40). Scaling up to our hardware limits, CAFE can leak as many as 2000 images in the VFL setting including 4 parties.

Evaluation Metrics To measure the data leakage performance, we introduce peak signal-to-noise ratio (PSNR) [8] value with mean squared error (MSE) defined in (15.30) and (15.31). A higher PSNR value of leaked data represents better performance of data recovery.

$$\text{MSE}_c(\mathbf{x}, \hat{\mathbf{x}}) = \frac{1}{HW} \sum_{i=1}^{H} \sum_{j=1}^{W} [\mathbf{x}_{ijc} - \hat{\mathbf{x}}_{ijc}]^2 \tag{15.30}$$

$$\text{PSNR}(\mathbf{x}, \hat{\mathbf{x}}) = \frac{1}{C} \sum_{c=1}^{C} \left[20 \log_{10}(\max_{i,j} \mathbf{x}_{ijc}) - 10 \log_{10}(\text{MSE}_c(\mathbf{x}, \hat{\mathbf{x}})) \right]. \tag{15.31}$$

Baseline Methods for Comparison We compare CAFE with three other baselines, (1) DLG [39], (2) DLG given labels (iDLG) [37], and (3) using cosine similarity to compare the real and fake gradients [7]. We implement the original DLG and our CAFE under the same model and optimization methods. We run the DLG on 50 single images, respectively, and compute the average rounds required to make the PSNR value of a single leaked image above 30. We also compute the expected round number per image leakage for our CAFE algorithm. Furthermore, we fix the batch size and compare the PSNR value obtained by CAFE with that of DLG. We also test the impact of given labels on CAFE by using the techniques in [37]. Moreover, we compare the performance of CAFE under different loss functions: (1) replacing the squared ℓ_2 norm term with the cosine similarity of two gradients (CAFE with cosine similarity) and (2) loss proposed in [7], which only contains the TV norm regularizer.

15.3.2 CAFE in HFL Settings

In the HFL setting, we use a neural network consisting of 2 convolutional layers and 3 fully connected layers. The number of output channels of the convolutional layers is 64 and 128, respectively. The number of nodes of the first two fully connected layers is 512 and 256. The last layer is the softmax classification layer. We assume that 4 parties are involved in HFL and each of them holds a dataset including 100

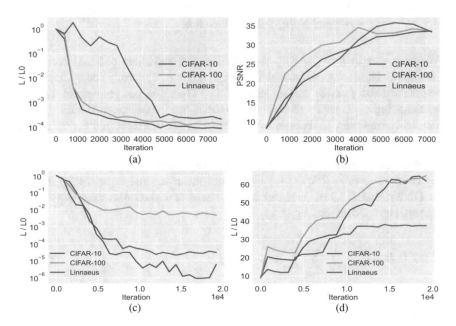

Fig. 15.8 CAFE loss ratio and PSNR curves. (**a**) HFL loss. (**b**) HFL PSNR. (**c**) VFL loss. (**d**) VFL PSNR

images. The batch size of each part in the training is 10, so there are 40 (10×4) images in total participating per round. For each experiment, we initialize the fake data using uniform distribution and optimize them for 800 epochs.

Figure 15.8a and b shows the CAFE loss curves and the PSNR curves on the three datasets in HFL cases. In the loss ratio curve, we set the ratio of current CAFE loss and the initial CAFE loss $\frac{\mathcal{L}(\theta, X^t)}{\mathcal{L}(\theta, X^0)}$ as label y. The PSNR values are always above 35 at the end of each CAFE attacking process, suggesting high data recovery quality (see Fig. 15.5 as an example). Figure 15.9 shows the attacking process of CAFE on Linnaeus. Under CAFE, PSNR reaches 35 at the 450th epoch where the private data are completely leaked visually.

Comparison with DLG Baseline In Table 15.1a, we set the batch ratio in CAFE as 0.1 and compare it with DLG under different batch sizes. Clearly, CAFE outperforms DLG thanks to our novel design of large-batch data leakage attack. As shown in Table 15.2a, DLG cannot obtain satisfactory results when the batch size increases to 40, while CAFE successfully recovers all the images. We also compare the algorithm performance in both given label and not given label cases.

From Table 15.3a and Fig. 15.10, recovery results on dataset with more categories are more likely to be effected if the labels are given. However, recoveries on datasets with few categories (10 or 5) have little influence.

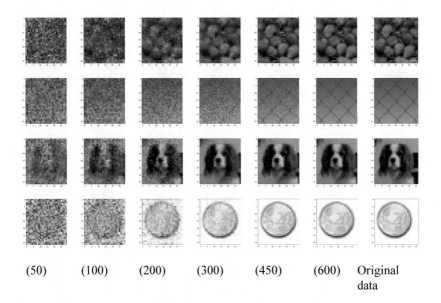

(50)   (100)   (200)   (300)   (450)   (600)   Original
data

Fig. 15.9 CAFE on Linnaeus (Epoch: 50, 100, 200, 300, 450, 600, Original data)

Table 15.1 CAFE vs DLG in speed

	Iters/Img		
	Datasets		
Batch size	CIFAR-10	CIFAR-100	Linnaeus
(a) Comparison of data leakage speed in HFL protocol. Lower round count is faster			
1(DLG)	284.4	266.9	366.7
10 × 4(CAFE)	9.50	6.00	9.50
20 × 4(CAFE)	6.75	3.86	4.75
30 × 4(CAFE)	4.83	3.41	3.17
40 × 4(CAFE)	3.75	3.75	2.375
(b) Comparison of data leakage speed in VFL protocol. Lower iteration count is faster			
1(DLG)	1530	1590	1630
10(DLG)	16.4	13.7	35.9
10(CAFE)	9.26	6.48	8.00
40(CAFE)	6.50	6.00	5.50
160(CAFE)	1.19	1.50	1.56

Comparison with Cosine Similarity   Table 15.4a shows that the PSNR values are still above 30 if we use cosine similarity instead of ℓ_2 norm. The slight drop in PSNR value may result from scaling ambiguity in cosine similarity. There is a performance gap between the loss of CAFE and the loss in [7], which validates the importance of our proposed auxiliary regularizers.

Table 15.2 CAFE vs DLG in performance

	PSNR		
	Datasets		
Algorithm	CIFAR-10	CIFAR-100	Linnaeus
(a) Comparison of leakage performance in HFL protocol. Higher PSNR is better. Batch size = 40			
CAFE	35.03	36.90	36.37
DLG	10.09	10.79	10.10
(b) Comparison of leakage performance in VFL protocol. Higher PSNR is better. Batch size = 40			
CAFE	66.20	66.33	38.72
DLG	34.67	45.12	29.49

Table 15.3 Impact by given labels

	PSNR		
	Datasets		
Setting	CIFAR-10	CIFAR-100	Linnaeus
(a) HFL			
Not given labels	35.03	36.90	36.37
Given labels	35.93	39.51	38.07
Number of categories	10	100	5
(b) VFL			
Not given labels	41.80	44.42	38.96
Given labels	40.20	40.29	39.50
Number of categories	10	100	5

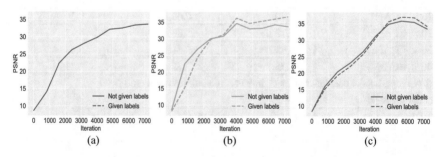

Fig. 15.10 Impact by given labels (HFL). (**a**) CIFAR-10. (**b**) CIFAR-100. (**c**) Linnaeus

15.3.3 CAFE in VFL Settings

We test the performance of CAFE on various factors. We slice one image into 4 small pieces. Each party holds one piece and the feature space dimension of each

Table 15.4 PSNR via loss

	PSNR		
	Datasets		
Loss	CIFAR-10	CIFAR-100	Linnaeus
(a) HFL (4 parties, batch ratio = 0.1, batch size 10 × 4)			
CAFE (15.27)	35.03	36.90	36.37
CAFE with cosine similarity	30.15	31.38	30.76
Loss in [7]	16.95	19.74	16.42
(b) VFL (4 parties, batch ratio = 0.1, batch size 40)			
CAFE (15.27)	66.20	66.33	38.72
CAFE with cosine similarity	30.96	43.68	34.90
Loss in [7]	12.76	10.85	10.46

piece is $16 \times 16 \times 3$. The model is composed of 2 parts. The first part consists of 2 convolutional layers and 3 fully connected layers for each part. The second part only consists of the softmax layer. In the training process, the pieces are sent into the first part, respectively, and turn to vectors as intermediate results. Parties then exchange their intermediate results, concatenate them, and put them into the second part. We set the batch size as 40 in VFL. Figure 15.8c and d shows the CAFE loss curves and the PSNR curves on the three datasets in VFL cases. The data recovery is even better than the results in HFL. The PSNR values of CIFAR-10 and CIFAR-100 rise higher than 40. By comparing with iDLG, we get the same conclusion as we discuss in the HFL part from Table 15.3b.

Comparison with DLG Baseline In Table 15.1b, we also set the batch ratio in CAFE as 0.1 and compare it with DLG with a single image and with a mini-batch (10 images) under different batch sizes. Clearly, CAFE still outperforms DLG. As shown in Table 15.2b, the PSNR values of CAFE keep 40 on average ahead of the ones of DLG.

Comparison with Cosine Similarity From Table 15.4b, we can conclude that the PSNR values still keep close to the ones by using CAFE. Scaling ambiguity in cosine similarity may also cause the drop in PSNR value. The performance gap between the loss of CAFE and the loss in [7] is much larger than the one in VFL, which indicates the utility of our auxiliary regularizers.

15.3.4   Attacking While Training in FL

Previous works have shown that DLG performs better on an untrained model than a trained one [7]. We also implement CAFE in the "attacking while learning" mode, in which the FL process is ongoing. When the network is training, the selected batch

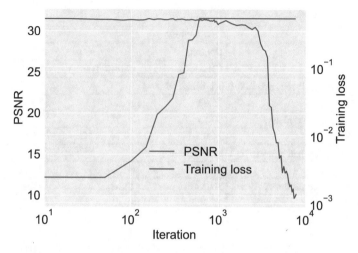

Fig. 15.11 PSNR and training loss curves

data and the parameters of the model change every round, which may cause the attack loss to diverge. To address this issue, for each real data batch, we compute the real gradients and optimize the corresponding fake data k times. We demonstrate on Linnaeus dataset, set $k = 10$, and stop CAFE after 1000 rounds (100 epochs). Figure 15.11 gives the curves of the training loss and the corresponding PSNR value. The PSNR value still can be raised to a relatively high value. It indicates that CAFE can be a practical data leakage attack in a dynamic training environment of FL and data can be recovered easier before the model is fully converged. It is mainly for the reason that the gradients are too small to recover data at the end of the training process.

15.3.5 Ablation Study

We test CAFE under different batch size, batch ratio, and with (without) auxiliary regularizers.

PSNR via Batch Size Table 15.5 shows that the PSNR values still keep above 30 when the batch size increases with a fixed number of parties and batch ratio. The result implies that the increasing batch size has little influence on data leakage performance of CAFE.

PSNR under Different Batch Ratio In HFL, 4 parties participate in the learning setting and we fix the amount of data held by each part as 500. In the VFL case, we implement CAFE on a total of 800 images. In Table 15.6, we change the batch ratio from 0.1 to 0.01 while keeping the trained epochs as 800. For both settings, the data leakage performance keeps at the same level.

Table 15.5 PSNR under Different Batch Ratio

	PSNR		
	Datasets		
Batch size	CIFAR-10	CIFAR-100	Linnaeus
(a) HFL (4 parties, batch ratio = 0.1)			
10 per party	35.03	36.90	36.37
20 per party	33.14	33.99	36.32
30 per party	32.31	33.21	35.96
40 per party	30.59	30.70	35.49
(b) VFL (4 parties, batch ratio = 0.2)			
8	41.80	44.42	39.96
40	59.51	65.00	41.37
80	57.20	63.10	43.66
160	54.74	64.75	38.72

Table 15.6 PSNR under Different Batch Ratio

	PSNR		
	Datasets		
Batch ratio	CIFAR-10 (HFL)	Linnaeus (HFL)	CIFAR-10 (VFL)
0.1	34.10	35.38	48.78
0.05	34.49	32.92	55.46
0.02	37.96	35.66	48.45
0.01	35.39	36.56	46.46

Impact of Auxiliary Regularizers Table 15.7 demonstrates the impact of auxiliary regularizers. From Fig. 15.12, adjusting the threshold ξ prevents images from being over blurred during the reconstruction process. TV norm can eliminate the noisy patterns on the recovered images and increase the PSNR. Images leaked without regularizing the Frobenius norm [33] of the difference between the internal representations $\mathcal{Z}$ and $\hat{\mathcal{Z}}$ may lose some details and cause the drop of PSNR.

Remarks We uncover the risk of *catastrophic data leakage in federated learning (CAFE)* through an algorithm that can perform large-batch data leakage with high data recovery quality. Extensive experimental results demonstrate that CAFE can recover large-scale private data from the shared aggregated gradients on both vertical and horizontal FL settings, overcoming the batch limitation problem in current data leakage attacks. Our advanced data leakage attack and its stealthy nature suggests practical data privacy concerns in FL and poses new challenges on future defenses.

Table 15.7 Effect of auxiliary regularizers

| | PSNR | | |
| | Datasets | | |
Algorithm	CIFAR-10	CIFAR-100	Linnaeus
(a) HFL (4 parties, batch size = 10 per part, 800 epochs)			
CAFE	35.03	36.90	36.37
CAFE ($\xi = 0$)	26.53	27.43	28.99
CAFE ($\beta = 0$)	23.19	22.09	31.67
CAEE ($\gamma = 0$)	25.41	18.14	24.17
CAEE ($\beta, \gamma = 0$)	20.42	16.50	20.33
(b) VFL (4 parts, batch size = 40, 800 epochs)			
CAFE	66.20	66.33	38.72
CAFE ($\xi = 0$)	45.35	36.86	37.37
CAFE ($\beta = 0$)	67.94	52.02	44.13
CAEE ($\gamma = 0$)	12.51	12.60	13.03
CAEE ($\beta, \gamma = 0$)	8.49	8.72	9.30

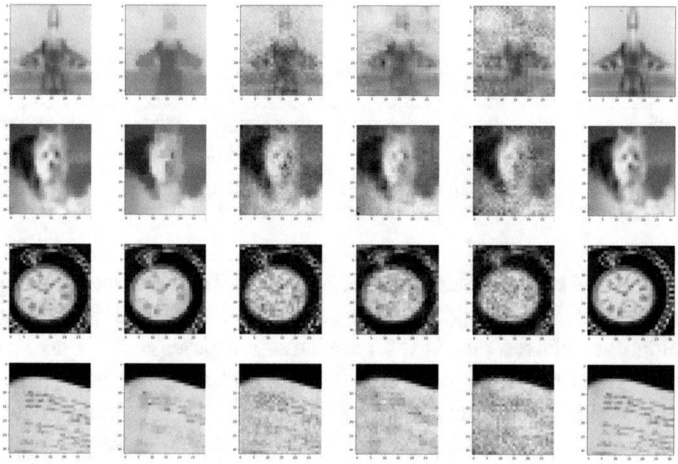

CIFAR-10

CIFAR-100

Linnaeus

Linnaeus

CAFE $\xi = 0$ $\beta = 0$ $\gamma = 0$ $\beta, \gamma = 0$ Original

Fig. 15.12 Effect of auxiliary regularizers

15.4 Concluding Remarks

In this chapter, we summarize our algorithm CAFE and give a discussion on open research directions.

15.4.1   Summary

In this chapter, we review the previous work on FL and potential attacking algorithms on it. We introduce a new algorithm named DLG for data attacking in FL. We improve the algorithm and propose CAFE as a new attacking method. By comparison, CAFE outperforms DLG when the training batch size is large. Moreover, we also add additional regularizers to help improve the algorithm performance. The simulation results indicate that CAFE can be a powerful attack in FL systems especially in the vertical protocols where a higher privacy protection level is required. The idea is simple yet effective.

15.4.2   Discussion

The assumption that the aggregator knows the sample indices may be strong in HFL settings. However, we have demonstrated that CAFE is an effective attacking method in VFL settings. To make the simulation results more convincing, more VFL simulations on other distributed datasets should be added. Our simulations on CAFE are only based on image data. Simulations on time serial or natural language processing models will also be added to show the attack performance. In CAFE, TV norm is designed only for image reconstruction. New regularizers should be designed to help recover other types of data in the attack algorithm.

It is common in FL that parties communicate locally trained weights instead of gradients [23]. Thus, the aggregator can derive the gradients of those parts of the model by using the change of the parameter. Even if the communication does not need to occur in each round, the aggregator can regard all the updates between two consecutive communications as a big round in which the batch size is the sum of the batch size in every single round which makes it possible to apply CAFE on other FL scenarios besides the ones we mentioned above. More simulations should be done to discuss the impact on algorithm performance in different FL scenarios.

Although [7] gives the complete proof of deriving the input of a fully connected layer with biases and in [5, 25], it is claimed that the input data can be reconstructed under some conditions. However, the paper cannot directly generalize this conclusion to the case where the input is a batch of data [24]. In our case, we assume the real internal representation is already known to us. From a new angle, it is also worth building new secure and privacy-preserving FL protocols [17, 18, 23, 30, 31] to prevent data leakage.

References

1. Beguier C, Tramel EW (2020) SAFER: Sparse secure aggregation for federated learning. arXiv, eprint:200714861
2. Chen T, Jin X, Sun Y, Yin W (2020) VAFL: a method of vertical asynchronous federated learning. In: International workshop on federated learning for user privacy and data confidentiality in conjunction with ICML
3. Cheng K, Fan T, Jin Y, Liu Y, Chen T, Yang Q (2019) Secureboost: A lossless federated learning framework. arXiv, eprint:190108755
4. Dwork C, Smith A, Steinke T, Ullman J, Vadhan S (2015) Robust traceability from trace amounts. In: 2015 IEEE 56th annual symposium on foundations of computer science, USA, pp 650–669
5. Fan L, Ng K, Ju C, Zhang T, Liu C, Chan CS, Yang Q (2020) Rethinking privacy preserving deep learning: How to evaluate and thwart privacy attacks. arXiv, eprint:200611601
6. Fredrikson M, Jha S, Ristenpart T (2015) Model inversion attacks that exploit confidence information and basic countermeasures. In: Proceedings of the 22nd ACM SIGSAC conference on computer and communications security. Association for Computing Machinery, New York, NY, pp 1322–1333
7. Geiping J, Bauermeister H, Dröge H, Moeller M (2020) Inverting gradients - how easy is it to break privacy in federated learning? In: Advances in neural information processing systems, vol 33, pp 16937–16947
8. Gonzalez RC, Woods RE (1992) Digital image processing. Addison-Wesley, New York
9. Guerraoui R, Gupta N, Pinot R, Rouault S, Stephan J (2021) Differential privacy and byzantine resilience in SGD: Do they add up? arXiv, eprint:210208166
10. Guo S, Zhang T, Xiang T, Liu Y (2020) Differentially private decentralized learning. arXiv, eprint:200607817
11. Hitaj B, Ateniese G, Pérez-Cruz F (2017) Deep models under the GAN: information leakage from collaborative deep learning. arXiv, eprint:170207464
12. Huang Y, Song Z, Chen D, Li K, Arora S (2020) TextHide: Tackling data privacy in language understanding tasks. In: The conference on empirical methods in natural language processing
13. Huang Y, Su Y, Ravi S, Song Z, Arora S, Li K (2020) Privacy-preserving learning via deep net pruning. arXiv, eprint:200301876
14. Li Z, Huang Z, Chen C, Hong C (2019) Quantification of the leakage in federated learning. In: International workshop on federated learning for user privacy and data confidentiality. West 118–120 Vancouver Convention Center, Vancouver
15. Li O, Sun J, Yang X, Gao W, Zhang H, Xie J, Smith V, Wang C (2021) Label leakage and protection in two-party split learning. arXiv, eprint:210208504
16. Liang G, Chawathe SS (2004) Privacy-preserving inter-database operations. In: 2nd Symposium on intelligence and security informatics (ISI 2004), Berlin, Heidelberg, pp 66–82
17. Liu R, Cao Y, Yoshikawa M, Chen H (2020) FedSel: Federated SGD under local differential privacy with top-k dimension selection. arXiv, eprint:200310637
18. Liu Y, Kang Y, Zhang X, Li L, Cheng Y, Chen T, Hong M, Yang Q (2020) A communication efficient vertical federated learning framework. arXiv, eprint:191211187
19. Long Y, Bindschaedler V, Wang L, Bu D, Wang X, Tang H, Gunter CA, Chen K (2018) Understanding membership inferences on well-generalized learning models. arXiv, eprint:180204889
20. Lyu L, Yu H, Ma X, Sun L, Zhao J, Yang Q, Yu PS (2020) Privacy and robustness in federated learning: Attacks and defenses. arXiv, eprint:201206337
21. McMahan HB, Moore E, Ramage D, y Arcas BA (2016) Federated learning of deep networks using model averaging. arXiv, eprint:160205629
22. Melis L, Song C, Cristofaro ED, Shmatikov V (2018) Inference attacks against collaborative learning. In: Proceedings of the 35th annual computer security applications conference. Association for Computing Machinery, New York, NY, pp 148–162

23. Niu C, Wu F, Tang S, Hua L, Jia R, Lv C, Wu Z, Chen G (2019) Secure federated submodel learning. arXiv, eprint:191102254

24. Pan X, Zhang M, Yan Y, Zhu J, Yang M (2020) Theory-oriented deep leakage from gradients via linear equation solver. arXiv, eprint:201013356

25. Qian J, Nassar H, Hansen LK (2021) On the limits to learning input data from gradients. arXiv, eprint:201015718

26. Scannapieco M, Figotin I, Bertino E, Elmagarmid A (2007) Privacy preserving schema and data matching. In: Proceedings of the ACM SIGMOD international conference on management of data, Beijing, pp 653–664

27. Shokri R, Shmatikov V (2015) Privacy-preserving deep learning. In: Proceedings of the 22nd ACM SIGSAC Conference on Computer and Communications Security, Association for Computing Machinery, New York, NY, CCS '15, pp 1310–1321

28. Shokri R, Stronati M, Song C, Shmatikov V (2017) Membership inference attacks against machine learning models. In: 2017 IEEE symposium on security and privacy (SP), pp 3–18

29. So J, Guler B, Avestimehr AS (2021) Byzantine-resilient secure federated learning. arXiv, eprint:200711115

30. So J, Guler B, Avestimehr AS (2021) Turbo-aggregate: Breaking the quadratic aggregation barrier in secure federated learning. arXiv, eprint:200204156

31. Sun L, Lyu L (2020) Federated model distillation with noise-free differential privacy. arXiv, eprint:200905537

32. Sun J, Li A, Wang B, Yang H, Li H, Chen Y (2020) Provable defense against privacy leakage in federated learning from representation perspective. arXiv, eprint:201206043

33. Trefethen LN, Bau D (1997) Numerical linear algebra. SIAM, Philadelphia

34. Wei W, Liu L, Loper M, Chow KH, Gursoy ME, Truex S, Wu Y (2020) A framework for evaluating gradient leakage attacks in federated learning. arXiv, eprint:200410397

35. Wei K, Li J, Ding M, Ma C, Su H, Zhang B, Poor HV (2021) User-level privacy-preserving federated learning: Analysis and performance optimization. arXiv, eprint:200300229

36. Yang Q, Liu Y, Chen T, Tong Y (2019) Federated machine learning: Concept and applications, vol 10. Association for Computing Machinery, New York, NY

37. Zhao B, Mopuri KR, Bilen H (2020) iDLG: Improved deep leakage from gradients. arXiv, eprint:200102610

38. Zhu J, Blaschko MB (2021) R-GAP: Recursive gradient attack on privacy. In: International conference on learning representations

39. Zhu L, Liu Z, Han S (2019) Deep leakage from gradients. In: Advances in neural information processing systems, Vancouver, pp 14774–14784

Chapter 16
Security and Robustness in Federated Learning

Ambrish Rawat, Giulio Zizzo, Muhammad Zaid Hameed, and Luis Muñoz-González

Abstract Federated learning (FL) has emerged as a powerful approach to decentralize the training of machine learning algorithms, allowing the training of collaborative models while preserving the privacy of the datasets provided by different parties. Despite the benefits, FL is also vulnerable to adversaries, similar to other machine learning (ML) algorithms in centralized settings. For example, just a single malicious or faulty participant in an FL task can entirely compromise the performance of the model when using unsecure implementations. In this chapter, we provide a comprehensive analysis of the vulnerabilities of FL algorithms to different attacks that can compromise their performance. We describe a taxonomy of attacks comparing the similarities and differences with respect to centralized ML algorithms. Then, we describe and analyze different families of existing defenses that can be applied to mitigate these threats. Finally, we review a set of comprehensive attacks that aim to compromise the performance and convergence of FL.

16.1 Introduction

Artificial intelligence (AI) and especially machine learning (ML) are at the core of the fourth industrial revolution. ML has become one of the main components of many systems and applications with success stories across different sectors, including healthcare [37], financial markets [10], or Internet of Things (IoT) [19]. The benefits of ML technologies are clear, as they allow the efficient automation of many processes and tasks by leveraging their capability to analyze a huge amount of data.

A. Rawat (✉) · G. Zizzo
IBM Research Europe, Dublin, Ireland
e-mail: ambrish.rawat@ie.ibm.com; giulio.zizzo2@ie.ibm.com

M. Z. Hameed · L. Muñoz-González
Imperial College, London, UK
e-mail: muhammad.hameed13@imperial.ac.uk; l.munoz-gonzalez@imperial.ac.uk

© The Author(s), under exclusive license to Springer Nature Switzerland AG 2022
H. Ludwig, N. Baracaldo (eds.), *Federated Learning*,
https://doi.org/10.1007/978-3-030-96896-0_16

363

Recently, federated learning (FL) has emerged as a promising approach for the development of distributed ML systems, allowing us to resolve challenges in some application domains. FL allows us to train a shared ML model from a federation of participants who use their own datasets to locally train a machine learning model while preserving the privacy of their datasets within the federation. In this approach, there is a central aggregator (server) that combines the information that controls the learning process and aggregates the information from the parties (clients) during the training of the ML model. These parties train the models locally using their own dataset and send the model updates back to the central aggregator in an iterative manner. In this way, during training, the data always remains with the party, keeping their datasets private.

Given current laws and privacy regulations such as the *General Data Protection Regulation* (GDPR) in the European Union, or the *Health Insurance Portability and Accountability Act* (HIPAA) in the US, FL offers an appealing alternative to build collaborative models across different institutions or companies in sensitive domains, such as healthcare or financial markets, by preserving the privacy of the party data. On the other hand, with the increasing computational capabilities of edge devices, including smartphones, sensors, and other IoT devices, FL also allows us to decentralize the training of the ML models and push the computation to edge devices. Thus, the data does not need to be collected and centralized, but edge devices contribute toward the shared FL model performing local computations using their own data.

Despite the benefits and the advantages of ML and FL technologies, there are still challenges and risks that need to be analyzed, understood, and mitigated. ML algorithms are known to be vulnerable to attackers. At training time, ML algorithms can be subject to poisoning attacks, where attackers can influence the training of the learning algorithm to manipulate and degrade its performance. This can be achieved, for example, by manipulating the data that is used to train the ML model. Attackers can also introduce *backdoors* during the training of the learning algorithm, so that the performance of the model is not altered for regular inputs, but a specific and unexpected behavior of the model is observed for inputs containing a *trigger* that activates the backdoor [12]. During deployment, ML algorithms are particularly vulnerable to *adversarial examples*, inputs specifically crafted by the attacker that contain a very small perturbation with respect to the original sample, that are designed to produce errors in the system [16].

These vulnerabilities of the learning algorithms are also present in FL. However, the mechanisms that attackers can leverage to compromise the learning algorithms are, in some cases, different to those where ML is applied to centralized data sources, requiring special consideration. In this chapter, we provide a comprehensive description of the different attacks that can be performed to compromise FL algorithms, including an extended taxonomy to model the attack surface compared to the taxonomies typically used for centralized learning algorithms. Using this taxonomy, we categorize these different sets of attacks as well as defenses that can be applied to mitigate them both at training and test time. This includes data and model poisoning attacks aiming to compromise the performance or the convergence of the FL algorithms, backdoors, and evasion attacks as exemplified by adversarial examples.

The rest of the chapter is organized as follows: In Sect. 16.2 we describe the threat model and present a taxonomy of attacks that can be performed against FL algorithms. Section 16.3, explains different defensive schemes capable of mitigating these threats. Section 16.4 provides a comprehensive description of different attack strategies that have been proposed to compromise FL algorithms, both at training and test time. Finally, Sect. 16.5 concludes the chapter.

16.1.1 Notation

In the rest of the chapter, we take classification models as the guiding example, but many principles transfer to other types of machine learning tasks. In a federated learning process, C parties with their individual data source, $\{D_i\}_{i=1}^{C}$, composed of (x, y) data–label pairs, seek to learn a common ML model f_w. During each training round, N parties participate by sending update vectors $\{v_i\}_{i=1}^{N}$ to the central aggregator. They obtain this update vector by optimizing for a common objective L with respect to their respective private data partition. The aggregator combines these updates, often by averaging, and broadcasts the corresponding vector to all C parties. For most of the discussions in the following section, we assume that all N parties participate in each training round.

16.2 Threats in Federated Learning

In this section, we present a threat model to describe the different threats and attacks possible against FL systems. This allows us to understand the vulnerabilities, providing a systematic framework to analyze the security aspects of FL.

For this, we rely on the frameworks originally proposed in [2, 18] and extended and revised in [25] for standard ML algorithms. Thus, we describe the threat model characterizing attacks according to the attacker's goal, capabilities to manipulate the data and influence the learning system, knowledge of the target system and data, as well as the attacker's strategy. Although some of these aspects are similar to those for standard learning algorithms, there are certain aspects of the threat model that are unique to FL scenarios which we discuss in the following sections.

16.2.1 Types of Attackers

Before contextualizing the threat model in similar terms to those in non-distributed ML algorithms, we need to define the specific types of attackers that are possible in FL scenarios. This differs from centralized ML algorithms, where the attacker is typically considered external to the system and aims to compromise or degrade the system's performance, produce errors in the system, or leak information about the target system or the data used to train the model. In other cases, the attackers can also manipulate the software or code implementations used to train the machine learning models. In addition to this, in FL, some of the parties (users) within the system can also behave maliciously.

Thus, in FL systems, we can categorize attackers as:

- **Outsiders:** similar to the case of centralized learning algorithms, outsiders are attackers that are not users (parties) of the platform. They can compromise the FL system at training time by poisoning the training datasets of benign parties to perform poisoning or backdoor attacks. At test time, they can exploit the weaknesses and blind spots of the resulting models to produce errors, e.g., with adversarial examples [16], or to extract some knowledge from the target model, e.g., membership inference attacks [28]. This category can also include attackers that are capable of intercepting and tampering with the communications between the central node and some of the parties of the FL platform.

- **Insiders:** this includes cases where one or several users (parties) of the FL platform are malicious. These attackers can also manipulate and degrade the performance of the system to gain some advantage with respect to other parties but have more freedom than outsiders to do so. For example, for poisoning the federated learning model, insiders can directly manipulate the parameters of the model sent to the aggregator. Insiders can also aim to leak information from the datasets used by the other users, e.g., with property inference attacks [17, 23, 28]. In cases where there are several insiders in the FL platform, as shown in Fig. 16.1, there are different possible scenarios depending on whether the attackers collude toward the same malicious objective.

16.2.2 Attacker's Capabilities

The capabilities of the attacker to compromise an FL system can be categorized in terms of the attacker's influence on the data, the model, and any additional constraints which limit the attacker such as the presence of defensive algorithms.

16.2.2.1 Attack Influence

According to the capabilities of the attacker to influence or compromise the ML model, attacks can be classified as:

- **Causative:** if the attacker can influence the learning algorithm by injecting or manipulating data used to train the learning algorithms or providing malicious

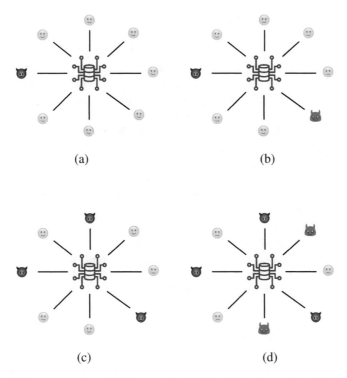

Fig. 16.1 Different scenarios of insider attackers in FL: (**a**) single attacker. (**b**) Several non-colluding attackers, i.e., the attackers have different objectives. (**c**) Group of colluding attackers. (**d**) Different groups of colluding attackers

information to manipulate the parameters of the system. These attacks are commonly referred to as *poisoning attacks*. *Byzantine* attacks [8] which send arbitrary updates to compromise model performance also qualify as causative within this categorization.

- **Exploratory:** the attacker cannot influence the training process but can attempt to exploit the weaknesses and blind spots of the system at test time or to extract information from the target system. Scenarios where attackers aim to produce errors in the target system are usually referred to as *evasion attacks*.

Poisoning attacks are an important threat in scenarios where the data collected to train the learning algorithms is untrusted. This is common in applications that collect data from humans who can act dishonestly or devices whose integrity can be at risk. In FL systems, poisoning attacks can be performed by both insiders and outsiders. In the case of outsiders, poisoning attacks can be achieved by injecting malicious data in the training datasets used by the participants, by compromising the integrity of software used by the participant or compromising communications between the participants and the central node. In this latter case, the attackers can perform stronger poisoning attacks via model poisoning [5].

At run-time, in evasion attacks, even if the data used for training the FL model is trusted and all the participants are honest, the attackers can probe the resulting model to produce intentional errors, for example, by crafting adversarial examples. On the other hand, there are other exploratory attacks that aim to compromise the privacy of the model, extracting or leaking information about the model and its training data. In this sense, similar to non-distributed ML settings, FL models can be vulnerable to *membership inference attacks*, where the attacker tries to assert if a data point has been used for training the learning algorithm, as described in Chap.. In this case, the difference in FL is that the attacker may not know which participant provided that data point. FL on the other hand leverages data from numerous participants, and the model is often trained with more data points compared to the centralized case, which increases the effort for the attacker to perform membership inference attacks [28]. In some settings, insider attackers can also perform *property inference attacks*, aiming to infer properties from the training data used by other participants [17, 23]. This can be achieved by examining the model updates during training. However, these attacks can only achieve a certain degree of success under very particular conditions: a limited number of participants and highly differentiated properties across the datasets of the participants.

16.2.2.2   Data Manipulation Constraints

The attacker's capabilities may be limited by the presence of constraints for the manipulation of the data or the parameters of the model in the case of model poisoning attacks. The attacker can also self-impose some constraints to remain undetected and perform stealthy attacks by, for example, crafting attack points that do not differ too much from benign data points. The manipulation constraints are also strongly related to the particular application domain. For example, in ML-based malware detection, the attacker's aim is to evade detection by manipulating the malware code, but those manipulations need to preserve the malicious functionality of the program [11, 32]. In contrast, in some computer vision applications, it is reasonable to assume that the attackers can manipulate every pixel in an image or every frame in a video.

In data poisoning attacks, the adversary can have different degrees of freedom to manipulate the data. For certain cases, the attacker may be in control of part of the labeling process (e.g., when using crowdsourcing). These are known as label flipping attacks. In other scenarios, even if the attacker is not in control of the labels assigned to the poisoning points, they can reliably estimate the label that will be assigned to the injected malicious points. For example, in a spam detection application, the attackers can assume that most of the malicious emails injected in the system will be labeled as spam.

When assessing the robustness of ML and FL algorithms to attacks, it is important to model realistic data constraints to better characterize worst-case scenarios, for example, through optimal attack strategies, where the attacker aims to maximize the damage on the target algorithm. However, it is important to consider appropriate detectability constraints; otherwise, the attacks can be trivial and can be easily detected with orthogonal methods, such as data pre-filtering or outlier detection [30, 31].

16.2.3   Attacker's Goal

The goal of the adversary can be categorized based on the type of security violation that the attacker seeks to achieve and the specificity of the attack, which can be described in terms of the number of data points affected by the attack, or on the type of errors to be produced in the system.

16.2.3.1   Security Violation

We can differentiate three different security violations against ML and FL systems:

- **Integrity violation:** when the attack evades detection without compromising the system's normal operation.
- **Availability violation:** when the attacker aims to compromise the functionality of the system.
- **Privacy violation:** when the adversary obtains private information about the target system, the data used for training, or the users of the system.

Integrity and availability violations depend upon the application to be deployed and the attacker's capabilities to influence the training of the learning algorithm. In this sense, in FL, for insider threats, the attackers can not only poison the learning algorithm but also prevent the algorithm to converge during its training. On the privacy side, as mentioned previously, FL models can be vulnerable to membership and property inference attacks.

16.2.3.2   Attack Specificity

This characteristic is defined by a continuum spectrum that describes the specificity of the attacker's intention ranging from targeted to indiscriminate attack scenarios:

- **Targeted Attacks:** where the attacker aims to degrade the performance of the system or to produce errors for a reduce set of target data points.
- **Indiscriminate Attacks:** where the attacker aims to degrade the system's performance or to produce errors in an indiscriminate fashion, i.e., affecting a broad set of cases or data points.

Different from the taxonomy originally proposed in [2, 18], in the research literature on adversarial examples, i.e., evasion attacks for specific inputs, the term *targeted attack* usually refers to the case where the attacker aims to evade the target model producing a specific type of error, whereas *untargeted attacks* refer to those attacks that just aim to produce errors regardless of the nature of the error. However, the related work on poisoning attacks follows the original taxonomy [2, 18], so that *indiscriminate poisoning attacks* are those that produce errors for a large set of inputs, and *targeted poisoning attacks* are those that produce errors on a reduced set of target inputs. However, the taxonomy in [2, 18] is limited to describe attacks depending on the nature of the errors. This limitation was addressed by Muñoz-González et al. [26], extending the taxonomy to categorize attacks according to the type of errors that the attacker wants to produce.

16.2.3.3 Error Specificity

As described in [26], in some cases, such as multi-class classification, depending on the nature of the errors that the attacker seeks to produce in the system, we can categorize the attacks as:

- **Error-generic:** when the adversary wants to produce errors in the target system regardless of the type of error to be produced.
- **Error-specific:** when the attacker aims to produce a specific type of errors in the system. This can be application dependent. In fact, depending on their capabilities, the attackers can be constrained on the type of errors that can be produced in the system.

While the categorization of targeted and indiscriminate attacks is based on specificity with respect to data samples, the error specificity characterizes the orthogonal dimension of quality of error—like an error-specific attacker could seek misclassification while an error-generic attacker might pursue more general objectives for system compromise. For example, in the context of data poisoning, an error-specific indiscriminate poisoning attack aims at maximizing the performance of the model over a large set of test inputs producing specific type of errors (e.g., classifying all the samples from all classes as samples from class "0"), whereas in the case of an error-generic indiscriminate attack, the adversary does not care about the nature of the errors produced in the system and just aims at maximizing the overall error of the model for a large set of inputs.

16.2.4 Attacker's Knowledge

The attacker's knowledge of the target FL system includes the following aspects:

- The datasets used by one or more participants
- The features used to train the learning algorithm and their range of valid values

- The learning algorithm, the objective function to be optimized, and the aggregation method used by the central node
- The parameters of the FL algorithm and the resulting model

Depending on how much the attacker knows about the previous points, we can differentiate two main scenarios: *perfect* and *limited* knowledge attacks.

16.2.4.1 Perfect Knowledge Attacks

These are scenarios where we assume that the attacker knows everything about the target system. Although this assumption can be unrealistic in most cases, perfect knowledge attacks are useful to assess the robustness and security of ML and FL algorithms in worst-case scenarios, helping to provide lower bounds in the performance of the algorithm for different attack's strength. Furthermore, they can be useful for model selection, by comparing the performance and robustness of different algorithms and architectures tested against these type of attacks.

16.2.4.2 Limited Knowledge Attacks

There is a broad range of possibilities to model attacks with limited knowledge. Typically, in the research literature, two main categories are considered:

- **Limited knowledge attacks with surrogate data:** this includes scenarios where the attacker knows the model used for the learning algorithm, the feature representation, the objective function, and the aggregation scheme used by the aggregator. However, the attackers do not have access to the training data, although they can have access to a surrogate dataset with similar characteristics to the dataset used to train the target learning algorithm. Then, the attacker can estimate the parameters of the targeted model by using this surrogate dataset, which can enable successful attacks depending on the quality of the surrogate dataset. In the case of FL, this is a reasonable assumption to model insider attackers. Such an adversary has access to the model information and their own dataset, but not to the datasets of the rest of the participants.
- **Limited knowledge attacks with surrogate models:** this category includes scenarios where the attackers have access to the dataset and the feature representation used by the target system, but they do not have access to the ML model, the objective function to be optimized, or the aggregation method used by the central node. In these cases, the attackers can train a surrogate model to estimate the behavior of the system. By crafting attacks against this surrogate model, the resulting malicious points are used to attack the real model. This strategy can be effective to achieve successful attacks, especially if the surrogate models are similar, as the vulnerabilities of different model architectures and learning algorithms are similar in some cases. This is commonly referred to as *attack transferability* and has been shown for both evasion [29] and poisoning attacks [26].

Although perfect knowledge attacks can be helpful to model worst-case scenarios for testing the robustness of many FL systems, a balance between realistic and worst-case scenarios should be considered in practical deployments. For example, in most cases, both insider and outsider attackers will not have access to the datasets from all the participants. Therefore, asserting robustness of FL algorithms against weaker adversaries can be a useful and well motivated threat model to investigate.

16.2.5   Attack Strategy

Attack strategies against both standard ML and FL systems can be formulated as an optimization problem capturing different aspects from the threat model. The attacker's goal can be characterized by an objective function evaluated on a set of predefined data points, which can be a specific set of target points or, for indiscriminate attacks, a representative set of the underlying data distribution used by the target system. This objective function typically helps the attacker to assess the effectiveness of an attack strategy. The objective function can also include specific constraints to prevent being detected by the defender. In Sect. 16.4, we will show a comprehensive set of attack strategies that can be used to compromise FL algorithms.

Finally, Table 16.1 summarizes the threat model presented in this section.

16.3   Defense Strategies

We now look at different defense strategies that have been devised to counter the different types of attacks described in the previous section. Designing a defense method incurs several challenges. To take one, it is essential that defense mechanisms preserve the model performance in the absence of malicious parties. The FL model assumptions may also affect the design strategy for defenses. The aggregator, for instance, may not have the ability to inspect model updates [9]. In this section we distill some broad themes that have been used for designing defenses for FL systems. First, we look at defenses developed for convergence attacks. Broadly speaking, these methods inspect the set of updates across all parties during each training round and use a filtering criterion with the aggregation. We then describe an alternate line of defenses which incorporate the update history for this process. A third category of defenses are based on redundancy between party data partitions.

It is worth noting that a large class of defenses developed for centralized systems naturally apply to federated settings. However, FL-specific scenarios do require specialized approaches which we discuss in the following section.

Table 16.1 Threat model in federated learning

Types of attackers	
Types of attackers	• **Outsiders:** attackers external to the platform • **Insiders:** attackers that are participating in the FL task
Attacker's capabilities	**Attack influence** • *Causative attacks:* the attacker can influence the learning algorithm (e.g., poisoning or backdoor attacks) • *Exploratory attacks:* the attacker can only manipulate data at test time (e.g., adversarial examples)
Attacker's goal	**Security violation** • *Integrity attacks* (e.g., backdoor attacks) • *Availability attacks* (e.g., poisoning attacks) • *Privacy violation* (e.g., property inference attacks) **Attack specificity** • *Targeted attacks:* focused on a specific set of cases or data points • *Indiscriminate Attacks:* target a broader set of cases or data points **Error specificity:** • *Error-generic attacks:* the attacker just aims to produce errors in the system, regardless of their nature • *Error-specific attacks:* the attacker aims to produce specific types of errors in the target system
Attacker's knowledge	• **Perfect knowledge:** the attacker knows everything about the target system • **Limited knowledge:** – *Surrogate dataset:* the attacker knows the target model but not the training dataset (or has partial knowledge of it) – *Surrogate model:* the attacker knows the training dataset but not the model (e.g., transfer attacks)

Backdoor attacks often include a subtask for which the adversary seeks high performance. The first lines of defenses against such attacks are implemented at the aggregator and assume that the updates for backdoor tasks would be outside the natural spread of benign updates. Two strategies that handle this perspective include *norm clipping* and *weak differential privacy*. For norm clipping, the central aggregator inspects the difference between the broadcasted global model and the received updates from the selected parties and clips the updates that exceed a pre-specified norm threshold [39]. On the other side, weak differential privacy approaches, as in [39, 43], add Gaussian noise with a small standard deviation to the model updates prior to the aggregation. Weak backdoor attacks can be easily countered with such defenses as the addition of Gaussian noise can neutralize the backdoor update.

Defenses that function against evasion attacks can similarly be employed in a federated setting. Adversarial training, [22] in which the defender trains against adversarial examples, is one such popular defense, and however this is a challenging training task and its difficulty increases in federated learning settings. For example, Shah et al. [36] observed that the performance against adversarial examples was strongly influenced by the amount of local computation conducted by the party, and Zizzo et al. [50] noted that the proportion of adversarial examples to clean data in a batch has a significant impact. The work in [41] also shows that robustness against affine distribution shifts (which can occur between parties in federated learning) can offer protection against adversarial examples. Effectively conducting adversarial training in a federated context remains an open problem, not only due to the underlying optimization difficulties but also attackers can interfere with the training process and create brittle models with misleading performance metrics for a defender [50].

16.3.1 Defending Against Convergence Attacks

For convergence attacks, we need to protect against adversaries who aim to degrade model performance in an unrestricted manner. A common attacker model to defend against in this scenario is a *Byzantine* attacker. This corresponds to a strong adversary who can send arbitrary model updates. Typically, in these attacks, the malicious model updates differ significantly from those sent by the benign parties and aim to produce a completely useless machine learning model, i.e., the performance of the resulting model is very poor. This can be achieved by, for example, sending random model updates adding noise with a very large variance to all the model's parameters. Blanchard et al. [8] showed that a single Byzantine adversary is enough to completely compromise a federated learning model when using standard aggregation methods, such as federated averaging.

This can be easily shown: for a set of party updates $\{v_k\}_{k=1}^{N}$, if the attacker aims for the global model to have a specific set of parameters w, and they control the party $k = N$, then the update required can be exactly computed as

$$w = \frac{1}{N} \sum_{k=1}^{N-1} v_k + \frac{1}{N} v_N \qquad (16.1)$$

$$v_N = Nw - \sum_{k=1}^{N-1} v_k. \tag{16.2}$$

Even without the knowledge of benign party updates, an attacker can trivially compromise the system. An attacker controlled party can send arbitrarily large updates which, when averaged with the benign parties, will break the model.

Thus, for practical FL deployments, it is essential to include mechanisms to filter out malicious (or faulty) model updates that can compromise the overall system's performance. This vulnerability has fostered research on robust aggregation methods aiming to detect and mitigate different types of poisoning attacks, including Byzantine adversaries.

16.3.1.1 Krum

Krum is one of the first algorithms proposed to defend against convergence attacks in FL [8]. A naive defender could try and filter out attackers by computing a score based on the squared distance between update i and all other received updates, to then select the update with the lowest score. This mechanism will however only tolerate a single Byzantine party. As soon as two Byzantine parties collude, then one Byzantine party can propose an update which shifts the barycenter of the benign party updates toward the other Byzantine update.

Krum solves this problem by being more selective in computing distance measures. Given N party updates $\{v_k\}_{k=1}^{N}$, Krum selects the update u which has the lowest squared Euclidean distance with respect to its $N - F - 2$ neighbors, where F is the allowable number of malicious parties in the system. We can express this as

$$s(i) = \sum_{i \to j} ||v_i - v_j||^2, \tag{16.3}$$

where we only sum over the $N - F - 2$ parties with the lowest squared distance. Krum requires that the number of malicious workers satisfies $2F + 2 < N$. We can see an example of Krum acting on a 2D set of updates in Fig. 16.2 where we only sum over the $N - F - 2$ parties with the lowest squared distance.

Although Krum can be effective to mitigate some attacks, especially Byzantine adversaries, it has been shown that this defense is not effective to mitigate other type of attacks, like label flipping attacks [27], or can be brittle against adaptive attacks targeting Krum [39]. Apart from this, the use of Krum slows the convergence of the FL algorithm, requiring more training rounds to achieve a good level of performance [8]. On the other side, Krum requires to compute the Euclidean distance between the model updates sent by all the parties participating at each training round, which can be computationally very demanding for scenarios where the number of parties is large.

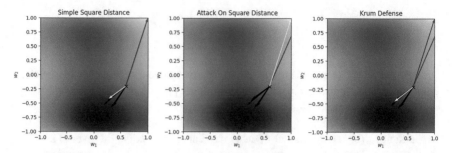

Fig. 16.2 Illustration of the Krum defense in a synthetic example with two parameters for the FL model. If we score each update based on its square distance to all other updates, then, by selecting the update which has the minimum score (white vector), we can successfully handle one Byzantine party. However, two malicious parties can collude, and now one update shifts the barycenter of all the supplied updates so that the original attacker update is selected. Note that this new Byzantine update (red line in the middle plot) is extremely large as it needs to counter the effect of all the benign parties. In fact, as we can see from the middle plot, it extends well beyond the w_1 and w_2 ranges we visualize. However, from the rightmost plot, if we apply Krum (which only considers the closest $N - F - 2$ parties), then many benign parties never have their score influenced by the malicious parties, and we can see that Krum reselects a benign party

The problems on the slow convergence of Krum can be partially mitigated with **Multi-Krum**, a straightforward variant of the algorithm where, instead of selecting a single update, we select the lowest scoring M updates so that the final update is given by

$$\frac{1}{M} \sum_i v_i^*, \tag{16.4}$$

where v^* is the set of the M lowest scoring party updates. This intuitively is interpolating between Krum and federated averaging, with M acting as a tunable parameter that a defender can set to prioritize convergence speed or robustness.

16.3.1.2 Median-Based Defenses

Methods based on the median form a broad family of defenses. If the number of malicious parties, F, is less than half of the total number of parties N, $F \leq \lceil \frac{N}{2} \rceil - 1$, then the median of a particular parameter must come from a benign party. Therefore, with this group of defenses, we are computing the median independently for every *dimension* of a parameter update. This is in contrast to Krum, which by using the squared Euclidean distance between two updates does not distinguish between the cases where updates differ significantly on only a few components, compared to when updates differ slightly on many components.

In its simplest form, we independently compute the median along every dimension and apply it to the global model as an update. However, there are a group of defenses that perform filtering around the median and then average the resulting parties. These are broadly referred to as *Trimmed Mean*-based defenses [24, 45, 49]. Concretely, the median for jth dimension in the N party updates $\{v_k\}_{k=1}^N$ is computed and a filtering operation is conducted. The remaining updates on each dimension are then averaged resulting in the final update vector. This is expressed as

$$w^{(j)} = \frac{1}{|U_j|} \sum_{i \in U_j} v_i^{(j)}, \tag{16.5}$$

where $|U_j|$ is the cardinality of the selected updates on dimension j. It is in the filtering step that the different Trimmed Mean algorithms differ. In particular,

- In [45], with $F \leq \lceil \frac{N}{2} \rceil - 1$, select the closest $N - F$ values to the median to average.
- In [24], with $N - 2F \geq 3$, only pick the nearest $N - 2F$ updates.
- Finally, for [49], with $F \leq \lceil \frac{N}{2} \rceil - 1$, remove the largest and smallest F updates on each dimension.

16.3.1.3 Bulyan

The Bulyan [24] defense seeks to combine the strengths of the previously discussed defenses. Krum has a shortcoming as it analyzes party updates based on the Euclidean distances of the local models across parties. Thus, adversaries can propose model updates which differ significantly on only a few parameters which will have little effect on the overall distance with respect to model updates from benign parties, but that can have a significant impact on the model performance. Bulyan thus computes a two-step process, in which Krum first produces a set of *likely* benign parties and then Trimmed Mean acts on this set derived from Krum. To be more precise,

- On the set of received party updates $V = \{v_i\}_{i=1}^N$, apply Krum which will select a single update.
- Add the update selected by Krum to a selection set S and remove the update from V.
- Apply the above two steps $N - 2F$ times. Thus, we are shrinking V and growing S by one update every iteration.
- Finally, apply Trimmed Mean on the resulting selection set S.

The Bulyan defense has robustness up to $N \geq 4F + 3$.

A different route is to directly limit the influence of the absolute value of any party's updates on the overall aggregation. One method for achieving this is to consider the sign of an update [4, 20]. In addition to limiting the influence of individual parties, it makes the communication between the parties and the aggregator much more efficient as only one-bit update is needed for every dimension in the update vector. Sign-based methods have been shown, under the assumption that updates are unimodal and symmetric about the mean, to be able to converge. Sign methods can also be viewed, as was done in [20], as a form of L_1 regularization. However, simple sign-based methods can be vulnerable to adaptive adversaries. Consider the algorithm in [4] in the following attack:

> *Example: Consider a system of 9 parties. The benign updates are modelled as coming from a Gaussian distribution $N(0.2, 0.15)$. We model 5 benign workers which generate updates $v_b = \{0.037, 0.4, 0.24, -0.026, 0.11\}$, which when signed have $v_b = \{1, 1, 1, -1, 1\}$. The attacker breaks the unimodal requirement and submits updates from 4 malicious parties of $v_m = \{-1, -1, -1, -1\}$ with a negative sign. Although the true update direction should be positive, the sum over all signed updates is now -1.*

16.3.1.4 Zeno

Should the aggregator have additional capabilities with access to the data itself, then further analysis can be conducted by examining the effect of the update on the model's performance on the aggregator data. This was examined in [48] which proposed the *Zeno* defense. Zeno produces a score s for every supplied gradient update v which indicates its reliability. The key idea here is to use the validation data to estimate the descent of the loss function value after a party update is applied. The score, s, is defined as

$$s = L(w, X) - L(w - \gamma v, X) - \rho||v||^2, \tag{16.6}$$

where w is the current parameter vector, γ is the learning rate at the aggregator, X represents samples of data drawn from the data distribution, and L is the loss function of the underlying machine learning task. The updates with the highest scoring s are averaged and used to update w. This can offer very strong defensive performance, and however the existence of an aggregator side dataset introduces additional requirements for the FL system.

16.3.2 Defenses Based on Parties' Temporal Consistency

In the previous section, we discussed defense aggregation methods that analyze a party's updates in each training round independently of their behavior during earlier rounds. This means a party's update in one training round does not affect its participation in the overall aggregation at later stages. Parties under the influence of attacks are likely to exhibit consistent malicious behavior across different training rounds and defense schemes can benefit from this knowledge by monitoring a party's temporal behavior during the training process. This insight can be employed for efficient and more accurate detection of malicious parties. Robust aggregation schemes based on these observations have been proposed in [27, 41], which either directly model the party's behavior during the training process or use a detection scheme to identify the parties sending malicious updates during the course of training. Furthermore, once the malicious parties are identified, they can be prevented from further participating in the training process which can result in reduced communication cost at the aggregator side.

16.3.2.1 Adaptive Model Averaging (AFA)

Muñoz-González et al. [27] propose an algorithm that relies on two components: (1) a robust aggregation rule to detect malicious model updates and (2) a probabilistic model using a hidden Markov model (HMM) that learns the quality of the model updates provided by each party during training and models their behavior.

The parameters of HMM are updated during each training round and implicitly incorporate the quality of update history for each party. They further use the HMM to detect a malicious party and then subsequently bar the malicious party from further participating in the training process. The proposed robust scheme aggregates the update at iteration $t + 1$ as

$$\sum_{k \in \mathcal{K}_t^g} \frac{p_{k_t} n_k}{P} v_k, \tag{16.7}$$

where p_{k_t} is the probability of party k providing a useful model update at iteration t and $P = \sum_{k \in \mathcal{K}_t^g} p_{k_t} n_k$, where n_k is the size of the dataset owned by party k. The set $\mathcal{K}_t^g \subset \mathcal{K}_t$ contains the parties that provide a good update according to the robust aggregation algorithm proposed in this chapter. For this, at each training round, AFA aims to detect malicious model updates iteratively using a distance-based algorithm (using cosine similarity or Euclidean distance). This detection algorithm is independent from the past contributions of the parties, to avoid situations where attackers are silent for some training rounds. At the start of the training process, updates from all parties are in the set of good updates. The aggregated model is estimated from Eq. 16.7 for given probabilities of the parties and the number of training data points provided. Then, the similarity of each party with respect to the aggregated model is calculated, and finally the mean, $\hat{\mu}$, and the median, $\bar{\mu}$, of all these similarity measures are calculated as shown in Fig. 16.3. Thereafter, each party's similarity score is compared to a threshold based on median score $\bar{\mu}$ and the standard deviation of the similarities. All model updates that are beyond that threshold (below or above depending on the position of the mean with respect

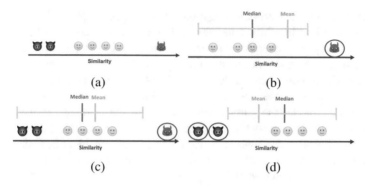

Fig. 16.3 (**a**) Considering that the benign parties are in majority during the training, they tend to send "similar" updates. (**b**) AFA calculates the median and mean of estimated similarity values of parties updates with the aggregated update. Parties whose similarity values are at a distance greater than a threshold based on these mean and median values of similarity can be identified easily in case of a single attacker. (**c–d**) AFA allows to detect different groups of attackers with different objectives by iteratively removing bad model updates at each training round

to the median) are considered as malicious. Then, the global model is recomputed and the procedure repeated until no model updates are considered as malicious. This iterative process allows to identify different types of attacks that can occur simultaneously, as described in Fig. 16.3. Finally, the probability p_{k_t} of each party is updated using the HMM accordingly for each party, depending on whether the model update at current training iteration was considered as malicious or not. If a party consistently sends malicious model updates, AFA includes a mechanism to block the user based on the beta posterior probability distribution used to model the parties' behavior.

Compared to Krum, AFA is more scalable, as it only needs to compute the similarity for each party model update with respect to the aggregated model, whereas Krum requires to compute the similarities among the model updates from all the parties. On the other side, compared to Krum and median-based aggregation rules, AFA enables the detection of the malicious parties and improves the communication efficiency by blocking parties that consistently send malicious model updates.

16.3.2.2 PCA

An alternative use of history was proposed in [41] to specifically combat against label flipping attacks. Given an update, for each output class, change (or delta) in the corresponding row in the final neural network layer is extracted and a history over many communication rounds is stored for each output class. From this history, the deltas are projected down to 2D via PCA, and the authors show that malicious and benign parties form well separated clusters.

16.3.2.3 FoolsGold

Along the same line, Fung et al. [15] devise a defense strategy based on comparison of historical updates between multiple parties. The algorithm works under the assumption that update from malicious parties tend to have similar and less diverse updates than those of honest parties. Cosine similarity is used to compare the histories of different participants and the party updates are rescaled to reflect the confidence before the subsequent aggregation.

16.3.2.4 LEGATO

Varma et al. [42] propose a fusion algorithm that can mitigate the effect of malicious gradients in various Byzantine attacks setting to train neural networks in FL. In particular, it analyzes the change of the norm of the gradient per layer and employs a dynamic gradient reweighing scheme based on layer-specific robustness computed based on the gradient analysis. Details about LEGATO can be found in Chap. 17.

16.3.3 Redundancy-Based Defenses

Thus far, the defenses we have discussed rely on improving the aggregation mechanism. However, an alternative line of proposed defensive methods function based on redundancy [13, 34, 38]. These defenses function by replicating data across several devices, so that each data partition is seen by at least R parties. If $R \geq 2S + 1$, where S is the number of malicious parties, then, by simple majority vote, the true update can be recovered. An example of this is illustrated in Fig. 16.4. The difficulty is that naively replicating data across R parties and having each party send R gradient updates corresponding to each replicated portion of data are computationally expensive. Thus, approaches have considered encoding all the gradients computed at each party. Then, the encoded representation is sent, and the individual gradients at the aggregator are then decoded [13]. Or, in [34], a hierarchical scheme was considered when combined with robust aggregation. More precisely, parties are assigned into groups and parties within the same group all perform the same redundant computation. The results from different groups are then hierarchically combined into a final model.

In general, redundancy-based defenses can be extremely strong and come with rigorous guarantees of the robustness offered. However, they have several significant drawbacks for application in federated (as opposed to *distributed*) learning. First, there is an inescapable communication overhead as the data will need to be replicated across devices. Second, there are privacy concerns with sharing data in such a manner. Data could potentially be anonymized prior to transmission (either by employing differential privacy or by other privacy mechanisms), and however, the risk might still be higher than not sharing data altogether.

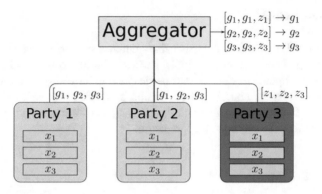

Fig. 16.4 Example of a simple redundancy-based defense. The two benign parties 1 and 2 each compute gradients g_{1-3} on data points x_{1-3}. Party 3 supplies arbitrary updates z_{1-3}. By a majority vote, the correct gradients g_{1-3} are used

16.4 Attacks

With a broad understanding of different threat models and defense strategies, we are now in a position for closer examination of specific attacks. The ability to supply arbitrary updates during model training allows an attacker to pursue a wide range of goals for data and model poisoning. Here, we categorize the types of attacks into three broad categories. First, convergence attacks which seek to indiscriminately degrade the global model performance on the underlying task. Second, in targeted attacks an adversary aims to produce errors for specific target data points or to introduce backdoors. For instance, a backdoor attacker can introduce a *key* (or trigger) into the data which will cause a machine learning model to always output an attacker chosen class when presented with the key, or alternatively backdoor task might consist in targeted misclassification for a subset of samples. Compared to targeted poisoning attacks, backdoors do not compromise the normal operation of the system, i.e., the performance of the resulting model is not affected for regular examples, and it only produces "unexpected" outputs for inputs that contain the key. Finally, we briefly discuss other attack strategies from centralized settings which naturally extend to federated setups.

Many of these attacks are specifically designed to counter certain defensive strategies. For scenarios where a defender is not employing any defense, it can be trivial to subvert a model undergoing federated learning [8]. An important dimension to consider for attack strategies is the amount of system compromise required in order to achieve the attack objective. Backdoor attacks, for instance, often require significantly lower compromise, with successful attacks needing as little as one malicious party. For cross-device setups, the frequency of attacks also affects their success rates. An attacker might control a fixed level of compromise for every selected quorum or might control a number of devices, a portion of which is selected every round of federated learning.

An alternative attack is for an adversary who crafts samples which expose the vulnerabilities of a deployed model at run-time. Machine learning models are known to be vulnerable to such *adversarial examples*. They represent indistinguishably perturbed inputs from a human standpoint that are misclassified with high confidence by a trained machine learning model [7, 16, 40]. Such attack vectors can be computed for both white-box and black-box scenarios and are even known to transfer across different models. Communication channels for model update sharing and broadcasting in federated learning could potentially expose additional surfaces for some of these white-box attacks, especially for insiders.

16.4.1 Convergence Attacks

For convergence attacks, an adversary seeks maximum damage to the model performance within the limits imposed by defensive aggregation schemes. According to the taxonomy in Sect. 16.2, these correspond to indiscriminate causative attacks, where the attackers can manipulate the parameters of the aggregated model by providing malicious local model updates, aiming to compromise the overall model's performance.

In this line, the simplest attacks that could be performed are Byzantine attacks, as the one proposed in [8], where malicious parties send model updates with very large values, which is enough to compromise vanilla aggregation methods, such as federated averaging. Another effective way to accomplish a convergence attack is to aim for the aggregation schemes to select an update with the sign that is opposite to the true update direction. This line of research has been examined in [14, 47]. If the secure aggregation scheme being targeted is using a median-based defense, the developed attacks in [14, 47] are similar. The strategy exploits benign parties that may supply updates with the opposite sign to the mean update of the benign parties. The attacker can force their selection either by supplying updates that are larger than any benign client, thereby trying to force the selection of a positive update, or by supplying malicious updates that are less than any of the benign parties, thus trying to select a negative direction.

Example: If the benign updates $V = \{-0.2, 0.2, 0.5\}$, then the true update mean $\mu = 0.167$. A simple median-based aggregation on this set would yield 0.2. However, if the attacker supplies updates smaller than $min(V)$, the selection of a negative gradient can be obtained. The attacker submits $V_{attacker} = \{-1, -1\}$; now with the combined update set being $V = \{-1, -1, -0.2, 0.2, 0.5\}$, the median selects -0.2.

For Krum, the two methodologies [14, 47] differ more substantially. Both methods try to deviate the chosen update vector toward the opposite sign of the true update mean μ. In [47], the formulation is similar to an attack on median-based defenses with the malicious update being

$$v_{\text{attacker}} = -\epsilon\mu \tag{16.8}$$

while in [14] the malicious update was formed via

$$v_{\text{attacker}} = w - \lambda s \tag{16.9}$$

where w is the global model, s is the sign of the direction the parameters should change by with only benign parties, and λ is our perturbation parameter.

In both cases, we would like to maximize the deviation ϵ or λ while still being selected by Krum. With [47], the deviation was manually set to a reasonably small number as to have a high selection chance. Conversely, in [14], the maximum value of λ was determined by running a binary search.

Example: With the benign updates of $V = \{0.0, 0.1, 0.25, 0.35, 0.5, 0.65\}$, we want a negative update that is selected by Krum. If the attacker controls 3 malicious parties, we search over $-\lambda$ in a simple grid search and see that $\lambda = 0.21$ is selected. Therefore, the attacker supplies $V_{\text{attacker}} = \{-0.21, -0.21, -0.21\}$. The updates as seen by the aggregator are $\{-0.21, -0.21, -0.21, 0.0, 0.1, 0.25, 0.35, 0.5, 0.65\}$ with the minimum Krum score belonging to a party that supplied -0.21.

Neither of those attacks considered Bulyan as a defensive method, which was instead tackled in [3]. The key observation that the authors exploited in their attack is that malicious updates can still cause significant harm by hiding in the natural spread of benign updates. The benign updates are modelled following a normal distribution with mean μ and variance σ. Then, the attacker submits updates of the form $\mu + k\sigma$. By setting k to the appropriate value, we can ensure that there are benign party updates that lie further away from the mean then the malicious updates. These parties support the selection of the malicious updates, which are selected with a high degree of probability by a robust aggregation algorithm. We can see an example for this attack in a 2D case in Fig. 16.5.

16.4.2 Targeted Model Poisoning

Having examined convergence-based attacks, we now turn our attention to model poisoning attacks which aim to be more specific in their objective. In particular, this

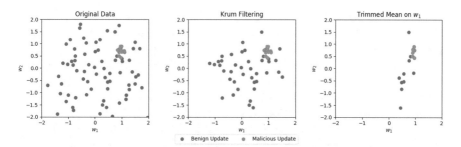

Fig. 16.5 Example illustration for the attack in [3] targeting Bulyan. Going from the plots left to right: we initially start with the distribution of two model parameters, w_1 and w_2, submitted by the benign parties (blue dots). The attackers submit model updates (orange dots) offset to the true mean, but still within the update variance. Then, in the middle plot, we apply the initial Krum filtering and we can see many malicious updates are still present. In the final plot, we apply Trimmed Mean on w_1 and a large amount of adversarial updates are included in the final averaging for w_1. An equivalent operation is also then done for w_2

involves attacks-based training a model on data that has had its feature manipulated through the insertion of backdoors or targeted label flipping attacks where particular data points are mislabelled. By performing one of these attacks, an adversary can force a model to learn attacker chosen correlations and therefore misclassify certain points at test time. We should note that this type of misclassification differs from the case of *adversarial examples*, as this results from explicit manipulations during the training process.

Both backdoor or label flipping attacks involve training models on manipulated data. One manner in which this can be achieved is if the adversary is able to tamper with the data collection process of the benign parties in a federated learning system. Thus, if manipulated data can be given to the benign parties, then the model learned through the federated learning process can be vulnerable. However, the more commonly modelled attack vector is that the adversary joins a federated learning system controlling one or more parties. The adversary then sends corrupted updates to the aggregator. This can be considered a stronger adversarial model compared to just poisoning the data that benign parties have access to, as the adversary has control over the update vector and its participation rate, and can even collude with other malicious parties to improve the attack success rate [5].

Should the adversary pursue their attack via label flipping, then the features of particular training data samples are left unchanged, but their associated labels are altered. An example of this in practice is changing the labels of all the green cars in a dataset to an attacker chosen class [1]. The model will then learn to associate green cars with the attacker class, rather than their original label. Label flipping attacks have been explored in a wide range of works [6, 15, 21, 30, 41, 44]. For label flipping attacks, it has been shown that attacking the model in the later part of training near convergence is more successfully compared to attacking the model during initial stage of training [41].

On the other hand, an adversary can mount backdoors by manipulating features, like certain pixels in the case of images, and also changing the label of the data point. Thus, a model will learn to associate the backdoor with a particular label and ignore the rest of the features in a data point if a backdoor is present. Although this is the most common attack method by which backdoors are inserted, clean label backdoor attacks in which the label is not altered are also possible [35].

Other nuances that can affect a backdoor attack depend also on the total proportion of samples in the training set that the attacker controls and wishes to affect. To continue our running example with misclassification of green cars to an attacker class, the attack will be easier if the attacker is able to control *all* the green cars in the dataset, rather than just a portion of them [39].

When attempting backdoor attacks, the challenge for an adversary depends on both:

1. The complexity of the target subtask, as an adversary might require varying numbers of malicious parties and potentially high participation frequency if the subtask has a high degree of complexity.
2. And the robust aggregation methods and anomaly detectors at the aggregator end, which need to be accounted for wen fabricating the malicious updates so as to circumvent such defenses.

In the simplest case where the aggregator uses federated averaging as the update rule, if the attacker sends their updates after training their local model on the backdoor task then as the number of parties that the attacker controls can be small in comparison to the total number of parties participating in an FL round then the backdoor updates can be cancelled out. To make these attacks effective, the work of [1] builds a strategy based on the observation that near convergence the updates sent by honest parties tend to be similar. An adversary can take advantage of this and rescale their update to ensure that the backdoor survives the eventual aggregation, thereby successfully replacing the global model with the malicious one. Specifically, with a global model w_t on round t, the attacker replaces it with their corrupted model w_{corrupt} by submitting v computed via

$$v_{\text{attacker}} \leftarrow \gamma \left(w_{\text{corrupt}} - w_t \right) + w_t \tag{16.10}$$

where γ is a scaling parameter. Independently, the work of [5] also arrives a similar rescaling strategy (explicit boosting) that accounts for the scaling at aggregation.

Example: Near convergence, a global model with parameter value of $\{4.03\}$, might receive update from honest parties as $0.08, 0.083, 0.09$. The malicious model might seek to replace the parameter value with $\{3.98\}$. Assuming that the aggregator will combine the updates with a learning rate of 0.015, the adversary in this case supplies update as $\frac{1}{0.015}(3.98 - 4.03) + 4.03 = 0.696$ as opposed to -0.05.

In order to supply updates that fall within the natural spread of updates received from non-malicious parties, an adversary can include additional constraints. For instance, Bhagoji et al. [5] proposes to include additional loss terms corresponding to benign training samples and regularizes the current update to be as close to the combined update from benign parties in the previous communication round. Similarly, Wang et al. [43] considers adversaries that employ projected gradient descent where the intermediate parameter states are periodically projected to an ϵ-ball around the previously received global update. Alternatively, rather than having the same backdoor key on all the data, in [46], the key is split between each malicious party according to a decomposition rule. Thus, each malicious party only inserts a part of the backdoor key into their data. The sum of all the key fragments is equal to the full backdoor key. Testing on LOAN and three image datasets shows that this approach yields better attack success rates as well as being more stealthy against FoolsGold [15] and RFA [33].

16.5   Conclusion

In this chapter, we have discussed the security of FL systems. In a similar fashion to standard, non-distributed ML systems, FL is vulnerable to attacks both at training and test time. For example, FL algorithms can be completely compromised during training just by the presence of one single malicious participant when using standard aggregation methods. Thus, the analysis of robust methods for FL is critical for the use of this technology in most practical settings.

In this chapter, we provide a comprehensive overview of different attack strategies and approaches to defend and mitigate them. However, some of the vulnerabilities of FL still need to be better understood and defending against some type of attacks remains an open research challenge. In this sense, it is also necessary to characterize and analyze further different trade-offs present in the design of FL systems, for example, a trade-off among performance, robustness and data heterogeneity or among performance, robustness, and privacy, just to cite some.

References

1. Bagdasaryan E, Veit A, Hua Y, Estrin D, Shmatikov V (2020) How to backdoor federated learning. In: Chiappa S, Calandra R (eds) The 23rd international conference on artificial intelligence and statistics, AISTATS 2020, 26–28 August 2020, Online [Palermo, Sicily, Italy], Proceedings of machine learning research. PMLR, vol 108, pp 2938–2948
2. Barreno M, Nelson B, Joseph AD, Tygar JD (2010) The security of machine learning. Mach Learn 81(2):121–148
3. Baruch G, Baruch M, Goldberg Y (2019) A little is enough: Circumventing defenses for distributed learning. In: Wallach H, Larochelle H, Beygelzimer A, d'Alché-Buc F, Fox E, Garnett R (eds) Advances in neural information processing systems 32, pp 8635–8645. Curran

Associates. http://papers.nips.cc/paper/9069-a-little-is-enough-circumventing-defenses-for-distributed-learning.pdf

4. Bernstein J, Zhao J, Azizzadenesheli K, Anandkumar A (2018) signSGD with majority vote is communication efficient and fault tolerant. Preprint. arXiv:1810.05291

5. Bhagoji AN, Chakraborty S, Mittal P, Calo S (2019) Analyzing federated learning through an adversarial lens. In: International conference on machine learning. PMLR, pp 634–643

6. Biggio B, Nelson B, Laskov P (2012) Poisoning attacks against support vector machines. In: Proceedings of the 29th international conference on machine learning, ICML 2012, Edinburgh, Scotland, June 26–July 1, 2012. icml.cc/Omnipress. http://icml.cc/2012/papers/880.pdf

7. Biggio B, Corona I, Maiorca D, Nelson B, Srndic N, Laskov P, Giacinto G, Roli F (2013) Evasion attacks against machine learning at test time. In: Blockeel H, Kersting K, Nijssen S, Zelezný F (eds) Machine learning and knowledge discovery in databases - European conference, ECML PKDD 2013, Prague, September 23–27, 2013, Proceedings, Part III, Lecture notes in computer science, vol 8190. Springer, pp 387–402

8. Blanchard P, Guerraoui R, Stainer J et al (2017) Machine learning with adversaries: Byzantine tolerant gradient descent. In: Advances in neural information processing systems, pp 119–129

9. Bonawitz K, Ivanov V, Kreuter B, Marcedone A, McMahan HB, Patel S, Ramage D, Segal A, Seth K (2017) Practical secure aggregation for privacy-preserving machine learning. In: Thuraisingham BM, Evans D, Malkin T, Xu D (eds) Proceedings of the 2017 ACM SIGSAC conference on computer and communications security, CCS 2017, Dallas, TX, October 30–November 03, 2017. ACM, pp 1175–1191

10. Buehler H, Gonon L, Teichmann J, Wood B (2019) Deep hedging. Quant Financ 19(8):1271–1291

11. Castro RL, Muñoz-González L, Pendlebury F, Rodosek GD, Pierazzi F, Cavallaro L (2021) Universal adversarial perturbations for malware. CoRR abs/2102.06747. https://arxiv.org/abs/2102.06747

12. Chen X, Liu C, Li B, Lu K, Song D (2017) Targeted backdoor attacks on deep learning systems using data poisoning. Preprint. arXiv:1712.05526

13. Chen L, Wang H, Charles Z, Papailiopoulos D (2018) Draco: Byzantine-resilient distributed training via redundant gradients. In: International conference on machine learning. PMLR, pp 903–912

14. Fang M, Cao X, Jia J, Gong N (2020) Local model poisoning attacks to byzantine-robust federated learning. In: 29th {USENIX} security symposium ({USENIX} Security 20), pp 1605–1622

15. Fung C, Yoon CJ, Beschastnikh I (2018) Mitigating sybils in federated learning poisoning. Preprint. arXiv:1808.04866

16. Goodfellow IJ, Shlens J, Szegedy C (2015) Explaining and harnessing adversarial examples. In: International conference on learning representations

17. Hitaj B, Ateniese G, Pérez-Cruz F (2017) Deep models under the GAN: information leakage from collaborative deep learning. In: Thuraisingham BM, Evans D, Malkin T, Xu D (eds) Proceedings of the 2017 ACM SIGSAC conference on computer and communications security, CCS 2017, Dallas, TX, October 30–November 03, 2017. ACM, pp 603–618

18. Huang L, Joseph AD, Nelson B, Rubinstein BIP, Tygar JD (2011) Adversarial machine learning. In: Chen Y, Cárdenas AA, Greenstadt R, Rubinstein BIP (eds) Proceedings of the 4th ACM workshop on security and artificial intelligence, AISec 2011, Chicago, IL, October 21, 2011. ACM, pp 43–58

19. Hussain F, Hussain R, Hassan S.A, Hossain E (2020) Machine learning in IoT security: Current solutions and future challenges. IEEE Commun Surv Tutorials 22(3):1686–1721

20. Li L, Xu W, Chen T, Giannakis GB, Ling Q (2019) RSA: Byzantine-robust stochastic aggregation methods for distributed learning from heterogeneous datasets. In: Proceedings of the AAAI conference on artificial intelligence, vol 33, pp 1544–1551

21. Liu Y, Ma S, Aafer Y, Lee W, Zhai J, Wang W, Zhang X (2018) Trojaning attack on neural networks. In: 25th Annual network and distributed system security symposium, NDSS 2018, San Diego, California, February 18–21, 2018. The Internet Society

22. Madry A, Makelov A, Schmidt L, Tsipras D, Vladu A (2018) Towards deep learning models resistant to adversarial attacks. In: International conference on learning representations. https://openreview.net/forum?id=rJzIBfZAb
23. Melis L, Song C, Cristofaro ED, Shmatikov V (2019) Exploiting unintended feature leakage in collaborative learning. In: 2019 IEEE symposium on security and privacy, SP 2019, San Francisco, CA, May 19–23, 2019. IEEE, pp 691–706
24. Mhamdi EME, Guerraoui R, Rouault S (2018) The hidden vulnerability of distributed learning in Byzantium. Preprint. arXiv:1802.07927
25. Muñoz-González L, Lupu EC (2019) The security of machine learning systems. In: AI in cybersecurity. Springer, pp 47–79
26. Muñoz-González L, Biggio B, Demontis A, Paudice A, Wongrassamee V, Lupu EC, Roli F (2017) Towards poisoning of deep learning algorithms with back-gradient optimization. In: Thuraisingham BM, Biggio B, Freeman DM, Miller B, Sinha A (eds) Proceedings of the 10th ACM workshop on artificial intelligence and security, AISec@CCS 2017, Dallas, TX, November 3, 2017. ACM, pp 27–38
27. Muñoz-González L, Co KT, Lupu EC (2019) Byzantine-robust federated machine learning through adaptive model averaging. Preprint. arXiv:1909.05125
28. Nasr M, Shokri R, Houmansadr A (2019) Comprehensive privacy analysis of deep learning: Passive and active white-box inference attacks against centralized and federated learning. In: 2019 IEEE symposium on security and privacy, SP 2019, San Francisco, CA, May 19–23, 2019. IEEE, pp 739–753
29. Papernot N, McDaniel PD, Goodfellow IJ (2016) Transferability in machine learning: from phenomena to black-box attacks using adversarial samples. CoRR abs/1605.07277. http://arxiv.org/abs/1605.07277
30. Paudice A, Muñoz-González L, György A, Lupu EC (2018) Detection of adversarial training examples in poisoning attacks through anomaly detection. CoRR abs/1802.03041. http://arxiv.org/abs/1802.03041
31. Paudice A, Muñoz-González L, Lupu EC (2018) Label sanitization against label flipping poisoning attacks. In: Alzate C, Monreale A, Assem H, Bifet A, Buda TS, Caglayan B, Drury B, García-Martín E, Gavaldà R, Kramer S, Lavesson N, Madden M, Molloy I, Nicolae M, Sinn M (eds) ECML PKDD 2018 Workshops - Nemesis 2018, UrbReas 2018, SoGood 2018, IWAISe 2018, and Green Data Mining 2018, Dublin, September 10–14, 2018, Proceedings, Lecture Notes in Computer Science, vol 11329. Springer, pp 5–15
32. Pierazzi, F, Pendlebury, F, Cortellazzi, J, Cavallaro, L (2020) Intriguing properties of adversarial ML attacks in the problem space. In: 2020 IEEE symposium on security and privacy, SP 2020, San Francisco, CA, May 18–21, 2020. IEEE, pp 1332–1349
33. Pillutla VK, Kakade SM, Harchaoui Z (2019) Robust aggregation for federated learning. CoRR abs/1912.13445. http://arxiv.org/abs/1912.13445
34. Rajput S, Wang H, Charles Z, Papailiopoulos D (2019) Detox: A redundancy-based framework for faster and more robust gradient aggregation. Preprint. arXiv:1907.12205
35. Shafahi A, Huang WR, Najibi M, Suciu O, Studer C, Dumitras T, Goldstein T (2018) Poison frogs! targeted clean-label poisoning attacks on neural networks. Preprint. arXiv:1804.00792
36. Shah D, Dube P, Chakraborty S, Verma A (2021) Adversarial training in communication constrained federated learning. Preprint. arXiv:2103.01319
37. Shen L, Margolies LR, Rothstein JH, Fluder E, McBride R, Sieh W (2019) Deep learning to improve breast cancer detection on screening mammography. Sci Rep 9(1):1–12
38. Sohn Jy, Han DJ, Choi B, Moon J (2019) Election coding for distributed learning: Protecting signSGD against byzantine attacks. Preprint. arXiv:1910.06093
39. Sun Z, Kairouz P, Suresh AT, McMahan HB (2019) Can you really backdoor federated learning? Preprint. arXiv:1911.07963
40. Szegedy C, Zaremba W, Sutskever I, Bruna J, Erhan D, Goodfellow IJ, Fergus R (2014) Intriguing properties of neural networks. In: Bengio Y, LeCun Y (eds) 2nd International conference on learning representations, ICLR 2014, Banff, AB, April 14–16, 2014, Conference Track Proceedings. http://arxiv.org/abs/1312.6199

41. Tolpegin V, Truex S, Gursoy ME, Liu L (2020) Data poisoning attacks against federated learning systems. In: European symposium on research in computer security. Springer, pp 480–501
42. Varma K, Zhou Y, Baracaldo N, Anwar A (2021) Legato: A layerwise gradient aggregation algorithm for mitigating byzantine attacks in federated learning. In: 2021 IEEE 14th international conference on cloud computing (CLOUD)
43. Wang H, Sreenivasan K, Rajput S, Vishwakarma H, Agarwal S, Sohn Jy, Lee K, Papailiopoulos D (2020) Attack of the tails: Yes, you really can backdoor federated learning. Preprint. arXiv:2007.05084
44. Xiao H, Xiao H, Eckert C (2012) Adversarial label flips attack on support vector machines. In: Raedt LD, Bessiere C, Dubois D, Doherty P, Frasconi P, Heintz F, Lucas PJF (eds) ECAI 2012 - 20th European conference on artificial intelligence. Including prestigious applications of artificial intelligence (PAIS-2012) System demonstrations track, Montpellier, August 27–31, 2012, Frontiers in artificial intelligence and applications, vol 242. IOS Press, pp 870–875
45. Xie C, Koyejo O, Gupta I (2018) Generalized byzantine-tolerant SGD. Preprint. arXiv:1802.10116
46. Xie C, Huang K, Chen PY, Li B (2019) Dba: Distributed backdoor attacks against federated learning. In: International conference on learning representations
47. Xie C, Koyejo O, Gupta I (2019) Fall of empires: Breaking byzantine-tolerant SGD by inner product manipulation. In: Globerson A, Silva R (eds) Proceedings of the thirty-fifth conference on uncertainty in artificial intelligence, UAI 2019, Tel Aviv, Israel, July 22–25, 2019. AUAI Press, p 83. http://auai.org/uai2019/proceedings/papers/83.pdf
48. Xie C, Koyejo S, Gupta I (2019) Zeno: Distributed stochastic gradient descent with suspicion-based fault-tolerance. In: International conference on machine learning. PMLR, pp 6893–6901
49. Yin D, Chen Y, Ramchandran K, Bartlett P (2018) Byzantine-robust distributed learning: Towards optimal statistical rates. Preprint. arXiv:1803.01498
50. Zizzo G, Rawat A, Sinn M, Buesser B (2020) Fat: Federated adversarial training. Preprint. arXiv:2012.01791

Chapter 17
Dealing with Byzantine Threats to Neural Networks

Yi Zhou, Nathalie Baracaldo, Ali Anwar, and Kamala Varma

Abstract Messages exchanged between the aggregator and the parties in a federated learning system can be corrupted due to machine glitches or malicious intents. This is known as a Byzantine failure or Byzantine attack. As such, in many federated learning settings, replies sent by participants may not be trusted fully. A set of competitors may work collaboratively to detect fraud via federated learning where each party provides local gradients that an aggregator uses to update a global model. This global model can be corrupted when one or more parties send malicious gradients. This necessitates the use of robust methods for aggregating gradients that mitigate the adverse effects of Byzantine replies. In this chapter, we focus on mitigating the Byzantine effect when training neural networks in a federated learning setting with a focus on the effect of having parties with highly disparate training datasets. Disparate training datasets or non-IID datasets may take the form of parties with imbalanced proportions of the training labels or different ranges of feature values. We introduce several state-of-the-art robust gradient aggregation algorithms and examine their performances as defenses against various attack settings. We empirically show the limitations of some existing robust aggregation algorithms, especially under certain Byzantine attacks and when parties admit non-IID data distributions. Moreover, we show that LayerwisE Gradient AggregaTiOn (LEGATO) is more computationally efficient than many existing robust aggregation algorithms and more generally robust across a variety of attack settings.

Y. Zhou (✉) · N. Baracaldo · A. Anwar
IBM Research – Almaden, San Jose, CA, USA
e-mail: yi.zhou@ibm.com; baracald@us.ibm.com; ali.anwar2@ibm.com

K. Varma
University of Maryland, College Park, MD, USA
e-mail: kvarma@umd.edu

17.1 Background and Motivation

In this section, we aim to provide more background information about non-IID party distributions and their associated challenges one may face when dealing with Byzantine threats.

As discussed in the introduction chapter, in a federated learning system, parties will collect and maintain their own training datasets and no training data samples will leave their owners. Parties therefore collect their training data samples from different data sources. For example, in a federated learning task involving collaboration among multiple hospitals, the local training data distribution may be affected by the speciality of the hospital, which might lead to heterogeneous local distribution. For instance, pediatric medical centers have more younger patients and general hospitals have more adult patients. Another example will be considering a federated learning system with multiple cellphones as parties trying to train an image recognition model using the photos stored with their phones to identify animals. Cellphone owners having a dog as a pet would have more dog pictures in their local training datasets, whereas cat owners would have more cat pictures.

As one of the key features for federated learning, heterogeneous local data distribution has raised challenges even for basic federated learning settings. It significantly affects the global model's performance and hence has motivated a new line of research addressing this issue, see, e.g., [6, 10, 18], and the references therein. However, the situation is even more complicated in a real-world scenario since there can be Byzantine threats presented in a federated learning system.

17.1.1 Byzantine Threats

Our goal in this section is to formally define Byzantine threats potentially present in a federated learning system. In particular, we discuss Byzantine failures and Byzantine attacks, which both aim at diminishing the final global model's performance.

We formally define *Byzantine failures* [8] as scenarios in a federated learning system when one or more parties have malfunctions in their computing devices or have encountered communication errors and consequently send corrupted information to the aggregator, which may compromise the quality of the global model. As shown in Fig. 17.1, consider a federated learning system with n parties where at t-round the reply from party P_2 received by the aggregator is corrupted as colored in red. Such a Byzantine failure may occur unintentionally or may be executed in the form of an attack, which we refer to as a *Byzantine attack*. Malicious entities intentionally corrupt an FL system by providing strategically dishonest replies. This powerful attack can be performed by only a single party like P_2 in Fig. 17.1 and can adversely affect indefinite numbers of models that are used in sensitive areas such as healthcare, banking, personal devices, and so on. We refer the party unintentionally sending inaccurate replies or intentionally performing *Byzantine*

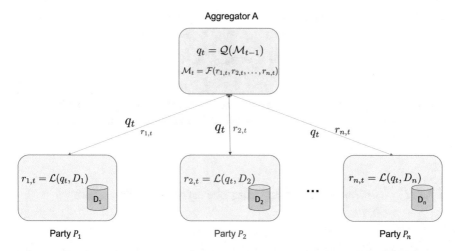

Fig. 17.1 A federated learning system with n parties where the second party's reply encounters Byzantine failure

attacks as a *Byzantine party*. The goal of Byzantine attacks is to produce a final global model with poor performance, usually referring to bad accuracy and F1 scores, and hence breaking the FL collaboration. A class of robust aggregation algorithms (a.k.a. robust fusion algorithms) have been introduced to defend against Byzantine attacks, in particular, mitigating the bad effects of those attacks and ensuring a final global model with good performance.

Now, we are ready to formally introduce two well-known Byzantine attacks that can happen in a federated learning system.

Gaussian Attack This type of Byzantine attack, first introduced in [20], can be performed by a single party in a federated learning system and does not need any collaboration among Byzantine parties. Parties performing this attack will randomly sample their replies from a Gaussian distribution, $\mathcal{N}(\mu, \sigma^2 * I)$, and its probability density function is $\frac{1}{\sigma} \frac{\exp\{-(x-\mu)^2/2\sigma^2\}}{2\pi}$, where μ and σ are the mean and the standard deviation respectively, regardless of their local training datasets.

Fall of Empires Attack Proposed in [22], it is designed to break robust aggregation algorithms like Krum [1] and coordinate-wise median (CWM) [23]. It requires a minimum number of Byzantine parties, which depends on the robust algorithms, to collaboratively construct the attack. Moreover, it belongs to the class of *perfect knowledge attacks* defined in Chap. 16, Security and Robustness in Federated Machine Learning, which assumes that the Byzantine parties know the replies sent by the honest parties. Let v_i, $i = 1, ..., m$, be the replies sent by m honest parties. The malicious replies, u_j, sent by the Byzantine parties who perform the Fall of

Empires attack can be formulated as

$$u_1 = u_2 = \ldots = u_{n-m} = -\frac{\epsilon}{m}\sum_{i=1}^{m} v_i, \qquad (17.1)$$

where ϵ is a real non-negative factor that depends on the lower and upper bounds of the distances between each reply. A detailed definition of this attack can be found in Chap. 16.

17.1.2 Challenges of Mitigating the Effects of Byzantine Threats

The aforementioned Byzantine attacks are very powerful. We can see from Fig. 17.2 that Gaussian attacks with larger σ values create more damage to the global model's performance, i.e., resulting in a global model with poorer test accuracy.

Krum and CWM are proven to successfully identify the malicious Byzantine parties in the setting where parties have independent identically distributed (IID) local datasets (see Fig. 17.3). Besides the IID scenario, we also examine their performances as defenses against Byzantine attacks when parties have heteroge-

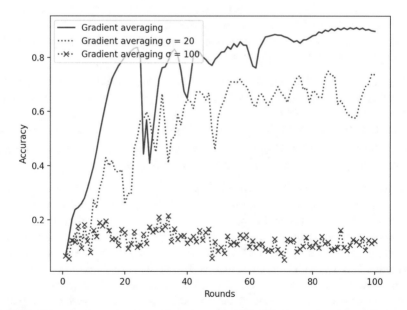

Fig. 17.2 This experiment involves 25 parties with 1000 IID data points from MNIST. We compare the performance of gradient averaging where 4 Byzantine parties execute a Gaussian attack with $\mu = 0$ and varying σ. Here, we use a mini-batch size of 50 and a learning rate of 0.05

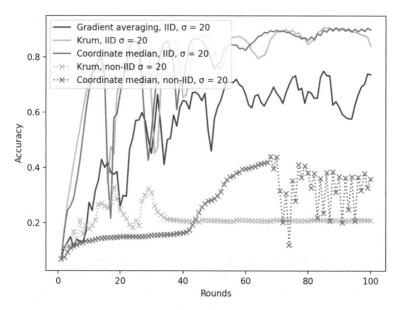

Fig. 17.3 This experiment involves 25 parties with 1000 data points randomly sampled from MNIST for an IID setting and 1000 points containing only 1 label per party for a non-IID setting, respectively. We compare the performance of gradient averaging, Krum, and coordinate-wise median where 4 Byzantine parties execute a Gaussian attack with $\mu = 0$ and $\sigma = 20$. Here we use a mini-batch size of 50 and a learning rate of 0.05

neous data distributions, which is a characteristic of FL [12]. From Fig. 17.3, we can see that they failed significantly when parties have heterogeneous local datasets. It can also be shown that even without any Byzantine attacks these robust aggregation algorithms cannot guarantee a global model with good performance under heterogeneous party data distributions. We will cover the details of these two robust algorithms in Sects. 17.2.3 and 17.2.4.

So far, we have shown empirically that several well-known fusion algorithms cannot defend against Byzantine attacks, especially in the non-IID setting. We want to conclude this section with opportunities and challenges we may have in this area.

Challenge One Can a robust aggregation algorithm smartly detect whether a party sending a "different" reply is a malicious party or a benign party with an unrepresentative local data distribution?

Challenge Two Can a robust aggregation algorithm train a global model under the non-IID local data distribution setting with reasonable performance?

Challenge Three Can a robust aggregation algorithm utilize all the information collected from parties to diagnose the parties' behaviors?

Different layers of a neural networks have different functionality and may behave different toward different input data samples. We may use this fact when dealing with the training of neural networks in a federated learning setting under Byzantine threats. We will propose a promising approach to largely mitigate the Byzantine threats, especially in the non-IID setting, for training neural networks in a federated learning system. But, we will first discuss gradient averaging, Krum, and CWM from a theoretical perspective and point out the drawbacks of these algorithms in our next section.

17.2 Gradient-Based Robustness

In this section, we briefly introduce a class of popular algorithms widely used for training neural networks in federated learning systems. In particular, parties will share gradients as model updates computed on their local datasets and the aggregator will use the collected gradients to updates the global model.

We first start with the basic gradient averaging algorithm in Sect. 17.2.1, and then, after introducing the threat model, we move on to gradient-based robust algorithms. Existing approaches to perform robust gradient aggregation can be classified into two different directions: (1) robust statistics and (2) anomaly detection. Examples of algorithms using robust statistics include CWM, trimmed mean [23], and their variants [2, 3, 7, 15, 20]. Krum [1] and other algorithms, e.g., [5, 13, 14, 16, 19, 21], exploit certain metrics like the ℓ_2 distance of the collected gradients to perform anomaly detection. In this section, we will take a close look at CWM (Sect. 17.2.3) and Krum (Sect. 17.2.4) since they are the most popular state-of-the-art robust aggregation algorithms in the literature.

17.2.1 Gradient Averaging

Very similar to FedAvg proposed in [12], the gradient averaging algorithm is a "distributed" version of (mini-batch) stochastic gradient descent. Instead of sharing local model weights like in FedAvg, the gradient averaging algorithm requires parties to share gradients. In particular, the gradient averaging algorithm requires the aggregator to request the gradient information from all registered parties at each round, which they compute based on their local datasets and the global model's weights, and then a simple averaging aggregation is performed over the collected gradients. The global model's weights are then updated by a gradient descent step using the resulting aggregated gradient and a predefined learning rate. Algorithm 17.1 shows the basic algorithmic schema of the gradient averaging algorithm.

It should be pointed out that in Line 7 of Algorithm 17.1, there is a simple average of the collected gradients, and however, one can employ a more involved aggregation step, for example, performing a weighted average of collected gradients

---

Algorithm 17.1 The gradient averaging algorithm

Input: Initial weight vector w_0, maximum global round K, a learning rate policy $\{\eta_k\}$, training batch-size B, and (Optional) a target accuracy ϵ.

 1: **procedure:** Aggregator (w_0, K, η, ϵ)
 2:     **for** round $k = 1, \ldots, K$ **do**
 3:         **if** (Optional) current global model reaches the target accuracy ϵ **then**
 4:             **return** w_k
 5:         **else**
 6:             Query party $i \in \mathcal{P}$ with the current global model weights w_{k-1} for its current gradient g_k^i.
 7:             $w_k = w_{k-1} - \eta_k \frac{1}{|\mathcal{P}|} \sum_{i \in \mathcal{P}} g_k^i$
 8:         **end if**
 9:     **end for**
10:     **return** w_K
11: **end procedure**
12: **procedure:** Party (B, w)
13:     Initialize the local model with $w_0 = w$
14:     $g = \nabla \ell(w_0; B)$ {ℓ denotes the loss function.}
15:     **return** g
16: **end procedure**

---

where the weights depend on the parties' local training dataset size. This is the same weighing approach used by FedAvg in [12].

Recently, the researchers have developed more robust aggregation methods to fuse the collected information from parties in an FL system. A majority of them are motivated by potential Byzantine threats presented in the FL system. We will discuss several state-of-the-art robust gradient aggregation algorithms in the following sections.

17.2.2 Threat Model

We first describe the threat model we assume throughout the rest of the chapter. We assume that the aggregator is honest and follows a provided procedure to detect malicious or erroneous gradients received and aggregate the gradients to update the global model iteratively during the training process. Parties may be dishonest and may try to provide gradient updates designed to evade detection. We also assume that malicious parties may collude and can obtain access to gradients of other parties to perform those attacks. Under this threat model, we will assess all fusion algorithms discussed within this chapter.

17.2.3 Coordinate-Wise Median

Coordinate-wise median [23] (cf. Algorithm 17.2) is very similar to gradient averaging except for Line 6 where instead of computing the average of the collected gradients, it computes the coordinate-wise median of all collected gradients. From [23], the definition of coordinate-wise median is as the following:

> For vectors $x^i \in \mathbb{R}^d, i \in [m]$, the coordinate-wise median $g := \mathrm{med}\{x^i : i \in [m]\}$ is a vector with its k-th coordinate being $g_k = \mathrm{med}\{x_k^i : i \in [m]\}$ for each $k \in [d]\}$, where med is the usual (one-dimensional) median.

Algorithm 17.2 The coordinate-wise median algorithm

Input: Initial weight vector w_0, maximum global round K, a learning rate policy $\{\eta_k\}$, training batch-size B, and (Optional) a target accuracy ϵ.

1: **procedure:** Aggregator (w_0, K, η, ϵ)
2:     **for** round $k = 1, \ldots, K$ **do**
3:         **if** (Optional) current global model reaches the target accuracy **then return** w_k
4:         **else**
5:             Query party $i \in \mathcal{P}$ with the current global model weights w_{k-1} for its current gradient g_k^i.
6:             $g_k = \mathrm{med}\{g_k^i : i \in \mathcal{P}\}$
7:             $w_k = w_{k-1} - \eta_k g_k$
8:             **return** w_K
9:         **end if**
10:    **end for**
11: **end procedure**
12: **procedure:** Party (B, w)
13:     Initialize the local model with $w_0 = w$
14:     $g = \nabla \ell(w_0; B)$ {ℓ denotes the loss function.}
15:     **return** g
16: **end procedure**

CWM requires the following assumptions for each loss function f_i and the overall generalization loss function F. It also requires the gradient of the loss function f_i of party $i \in \mathcal{P}$.

Assumption 1 (Smoothness of f_i and F) For any training data sample, assume the k-th coordinate of the partial derivative of f_i with respect to the weight vector w is L_k-Lipschitz for each $k \in [d]$ and each loss function f_i is L-smooth. Also assume that the generalization loss function $F(\cdot) := \mathbb{E}_{z \sim \mathcal{D}}[f(w; z)]$ is L_F-smooth, where $\mathcal{D}$ denotes the overall data distribution across parties.

Assumption 2 (Bounded Variance of Gradients) For any $w \in W$, $\text{Var}(\nabla f_i(w)) \leq V^2$, $\forall i \in \mathcal{P}$.

Assumption 3 (Bounded Skewness of Gradients) For any $w \in W$, $\|\gamma(\nabla f_i(w))\|_\infty \leq S$, $\forall i \in \mathcal{P}$, where $\gamma(X) := \frac{\mathbb{E}[X - \mathbb{E}(x)]^3}{\text{Var}(X)^{3/2}}$ is the absolute skewness of the vector X.

These assumptions are satisfied usually in the case where the parties' local data distribution is IID and the potential noise added during the data collection process is also IID. However, it will not hold for the non-IID case where parties have heterogeneous data resources.

17.2.4 Krum

In [1], Krum chooses party gradients to aggregate using the choice function $\text{Kr}(X_1, X_2, \ldots, X_n)$, which is defined as follows. For each party i, let us denote $s(i) = \sum_{i \to j, i \neq j} \|X_i - X_j\|^2$, where $i \to j$ denotes the set of vectors that are the $n - t - 2$ closest vectors to X_i and t is the maximum number of Byzantine parties in the federated learning system. Additionally, $\text{Kr}(X_1, X_2, \ldots, K_n) := X_{i*}$, where $s(i^*) \leq s(i)$, $\forall i \in \mathcal{P}$.

Algorithm 17.3 The Krum algorithm

Input: Initial weight vector w_0, maximum global round K, a learning rate policy $\{\eta_k\}$, training batch-size B, and (Optional) a target accuracy ϵ.

1: **procedure:** Aggregator (w_0, K, η, ϵ)
2:     **for** round $k = 1, \ldots, K$ **do**
3:         **if** (Optional) current global model reaches the target accuracy **then return** w_k
4:         **else**
5:             Query party $i \in \mathcal{P}$ with the current global model weights w_{k-1} for its current gradient g_k^i.
6:             $g_k = \text{Kr}(g_k^1, g_k^2, \ldots, g_k^{|\mathcal{P}|})$
7:             $w_k = w_{k-1} - \eta_k g_k$
8:             **return** w_K
9:         **end if**
10:     **end for**
11: **end procedure**
12: **procedure:** Party (B, w)
13:     Initialize the local model with $w_0 = w$
14:     $g = \nabla \ell(w_0; B)$ {ℓ denotes the loss function.}
15:     **return** g
16: **end procedure**

As shown in Algorithm 17.3, assuming that the maximum number of Byzantine parties is $0 \leq t \leq n$, Krum [1] can evade Byzantine attacks. Specifically, Krum enables the aggregator to only use a single party's gradients to update the global model in a global training round, rather than computing an aggregated gradient that incorporates all parties' gradients. The Krum choice function is called in Line 6 to choose a party at every round which is deemed the most trustworthy by having the smallest ℓ_2 distance with respect to all other parties' gradients. In other words, its local gradient is the most similar to all other local gradients. Since the Krum choice function requires the summation of at least $n - t - 2$ number of closest distances for each party, Krum enforces a limitation of minimum number of parties in a federated learning system to be at least 5. Therefore, it will not work for small federations of less than 5 parties. Moreover, similar to CWM, Krum assumes that the gradient vectors g_k^i proposed by the benign parties are IID random vectors, which might not be true for the case where parties have non-IID local data distributions. The computational complexity required by Krum is also much higher than gradient averaging which is only $O(dn)$.

The time complexity of the Krum Function $\text{Kr}(X_1, \ldots, X_n)$, where $X_1, \ldots, X_n$ are d-dimensional vectors, is $O(n^2(d + \log n))$.

We can conclude that all the abovementioned algorithms reject gradients that are assumed to be Byzantine under the assumption that a collected gradient is dishonest if it is comparatively more distant than the other gradients are to each other. One weakness of all aforementioned solutions is that they require bounds on the variance of the honest gradients [4], which has been shown unknown in practice and has been exploited through attack strategies such as the Fall of Empires attack [22].

17.3   Layerwise Robustness to Byzantine Threats

As we conclude from Sect. 17.2, many robust aggregation algorithms utilize some form of "outlier detection" to identify and filter out potential Byzantine gradients based on the assumption that in comparison to Byzantine gradients honest gradients will be "similar." However, it is not hard to find out that the similarity among gradients is closely related to the individual party's local distribution. Let us use a simple example to illustrate this statement. We can see in Fig. 17.4 that there are two clusters which correspond to two classes in a classification problem, blue and orange. In the case where each party has only one class label in its local training dataset, the gradient sent by a party will pull the gradient descent direction toward its own local data cluster (distribution). The relative similarity of gradients will decrease as the similarity of their local distributions decreases. For simplicity, we now refer to parties owning the blue class as the blue team and those owning orange class as the orange team. Once the robust aggregation algorithm, such as Krum, has decided the gradients sent by the blue team are benign, which often happens when there are more parties in the blue team in the FL system, it will eliminate gradients

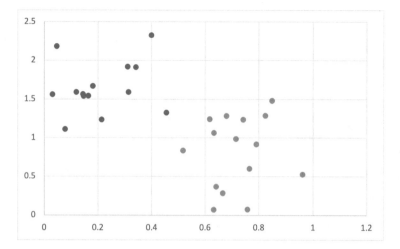

Fig. 17.4 Two-dimensional data distribution with two clusters

Table 17.1 List of trainable
layers of a simple CNN

Order	Layer	Parameters	Dimension
1	conv1	Weights	288
		Biases	32
2	conv2	Weights	18, 432
		Biases	64
3	dense1	Weights	1, 179, 648
		Biases	128
4	dense2	Weights	1280
		Biases	10

sent by parties from the orange team. This is because these gradients lie far away from replies sent by the blue team and are considered "malicious." In this case, the global model will always be updated based on gradients sent by the blue team and hence predict accurately for blue class, but not for the orange class since the model never gets to see gradients computed based on data samples from the orange class.

Instead of eliminating any gradient that seems to be malicious, we consider the problem of mitigating Byzantine threats via evaluating the reaction of a neural network layer when under certain Byzantine attacks. We hence conduct a preliminary study to verify our conjecture that different layers of a neural network will behave differently under a Byzantine attack. In this study, we construct an FL system with ten parties to collaboratively train a convolutional neural network (CNN) with two convolutional layers and dense layers, which we refer to as conv1, conv2, dense1, and dense2. We list the trainable layer weights and biases in Table 17.1. Notice that we separate the weights and biases from the same layer, since based on our experiment results (see Fig. 17.5), they behave differently toward

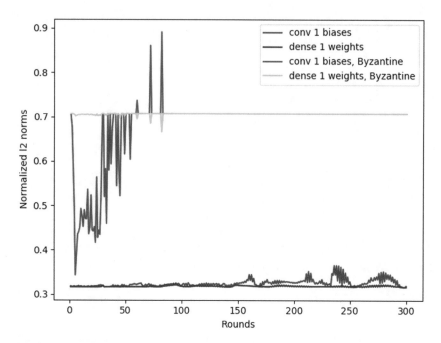

Fig. 17.5 Normalized ℓ_2 norms associated with the biases of the first convolutional layer and the weights of the first dense layer. This experiment involves ten workers with 1000 IID data points from MNIST. We compare results with and without two Byzantine workers executing a Gaussian attack with $\mu = 0$ and $\sigma = 200$

Gaussian attacks. Each party has 1000 data points randomly drawn from the MNIST dataset [9] and thereby the parties' local distributions are IID.

Before we discuss the details of our experimental study and key observations from this study, we first provide the notation used throughout the rest of this chapter.

Notation We use n to represent the total number of parties in an FL system. To denote the d-dimensional gradients of party p for layer l at global training round k, we use $g_{p,l}^k \in \mathbb{R}^d$. We use L to denote the total number of layers in a neural network. To denote the complete set of gradients of a party p at round k, we use G_p^k. Finally, we use $\mathcal{G}^k$ to represent a list of collected party gradient vectors at round k.

Since the parties' local data distribution is IID, we expect local gradients sent by each party to be relatively similar. Therefore, in our study, we use the ℓ_2 norm to measure the difference among them. In particular, at round k, we compute the normalized ℓ_2 norm of a specific layer, P_l^k as follows:

$$P_l^k \leftarrow \frac{\| [g_{1,l}^k, g_{2,l}^k, \ldots, g_{n,l}^k] \|_2}{\sum_{p=1}^{n} x_p}, \tag{17.2}$$

where x_p is the norm of the current gradient of the party p, i.e., $x_p = \| < g_{p,0}^k, g_{p,1}^k, \ldots, g_{p,L}^k > \|_2$, and $[g_{1,l}^k, g_{2,l}^k, \ldots, g_{n,l}^k]$ denotes a matrix with the p-th row being the l-th layer of party p's current gradient. We compute these normalized ℓ_2 norms across all of ten total parties at each round and compare the results between the layer weights and biases over time. We run our experiment in two settings, one of which is a normal FL training without any attack and the other is an FL training with two out of ten Byzantine parties. These Byzantine parties execute a Gaussian attack by sending gradients that are randomly sampled from a Gaussian distribution, $N(0, 200^2 I)$, which we take from [20].

Figure 17.5 compares the results for two layer parameters' gradients under non-attack and Byzantine attack scenarios. In absence of the Byzantine attack, the bias gradients from the first convolutional layer have similar ℓ_2 norms to the weight gradients from the first dense layer until around round 150. Then, the convolutional bias gradients' ℓ_2 norms become clearly larger and more varied across rounds than the weights of the first dense layer. However, even after 150 rounds, we can see that the normalized ℓ_2 norm does not vary too much when there are no Byzantine parties in the FL system. And hence we confirm our previous statement that local gradients sent by each party tend to be relatively similar even at the layer level when they have IID local data distribution.

We further study the similarity at a layer level with the inclusion of Byzantine gradients. In the Byzantine case, the convolutional bias norms start to even more decisively exceed the dense weight norms even earlier in training. While the variance of the dense layer ℓ_2 norms across rounds is comparably small in both settings, the variance in the convolutional layer norms is significantly amplified with the addition of Byzantine workers. These patterns demonstrate the fact that the gradient variance imposed by the Byzantine workers is more drastically affecting the convolutional bias gradients than the dense weights. They also have larger and more varied norms across rounds in the attack and non-attack settings separately. Therefore, we conclude more generally that the layers whose gradient norms vary more across rounds have greater inherent vulnerability that is more intensely exploited by Byzantine attacks.

17.4 LEGATO: Layerwise Gradient Aggregation

We introduce a new robust gradient aggregation method, which utilizes a layerwise robustness factor to reweigh collected gradients, for training neural networks in a federated learning setting. It is inspired by the observations of the preliminary study in Sect. 17.3. This new robust algorithm is called LayerwisE Gradient AggregatTiOn (LEGATO) [17]. The goal of LEGATO is to be able to mitigate the effect of erroneous or malicious gradients while preserving potential useful information collected from parties owning rare training data samples.

Algorithm 17.4 Federated learning with LEGATO

Input: Initial weight vector w_0, maximum global round K, a learning rate policy $\{\eta_k\}$, training batch-size B.

 1: **procedure:** Aggregator $(w_0, K, \{\eta_k\})$

 2:    **for** round $k = 1, \ldots, K$ **do**

 3:       $\mathcal{G}^k \leftarrow$ new list

 4:       Query party $p \in \mathcal{P}$ with the current global model weights w_{k-1} for its current gradient G_p^k and add it to $\mathcal{G}^k$

 5:       $\mathcal{G}_{agg}^k = \text{LEGATO}(\mathcal{G}^k)$ {Aggregates gradients as in Algorithm 17.5}

 6:       $w_k = w_{k-1} - \eta_k \mathcal{G}_{agg}^k$

 7:    **end for**

 8: **end procedure**

 9: **procedure:** Party (B, w)

10:    Initialize the local model with $w_0 = w$

11:    $g = \nabla \ell(w_0; B)$ {ℓ denotes the loss function.}

12:    **return** g

13: **end procedure**

17.4.1 LEGATO

Before we present the details about the LEGATO algorithm, we first describe a basic federated learning process in Algorithm 17.4. In this FL setting, we adopt the scheme of the mini-batch stochastic gradient descent method for parties to compute the local gradients and for the aggregator to update the global model via the aggregated gradients. In particular, an aggregator will iteratively request all parties in the system to send their current gradients computed based on the shared global model weights and their local datasets as shown in Line 4 of Algorithm 17.4. It then aggregates the collected gradients via some robust aggregation method, e.g., LEGATO, in Line 5 and update the global model via a gradient descent step using the aggregated gradient as shown in Line 6. It needs to be pointed out that we solely consider a synchronous setting, meaning the aggregator will wait until it receives responses from all queried parties. The aggregator also keeps a log of the most recent past gradients from all parties. At round k, the log is denoted by $GLog := [\mathcal{G}^{k-m}, \mathcal{G}^{k-m+1}, \ldots, \mathcal{G}^{k-1}]$, where m is the maximum size of the log.

LEGATO starts when the aggregator receives gradients from all parties at each round of training. First, the aggregator updates the gradient log, $GLog$, so that it contains the most recent m gradients collected from all parties. In Lines 7–11, the aggregator exploits the gradient log $GLog$ to compute the layerwise normalized ℓ_2 norms P_l^k for layer l at round k in the same way as (17.2). It then computes the reciprocal of the standard deviation of these norms across all logged rounds as a robustness factor that is assigned to each layer and normalized across all layers (see Lines 13–14). These steps are inspired by the observations from the experimental

---

Algorithm 17.5 LEGATO. An aggregation algorithm to aggregate gradients at round k

---

Input: Current round number k, a list of current parties' gradients $\mathcal{G}^k = [G_1^k, G_2^k, \ldots, G_n^k]$, and a log of recent past party's gradients with maximum size m $GLog := [\mathcal{G}^{k-m}, \mathcal{G}^{k-m+1}, \ldots, \mathcal{G}^{k-1}]$.

1: **procedure: Aggregator)**$(k, \mathcal{G}^t)$
2:    **if** $k = 1$ **then**
3:       Initialize an empty log $GLog$
4:    **else**
5:       **UpdateGradientLog**$(GLog, \mathcal{G}^t)$
6:    **end if**
7:    **for** $\mathcal{G}^k$ in $GLog$ and each party p in $\mathcal{P}$ **do**
8:       $x_p = \| G_p^k \|_2$
9:    **end for**
10:   **for** each layer l **do**
11:      $P_l^k \leftarrow \frac{\|[g_{1,l}^k, g_{2,l}^k, \ldots, g_{n,l}^k]\|_2}{\|[x_0, x_1, \ldots, x_n]\|_1}$
12:   **end for**
13:   **for** each layer l **do**
14:      $w_l \leftarrow \text{Normalize}(\frac{1}{\sqrt{\text{Var}(P_l^1, \ldots, P_l^k)}})$
15:   **end for**
16:   **for** p in $\mathcal{P}$ and each layer l **do**
17:      $G_{p,l}^* \leftarrow w_l G_{p,l}^k + \frac{1-w_l}{m-1} \sum_{j=1}^{m-1} G_{p,l}^{k-j}$
18:   **end for**
19:   **return** $\sum_{p \in \mathcal{P}} G_{p,l}^*$
20: **end procedure**
21: **procedure: UpdateGradientLog**$(GLog, \mathcal{G}^k)$
22:   $GLog \leftarrow GLog + \mathcal{G}^k$
23:   **if** $len(GLog) > m$ **then**
24:      $GLog \leftarrow GLog[1 :]$
25:   **end if**
26: **end procedure**

---

study conducted in Sect. 17.3 that the less the ℓ_2 norms vary across rounds, the more robust a layer is. In Line 17, each party's gradient information is updated as a weighted sum of the average of the party's historic gradients and its current gradient vector. The weights are chosen as a function of the robustness factor per layer that allows the updates of less robust layers to rely more heavily on the average of past gradients. Finally, all of these reweighed gradients are averaged across all parties at Line 19, and the result is used as round k's aggregated gradient, $\mathcal{G}_{agg}^t$.

LEGATO is robust and also stretches the aggregator's ability to utilize as much of the information provided by the parties as possible. On the one hand, this reweighing strategy in Line 19 is ultimately dampening the noise across gradients that may be a result of a Byzantine attack, with the goal of mitigating the Byzantine gradients' effect of pulling the aggregated gradient away from an honest value. On the other hand, the fact that it applies this dampening step at the most vulnerable layers allows reliable layers' current gradients, which are most accurate, to still weigh heavily into the aggregated gradient, and hence limits the sacrifice of convergence speed that could result from using past gradients. Furthermore, the online robustness factor computation allows LEGATO to generalize to a variety of model architectures because it adopts online factor assignments rather than relying on an absolute quantification of robustness. As is evidenced in [24], the knowledge of layerwise robustness varies between architectures, so an online and model-agnostic approach is more desirable.

17.4.2 Complexity Analysis of LEGATO

Due to its simplicity, gradient averaging is one of the most efficient gradient aggregation algorithms. It has computational complexity that is linear in terms of n. Other robust algorithms that utilize robust statistics have similar computational complexity, for example, coordinate-wise median.

Recall that we assume that there are n parties in a federated learning system and each neural network layer has at most d dimension. Proposition 17.1 states that LEGATO has time complexity $O(dn + d)$, which is also linear in n. This is a crucial improvement that LEGATO has over state-of-the-art robust algorithms such as Krum [1] and Bulyan [13] whose time complexities are $O(n^2)$.

Proposition 17.1 *LEGATO has time complexity $O(dn + d)$.*

Proof First, for each logged round at each layer, the aggregator computes the ℓ_2 norms of each worker's gradients, which has time complexity $O(dn)$. Then, for each logged round at each layer, the aggregator computes the ℓ_2 norm of the matrix of the party's gradients and normalizes it by dividing by the ℓ_1 norm of all parties' ℓ_2 gradient norms, which adds time complexity of $O(dn)$. The step that computes the standard deviation across these normalized round ℓ_2 norms for each layer incurs a time complexity of $O(dn)$. Next, the aggregator normalizes the reciprocals of the standard deviations corresponding to each layer induces $O(d)$. It then computes the average of logged gradients from each worker at each layer, which is $O(dn)$. Lastly, workers computes a weighted average of the gradients with the weight vector assigned by the workers, ends up with a time complexity of $O(dn)$.

Note that iterating through each individual layer of gradients does not add a factor of l to the time complexity because d encompasses all gradients at all layers. Therefore, iterating through all gradients by layer is $O(d)$ and iterating through all gradients in a flattened vector is also $O(d)$.

In conclusion, summing up the time complexity of all steps, the resulting time complexity of $O(dn + d)$ follows immediately.   □

LEGATO's space complexity is formalized in the following proposition.

Proposition 17.2 *LEGATO has space complexity $O(dmn)$, where m is the selected log size for LEGATO.*

Proof *LEGATO's maintenance of a gradient log stores m past gradients from each party which introduces a space requirement of $O(dmn)$.* □

17.5 Comparing Gradient-Based and Layerwise Robustness

In this section, we evaluate LEGATO's performance against Krum [1] and coordinate-wise median (CWM) [23] under a variety of non-attack and attack settings in Sects. 17.5.1 and 17.5.2, respectively. We also demonstrate the potential advantages of LEGATO over gradient averaging (see Algorithm 17.1) in an overparameterized neural network setting.

Through numerical experiments, we can make the following claims:

- In settings without strictly bounded honest gradient variance, which we demonstrate through experiments with non-IID data, LEGATO is more robust than Krum and CWM.
- LEGATO has the best performance of the three algorithms in the absence of a Byzantine attack in both IID and non-IID settings.
- Considering two attack settings we test against LEGATO is generally the most robust of the three robust aggregation algorithms.

Throughout this section, we use the IBM Federated Learning library [11] to conduct our numerical experiments. We use a standard datasets: MNIST handwritten digits for our experiments. All experiments train a global model in an FL system with 25 parties where each party's training data points are randomly sampled either across the entire dataset (in the IID setting) or across only one class each (in the non-IID setting, the same as in [25]). Parties use mini-batch technique to compute the gradient based on its local dataset before responding to the aggregator's request at every global training round. The global model for MNIST is a simple CNN with two convolutional layers, one dropout and two dense layers. We use 10 as the size of the log/history of past gradients. Considering we use a nearly balanced global test set to evaluate the global model's performance, we use accuracy across global training rounds as a performance metric, which would be comparable to F1 score in these experiments.

17.5.1 Dealing with Non-IID Party Data Distributions

Many robust aggregation algorithms assume that gradients collected from the honest parties are bounded (see the details in Sects. 17.2.3 and 17.2.4). However, this may

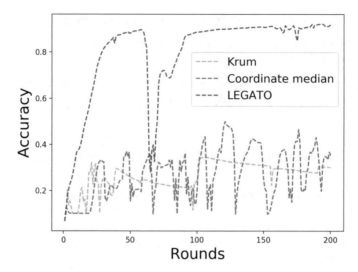

Fig. 17.6 MNIST dataset, non-IID setting, where each party has 1000 training images. All algorithms use mini-batch size of 50 and a learning rate of 0.05

not be necessarily true in the non-IID setting where each party only has some labels, in particular one label in our experimental setting, in its local training dataset.

In Fig. 17.6, we can see that Krum and CWM perform poorly in the non-IID setting without any attack presented. In fact, they falsely identify the benign gradients from parties owning different local distributions as Byzantine replies. In contrast, LEGATO, independent of the aforementioned assumption, performs fairly well in such non-IID setting.

17.5.2 Dealing with Byzantine Failures

In this section, we consider the case where there are possible Byzantine attacks in the FL system. In particular, we evaluate the performances of LEGATO, Krum, and CWM under the following Byzantine attacks:

Fall of Empires For Krum, we create 11 Byzantine parties out of a total of 25 parties in the FL system, and we use $\epsilon = 0.001$, following a similar attack setting as in [22].

Gaussian Attacks In our experiments, we set $\mu = 0$ and vary the selection of σ to be either relatively large or relatively small.

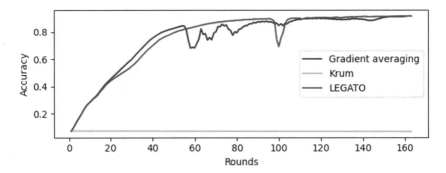

Fig. 17.7 MNIST dataset, IID setting, where each party has 1000 training images. All algorithms use mini-batch size of 50 and a learning rate of 0.05

17.5.2.1 Defense Against Fall of Empires

We first study the case when the Fall of Empires attack is presented. Figure 17.7 shows results for robust algorithms against the Fall of Empires attack for training on MNIST dataset. As it can be seen in the figure, the Fall of Empires attack is effective against Krum, but not against LEGATO or gradient averaging. The reason behind this is that all Byzantine gradients in this attack are the same and hence form a large cluster. Therefore, Krum identifies these Byzantine parties' gradients as being more trustworthy and will use them to update the global model. The Fall of Empires attack takes advantage of Krum's reliance on the assumption that honest gradients are sufficiently clumped [4] and a similar effect is shown for the CWM algorithm.

17.5.2.2 Defense Against Gaussian Attacks

Let us investigate the performances of three robust algorithms under Gaussian attacks with different magnitudes of variance.

As we can see from Fig. 17.8, Krum and CWM perform almost similarly for both cases where the variance of the Byzantine attacks varies from 20 to 100. They cannot distinguish between Byzantine gradients and the benign gradients sent from parties owning different local distributions. Recall that both Krum and CWM require the assumption that benign gradients look "alike"; in other words, the gradients sent from the parties are supposed to be IID. Our experimental results demonstrate that LEGATO is independent of this assumption. LEGATO is able to obtain better model test accuracy than Krum and CWM, despite being affected by Byzantine parties. In particular, when the Gaussian attack uses a small variance, $\sigma = 20$, LEGATO achieves a significant improvement in model accuracy compared to Krum and CWM. It is also slightly better than gradient averaging, especially after 200 rounds. When $\sigma = 100$, LEGATO significantly outperforms Krum and gradient

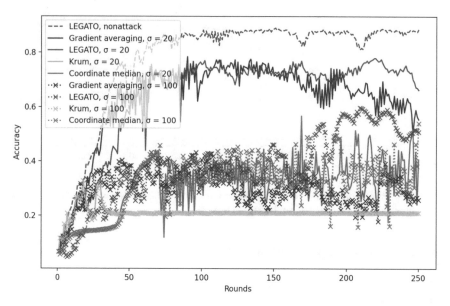

Fig. 17.8 MNIST dataset, non-IID setting, where each party has 1000 training images. For Gaussian attacks, 4 out of the 25 parties are Byzantine and randomly sample gradients from $N(0, \sigma^2 I)$. For all cases, the mini-batch size is 50 and the learning rate is 0.03

averaging, but only slightly exceeds CWM in terms of model accuracy after 175 rounds because the high standard deviation exploits a vulnerability LEGATO has to extreme outliers.

17.5.3 Dealing with Overparameterized Neural Networks

Increasing the overparameterization of the model is a condition worth studying due to its common existence in practice. In this section, we investigate a special neural network architecture where the neural network to be trained is overparametrized. Specifically, the global model possesses more convolutional layers and is able to capture more information and/or noises during the training process. Our goal is to observe if LEGATO is more robust than gradient averaging in this special case.

Figure 17.9 shows that the performance gap of LEGATO and gradient averaging widens when more convolutional layers are added to the model architecture. As the number of parameters increases, neural networks become more susceptible to learning the Gaussian noise in the same way that they would become more susceptible to learning noise in training data and over-fitting in non-attack settings. Since LEGATO exploits the layerwise robustness factors to aggregate the collected gradients, it is able to dampen the gradient oscillations, which resembles the effect of regularization methods for reducing over-fitting. It is worth mentioning that the

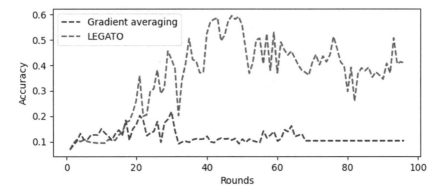

Fig. 17.9 Comparing the accuracy of gradient averaging and LEGATO for training a CNN with eight convolutional model layers instead of the well-known architecture with two convolutional layers. All algorithms use a mini-batch size of 50 and a learning rate of 0.05. Four of twenty-five parties are Byzantine parties executing a Gaussian attack where gradients are randomly sampled from $N(0, 20^2 I)$

overall performances of LEGATO and gradient averaging drop comparing to those training a model with two convolutional layers because overparameterized neural networks are more vulnerable to Gaussian attacks.

17.5.4 Effectiveness of the Log Size

In this section, we examine the effect of the log size for the LEGATO algorithm, in particular, how the log size affects LEGATO's ability to defend against Byzantine threats.

From Fig. 17.10, we can see that the choice of log size does not affect the model's performance a lot in the non-attack and Gaussian attack with small variance cases, but in the case of Gaussian attack with high variance, the selection of log size is quite crucial. It is reasonable that when log size is as small as five, the performance of LEGATO will lean toward gradient averaging, while as the log size increases, the LEGATO algorithm becomes more conservative to make a move toward the new gradient direction. Therefore, when the log size is as large as 30, it will converge very slowly and even oscillate around a certain point for a long time.

17.6 Conclusion, Open Problems, and Challenges

We have shown that a layerwise gradient-based robust algorithm (LEGATO) performs better than other gradient-based robust algorithms, e.g., Krum and CWM, for training neural networks in non-IID settings with Byzantine attacks, where a

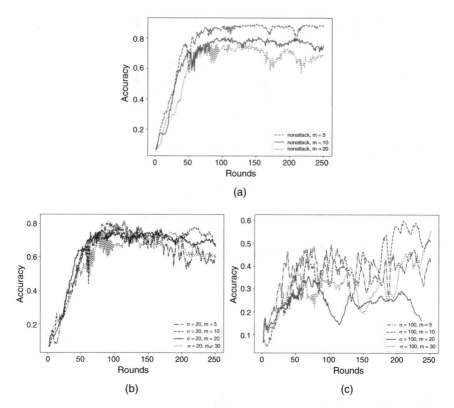

Fig. 17.10 MNIST dataset, non-IID setting, where each party has 1000 training images. For Gaussian attacks, 4 out of the 25 parties are Byzantine and randomly sample gradients from $N(0, \sigma * I)$. For all cases, the mini-batch size is 50 and the learning rate is 0.03. We vary the log size from 5 to 30. (**a**) Non-attack setting. (**b**) Gaussian attack, $\sigma = 20$. (**c**) Gaussian attack, $\sigma = 100$

party or a subset of parties aims to bring down the final model's performance by sending crafted malicious replies to the aggregator during the FL training process. Although LEGATO is a first successful attempt to mitigate Byzantine threats for training neural networks in federated learning, it is not a general algorithm that can be applied to train all types of machine learning models in an FL setting. Other models, like linear models, Support Vector Machines (SVMs), and XGBoost among others, do not possess a layered architecture similar to neural networks that can be utilized by LEGATO. One open problem in this field is determining a general robust aggregation algorithm that can be applied to mitigating Byzantine threats even under the non-IID local data distribution setting.

Moreover, as we can see in the case when the variance of the Gaussian attack is very high, the effectiveness of LEGATO is largely affected, since it does not reject any gradient information and hence is vulnerable to extreme outliers. Another open problem in this field is finding a good way to define and identify "extreme outliers" according to the problem context. In particular, the procedure to define and identify those "extreme outliers" should be data-aware, i.e., based on local data distributions from all parties. This is a very challenging problem since it will be hard to identify "extreme outliers" while maintaining some level of a party's local data privacy.

References

1. Blanchard P, El Mhamdi EM, Guerraoui R, Stainer J (2017) Machine learning with adversaries: byzantine tolerant gradient descent. Advances in Neural Information Processing Systems, p 30
2. Charikar M, Steinhardt J, Valiant G (2017, June) Learning from untrusted data. In: Proceedings of the 49th annual ACM SIGACT symposium on theory of computing, pp 47–60
3. Chen, Y., Su, L., & Xu, J. (2017). Distributed statistical machine learning in adversarial settings: Byzantine gradient descent. Proceedings of the ACM on measurement and analysis of computing systems, 1(2), 1–25
4. El-Mhamdi, E. M., Guerraoui, R., & Rouault, S. (2020). Distributed momentum for byzantine-resilient learning. arXiv preprint arXiv:2003.00010
5. Fung, C., Yoon, C. J., & Beschastnikh, I. (2018). Mitigating sybils in federated learning poisoning. arXiv preprint arXiv:1808.04866
6. Gao D, Liu Y, Huang A, Ju C, Yu H, Yang Q (2019) Privacy-preserving heterogeneous federated transfer learning. In: 2019 IEEE international conference on big data (Big Data). IEEE, pp 2552–2559
7. Krishnaswamy R, Li S, Sandeep S (2018, June) Constant approximation for k-median and k-means with outliers via iterative rounding. In: Proceedings of the 50th annual ACM SIGACT symposium on theory of computing, pp 646–659
8. Lamport L, Shostak R, Pease M (1982) The byzantine generals problem. ACM Trans Program Lang Syst 4(3):382–401. https://doi.org/10.1145/357172.357176
9. Lecun Y, Bottou L, Bengio Y, Haffner P (1998) Gradient-based learning applied to document recognition. Proc IEEE 86:2278–2324
10. Li T, Sahu AK, Zaheer M, Sanjabi M, Talwalkar A, Smith V (2018) Federated optimization in heterogeneous networks. Preprint. arXiv:1812.06127
11. Ludwig H, Baracaldo N, Thomas G, Zhou Y, Anwar A, Rajamoni S, Ong Y, Radhakrishnan J, Verma A, Sinn M et al (2020) IBM federated learning: an enterprise framework white paper v0. 1. Preprint. arXiv:2007.10987
12. McMahan B, Moore E, Ramage D, Hampson S, y Arcas, B. A. (2017, April) Communication-efficient learning of deep networks from decentralized data. In: Artificial intelligence and statistics, PMLR, pp 1273–1282
13. Guerraoui R, Rouault S (2018, July) The hidden vulnerability of distributed learning in byzantium. In: International conference on machine learning. PMLR, pp 3521–3530
14. Muñoz-González, L., Co, K. T., & Lupu, E. C. (2019). Byzantine-robust federated machine learning through adaptive model averaging. arXiv preprint arXiv:1909.05125
15. Pillutla, K., Kakade, S. M., & Harchaoui, Z. (2019). Robust aggregation for federated learning. arXiv preprint arXiv:1912.13445
16. Rajput S, Wang H, Charles ZB, Papailiopoulos DS (2019) DETOX: A redundancy-based framework for faster and more robust gradient aggregation. CoRR abs/1907.12205. http://arxiv.org/abs/1907.12205
17. Varma K, Zhou Y, Baracaldo N, Anwar A (2021) Legato: A layerwise gradient aggregation algorithm for mitigating byzantine attacks in federated learning
18. Wang H, Yurochkin M, Sun Y, Papailiopoulos D, Khazaeni Y (2020) Federated learning with matched averaging

19. Xia Q, Tao Z, Hao Z, Li Q (2019) Faba: An algorithm for fast aggregation against byzantine attacks in distributed neural networks. In: Proceedings of the twenty-eighth international joint conference on artificial intelligence, IJCAI-19. International joint conferences on artificial intelligence organization, pp 4824–4830. https://doi.org/10.24963/ijcai.2019/670

20. Xie, C., Koyejo, O., & Gupta, I. (2018). Generalized byzantine-tolerant sgd. arXiv preprint arXiv:1802.10116

21. Xie C, Koyejo S, Gupta I (2019, May) Zeno: distributed stochastic gradient descent with suspicion-based fault-tolerance. In: International conference on machine learning. PMLR, pp 6893–6901

22. Xie C, Koyejo O, Gupta I (2020, August) Fall of empires: breaking byzantine-tolerant sgd by inner product manipulation. In: Uncertainty in artificial intelligence. PMLR, pp 261–270

23. Yin D, Chen Y, Kannan R, Bartlett P (2018, July) Byzantine-robust distributed learning: towards optimal statistical rates. In: International conference on machine learning. PMLR, pp 5650–5659

24. Zhang, C., Bengio, S., & Singer, Y. (2019). Are all layers created equal?. arXiv preprint arXiv:1902.01996

25. Zhao, Y., Li, M., Lai, L., Suda, N., Civin, D., & Chandra, V. (2018). Federated learning with non-iid data. arXiv preprint arXiv:1806.00582

Part IV
Beyond Horizontal Federated Learning: Partitioning Models and Data in Diverse Ways

In this part of the book, we cover two enterprise use cases that do not comply with the *horizontal* federated learning setup that we have discussed this far. In particular, this part of the book focuses on two new paradigms *vertical federated learning* and *split learning*, where a single party does not have all the information to train a model by itself.

Vertical federated learning arises in enterprise settings where a single party does not have all the data required to train a model by itself. Therefore, algorithms for horizontal federated learning training cannot be applied in this setting. Chapter 18 provides an overview of vertical federated learning techniques and in particular discusses a novel and efficient approach based on functional encryption that considers some important enterprise requirements, including no peer-to-peer connections between participating parties.

Split learning is an alternative approach where multiple parties may train different parts of a model ensuring their data remains private. Multiple setups of the system are possible. Chapter 19 introduces this novel approach and associated algorithms in detail.

Chapter 18
Privacy-Preserving Vertical Federated Learning

Runhua Xu, Nathalie Baracaldo, Yi Zhou, Annie Abay, and Ali Anwar

Abstract Many federated learning (FL) proposals follow the structure of horizontal FL, where each party has all the necessary information to train a model available to them. However, in important real-world FL scenarios, not all parties have access to the same information, and not all have what is required to train a machine learning model. In what is known as *vertical* scenarios, multiple parties provide disjoint sets of information that, when brought together, can create a full feature set with labels, which can be used for training. Legislation, practical considerations, and privacy requirements inhibit moving all data to a single place or freely sharing among parties. Horizontal FL techniques cannot be applied to vertical settings. This chapter discusses the use cases and challenges of vertical FL. It introduces the most important approaches for vertical FL and describes in detail *FedV*, an efficient solution to perform secure gradient computation for popular ML models. FedV is designed to overcome some of the pitfalls inherent to applying existing state-of-the art techniques. Using FedV substantially reduces training time and the amount of data transfer and enables the use of vertical FL in more real-world use cases.

18.1 Introduction

Federated learning (FL) [21] has recently been proposed as a promising approach for enabling collaborative training of ML models. Parties—under the orchestration of a central node (an *aggregator*)—train together, without having to share any of their raw training data.

There are two types of FL approaches, *horizontal FL* and *vertical FL*, which differ based on how data is partitioned between parties. In *horizontal FL*, each party can access the entirety of its own feature set and labels, and hence each party can train a complete local model based on its dataset. The aggregator can query

R. Xu (✉) · N. Baracaldo · Y. Zhou · A. Abay · A. Anwar
IBM Research – Almaden, San Jose, CA, USA
e-mail: runhua@ibm.com; baracald@us.ibm.com; yi.zhou@ibm.com; anniek@ibm.com; ali.anwar2@ibm.com

each party for model updates and create a global model by fusing together model weights from the updates. Most of the literature about FL focuses on *horizontal FL*, addressing issues related to privacy and security [10, 32], system architecture [3, 19], and new learning algorithms [7, 37].

In contrast, *vertical FL* (VFL) refers to FL scenarios where individual parties do **not** have a complete set of features and labels. Thus, they cannot train a local model using their data. For instance, in a financial scenario involving a set of banks and a regulator, banks may want to collaboratively create an ML model using their data to flag accounts involved in money laundering. If several banks collaborate to find a common feature vector for the same set of clients, and a regulator provides labels showing which clients have committed money laundering, such fraud can be identified and mitigated. However, banks may not want to share their clients' account details, and in some cases, they are prevented from sharing data by regulations. As a result, such a money laundering detection scenario would benefit substantially from a VFL solution.

The critical steps of VFL include (1) *private entity resolution* and (2) *private vertical training*. In private entity resolution, parties' datasets need to be aligned to create the complete feature set while maintaining data privacy. Private vertical training focuses on training the global model in a privacy-preserving way. Various approaches, e.g., [5, 9, 12, 13, 26, 31, 34, 36], have been proposed to perform VFL. Most are model-specific and rely on particular privacy-preserving techniques, such as garbled circuits, secure multi-party computation (SMC), differential privacy noise perturbation, partially additive homomorphic encryption (HE), and inner-product functional encryption schemes.

In this chapter, we provide an overview of VFL, its challenges, and the state-of-the art. Section 18.2 provides an overview of the overall process required to train a model in VFL settings. In Sect. 18.3, we explain the inherent challenges of training using gradient descent according to the restrictions imposed by VFL. Then, we explore representative VFL solutions from *communication topology*, *system efficiency*, and *supported ML models* in Sect. 18.4. Finally, we present *FedV*, an efficient VFL solution in Sect. 18.5. We conclude the chapter in Sect. 18.6.

18.2   Understanding Vertical Federated Learning

VFL is a powerful approach that can help create ML models in many real-world scenarios where horizontal FL is not applicable, i.e., where a single entity does not have access to all the training features or labels. In the healthcare domain, different entities may collect diverse sets of data about patients that, if combined, can help achieve significant improvements in the prediction and diagnosis of the patients' health conditions. For instance, a sensor company may collect records of body sensors, including heart rate and sleep cycle data of a patient. Then, a hospital that keeps track of a patient's medical history (including labels) can use VFL by training a machine learning model that can assist in diagnosing the patient's health conditions more accurately. Another example use case is predicting the lifecycle of a product using information about its fabrication, transportation, and shelf life to determine if it will be returned due to malfunction. Linking information collected at the (1) factory where the device is produced, (2) during the transportation period, i.e., such as the monitored temperature and shaking, and (3) from stores including when the item was sold and whether it was returned could help this use case. Here, whether the product was sold or not and if it was returned at a point of time is a valuable label.

In all of these examples, different information is available at multiple locations and is owned by several organizations (parties) that do not share their information with each other. In VFL, it is often assume that a single party has access to the labels. However, there are three potential scenarios:

1. *A single party owns the labels and labels are private.* This case follows most of our previously discussed examples.
2. *A single party owns the labels, but they are not private and can be shared with all other parties.*
3. *All parties have a label for their data.* Labeling data is often costly and cumbersome. Hence, this use case is not common. Additionally, it opens the door for potential discrepancies in the labeling.

In this chapter, we focus on the first case that reflects most of the use cases we have encountered in real-world scenarios. We thus discuss VFL in settings where only one party has the class labels and wants to keep them private. We call *active party* the party owning the label, while we refer to other parties without labels *passive parties*.

18.2.1 Notation, Terminology and Assumptions

To help make the discussion more concrete, we now introduce the notation used throughout the rest of the chapter. Let $\mathcal{P} = \{p_i\}_{i \in \{1,...,n\}}$ be the set of n parties in VFL. Let $\mathcal{D}^{[X,Y]}$ be the training dataset across the set of parties $\mathcal{P}$, where $X \in \mathbb{R}^d$ represents the feature set and $Y \in \mathbb{R}$ represents the labels. The goal of VFL is to train a ML model $\mathcal{M}$ over the dataset $\mathcal{D}$ from the party set $\mathcal{P}$ without leaking any party's data.

In VFL, a dataset $\mathcal{D}$ is vertically distributed across multiple parties, where each local dataset $\mathcal{D}_{p_i}$ has all of the training data samples, but only a portion of the feature set. Figure 18.1 presents a simple representation of how $\mathcal{D}$ can be vertically partitioned among two parties. Following the scenarios explained before, feature sets and labels involved in the learning need to be kept private during the VFL process.

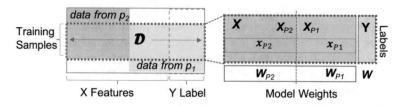

Fig. 18.1 Vertically partitioned data across parties. In this example, p_1 and p_2 have overlapping identifier features, and p_1 is the active party that has the labels

18.2.2 Two Phases of Vertical FL

When features are distributed among parties, two questions arise:

1. How can we correctly match data samples to corresponding feature values from different parties without leaking data?
2. How can we train a model with vertically partitioned data while complying with the FL privacy constraints?

When each party collects data independently, it is necessary to first identify common records and match the records. This process is called entity matching or entity resolution and requires that all parties have an identifier or a set of identifiers that can serve to determine when two of their records belong to the same sample. For example, in the case of multiple hospitals collecting data from patients, the social security number of each patient can serve to link patient's information collected at different hospitals. Private entity resolution is a set of techniques that have been developed to deal with record matching without revealing to parties the records that they share.

After the private entity resolution takes place, vertical training can start. Typically, an *aggregator* coordinates the entity resolution and training process among parties. In the following, we present in more detail the private entity resolution process and then highlight some of the challenges of training in vertical settings.

18.2.2.1 Phase I: Private Entity Resolution (PER)

One crucial requirement of entity resolution is ensuring the process does not leak private information about the parties' training data. An *honest-but-curious* or *adversary* party should not be able to infer the presence or absence of a specific data sample. To achieve privacy-preserving entity resolution, existing approaches, such as [14, 23], usually employ techniques or tools like bloom filters, random oblivious transfer, or private set intersection that help to match data without sacrificing the privacy guarantee [14, 23]. Among those approaches, a common assumption is that there exist public record identifiers, such as names, dates of birth, or universal identification numbers, that can be used to perform entity matching.

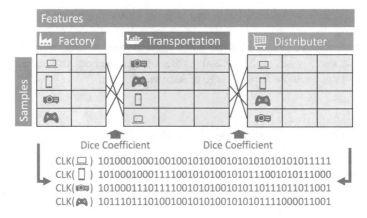

Fig. 18.2 Illustration of privacy-preserving entity resolution based on anonymous linking code

Figure 18.2 illustrates a common PER approach adopted by VFL techniques such as [13, 33]. This typical entity resolution approach employs an anonymous linking code technique called *cryptographic long-term key (CLK)*, and a corresponding matching method called the *Dice coefficient* [24]. First, each party generates a set of CLKs based on the identifiers of the local dataset and shares it with an *aggregator*, which matches the CLKs received and generates a permutation vector that each party uses to *shuffle* its local dataset. As a result, the shuffled local datasets are similarly ordered and ready to be used for private VFL training.

18.2.2.2   Phase II: Private Vertical Training

After the data has been matched and aligned according to PER, training can start. In vertical training, the aggregator coordinates the training process among parties. Each party trains a "partial model update" (i.e., partial gradient computation results or intermediate results) and employs privacy-preserving approaches to protect the "partial model update." In each training round, the aggregator orchestrates a specially designed training process that performs secure computations to ensure data is never moved or shared. In the following section, we overview the challenges of applying Gradient Descent in VFL.

18.3   Challenge of Applying Gradient Descent in Vertical FL

The Gradient Descent (GD) and its variants, e.g., [8, 11, 16, 17, 21] and the reference therein, have been the dominant approaches to train a model in (distributed) machine learning. In particular, GD can be applied to train non-tree-based traditional models including logistic regression, lasso, and support vector machines, as well as deep learning models.

As the subsets of feature set are distributed among different parties, it is not possible to directly adopt the normal gradient-based methods in vertical partition settings. Here, we show the challenge of such an adoption.

18.3.1 Gradient Descent in Centralized ML

GD method [22] represents a class of optimization algorithms that find the minimum of a target loss function; for example, a typical loss function can be defined as follows:

$$E_{\mathcal{D}}(\boldsymbol{w}) = \frac{1}{n} \sum_{i=1}^{n} \mathcal{L}(y^{(i)}, f(\boldsymbol{x}^{(i)}; \boldsymbol{w})) + \lambda R(\boldsymbol{w}), \qquad (18.1)$$

where $\mathcal{L}$ is the loss function, $y^{(i)}$ is the corresponding class label of data sample $\boldsymbol{x}^{(i)}$, $\boldsymbol{w}$ denotes the model parameters, f is the prediction function, and R is regularization term with coefficient λ. GD finds optimal $\boldsymbol{w}$ that minimizes Eq. (18.1) by iteratively moving in the direction of the local steepest descent, as defined by the negative of the gradient, i.e.,

$$\boldsymbol{w} \leftarrow \boldsymbol{w} - \alpha \nabla E_{\mathcal{D}}(\boldsymbol{w}), \qquad (18.2)$$

where α is the learning rate, and $\nabla E_{\mathcal{D}}(\boldsymbol{w})$ is the gradient computed at the current iteration. GD and its variants (i.e., SGD) have become common approaches to find optimal parameters (weights) of a ML model due to their simple algorithmic schemes [22].

18.3.2 Gradient Descent in Vertical FL

In a VFL setting, since $\mathcal{D}$ is vertically partitioned among parties, the gradient computation $\nabla E_{\mathcal{D}}(\boldsymbol{w})$ is more computationally involved than in a centralized ML setting. Considering the simplest case where there are only two parties, p_1 and p_2, in a VFL system, as illustrated in Fig. 18.1, and Mean Squared Loss (MSE) is used as the target loss function, i.e.,

$$E_{\mathcal{D}}(\boldsymbol{w}) = \frac{1}{n} \sum_{i=1}^{n} (y^{(i)} - f(\boldsymbol{x}^{(i)}; \boldsymbol{w}))^2. \qquad (18.3)$$

The gradient of $E_\mathcal{D}(\mathbf{w})$ is computed as

$$\nabla E_\mathcal{D}(\mathbf{w}) = -\frac{2}{n} \sum_{i=1}^{n} (y^{(i)} - f(\mathbf{x}^{(i)}; \mathbf{w})) \nabla f(\mathbf{x}^{(i)}; \mathbf{w}). \tag{18.4}$$

If we expand Eq. (18.4) and compute the result of the summation, we need to compute $-y^{(i)}\nabla f(\mathbf{x}^{(i)}; \mathbf{w})$ for $i = 1, \ldots n$, which requires feature information from both p_1 and p_2, and labels from p_1. And, clearly,

$$\nabla f(\mathbf{x}^{(i)}; \mathbf{w}) = [\partial_{\mathbf{w}_{p_1}} f(\mathbf{x}_{p_1}^{(i)}; \mathbf{w}); \partial_{\mathbf{w}_{p_2}} f(\mathbf{x}_{p_2}^{(i)}; \mathbf{w})]$$

does not always hold for any function f, since f may not be well-separable with respect to $\mathbf{w}$. For instance, when it applies in linear functions, such as

$$f(\mathbf{x}^{(i)}; \mathbf{w}) = \mathbf{x}^{(i)}\mathbf{w} = \mathbf{x}_{p_1}^{(i)}\mathbf{w}_{p_1} + \mathbf{x}_{p_2}^{(i)}\mathbf{w}_{p_2}, \tag{18.5}$$

Equation (18.4), namely, the partial gradient $\nabla E_\mathcal{D}(\mathbf{w})$ will be reduced as follows:

$$\frac{2}{n} \sum_{i=1}^{n} \left([(\mathbf{x}_{p_1}^{(i)}\mathbf{w}_{p_1} - y_{p_1}^{(i)} + \mathbf{x}_{p_2}^{(i)}\mathbf{w}_{p_2})\mathbf{x}_{p_1}^{(i)}; (\mathbf{x}_{p_1}^{(i)}\mathbf{w}_{p_1} - y_{p_1}^{(i)} + \mathbf{x}_{p_2}^{(i)}\mathbf{w}_{p_2})\mathbf{x}_{p_2}^{(i)}] \right).$$
$$\tag{18.6}$$

In such a case, this may lead to exposure of training data between two parties, due to the computation of some terms, as demonstrated in Eq. (18.6). For example, it may result in the leakage of the second party's input $\mathbf{x}_{p_2}^{(i)}$ to the first party p_1.

In short, under the VFL setting, the gradient computation at each training epoch requires either (i) the parties' collaboration, to exchange their "partial model update" with each other or (ii) parties sharing their data with the aggregator to compute the final gradient update. Therefore, any naive solutions will lead to a significant risk of privacy leakage, countering the initial goal of FL, to protect data privacy.

18.4   Representative Vertical FL Solutions

The studies and mining of data across vertically partitioned datasets is not a novel topic in machine learning literature. Several previous approaches, such as those in [26, 28, 29, 35], have been developed for distributed data mining, where methods are proposed to train specific ML models, such as support vector machines [35], logistic regressions [26], and decision trees [29]. Vaidya's survey [28] presents a summary of vertical data mining methods. These solutions place a higher premium on developing distributed data mining algorithms in a vertical context than on ensuring strong privacy in a variety of threat model scenarios and thus are not designed to prevent upcoming inference of private data threats. For instance, the privacy-preserving decision tree model in [29] has to reveal class distribution over the given attributes and thus trivially leaks private information.

Additionally, split learning [25, 30], a new paradigm to train deep learning, was recently proposed to train neural networks without sharing raw data. Split learning is also an alternative approach to achieve VFL [27]. However, the approach's primary objective is to horizontally or vertically partition the neural networks among parties, rather than to ensure strong privacy protections. At first glance, split learning appears to be a private process because only the intermediate computing results of the cut layer are exchanged among parties, rather than raw features or labels. However, it has been demonstrated that this gradient exchange method is vulnerable to inference attacks. For example, Li et al. describe how to uncover labels by utilizing the norm of the active party's exchanged gradients [18].

Numerous emerging inference attacks are proposed in [15, 18, 20], with a focus on vertical federated learning. The threats in these settings include, for example, label inference [18] and batch gradient's inference [15] during the training phase, and feature inference during the prediction stage [20]. To address the privacy leakage issue posed by those inference attacks in the context of VFL, a few emerging VFL solutions have incorporated privacy-preserving mechanisms into VFL, resolving those privacy inference attacks completely or partially. In this section, we compare and summarize those approaches from three perspectives: *communication topology*, *privacy-preserving offered*, and *supported ML models*, respectively.

18.4.1 Contrasting Communication Topology and Efficiency

One of the crucial challenges of VFL is computing the gradient descent in a vertical setting, as discussed in Sect. 18.3. For simplicity, let us take the linear function as an example. As demonstrated in Eq. (18.6), the key steps of computing gradients $\nabla E_{\mathcal{D}}(\boldsymbol{w})$ in VFL are as follows:

(i) Enabling all parties to collaboratively compute partial model prediction $u^{(i)}$ in a private manner:

$$u^{(i)} \leftarrow y^{(i)} - \sum_i x_{p_i}^{(i)} \boldsymbol{w}_{p_i}.$$

(ii) Allowing the aggregator compute gradient update $\nabla E_{\mathcal{D}}(\boldsymbol{w})$ without leaking private data:

$$\nabla E_{\mathcal{D}}(\boldsymbol{w}) \leftarrow -\frac{2}{n} \sum_{i=1}^{n} ([u^{(i)} x_{p_1}^{(i)}; u^{(i)} x_{p_2}^{(i)}]).$$

Figure 18.3 summarizes the five types of communication topologies used by existing VFL proposals to compute the vertical gradient update. The communication type includes *generic multi-party computation* communication, *peer-to-aggregator* communication, and (partial) *peer-to-peer* communication. Notably, the VFL solution utilizing generic garbled circuits, as illustrated in Fig. 18.3 Type (E), does not follow the previously stated SGD-based gradients computation and hence has a different and more complex architecture than the other four types. Except for a recent VFL framework [33], elaborated on in Sect. 18.5, most existing vertical solutions require at least two rounds of communication to compute a gradient descent step.

Table 18.1 presents VFL proposals contrasting the communication and interaction topology. The proposals presented in [13, 34] rely on fully peer-to-peer communication to compute $u^{(i)}$ and then each party p_i shares $u^{(i)} x_{p_i}$ to the aggregator to finalize the gradients update. The solutions in [12, 36] relax the

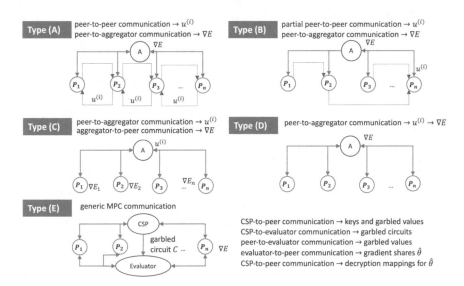

Fig. 18.3 Communication topology of existing VFL solutions

Table 18.1 Communication topology summary of VFL solutions

Proposal	Communication and interaction topology
Gascón et al. [9]	Multi-party computation interactions
Hardy et al. [13]	Full peer-to-peer + 1 round peer-to-aggregator
Yang et al. [34]	Full peer-to-peer + 1 round peer-to-aggregator
Gu et al. [12]	Partial peer-to-peer + 2 rounds peer-to-aggregator
Zhang et al. [36]	Partial peer-to-peer + 2 rounds peer-to-aggregator
Chen et al. [5]	2 rounds peer-to-aggregator
Wang et al. [31]	2 rounds peer-to-aggregator
Xu et al. [33]	1 round peer-to-aggregator

requirement of fully peer-to-peer communication by adopting tree-based partial peer-to-peer and peer-to-aggregator communications to compute $u^{(i)}$.

Another line of research [5, 31] focuses on asynchronous VFL using the coordinate descent (CD) approach rather than the standard gradient descent method. Instead of peer-to-peer communication, these systems rely on two rounds of peer-to-aggregator communication to coordinate immediate results and compute the final gradients update. The recent *FedV* framework [33] requires only one round of peer-to-aggregator communication.

18.4.2 Contrasting Privacy-Preserving Mechanisms and Their Threat Models

Existing solutions use a diverse set of privacy-preserving mechanisms. There are multiple applied techniques, including those based on mixed-protocol multi-party computation techniques and other cryptosystems like HE and functional encryption. In Table 18.2, we present the privacy-preserving approaches that each VFL solution employs. Gascón et al. [9] employ mixed traditional MPC techniques such as secret sharing, garbled circuits, and oblivious transfer. Approaches presented in [12, 36] adopt tree-based communication with random masking, and [5, 31] employ differential privacy mechanisms. The solutions presented in [13, 33, 34] rely on cryptosystems, such as partially additive homomorphic encryption (i.e., the Paillier cryptosystem) and inner-production functional encryption.

As shown in Table 18.2, there exist three types of computation: *garbled circuits*-based secure computation, computation over *ciphertext*, and *plaintext* where model updates are transmitted without encryption to the aggregator. Unfortunately, achieving privacy-preserving objectives while computing gradient descent negatively impacts computation, communication, and model performance. In particular, solutions based on tree-structured communication with random noise perturbation are constrained to scenarios where all parties are assumed to be trustworthy and non-colluding, limiting these approaches' applicability. Differential privacy often leads

Table 18.2 Privacy-preserving approach summary of VFL solutions

Proposal	Computation over	Privacy-preserving approach
Gascón et al. [9]	Garbled circuits	Mixed-protocol multi-party computation
Gu et al. [12]	Plaintext	Random mask + tree-structured communication
Zhang et al. [36]	Plaintext	Random mask + tree-structured communication
Chen et al. [5]	Plaintext	Gaussian DP perturbation
Wang et al. [31]	Plaintext	Gaussian DP perturbation
Hardy et al. [13]	Ciphertext	Cryptosystem (partially HE)
Yang et al. [34]	Ciphertext	Cryptosystem (partially HE)
Xu et al. [33]	Ciphertext	Cryptosystem (Functional Encryption)

Table 18.3 Entities and assumptions of representative VFL solutions

Proposal	Entities and assumptions
Gascón et al. [9]	Semi-honest, non-colluding crypto service provider (CSP) and evaluator
Hardy et al. [13]	Semi-honest parties and aggregator
Yang et al. [34]	Semi-honest parties and aggregator
Gu et al. [12]	Semi-honest and non-colluding parties
Zhang et al. [36]	Semi-honest and non-colluding parties
Chen et al. [5]	Semi-honest parties and aggregator
Wang et al. [31]	Semi-honest parties
Xu et al. [33]	Semi-honest aggregator, dishonest passive parties, and trusted crypto service

to less accurate models, while mixed-protocol MPC techniques require multiple rounds of communication for a simple function, incurring substantial transmission overheads; their cryptosystems also consume extra computational resources.

The different privacy mechanisms presented in Table 18.2 are designed to cover a variety of threat models and may require deployment of additional entities. A threat model defines how different entities in the system should behave in order to ensure the defined privacy goals are met. We summarize the threat models for the analyzed solutions in Table 18.3. A *semi-honest entity* is one that adheres to the protocol but may use the information produced during the operation to infer private data.[1]

Gascón et al. [9] use garbled circuits and an architecture with two additional semi-honest, non-colluding entities: a crypto service provider for generating crypto-related parameters and an evaluator to assist with secure protocol evaluation. Hardy et al. [13] and Yang et al. [34] rely on an additive homomorphic encryption cryptosystem to ensure privacy. Their architecture utilizes an aggregator to coordinate the VFL training process. They make the assumption that all parties and the aggregator are semi-honest.

Chen et al. [5] also employ an aggregator to coordinate the training process and assume semi-honest parties and aggregator. Wang et al. [31] propose an architecture where the active party acts as a coordinator of the entire process. Likewise, Gu et al. [12] and Zhang et al. [36] propose employing tree-based communication with random noise perturbation to preserve privacy among all semi-honest active and passive parties but indicate that parties do not collude. Xu et al. [33] deploy an honest-but-curious aggregator and dishonest passive parties, with the additional assumption that the aggregator and passive parties are not colluding. Additionally, they require a trustworthy crypto infrastructure capable of providing key services.

---

[1] The term *honest-but-curious* has the same meaning as the term *semi-honest* in the crypto community.

18.4.3 Contrasting Supported Machine Learning Models

Table 18.4 presents a summary of the ML models supported by existing VFL solutions. Logistic regression is the most popular supported model. For non-crypto solutions, only the proposal [5] can be applied to neural networks model. Homomorphic encryption-based approaches such as [13, 34] rely on Taylor approximation technique to transform the nonlinear loss function used by regression models to an approximated linear loss function, and hence those approaches may result in underperforming models. Only the *FedV* framework [33] can be used to train logistic regression models without requiring Taylor series approximation.

Existing techniques suffer from one or more limitations:

1. Most approaches apply to a limited pool of models. They require the use of the Taylor series approximation to train nonlinear ML models, such as logistic regression, which possibly weakens the model's performance and *cannot* be generalized to solve classification problems. Furthermore, the prediction and inference phases of these VFL solutions rely on approximation-based secure computation or noise perturbation. As such, these solutions cannot predict as accurately as a centralized ML model could.
2. Most approaches that use cryptosystems as part of the training process may substantially increase the training time, unless they reduce the number of communication rounds.
3. Most protocols require a large number of peer-to-peer communication rounds among parties, making it challenging to deploy them in systems that have poor connectivity, or where communication is limited to a few specific entities due to regulations, such as HIPAA.
4. Approaches such as [35] require sharing class distributions, which may lead to potential leakage of parties' private information.

Table 18.4 Supported machine learning models of VFL solutions. Solutions that do not rely on Taylor approximation are preferred

Proposal	Supported models with SGD training
Gascón et al. [9]	Linear regression
Hardy et al. [13]	Logistic regression (LR) with Taylor approximation
Yang et al. [34]	Taylor approximation-based LR with quasi-Newton method
Gu et al. [12]	SVM with kernels
Zhang et al. [36]	Logistic regression
Chen et al. [5]	DP noise-injected LR and neural networks
Wang et al. [31]	DP noise-injected LR
Xu et al. [33]	Linear models, LR, and SVM with kernels

18.5   *FedV*: An Efficient Vertical FL Framework

This section introduces the *FedV* framework, an efficient VFL solution. It significantly decreases the total number of communication interactions required to train models. *FedV* does not require peer-to-peer communication between parties and can be used to train a range of machine learning models using gradient-based algorithms such as stochastic gradient descent and its derivatives. *FedV* achieves these benefits by orchestrating various non-interactive functional encryption techniques, which accelerates the training process in comparison to state-of-the-art approaches. Additionally, FedV supports multiple parties and enables parties to drop out and rejoin without the requirement for dynamic re-keying, while other approaches do not.

18.5.1   Overview of FedV

The *FedV* framework consists of three components: an *aggregator*, a group of *parties*, and an *optional third-party authority (TPA)* crypto-infrastructure for functional encryption that depends on the specific employed cryptosystem. The *aggregator* orchestrates the private entity resolution process and oversees the parties' training. Each *party* has a training dataset with a subset of features and wishes to collectively train a global model. There are two sorts of parties: one active party that has a partial feature set and class labels (denoted by p_1 in the rest of the section), and several passive parties that own only the partial feature set.

Algorithm 18.1 *FedV* framework

Input: batch size s, *maxEpochs*, and total batches per epoch S, total number of features d.

Output: TPA initializes cryptosystems with keys and a secret seed r to each party.
  Party:
    Re-shuffle samples using entity resolution vector $(\pi_1, \ldots, \pi_n)$
    Use r to generate its one-time password chain for batch selection
  Aggregator:
    $w \leftarrow$ random initialization
    foreach epoch in *maxEpochs* **do**
      $\nabla E(w) \leftarrow$ FedV-SecGrad($epoch, s, S, d, w$)
      $w \leftarrow w - \alpha \nabla E(w)$
    end for
    return w

Algorithm 18.1 illustrates the generic operations of *FedV*. To begin, the system is initialized by creating the necessary cryptographic keys. Following that, a PER process is performed to align the training data samples across all parties, in which each party receives an entity resolution vector, π_i, and shuffles its local data samples appropriately.

For each training epoch, the proposed *Federated Vertical Secure Gradient Descent (FedV-SecGrad)* approach is used to securely compute the gradient update, which is then used to guide the training process forward. *FedV-SecGrad* is a two-phased secure aggregation operation that enables the computation of gradients as stated in Sect. 18.2 and requires the parties to conduct sample and feature dimension encryption, as well as transmit ciphertext to the aggregator.

The aggregator then generates a *fusion weight vector* based on each party's assigned weight and submits it to the TPA in order to request the functional decryption key. For instance, when the aggregator receives two encrypted inputs ct_1 and ct_2 from parties p_1 and p_2, respectively, it constructs a fusion weight vector (w_1, w_2), each element of which corresponds to the party's assigned weight. Following that, the aggregator transmits it to the TPA. The TPA provides the aggregator with the functional decryption key for computing the inner product between (ct_1, ct_2) and (w_1, w_2). Remain aware that the TPA does not have access to ct_1, ct_2, or the aggregated result. Additionally, keep in mind that the fusion weight vectors include no secret information; they just contain the weights required to aggregate ciphertext received from parties.

A curious aggregator may try to manipulate a fusion weight vector to infer private gradient data. To eliminate this potential inference threat, once the TPA receives a fusion weight vector, a special module called the *Inference Prevention Module (IPM)* inspects the vector to ensure it does not isolate any replies according to a pre-specified aggregation policy. If the IPM determines that the fusion weight vectors are not manipulated by a curious aggregator, the TPA supplies the aggregator with the functional decryption key.

Notably, FedV is also compatible with TPA-free FE schemes in which the functional decryption key is generated by all parties collaboratively. The IPM can also be deployed at each party in this circumstance. The aggregator then obtains the result of the corresponding inner product via decryption, which is performed with the use of the functioning decryption key. As a result, the aggregator can acquire the exact gradients to update the model.

18.5.2   FedV Threat Model and Assumptions

FedV's primary objective is to train a machine learning model while securing the privacy of the features provided by each party. Thus, *FedV* guarantees the privacy of the input, and the adversary's objective is to infer the features of the parties.

FedV considers an honest-but-curious aggregator, who follows the algorithms and protocols appropriately but may attempt to acquire private information from aggregated model updates. *FedV* also assumes a small number of dishonest parties who may attempt to infer the private information of the honest parties. In real-world applications, the aggregator is typically run by large businesses, making it more difficult for adversaries to manipulate the protocol without being noticed. Dishonest parties may collude with each other to try to obtain features from other participants. We presumptively exclude potential collusion between the aggregator and the parties in *FedV*.

A TPA may be used to facilitate functional encryption. However, FedV is also compatible with functional encryption systems that do not use TPA, such as in [1, 6]. In a cryptosystem that makes use of a TPA, the TPA must be completely trusted by the other entities in the system in order to provide the aggregator and party with corresponding keys. In real-world circumstances, various sectors already have entities capable of acting as TPAs. For instance, the Federal Reserve System typically fulfills this role in the banking business. In other areas, TPAs can be managed by third-party corporations such as consulting firms.

Finally, FedV does not explicitly cover denial of service attacks or backdoor attacks [2, 4] in which parties attempt to force the final model to generate a targeted misclassification.

18.5.3 Vertical Training Process: FedV-SecGrad

FedV-SecGrad supports a variety of machine learning models, including logistic regression, SVM, and others. Formally, it can encompass any prediction function that can be written as

$$f(\boldsymbol{x}; \boldsymbol{w}) := g(\boldsymbol{w}^\mathsf{T}\boldsymbol{x}), \tag{18.7}$$

where $g : \mathbb{R} \to \mathbb{R}$ is a differentiable function, and $\boldsymbol{x}$ and $\boldsymbol{w}$ denote the feature vector and the model weights, respectively.

If g is the identity function, then f is simplified to a linear model; otherwise, it defines a subclass of nonlinear machine learning models. When g is the sigmoid function, for example, the defined machine learning objective is a logistic classification/regression model. The following portion of this section contains a more in-depth discussion.

18.5.3.1 *FedV-SecGrad* for Linear Models

Suppose that we use mean-squared loss as the loss function and g is the identity function, for simplicity, $g(\boldsymbol{w}^\mathsf{T}\boldsymbol{x}) = \boldsymbol{w}^\mathsf{T}\boldsymbol{x}$. In the case of linear models, the target loss is denoted as

$$E(\boldsymbol{w}) = \frac{1}{2n}\sum_{i=1}^{n}(y^{(i)} - \boldsymbol{w}^{\mathsf{T}}\boldsymbol{x}^{(i)})^2. \tag{18.8}$$

Following that, the computation of gradients over vertically partitioned data in *FedV-SecGrad* is denoted as

$$\nabla E(\boldsymbol{w}) = -\frac{2}{n}\sum_{i=1}^{n}(y^{(i)} - \boldsymbol{w}^{\mathsf{T}}\boldsymbol{x}^{(i)})\boldsymbol{x}^{(i)}. \tag{18.9}$$

Then, the secure computation can be simplified to two types of operations: feature dimension aggregation and sample/batch dimension aggregation. *FedV-SecGrad* performs these two actions using a *two-phase secure aggregation (2Phase-SA)* technique. *Feature dimension secure aggregation* securely aggregates a batch of training data from all parties via grouping across features, to acquire the value of $y^{(i)} - \boldsymbol{w}^{\mathsf{T}}\boldsymbol{x}^{(i)}$ for each data sample. Next, *Sample dimension secure aggregation* can securely aggregate one party's training data via grouping across samples with the weight of $y^{(i)} - \boldsymbol{w}^{\mathsf{T}}\boldsymbol{x}^{(i)}$ for each sample, to obtain the batch gradient $\nabla E(\boldsymbol{w})$ as illustrated in Eq. (18.9). The interaction between the parties and the aggregator is one-way and requires only one round of message.

To explain *FedV-SecGrad*, we will assume the simple example of two parties, where p_1 is the active party and p_2 is the passive party. Recall that the training batch size is s and the total number of features is d. The current training batch samples for p_1 and p_2 can be denoted as

$$\mathcal{B}_{p_1}^{s\times m} : \{y^{(i)}; x_{k_1}^{(i)}\}_{1\leq k_1\leq m}^{1\leq i\leq s}, \tag{18.10}$$

$$\mathcal{B}_{p_2}^{s\times(d-m)} : \{x_{k_2}^{(i)}\}_{m+1\leq k_2\leq d}^{1\leq i\leq s}, \tag{18.11}$$

where, to be more precise, active party p_1 has labels $y^{(i)}$ and partial features indexed from 1 to m, whereas passive party p_2 only has partial features indexed from $m+1$ to d. Then, two distinct types of secure aggregation are performed:

- *Feature Dimension Secure Aggregation.* The objective of feature dimension secure aggregation is to securely aggregate the sum of multiple parties' *partial model prediction*, denoted as

$$\boldsymbol{w}_{p_1}^{\mathsf{T}}\boldsymbol{x}_{p_1}^{(i)} = w_1 x_1^{(i)} + w_2 x_2^{(i)} + \ldots + w_m x_m^{(i)}, \tag{18.12}$$

$$\boldsymbol{w}_{p_2}^{\mathsf{T}}\boldsymbol{x}_{p_2}^{(i)} = w_{m+1} x_{m+1}^{(i)} + w_{m+2} x_{m+2}^{(i)} + \ldots + w_d x_d^{(i)}, \tag{18.13}$$

without disclosing the inputs $\boldsymbol{x}_{p_1}^{(i)}$ and $\boldsymbol{x}_{p_2}^{(i)}$ to the aggregator. Taking the sth data sample in the batch as an example, the aggregator is capable of securely aggregating $\boldsymbol{w}_{p_1}^{\mathsf{T}}\boldsymbol{x}_{p_1}^{(s)} - y^{(s)} + \boldsymbol{w}_{p_2}^{\mathsf{T}}\boldsymbol{x}_{p_2}^{(s)}$. For this purpose, the active party and all other passive parties perform slightly different pre-processing steps before invoking *FedV-SecGrad*.

The active party, p_1, directly subtracts $\boldsymbol{w}_{p_1}^{\mathsf{T}} \boldsymbol{x}_{p_1}^{(i)}$ with labels y to obtain $\boldsymbol{w}_{p_1}^{\mathsf{T}} \boldsymbol{x}_{p_1}^{(s)} - y^{(s)}$ as its "partial model prediction." For the passive party p_2, its "partial model prediction" is defined by $\boldsymbol{x}_{p_2}^{(i)} \boldsymbol{w}_{p_2}$. Each party p_i encrypts its "partial model prediction" using the multi-input functional encryption (MIFE) algorithm with its encryption key $\mathrm{sk}_{p_i}^{\mathrm{MIFE}}$ and sends it to the aggregator. Once the aggregator receives the partial model predictions, it prepares a fusion weight vector $\boldsymbol{v}_{\mathcal{P}}$ of size equal to the number of parties involved in the aggregation (as explained in Sect. 18.5.1) and sends it to the TPA, to request a functional decryption key $\mathrm{dk}_{\boldsymbol{v}_{\mathcal{P}}}^{\mathrm{MIFE}}$. With the received key $\mathrm{dk}_{\boldsymbol{v}_{\mathcal{P}}}^{\mathrm{MIFE}}$, the aggregator can obtain the sum of $\boldsymbol{w}_{p_1}^{m \times 1} \mathcal{B}_{p_1}^{s \times m} - \boldsymbol{y}^{1 \times s}$ and $\boldsymbol{w}_{p_2}^{(d-m) \times 1} \mathcal{B}_{p_2}^{s \times (d-m)}$ through one-step decryption.

- *Sample Dimension Secure Aggregation.* The purpose of the sample dimension secure aggregation is to securely aggregate the batch gradients. For instance, for feature weight w_1 from p_1, the aggregator is able to securely aggregate "partial gradient updates" via *sample dimension secure aggregation*, denoted as

$$\nabla E(w_1) = \sum_{k=1}^{s} x_1^{(k)} u_k, \qquad (18.14)$$

where *aggregated model prediction* $u_k = \boldsymbol{w}_{p_1}^{\mathsf{T}} \boldsymbol{x}_{p_1}^{(k)} - y^{(k)} + \boldsymbol{w}_{p_2}^{\mathsf{T}} \boldsymbol{x}_{p_2}^{(k)}$ is the aggregation result of previous *feature dimension secure aggregation*. This secure aggregation protocol requires the party to encrypt its batch samples with its public key $\mathrm{pk}^{\mathrm{SIFE}}$ using the single-input functional encryption (SIFE) cryptosystem. The aggregator then requests a functional decryption key $\mathrm{dk}_{\boldsymbol{u}}^{\mathrm{SIFE}}$ from the TPA using the results of the feature dimension secure aggregation, i.e., the aggregated model prediction $\boldsymbol{u}$ as discussed above. Following that, the aggregator is able to decrypt the ciphertext and obtain the batch gradient $\nabla E(\boldsymbol{w})$ using the functional decryption key $\mathrm{dk}_{\boldsymbol{u}}^{\mathrm{SIFE}}$.

The protocol described above is simple to generalize to n parties. In this situation, the fusion vector $\boldsymbol{v}$ can be specified as a binary vector with n elements, with 1 indicating that the aggregator has received responses from the appropriate party and 0 indicating that it has not.

18.5.3.2   *FedV-SecGrad* for Nonlinear Models

FedV-SecGrad requires the active party to share plaintext labels with the aggregator for nonlinear models. Due to the fact that g is not the identity function and may be nonlinear, the associated gradient computation does not involve exclusively linear operations. We will briefly discuss the extension of logistic and SVM models here. Additional information is available in [33].

- *Logistic Models.* Here, following the generic equation (18.7), the prediction function is described as

$$f(x; w) = \frac{1}{1 + e^{-w^\mathsf{T} x}}, \tag{18.15}$$

where $g(\cdot)$ is the sigmoid function, i.e., $g(z) = \frac{1}{1+e^{-z}}$. Assuming we are working on a classification problem and employing cross-entropy loss, the gradient computation over a mini-batch $\mathcal{B}$ of size s can be expressed as

$$\nabla E_\mathcal{B}(w) = \frac{1}{s} \sum_{k \in \mathcal{B}} (g(w^\mathsf{T} x^{(k)})) - y^{(k)}) x^{(k)}. \tag{18.16}$$

As explained above, the aggregator is capable of securely acquiring $z^{(k)} = w^\mathsf{T} x^{(k)}$ via the feature dimension SA procedure. Following that, it can compute the aggregated *model prediction* $u_k = g(z) - y^{(k)}$ using the shared labels. Finally, *sample dimension SA* is applied to compute $\nabla E_\mathcal{B}(w) = \sum_{i \in \mathcal{B}} u_i x^{(i)}$.

FedV-SecGrad also offers an alternate strategy in circumstances where label sharing is prohibited, by transferring the logistic calculation to linear computation via Taylor approximation, as several existing VFL solutions [13, 34] do.

- *SVM Models.* SVM with kernel is typically employed when the data is not separable linearly. The linear SVM model employs a squared hinge loss function with the goal of minimizing the following function:

$$\frac{1}{n} \sum_{k \in \mathcal{B}} \left(\max(0, 1 - y^{(k)} w^\mathsf{T} x^{(k)}) \right)^2. \tag{18.17}$$

Then, the gradient computation over a mini-batch $\mathcal{B}$ of size s can be described as

$$\nabla E_\mathcal{B}(w) = \frac{1}{s} \sum_{k \in \mathcal{B}} -2 y^{(k)} (\max(0, 1 - y^{(k)} w^\mathsf{T} x^{(k)})) x^{(k)}. \tag{18.18}$$

With the provided labels and acquired $w^\mathsf{T} x^{(k)}$ as explained above, the aggregator is capable of securely computing the following aggregated *model prediction* value:

$$u_k = -2 y^{(k)} \max(0, 1 - y^{(k)} w^\mathsf{T} x^{(k)}). \tag{18.19}$$

Following that, the partial gradients $\nabla E_\mathcal{B}(w) = \frac{1}{s} \sum_{k \in \mathcal{B}} u_k x^{(k)}$ can be updated as the same *simple dimension secure aggregation* approach.

Additionally, for the case of SVM with nonlinear kernels, assuming the prediction function is

$$f(x; w) = \sum_{i=1}^{n} w_i y_i k(x_i, x),\qquad(18.20)$$

where $k(\cdot)$ denotes the corresponding kernel. Nonlinear kernel functions, such as polynomial kernel $(x_i^{\mathsf{T}} x_j)^d$, and sigmoid kernel $tanh(\beta x_i^{\mathsf{T}} x_j + \theta)$ (β and θ are kernel coefficients), are based on inner-product computation, which is supported by our *feature dimension secure aggregation* and *sample dimension secure aggregation* protocols. These kernel matrices can be generated prior to the start of the training procedure. The objective stated previously for SVMs with nonlinear kernels will be reduced to an SVM with linear kernels using the pre-computed kernel matrix. Then, the gradient computation procedure for these SVM models will be reduced to that of a conventional linear SVM, which *FedV-SecGrad* can plainly support.

18.5.4   Analysis and Discussion

We present the performance comparison of *FedV*, *HE-VFL* by *Hardy et al.* [13], and centralized LR. *HE-VFL* is a VFL solution with secure protocols developed using additive homomorphic encryption (HE), and centralized LR is a centralized (non-FL) logistic regression with and without Taylor series approximation. It is denoted by the acronyms *centralized LR and centralized LR(approx.)*. Correspondingly, *FedV* also trained two models: a logistic regression model, referred as *FedV*, and a logistic regression model with Taylor series approximation, which reduces the logistic regression model to a linear model, referred as *FedV(approx.)*.

The test accuracy and training time for each approach to logistic regression training on several datasets are shown in Fig. 18.4. *FedV* and *FedV(approx.)*, *HE-VFL* and *centralized* have comparable accuracy across all four datasets. With 360 total training epochs, *FedV*'s and *FedV(approx.)* cut training time for a set of chosen

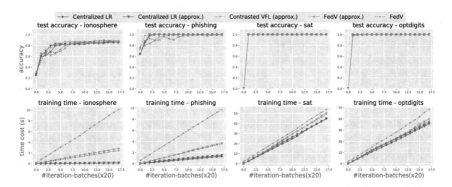

Fig. 18.4 Model accuracy and training time comparisons for logistic regression with two parties

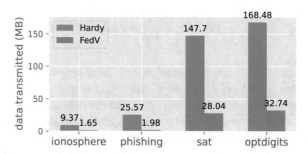

Fig. 18.5 Total data transmitted while training a LR model over 20 training epochs with two parties

datasets. The variance in training time reduction across datasets is due to the fact that data sample sizes and model convergence speed are varied. As illustrated in Fig. 18.5, *FedV* also improves with respect to data transmission efficiency as compared to additive homomorphic encryption-based VFL solution; this is because *FedV* depends entirely on non-interactive secure aggregation methods and does not require the multiple communication rounds.

In some applications, parties may suffer connectivity issues that temporarily prevent them from communicating with the aggregator. The ability to easily recover from such disturbances, without losing other parties' computations, would help reduce training time. *FedV* dynamically permits a restricted number of inactive parties to leave and rejoin throughout the training phase. This is achievable because *FedV* does not require sequential peer-to-peer communication between parties nor does it require re-keying procedures in the event of a party failure. To overcome failure situations, *FedV* enables the aggregator to set the corresponding element in fusion weight vector to zero.

18.6 Conclusions

The majority of existing privacy-preserving FL systems are limited to datasets that are horizontally partitioned. In some real-world FL scenarios, datasets may be partitioned vertically, which means that not all parties have access to the same feature set. As a result, parties cannot independently train a complete local model, as done in horizontal FL. To overcome this limitation, several VFL solutions have been proposed. In this chapter, we have reviewed these solutions and demonstrated differences in communication, computation requirements, number of entities required, and trust assumptions. These differences make the solutions suitable for different types of FL scenarios. We also presented *FedV*, an approach that substantially reduces the training time and data transfer by eliminating all communication between parties.

References

1. Abdalla M, Benhamouda F, Kohlweiss M, Waldner H (2019) Decentralizing inner-product functional encryption. In: IACR international workshop on public key cryptography. Springer, pp 128–157
2. Bagdasaryan E, Veit A, Hua Y, Estrin D, Shmatikov V (2018) How to backdoor federated learning. Preprint. arXiv:1807.00459
3. Bonawitz K, Eichner H, Grieskamp W, Huba D, Ingerman A, Ivanov V, Kiddon C, Konecny J, Mazzocchi S, McMahan HB et al (2019) Towards federated learning at scale: System design. Preprint. arXiv:1902.01046
4. Chen B, Carvalho W, Baracaldo N, Ludwig H, Edwards B, Lee T, Molloy I, Srivastava B (2018) Detecting backdoor attacks on deep neural networks by activation clustering. Preprint. arXiv:1811.03728
5. Chen T, Jin X, Sun Y, Yin W (2020) Vafl: a method of vertical asynchronous federated learning. Preprint. arXiv:2007.06081
6. Chotard J, Sans ED, Gay R, Phan DH, Pointcheval D (2018) Decentralized multi-client functional encryption for inner product. In: International conference on the theory and application of cryptology and information security. Springer, pp 703–732
7. Corinzia L, Buhmann JM (2019) Variational federated multi-task learning. Preprint. arXiv:1906.06268
8. Fang C, Li CJ, Lin Z, Zhang T (2018) Spider: Near-optimal non-convex optimization via stochastic path integrated differential estimator. Preprint. arXiv:1807.01695
9. Gascón A, Schoppmann P, Balle B, Raykova M, Doerner J, Zahur S, Evans D (2016) Secure linear regression on vertically partitioned datasets. IACR Cryptology ePrint Archive 2016, 892
10. Geyer RC, Klein T, Nabi M (2017) Differentially private federated learning: A client level perspective. Preprint. arXiv:1712.07557
11. Ghadimi S, Lan G (2013) Stochastic first-and zeroth-order methods for nonconvex stochastic programming. SIAM J Optim 23(4):2341–2368
12. Gu B, Dang Z, Li X, Huang H (2020) Federated doubly stochastic kernel learning for vertically partitioned data. In: Proceedings of the 26th ACM SIGKDD international conference on knowledge discovery & data mining, pp 2483–2493
13. Hardy S, Henecka W, Ivey-Law H, Nock R, Patrini G, Smith G, Thorne B (2017) Private federated learning on vertically partitioned data via entity resolution and additively homomorphic encryption. Preprint. arXiv:1711.10677
14. Ion M, Kreuter B, Nergiz AE, Patel S, Raykova M, Saxena S, Seth K, Shanahan D, Yung M (2019) On deploying secure computing commercially: Private intersection-sum protocols and their business applications. IACR Cryptol. ePrint Arch. 2019, 723
15. Jin X, Du R, Chen PY, Chen T (2020) Cafe: Catastrophic data leakage in federated learning. OpenReview - Preprint
16. Kingma DP, Ba J (2014) Adam: A method for stochastic optimization. Preprint. arXiv:1412.6980
17. Lan G, Lee S, Zhou Y (2020) Communication-efficient algorithms for decentralized and stochastic optimization. Math Program 180(1):237–284
18. Li O, Sun J, Yang X, Gao W, Zhang H, Xie J, Smith V, Wang C (2021) Label leakage and protection in two-party split learning. Preprint. arXiv:2102.08504
19. Ludwig H, Baracaldo N, Thomas G, Zhou Y, Anwar A, Rajamoni S, Ong Y, Radhakrishnan J, Verma A, Sinn M, et al (2020) IBM federated learning: an enterprise framework white paper v0. 1. Preprint. arXiv:2007.10987
20. Luo X, Wu Y, Xiao X, Ooi BC (2021) Feature inference attack on model predictions in vertical federated learning. In: 2021 IEEE 37th international conference on data engineering (ICDE). IEEE, pp 181–192
21. McMahan HB, Moore E, Ramage D, Hampson S et al (2016) Communication-efficient learning of deep networks from decentralized data. Preprint. arXiv:1602.05629

22. Nesterov Y (1998) Introductory lectures on convex programming volume I: Basic course. Lecture Notes 3(4):5
23. Nock R, Hardy S, Henecka W, Ivey-Law H, Patrini G, Smith G, Thorne B (2018) Entity resolution and federated learning get a federated resolution. Preprint. arXiv:1803.04035
24. Schnell R, Bachteler T, Reiher J (2011) A novel error-tolerant anonymous linking code. German Record Linkage Center, Working Paper Series No. WP-GRLC-2011-02
25. Singh A, Vepakomma P, Gupta O, Raskar R (2019) Detailed comparison of communication efficiency of split learning and federated learning. Preprint. arXiv:1909.09145
26. Slavkovic AB, Nardi Y, Tibbits MM (2007) Secure logistic regression of horizontally and vertically partitioned distributed databases. In: Seventh IEEE international conference on data mining workshops (ICDMW 2007). IEEE, pp. 723–728
27. Thapa C, Chamikara MAP, Camtepe S (2020) Splitfed: When federated learning meets split learning. Preprint. arXiv:2004.12088
28. Vaidya J (2008) A survey of privacy-preserving methods across vertically partitioned data. In: Privacy-preserving data mining. Springer, pp 337–358
29. Vaidya J, Clifton C, Kantarcioglu M, Patterson AS (2008) Privacy-preserving decision trees over vertically partitioned data. ACM Trans Knowl Discov Data (TKDD) 2(3):14
30. Vepakomma P, Gupta O, Swedish T, Raskar R (2018) Split learning for health: Distributed deep learning without sharing raw patient data. Preprint. arXiv:1812.00564
31. Wang C, Liang J, Huang M, Bai B, Bai K, Li H (2020) Hybrid differentially private federated learning on vertically partitioned data. Preprint. arXiv:2009.02763
32. Xu R, Baracaldo N, Zhou Y, Anwar A, Ludwig H (2019) Hybridalpha: An efficient approach for privacy-preserving federated learning. In: Proceedings of the 12th ACM workshop on artificial intelligence and security. ACM
33. Xu R, Baracaldo N, Zhou Y, Anwar A, Joshi J, Ludwig H (2021) Fedv: Privacy-preserving federated learning over vertically partitioned data. Preprint. arXiv:2103.03918
34. Yang, K, Fan T, Chen T, Shi Y, Yang Q (2019) A quasi-newton method based vertical federated learning framework for logistic regression. Preprint. arXiv:1912.00513
35. Yu H, Vaidya J, Jiang X (2006) Privacy-preserving SVM classification on vertically partitioned data. In: Pacific-Asia conference on knowledge discovery and data mining. Springer, pp 647–656
36. Zhang Q, Gu B, Deng C, Huang H (2021) Secure bilevel asynchronous vertical federated learning with backward updating. Preprint. arXiv:2103.00958
37. Zhao Y, Li M, Lai L, Suda N, Civin D, Chandra V (2018) Federated learning with non-IID data. Preprint. arXiv:1806.00582

Chapter 19
Split Learning: A Resource Efficient Model and Data Parallel Approach for Distributed Deep Learning

Praneeth Vepakomma and Ramesh Raskar

Abstract Resource constraints, workload overheads, lack of trust, and competition hinder the sharing of raw data across multiple institutions. This leads to a shortage of data for training state-of-the-art deep learning models. Split Learning is a model and data parallel approach of distributed machine learning, which is a highly resource efficient solution to overcome these problems. Split Learning works by partitioning conventional deep learning model architectures such that some of the layers in the network are private to the client and the rest are centrally shared at the server. This allows for training of distributed machine learning models without any sharing of raw data while reducing the amount of computation or communication required by any client. The paradigm of split learning comes in several variants depending on the specific problem being considered at hand. In this chapter we share theoretical, empirical, and practical aspects of performing split learning and some of its variants that can be chosen depending on the application of your choice.

19.1 Introduction to Split Learning

Federated learning [1] is a data parallel approach where the data is distributed while every client that is part of a training round trains the exact same model architecture using its own local data. The server that could potentially be a powerful computational resource in the real world ends up performing a relatively easier computation, which is that of performing a weighted average of the weights learnt by each of the clients. In the real world, there often exist clients that are relatively resource constrained in comparison to a server.

P. Vepakomma (✉) · R. Raskar
Massachusetts Institute of Technology, Cambridge, MA, USA
e-mail: vepakom@mit.edu; raskar@mit.edu

Split learning [2, 3] caters to this realistic setting by splitting the model architecture across layers such that each client maintains the weights up to an intermediate layer known as the split layer. The rest of the layers are held at the server.

Benefits and Limitations This approach not only reduces the computational work that is to be performed at any client, but it also reduces the size of communication payloads required to be sent during the distributed training. This is because it only requires activations from just one layer (split layer) to be sent to the server from any client during the forward propagation step. At the same time gradients from only one layer (the layer after the split layer) need to be sent by the server to the client during the backpropagation step. In terms of model performance, we empirically observe that the convergence of SplitNN remained much faster than federated learning and large batch synchronous stochastic gradient descent [4]. That said, it requires a relatively larger overall communication bandwidth when training over a smaller number of clients although it ends up being much lower than other methods in settings with large number of clients. Advanced neural network compression methods such as [5–7] can be used to reduce the communication load. The communication bandwidth can also be traded for computation on client by allowing for more layers at client to represent further compressed representations.

Sharing of activations from intermediate layers as in split learning is also relevant in distributed learning approaches of local parallelism [8], features replay [9], and divide and conquer quantization [10]. This is as opposed to weight sharing as is done in federated learning.

19.1.1 Vanilla Split Learning

In this method each client trains the network up to a certain layer known as the split layer and sends the weights to server (Fig. 19.1). The server then trains the network for rest of the layers. This completes the forward propagation. Server then generates the gradients for the final layer and back-propagates the error until the split layer. The gradient is then passed over to the client. The rest of the back-propagation is completed by the client. This is continued till the network is trained. The shape of the split could be arbitrary and not necessarily, vertical. In this framework as well there is no explicit sharing of raw data.

19.1.1.1 Synchronization Step

After each client finishes its epoch, the next client that is in queue to finish its epoch receives the local weights (weights up to the split layer) from the previous client as its initialization for its epoch.

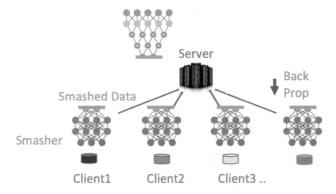

Fig. 19.1 Split learning setup with multiple clients and a server with dotted green line showing the split between the client's share of layers and the server's share of layers. Activations from only the split layer (last layer of client) are shared during forward propagation and gradients from only first layer of server are shared with client during backpropagation

19.1.1.2 Relaxing Synchronization Requirements

This additional communication between clients can be avoided via approaches like BlindLearning [11] for split learning, which is based on using a loss function that is an average of losses obtained by the forward propagations completed by each client. Similarly, the communication and synchronization requirements are further reduced via splitFedv1 [12], splitFedv2 [12], and splitFedv3 [13], which are hybrid approaches of split learning and federated learning. A hybrid approach that improves upon latencies is provided in [14].

19.2 Communication Efficiency [15]

In this section we describe our calculations of the communication efficiency for both of the distributed learning setups of split learning and federated learning. For analyzing the communication efficiency, we consider the amount of data transferred by every client for the training and client weight synchronization since rest of the factors affecting the communication rate is dependent on the setup of training cluster and is independent of the distributed learning setup. We use the following notation to mathematically measure the communication efficiencies.

Notation K = # clients, N = # model parameters, p = total dataset size, q = size of the split layer, η = fraction of model parameters (weights) with client, and therefore $1 - \eta$ is fraction of parameters with server.

In Table 19.1 we show the communication required per client per one epoch as well as total communication required across all clients per one epoch. As there are K clients, when size of the training dataset across each client is the

Algorithm 19.1 `SplitNN`. The K clients are indexed by k; B is the local minibatch size, and η is the learning rate

---

Server executes at round $t \geq 0$:
  for each client $k \in S_t$ **in parallel do**
    $\mathbf{A}_t^k \leftarrow$ ClientUpdate(k, t)
    Compute $\mathbf{W}_t \leftarrow \mathbf{W}_t - \eta \nabla \mathcal{L}(\mathbf{W}_t; \mathbf{A}_t)$
    Send $\nabla \mathcal{L}(\mathbf{A}_t; \mathbf{W}_t)$ to client k for ClientBackprop(k, t)
  end for

ClientUpdate(k, t):   *// Run on client k*
  $\mathbf{A}_t^k = \phi$
  for each local epoch i from 1 to E **do**
    for batch $b \in \mathcal{B}$ **do**
      Concatenate $f(b, \mathbf{H}_t^k)$ to $\mathbf{A}_t^k$
    end for
  end for
  return $\mathbf{A}_t^k$ to server

ClientBackprop$(k, t, \nabla \mathcal{L}(\mathbf{A}_t; \mathbf{W}_t))$:   *// Run on client k*
  for batch $b \in \mathcal{B}$ **do**
    $\mathbf{H}_t^k = \mathbf{H}_t^k - \eta \nabla \mathcal{L}(\mathbf{A}_t; \mathbf{W}_t; b)$
  end for

Table 19.1 Communication per client and total communication for the distributed learning setup as measured by the data transferred by all of the nodes in the learning setup

Method	Communication per client	Total Comm.
Split learning (client weight sharing)	$(p/K)q + (p/K)q + \eta N$	$2pq + \eta N K$
Split learning (no client weight sharing)	$(p/K)q + (p/K)q$	$2pq$
Federated learning	$2N$	$2KN$

same, there would be p/K data records per client in split learning. Therefore during forward propagation the size of the activations that are communicated per client in split learning is $(p/K)q$ and during backward propagation the size of gradients communicated per client is also $(p/K)q$. In the vanilla split learning case where there is client weight sharing, passing on the weights to next client would involve a communication of ηN. In federated learning the communication of weights/gradients during upload of individual client weights and download of averaged weights are both of size N each.

Table 19.2 Computation load (per client), total communication load, and latency required for one global round

Methods	Comp.	Comm.	Latency
FL	$|D|\|\mathbf{w}\|$	$2|\mathbf{w}|K$	$\frac{2|\mathbf{w}|K}{R} + \frac{|D|\|\mathbf{w}\|}{P_C}$
SplitFed	$\alpha|D|\|\mathbf{w}\|$	$(2q|D| + 2\alpha|\mathbf{w}|)K$	$\frac{(2q|D|+2\alpha|\mathbf{w}|)K}{R} + \frac{\alpha|D|\|\mathbf{w}\|}{P} + \frac{(1-\alpha)|D|\|\mathbf{w}\|K}{P_S}$
Hybrid	$\alpha|D|\|\mathbf{w}\|$	$(q|D| + 2\alpha|\mathbf{w}|)K$	$\frac{(q|D|+\alpha|\mathbf{w}|)K}{R} + \frac{\alpha\beta|D|\|\mathbf{w}\|}{P_C} +$ $\max\left(\frac{\alpha|\mathbf{w}|K}{R} + \frac{\alpha(1-\beta)|D|\|\mathbf{w}\|}{P_C}, \frac{(1-\alpha)|D|\|\mathbf{w}\|K}{P_S} \right)$

19.3 Latencies

Depending on the computing power constraints of the client and server, latencies in computation need to be minimized while keeping the communication efficiency to be high. To that effect, [14] provides analytical comparison of latencies of vanilla split learning, splitFed, and the approach proposed in [14]. They consider the following notation of model size being $|w|$, proportion of weights on client αw and server $(1 - \alpha)|w|$, computing powers of client P_C and that of server P_S, and uplink and downlink transmission rates of R and K being the number of clients. The time required for forward propagation is modeled as $\frac{\beta|D|\|w\|}{P}$ and time required for backward propagation as $\frac{(1-\beta)|D|\|w\|}{P}$. With this notation [14] gives the following Table 19.2 comparing the latencies, and resource efficiencies of federated learning, splitFed, and the recent hybrid method between split learning and federated learning in [14].

19.4 Split Learning Topologies

19.4.1 Versatile Configurations

In addition to the discussed vanilla split learning and its variants that require lesser synchronization, there are other topologies in which split learning could be used as described below.

1. **U-shaped configuration for split learning without label sharing [3, 16]:** The other two configurations described in this section involve sharing of labels although they do not share any raw input data with each other. We can completely mitigate this problem by a U-shaped configuration that does not require any label sharing by clients. In this setup we wrap the network around at end layers of server's network and send the outputs back to client entities as seen in Fig. 19.2b. While the server still retains a majority of its layers, the clients generate the gradients from the end layers and use them for backpropagation without sharing the corresponding labels. In cases where labels include highly

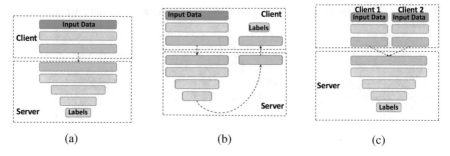

Fig. 19.2 Split learning configuration for health shows raw data is not transferred between the client and server health entities for training and inference of distributed deep learning models with SplitNN. (**a**) Simple vanilla split learning. (**b**) Split learning without label sharing. (**c**) Split learning for vertically partitioned data

sensitive information like the disease status of patients, this setup is ideal for distributed deep learning.

2. **Vertically partitioned data for split learning [17]:** This configuration allows for multiple institutions holding different modalities of patient data to learn distributed models without data sharing. In Fig. 19.2c, we show an example configuration of SplitNN suitable for such multi-modal multi-institutional collaboration. As a concrete example we walk through the case where radiology centers collaborate with pathology test centers and a server for disease diagnosis. As shown in Fig. 19.2c radiology centers holding imaging data modalities train a partial model up to the split layer. In the same way the pathology test center having patient test results trains a partial model up to its own split layer. The outputs at the split layer from both these centers are then concatenated and sent to the disease diagnosis server that trains the rest of the model. This process is continued back and forth to complete the forward and backward propagations in order to train the distributed deep learning model without sharing each other's raw data.

3. **Extended vanilla split learning:** As shown in Fig. 19.3a we give another modification of vanilla split learning where the result of concatenated outputs is further processed at another client before passing it to the server.

4. **Configuration for multi-task split learning:** As shown in Fig. 19.3b, in this configuration multi-modal data from different clients is used to train partial networks up to their corresponding split layers. The outputs from each of these split layers are concatenated and then sent over to multiple servers. These are used by each server to train multiple models that solve different supervised learning tasks.

5. **Tor [18] like configuration for multi-hop split learning:** This configuration is an analogous extension of the vanilla configuration. In this setting multiple clients train partial networks in sequence where each client trains up to a split layer and transfers its outputs to the next client. This process is continued as

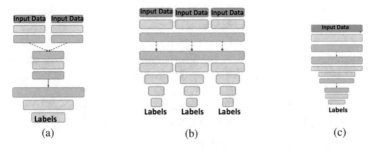

Fig. 19.3 Split learning configuration for health shows raw data is not transferred between the client and server health entities for training and inference of distributed deep learning models with SplitNN. (**a**) Extended vanilla. (**b**) Multi-task output with vertically partitioned input. (**c**) 'Tor' [18]

shown in Fig. 19.3c as the final client sends its activations from its split layer to a server to complete the training.

We would like to note that although these example configurations show some versatile applications for SplitNN, they are by no means the only possible configurations.

19.4.2 Model Selection with ExpertMatcher [19]

In some scenarios a powerful server hosts a repository of multiple proprietary models that it would like to use in a machine learning as a service (MLaaS) business model via a prediction API. The proprietary models cannot be offered to be downloaded by the client. At the same time, the clients often have sensitive datasets that it would like to obtain predictions for. In this setup, arises the problem of matching the right model(s) from the server's repository with respect to the dataset held by the client. ExpertMatcher is such a model-selection architecture based on the U-shaped boomerang split learning topology.

19.4.3 Implementation Details

We assume that we have K pre-trained expert networks on the centralized server, each of these networks has its corresponding pre-trained unsupervised representation learning models (we consider autoencoders (AE) in this example) ϕ_K trained on a task-specific dataset. Given the dataset on which the AE was trained on, we extract the encoded representations of the whole dataset and compute an average representation of the dataset $\mu_k \in \mathbb{R}^d, k \in \{1, \ldots, K\}$, where d is the feature dimension. Assuming the dataset consist of N object classes, we also compute the average representation of each class in the dataset $\mu_k^n \in \mathbb{R}^d$, $n \in \{1, \ldots, N\}$, $k \in \{1, \ldots, K\}$.

The clients (Client A and Client B) utilize a similar approach as the server, where the clients train their unique AE's one for each pth and qth datasets, Client A: $p \in \{1, \ldots, P\}$ and Client B: $q \in \{1, \ldots, Q\}$. Let us assume for Client A, the intermediate features extracted from a hidden layer for a sample X_p^1 are given as $x_p^1 = \phi_p^1(X_p^1)$, and similarly, for Client B it is $x_q^2 = \phi_q^2(X_q^2)$. For brevity, we denote the intermediate representation coming from any client as x'.

We would like to first explain the notion of coarseness or fineness of labels, by which we mean that the classes in the data are separated by high-level (coarse) or low-level (fine) semantic categories. As an example, classes that separate dogs from cats are coarse categories, while classes that separate different types of dogs are fine categories. In the concept of ExpertMatcher, for coarse assignment (CA) of clients data: To the encoded representation x', we assign a server AE, $k^* \in \{1, \ldots, K\}$, that has maximum similarity of x' with μ_k; see Fig. 19.4.

For fine-grained (FA) assignment of clients data: To the encoded representation x', we assign an expert network M_n, $n \in \{1, \ldots, N\}$, that has the maximum similarity of x'_{k*} with μ_k^n. The choice of similarity used for assignment depends on the user. Cosine similarity, distance correlation, information theoretic measures, Hilbert–Schmidt independence criterion, maximum mean discrepancy, kernel target alignment, and integral probability metrics are just a few possibilities for the similarity metric.

Finally after the assignment of the given sample to the model, one can easily train a SplitNN type architecture [3].

In the current setup, a weak level of privacy is preserved as the server does not have access to the raw client's data, but rather a very low dimensional encoded representation.

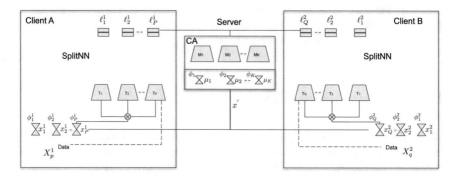

Fig. 19.4 Pipeline of ExpertMatcher with encoded representations at the client and server for automatic model selection from a repository of server models based on their relevance to query data set hosted by the client. The matching is done based on encoded intermediate representations

Note that there is a shortcoming of this approach. If the server has no AE model dedicated to the client data, the wrong assignment of client data takes place because of maximum cosine similarity criteria—this can be resolved by adding an additional model on the server that performs a binary classification: if the client data matches the server data or not.

19.5   Collaborative Inference with Split Learning

As organizations are able to train ultra-large machine learning models on huge datasets with massive computing resources, it opens up a new set of problems for external clients that intend to predict with these models. The client would not like to download these large models in their entirety on-device given that they often have billions of parameters. Predicting with these models is computationally resource intensive to solely be performed on-device. This opens up the problem of private collaborative inference (PCI) where the model is split across the client and server (Table 19.3).

The clients' data is private and therefore the activations that are communicated in this setting need to be formally privatized to prevent membership inference and reconstruction attacks. There has been considerable work in the alternate setting, where the server intends to privately share the weights of a trained model. In this setting of PCI, the privacy considered is with regard to the server's own data. The setting of PCI is instead relatively new, as it requires private sharing of activations during private inference with regard to client's own private data as opposed to private sharing of weights after private training with regard to server's data. This requires innovations at the intersection of distributed machine learning based on activation sharing as opposed to weight sharing and formal privacy.

19.5.1   Preventing Reconstruction Attacks in Collaborative Inference

The client's data records on which the predictions need to be obtained are private and therefore the model's intermediate representations (or activations) that are com-

Table 19.3 Differences between the settings of private collaborative inference and private distributed model training

	Private collaborative inference	Private model training
Communication payload	Intermediate activations	Weights
Inference mechanism	Model inference is distributed	Client downloads model
Privatized entity	Query sample	Training data

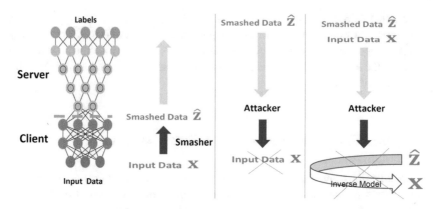

Fig. 19.5 The setting of reconstruction attacks from intermediate activations in the context of split learning

municated in this setting of PCI need to be desensitized to prevent reconstruction attacks (Fig. 19.5). Privacy-preserving machine learning has not reached its AlexNet moment from an architectural perspective. The field has made rapid strides on formal privacy mechanisms like DP-SGD [24] and its variants. There is still a lot of room for improving the current tradeoffs of privacy vs. utility in these methods to make them amenable to many production use cases. We now describe some advances in activation sharing for (a) preventing membership inference attacks with respect to training data and (b) in preventing reconstruction attacks of prediction query data in the setting of PCI.

19.5.1.1 Channel Pruning

The work in [20] shows that learning a pruning filter to selectively prune out channels in the latent representation space at the split layer helps in empirically preventing various state-of-the-art reconstruction attacks during the prediction step in the setting of PCI (Fig. 19.6).

19.5.1.2 Decorrelation

The key idea here is to reduce information leakage by adding an additional loss term to the commonly used classification loss term, categorical cross-entropy. The information leakage reduction loss term we use is distance correlation, a powerful measure of non-linear (and linear) statistical dependence between random variables. The distance correlation loss is minimized between raw input data and the output of any chosen layer whose outputs need to be communicated from the client to another untrusted client or untrusted server. This setting is crucial to some popular forms of distributed machine learning that require sharing of activations from an intermediate

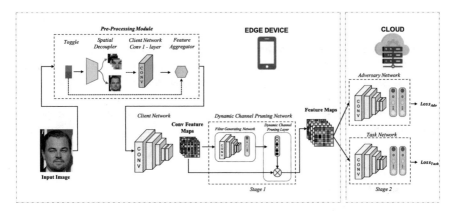

Fig. 19.6 Reference [20] shows that learning a pruning filter to selectively prune out channels in the latent representation space at the split layer helps in empirically preventing various state-of-the-art reconstruction attacks during the prediction step in the setting of PCI

layer. This has been motivated under the "activation sharing" subsection in the motivation section.

Optimization of this combination of losses helps ensure the activations resulting from the protected layer have minimal information for reconstructing raw data while still being useful enough to achieve reasonable classification accuracies upon post-processing. The quality of preventing reconstruction of raw input data while maintaining reasonable classification accuracies is qualitatively and quantitatively substantiated in the experiments section. The joint minimization of distance correlation with cross entropy leads to a specialized feature extraction or transformation such that it is imperceptible in leaking information about the raw dataset with respect to both the human visual system and more sophisticated reconstruction attacks.

19.5.1.3 Loss Function

The total loss function for n samples of input data $\mathbf{X}$, activations from protected layer $\mathbf{Z}$, true labels $\mathbf{Y}_{true}$, predicted labels $\mathbf{Y}$, and scalar weight α is given by:

$$\alpha DCOR(\mathbf{X}, \mathbf{Z}) + (1 - \alpha)CCE(\mathbf{Y}_{true}, \mathbf{Y}). \tag{19.1}$$

19.5.2 Differential Privacy for Activation Sharing

Arachchige et al. [21] provides a differentially private mechanism for sharing activations post a flattening layer obtained after the convolutional and pooling layers. These flattened outputs are binarized and a "utility enhanced randomization" mechanism inspired by RAPPOR is applied to create a differentially private binary representation. These are then communicated to the server where fully connected layers act on them to generate the final predictions. The work in [22] provides a differentially private mechanism for supervised manifold embeddings of features extracted from deep networks in order to perform image retrieval tasks from a database on the server. The work in [23] looks at preventing leakage of information about the labels in the context of split learning. They provide defenses to prevent from norm and hint attacks for revealing label information.

19.6   Future Work

There are several aspects to study with regard to distributed machine learning methods like split learning and federated learning. These include issues of resource efficiency, privacy, convergence, non-homogeneity in real-world data, latency of training, collaborative inference, straggler clients, topologies of communication, attack testbeds, and so forth making this field an active area of current research.

References

1. Konečný J, McMahan HB, Yu FX, Richtárik P, Suresh AT, Bacon D (2016) Federated learning: Strategies for improving communication efficiency. Preprint. arXiv:1610.05492
2. Gupta O, Raskar R (2018) Distributed learning of deep neural network over multiple agents. J Netw Comput Appl 116:1–8
3. Vepakomma P, Gupta O, Swedish T, Raskar R (2018) Split learning for health: Distributed deep learning without sharing raw patient data. Preprint. arXiv:1812.00564
4. Chen J, Pan X, Monga R, Bengio S, Jozefowicz R (2016) Revisiting distributed synchronous SGD. Preprint. arXiv:1604.00981
5. Lin Y, Han S, Mao H, Wang Y, Dally WJ (2017) Deep gradient compression: Reducing the communication bandwidth for distributed training. Preprint. arXiv:1712.01887
6. Han S, Mao H, Dally WJ (2015) Deep compression: Compressing deep neural networks with pruning, trained quantization and huffman coding. Preprint. arXiv:1510.00149
7. Louizos C, Ullrich K, Welling M (2017) Bayesian compression for deep learning. Preprint. arXiv:1705.08665
8. Laskin M, Metz L, Nabarro S, Saroufim M, Noune B, Luschi C, Sohl-Dickstein J, Abbeel P (2020) Parallel training of deep networks with local updates. Preprint. arXiv:2012.03837
9. Huo Z, Gu B, Huang H (2018) Training neural networks using features replay. Preprint. arXiv:1807.04511

10. Elthakeb AT, Pilligundla P, Mireshghallah F, Cloninger A, Esmaeilzadeh H (2020) Divide and conquer: Leveraging intermediate feature representations for quantized training of neural networks. In: International conference on machine learning. PMLR, pp 2880–2891

11. Gharib G, Vepakomma P (2021) Blind learning: An efficient privacy-preserving approach for distributed learning. In: Workshop on split learning for distributed machine learning (SLDML'21)

12. Thapa C, Chamikara MAP, Camtepe S (2020) Splitfed: When federated learning meets split learning. Preprint. arXiv:2004.12088

13. Madaan H, Gawali M, Kulkarni V, Pant A (2021) Vulnerability due to training order in split learning. Preprint. arXiv:2103.14291

14. Han DJ, Bhatti HI, Lee J, Moon J (2021) Han DJ, Bhatti HI, Lee J, Moon J (2021) Accelerating federated learning with split learning on locally generated losses. In: ICML 2021 workshop on federated learning for user privacy and data confidentiality. ICML Board

15. Singh A, Vepakomma P, Gupta O, Raskar R (2019) Detailed comparison of communication efficiency of split learning and federated learning. arXiv:1909.09145

16. Poirot MG, Vepakomma P, Chang K, Kalpathy-Cramer J, Gupta R, Raskar R (2019) Split learning for collaborative deep learning in healthcare. Preprint. arXiv:1912.12115

17. Ceballos I, Sharma V, Mugica E, Singh A, Roman A, Vepakomma P, Raskar R (2020) SplitNN-driven vertical partitioning. Preprint. arXiv:2008.04137

18. Dingledine R, Mathewson N, Syverson P (2004) Tor: The second-generation onion router. Technical report, Naval Research Lab Washington DC

19. Sharma V, Vepakomma P, Swedish T, Chang K, Kalpathy-Cramer J, Raskar R (2019) Expertmatcher: Automating ML model selection for clients using hidden representations. Preprint. arXiv:1910.03731

20. Singh A, Chopra A, Garza E, Zhang E, Vepakomma P, Sharma V, Raskar R (2020) Disco: Dynamic and invariant sensitive channel obfuscation for deep neural networks. Preprint. arXiv:2012.11025

21. Arachchige PCM, Bertok P, Khalil I, Liu D, Camtepe S, Atiquzzaman M (2019) Local differential privacy for deep learning. IEEE Internet Things J 7(7):5827–5842

22. Vepakomma P, Balla J, Raskar R (2021) Differentially private supervised manifold learning with applications like private image retrieval. Preprint. arXiv:2102.10802

23. Li O, Sun J, Yang X, Gao W, Zhang H, Xie J, Smith V, Wang C Label leakage and protection in two-party split learning. Preprint. arXiv:2102.08504

24. Abadi M, Chu A, Goodfellow I, McMahan HB, Mironov I, Talwar K, Zhang L (2016) Deep learning with differential privacy. In: Proceedings of the 2016 ACM SIGSAC conference on computer and communications security, pp 308–318

Part V
Applications

This part sheds more light on multiple enterprise use cases of federated learning. This includes financial services and healthcare, two domains in which federated learning has been adopted early due to their high degree of privacy regulation. We also discuss retail and telecommunications and the needs of these domains.

This part starts with the quite formal discussion of applications of FL in their respective domains but also extends to experiences shared by an FL platform provider and to opportunities lined out for the telecommunications industry.

Chapter 20 reviews an application of federated learning for *financial crimes detection*, in particular anti-money laundering. This is a typical example of collaboration between different institutions in a non-competitive field. Financial regulators require financial institutions to take measure to combat fraud and money laundering due to the social cost incurred by drug trade, people trafficking, and tax evasion. Chapter 21 introduces an interesting case in which federated learning is used for financial portfolio management. This chapter illustrates the use of reinforcement learning in a federated setting.

Chapter 22 provides an interesting example of using federated learning in medical imaging for both segmentation and classification tasks. Chapter 23 addresses the practical application of federated learning in the medical field, also in the context of COVID-19. The chapter is written by an author from Persistent Systems, which offers federated learning services to clients in the medical field. The chapter provides guidance for the development of FL platforms in healthcare based on practical experiences.

Chapter 24 introduces a retail use case and federated learning algorithms for a product recommender system. This is a good example of how to avoid aggregating data in a large-scale retail network, avoiding privacy risk and regulatory scrutiny. It outlines insights into practical issues such as the strong imbalance of data sets between stores.

Lastly, Chap. 25 discusses potential use cases in the telecommunication industry, one of the potentially largest users of federated learning.

Chapter 20
Federated Learning for Collaborative Financial Crimes Detection

Toyotaro Suzumura, Yi Zhou, Ryo Kawahara, Nathalie Baracaldo, and Heiko Ludwig

Abstract Mitigating financial crime risk (e.g., fraud, theft, money laundering) is a large and growing problem. In some way it touches almost every financial institution, as well as many individuals, and in some cases, entire societies. Advances in technology used in this domain, including machine learning-based approaches, can improve upon the effectiveness of financial institutions' existing processes. However, a key challenge that most financial institutions continue to face is that they address financial crimes in isolation without any insight from other firms. Where financial institutions address financial crimes through the lens of their own firm, perpetrators may devise sophisticated strategies that may span across institutions and geographies. In this chapter, we describe a methodology to share key information across institutions by using a federated graph learning platform that enables us to train more accurate detection models by leveraging federated learning as well as graph learning approaches. We demonstrate that our federated model outperforms a local model by 20% with the UK FCA TechSprint data set. This new platform opens up the door to efficiently detect global money laundering activity.

20.1 Introduction: Financial Crimes Detection

Financial crime [3–6, 8, 14] is a broad and growing class of criminal activity involving the misuse, misappropriation, or misrepresentation of entities with monetary value. Common subclasses of financial crime include theft, fraud, and money

T. Suzumura
The University of Tokyo, Tokyo, Japan
e-mail: suzumura@acm.org

Y. Zhou (✉) · N. Baracaldo · H. Ludwig
IBM Research – Almaden, San Jose, CA, USA
e-mail: yi.zhou@ibm.com; baracald@us.ibm.com; hludwig@us.ibm.com

R. Kawahara
IBM Research, Tokyo, Japan
e-mail: RYOKAWA@jp.ibm.com

© The Author(s), under exclusive license to Springer Nature Switzerland AG 2022   455
H. Ludwig, N. Baracaldo (eds.), *Federated Learning*,
https://doi.org/10.1007/978-3-030-96896-0_20

laundering (i.e., obscuring the true origin of monetary entities to evade regulations or avoid taxes). The monetary value of such crimes can range from tens of dollars to tens of billions of dollars. However, the overall negative consequences of such crimes extend far beyond their monetary value. In fact, the consequences may even be societal in scope, such as in cases of terrorist financing or large-scale frauds that topple major institutions and governments.

In response, regulators require efforts to combat money laundry from financial institutions. Financial institutions spend substantial resources to develop compliance programs and infrastructures in order to combat financial crimes. Managing financial crime risk presents challenges due to the scale of the effort (large banks may have upwards of 100 million customers or more, which together generate billions of transactions that must be screened) and the availability of data (when transactions cross bank or country boundaries, little may be known about the remote counterparty). Current technology employed to assist with these processes focuses on the identification of anomalies and known patterns of malfeasance. However, usually also a large number of false positive alerts are created in the process. These alerts then require further (often manual) review to parse out suspicious behavior from valid financial activity that is inadvertently picked up by the models (referred to as "false positives").

20.1.1 Combating Financial Crimes with Machine Learning and Graph Learning

Recently, financial institutions have been exploring the use of machine learning techniques to augment existing transaction monitoring capabilities. Machine learning techniques offer a promising capability to identify suspicious activity from an incoming stream of transactions, as well as to filter the false positives from the alerts generated by current technology, thereby making existing processes more efficient and ultimately more effective. These machine learning techniques rely on a set of features generated from knowledge about the transacting parties, from individual and aggregate transaction metrics, and from the topology of party-to-party relationships derived from static knowledge and transactional history. Topological features are computed from graph embeddings or from the results of traditional graph algorithms such as PageRank [13], count of suspicious parties within an egonet network. An ego network consists of a focal node ("ego") and the nodes to whom ego is directly connected to (these are called "alters") plus the ties, if any, among the alters.

Overall, this approach has been shown to have a positive effect when evaluated against a ground truth determined by currently deployed methods. In one such evaluation, false positives were reduced by 20–30%.

20.1.2 Need for Global Financial Crimes Detection and Contributions

Notwithstanding the value of leveraging machine learning in the context of transaction monitoring, financial institutions are limited to identifying suspicious activity as it pertains to their organization. This presents a conundrum since bad actors are increasingly sophisticated with their techniques that often span across organizations and geographies (i.e., many use multiple banks to launder money). Financial institutions are realizing that without looking at data across multiple organizations, it would be impossible to detect a portion of suspicious activity. Regulatory requirements, data privacy concerns, as well as commercial competitiveness, all pose challenges to explicit sharing in information among financial institutions. Given the challenge at hand, an innovative solution is needed to detect suspicious activities across organizations. This chapter uses an approach to combine federated graph learning across parties with a federated machine learning approach to facilitate the collaboration of multiple financial institutions in training better detection models for money laundering. Its main contributions are:

- A federated graph learning platform detects global financial crime activities across multiple financial institutions.
- It demonstrates the effectiveness of federated graph learning as a tool to help identify financial crime during the TechSprint hosted by the United Kingdom's Financial Conduct Authority (FCA) in 2019, using the data set and use cases provided by the FCA.
- We combine federated learning with graph learning as a means to detect potential financial crimes and share typologies across multiple financial institutions for which money laundry detection is a non-competitive activity.

The rest of this chapter is organized as follows: We outline our core technologies, including preliminaries for Graph Learning. This is followed by the overall architecture and our federated graph learning capabilities. We then provide an overview of an implementation and evaluate using the data set provided by the UK FCA for its TechSprint. Finally we describe concluding remarks and future directions.

20.2 Graph Learning

This section describes the underlying technologies, along with relevant prior art, that constitute our platform—including graph learning or machine learning techniques to detect financial crimes.

Graph learning is defined as a type of machine learning that utilizes graph-based features to add richer context to data by first linking that data together as a graph structure and then deriving features from different metrics on the graph. Various graph features can be defined by exploiting a set of graph analytics such as connectivity, centrality, community detection, and pattern matching. Graph features can also be combined with non-graph features (e.g., features on attributes for a specific data point). Once a set of features including graph features and non-graph features are defined, a problem can then be formulated as a supervised machine learning problem (assuming that label data is provided). However, if label data is not provided, it can be approached as an unsupervised machine learning problem so that we can apply clustering (e.g., k-means) or outlier detection (e.g., LoF or DBScan). Recently, there have been many advances in scalable graph computation for billion-scale or even trillion-scale graphs [7, 16]. Hence, it is reasonable to expect that this approach would remain practical, even for large graphs.

The paper of Akoglu et al. [2] provides a good review of prior art about graph-based approaches for general anomaly detection problems. Molloy et al. [11] use PageRank-based features for fraud detection. The papers [9] and [17] explore graph embedding methods for financial crime detection applications such as anti-money laundering (AML). Recently research communities are also exploring the use of neural networks to compute graph embeddings without determining pre-defined graph topologies as graph features. However, the lack of explainability of the black-box model of neural network presents adoption challenges for financial institutions that have stringent model validation processes that hinge on explainability of the decisions.

Notably these prior works focus more on local graph features, while the approach in this chapter is focused on global graph features spanning multiple financial institutions.

20.3 Federated Learning for Financial Crimes Detection

In this section we propose a new platform that enables us to capture complex global money laundering activities spanning multiple financial institutions as opposed to current AML (Anti-Money Laundering) systems that only look at transactions at a single bank. The proposed federated learning system is comprised of 3 steps: First, we compute local features, then, we compute global graph features, and finally we perform federated learning over computed features. Subsequent sections describe each step.

20.3.1 Local Feature Computation

We firstly compute local features for each financial institution. As local features, we can firstly compute demographic features of customers such as account types (individual or business), business types, countries, account opening date, and some risk flags based on "Know Your Customer" (KYC) attributes. KYC refers to onboarding process when a client opens a bank account to prove his or her identity by providing his/her personal information such as driver information, etc. We then compute various statistical features on transaction behaviors such as min, max, average, mean, and standard deviation for transaction of various types such as international wire, domestic wire, credit, cash, check, and so forth. We can also compute graph features such as egonet, pagerank, and degree distribution, similarly to the practice in a single bank case.

20.3.2 Global Feature Computation

As a next step, we compute global features that provide global context related to suspicious activities among multiple financial institutions—using a privacy-preserving graph computation framework called GraphSC [12]. Global graph features are mainly computed using graph analytics over the entire graph of transactions and a party relationship graph—while each party does not have to reveal their graph to other parties. Graph features include 1 hop/2 hop egonets, cycle and temporal cycle, betweeness centrality, community detection, and so forth. The advantage of using global features over local graph features is if we can create richer and denser graph by assembling sub-graphs from multiple graphs, then the graph features should be more effective since you can also acquire contexts from other financial institutions as to which bank accounts may be associated with bad actors.

For computing global graph features, we need to take privacy into account, so as not to disclose any sensitive information from each financial institution. For instance, if there is a cycle of transactions consisting of 3 accounts in two different financial institutions—starting from an account A in Financial Institution X to an account B in Financial Institution Y, and to an account C in Financial Institution Y. A challenge is that a transaction from B to C in Financial Institution Y cannot be revealed to Financial Institution X. Thus, one of the requirements is to design and implement a secure protocol that allows Financial Institution X to send an inquiry to Financial Institution Y to ask whether there is a transaction between B and C—without letting Financial Institution Y to reveal sensitive information. GraphSC [12] is one of such secure graph computation frameworks, and we implemented some graph features such as temporal cycle features based on this approach.

20.3.3 Federated Learning

Next, we build a federated learning model using local features and global features that we describe in the previous sections. It uses one of federated learning platforms [10], which implements a centralized federated learning approach. Data owners share model updates with a central server, the aggregator, which does not have access to the data of any of the parties. This central server is hosted by a third party such as a Financial Intelligence Unit (FIU), a common type of financial crimes watchdog in many mature financial markets. To further protect the privacy, even the model updates shared with the aggregator can be strictly secured via privacy-preserving techniques such as differential privacy, secure multi-party computation, or different encryption techniques.

We target a scenario where different financial institutions collaborate together to train a model that can more accurately predict suspicious money laundering efforts. In our setup, each financial institution trains on its local data and shares the model parameters of the trained model with the central aggregator. The aggregator then fuses all of the model parameters and generates a global model whose weights will be sent back to all the collaborating banks to reinitialize their local model for another round of local training. This process is repeated for a set number of rounds or until desired model accuracy is achieved.

The framework [10] that we used is a framework designed for federated learning in an enterprise environment. It provides a basic fabric for federated learning on which advanced features, such as differential privacy and secure multiparty computation, can be added. It is agnostic to the specific machine learning platform used and supports different learning topologies, e.g., a shared aggregator, and protocols.

20.4   Evaluation

In this section we describe how federated learning can help improving model accuracy for AML problems across multiple financial institutions.

20.4.1   Data Set and Graph Modelling

For the evaluation of the approach described in this chapter, we use the data set provided by the FCA (Financial Conduct Authority) of the United Kingdom, who hosted a TechSprint in 2019 [1]. The data set is a simulated—but realistic—data set comprised of data from 6 financial institutions in the UK and reflects real-world statistical distributions and well-known suspicious patterns.

The data set spans 2 years of activity and includes customer profile, transactions, customer relationship data, an indicator of suspicious activity alerts, and an indication of whether the customer relationship is terminated over suspicion of misconduct. We use the last item as a form of ground truth for suspicious activity.

On the basis of this data set, we build two types of graphs: one called a transaction graph where a vertex represents a bank account and an edge represents a money transfer. Another graph is called party relationship graph where a vertex represents a customer and an edge represents a social relationship between customers such as family.

20.4.2   Graph Features for Party Relationship Graph

As a preliminary evaluation, we firstly focus here only on the party relationship graph, which consists of social relationships between bank accounts. For example, a person who owns a company can use both his or her personal account and the company's business account. In this case, those two bank accounts can be related through the owner. This kind of relations could be an important indicator of a financial crime because a criminal might use an account indirectly through the relationship (e.g., ownership of a company) to send his or her private money to obscure the true source or beneficiary (i.e., the layering).

Here, we assume that a bank has the following information for each customer:

- Customer profile (e.g., account ID, name, date of birth, nationality, etc.)
- Related party profiles (e.g., name, date of birth, etc.)
- Relations between the customer and the related parties (e.g., director, owner, family, etc.)
- Customer risk intelligence (e.g., past Suspicious Activity Report (SAR) flags, financial crime exit markers)

Such information is obtained in the course of the KYC (Know-Your-Customer) processes (when onboarding new customers or in performing periodic reviews of existing customers) or when performing detailed investigations of AML alerts. The related parties may or may not be a customer of a financial institution and could include the customer itself. If multiple customers have relationships with a common related party, this indicates that the accounts of the customers might be affected by a single party and thus could work in a coordinated manner.

Since there are many financial institutions in the market, one needs to consider the case of an individual having accounts in multiple financial institutions. Similarly, the same related party could appear in the data of multiple financial institution. To reveal the connection between accounts across the financial institution boundaries, one needs to go through the process of "entity resolution" to draw the connections between the customer profiles and related party profiles. That is, one needs to identify the profiles that correspond to the same entity by comparing the attributes such as the names, addresses, or identification numbers.

Here, we applied the following simple rule for the entity resolution:

- Individual customer or party: (full name, date of birth, and nationality are equal) OR (ID document type, ID document number, and nationality are equal)
- Business customer or party: (full name, date of incorporation, and country of incorporation are equal) OR (company registration type, company registration number, and country of incorporation are equal)

However, in practice, entity resolution presents many challenges due to the existence of typos, document quality issues, OCR errors, or fluctuations in conversions of non-Latin characters. There are a number of commercial products that address this

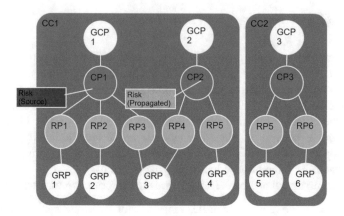

Fig. 20.1 Relation between customers and related parties

challenge in contexts where raw data can be shared; however, performing entity resolution under privacy-preserving constraints remains an area of future work.

Once the entity resolution for the customers and their related parties has been performed, one will get a graph of those entities, as shown in Fig. 20.1. In the figure, CP1, CP2, $\cdots$ are the customers, each of which has an account, RP1, RP2, $\cdots$ are their related parties, GRP1, GRP2, $\cdots$ and GCP1, GCP2, $\cdots$ are the grouping IDs issued in the course of the entity resolution. Edges between the customers and the related parties are the social relations, and edges between the grouping IDs and the customers or related parties are created if those parties are identified as belonging to the same entity by the entity resolution.

Since the connected accounts (customers) in the graph are possible collaborators, we think that the risk of being involved in money laundering is shared among the accounts. From this hypothesis, we compute each customer's features based on the statistics in each (weakly) connected component in the graph. In the current implementation, the following features are used:

- Number of customers who have alerted by a transaction monitoring system in the past within the connected component
- Number of customers who have SAR flags in the past within the connected component
- Number of customers who have financial crime exit markers in the past within the connected component
- Number of the nodes within the connected component

The status of the risk flags (SAR, financial crime exit marker, etc.) can be obtained from the customer risk intelligence data as mentioned in a previous paragraph in this section. Please note that the risk flag status of a customer is often used as a target variable in a machine learning-based prediction/classification task of financial crimes. In such cases, those features must not include the status of the risk flag of the

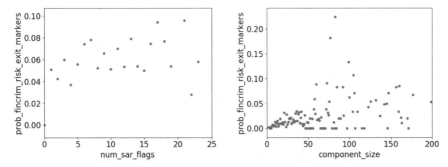

Fig. 20.2 Conditional probability of a customer's fincrime exit marker being flagged as a function of a feature. Left: the feature is the number of SAR-flagged customers in the same connected component. Right: the feature is the number of nodes in the connected component

customer in question and must contain the information from only other customers when those are used as a training or a testing data set.

Our preliminary analysis on the synthetic data set used in the course of the TechSprint is shown in Fig. 20.2. Here, we assume that the financial crime exit marker of a customer is the target variable to be predicted for the supervised machine learning setting. It shows a positive correlation between the probability of a customer being flagged with a fincrime exit marker and the number of customers who have SAR flags (left) within the same connected component and the number of nodes (including the customers, related parties, and the grouping IDs) within the same connected component. This result indicates that these values can be used as features for detecting money laundering with other features.

20.4.3   Model Accuracy

Here we show the evaluation result of our federated graph learning using the TechSprint data. We compute local features including transaction-based features, global graph features in party relationship graph defined in the previous section, and global graph features in transaction graph, and local transaction features.

With regard to the platform setting, in an ideal federated learning environment, each financial institution will perform its local training on its own server or virtual machine and communicate with the aggregator, which can be hosted by a third party (e.g., government agency) or by one of the banks, after each local training period is performed. However, due to the limited resources provided by during the TechSprint, we only had one host and hence needed to simulated 6 processes representing 6 UK banks' local training processes and one process was used as a proxy for the role of the aggregator using our federated learning framework [15]. We have trained several types of machine learning models, for example, ℓ-1 regularized logistic regression, ℓ-2 regularized linear support vector machine (SVM), a decision tree, and a simple neural network.

We found similar performance results (with less than 10% difference in testing accuracy and F1 scores) for these machine learning models. Therefore, we only report the results for the neural network, which is composed of two dense layers of sigmoid units and a sigmoid layer with binary cross-entropy loss. As previously noted, explainability for neural networks is still an ongoing research area, so we could leverage existing work or use other machine learning models for the current financial regulation policy that requires transparency and explainability in the machine learning models that are used. Since the data set that was provided is highly imbalanced with only around 5% bank accounts containing labels on whether they were filed as Suspicious Activity Reports (SAR) and around 0.4% are labeled as financial criminals, we exploited the under-sampling strategy in the majority label class (i.e., the clean bank accounts, to create balance training data sets for all financial institutions). We then trained local models and the aggregated model all based on the balanced training data sets.

In Table 20.1, we provide the results of local models trained on each financial institution's transaction records. We observed that since the local test sets are balanced, the test accuracy and F1 score are the same, which seems to demonstrate good performance of the local models. However, if we test the local trained model against account records from all financial institutions, we can see that F1 scores drop significantly due to the imbalanced nature of the test set and the test accuracy also drops a bit. Moreover, if we add graph features that we described in the previous section into our training features, we see improvements in both test accuracy and F1 scores as shown in Table 20.2. From Table 20.3, we can conclude from the results that by training an aggregated model collaboratively via federated learning, all financial institutions can benefit from the aggregated model without sacrificing their data privacy. This also includes significantly reducing costs related to processing false positives of money laundry.

The improvement in accuracy and F1 means a more accurate flagging of cases has the potential to significantly reduce the compliance cost for banks, reducing the need for follow-up.

Table 20.1 Centralized local models trained on transaction features

	BWBAGB	PCOBGB	NUBAGB	HCBGGB	GVBCGB	FOCSGB
Local test set (Accuracy/F1)	0.971/0.971	0.976/0.976	0.984/0.984	0.982/0.982	0.967/0.966	0.988/0.988
All record test set (Accuracy/F1)	0.956/0.550	0.956/0.550	0.953/0.546	0.957/0.551	0.960/0.550	0.962/0.552

Table 20.2 Local models trained on transaction and graph features

	BWBAGB	PCOBGB	NUBAGB	HCBGGB	GVBCGB	FOCSGB
Local test set (Acc/F1)	0.996/0.996	0.997/0.997	0.997/0.997	0.996/0.996	0.990/0.990	1/1
All record test set (Acc/F1)	0.994/0.761	0.994/0.769	0.990/0.692	0.995/0.766	0.995/0.764	0.995/0.765

Table 20.3 Federated model trained on transaction and graph features

	Aggregated model
Accuracy	**0.995**
F1	**0.769**

20.5   Concluding Remarks

In this chapter we described a novel framework that enables us to better identify patterns of suspicious activity by sharing insights across multiple financial institutions without sharing any raw data between each financial institution or a third party. This was made possible by combining graph learning techniques with federated learning. We described the overall architecture and prototypical implementation. We demonstrated that the federated learning model using multiple financial institutions outperformed a local model by 20% based on a data set from the 2019 FCA TechSprint. We believe that this capability lays the foundation to pilot these techniques on real-world data and scenarios.

Federated learning combined with graph analysis provides a good approach for financial institutions to collaborate identifying patterns of financial fraud, in particular patterns of money laundering. If privacy and security of customers' transaction data can be guaranteed, financial institutions can collaborate in this non-competitive area or be compelled to do so by regulators. Reduced rates of false positives provide savings of regulatory costs to the financial sector, but the reduction of money laundry has large societal benefits: Drug, arms, and sex trafficking as well as reduced terrorism financing becomes more difficult and is hopefully reduced. Moreover, this form of collaboration can be extended to other industries that face fraud or crime in a similar way and can address this in a non-competitive way if secrecy and privacy is maintained. Detecting counterfeit goods or retail theft are good candidates.

References

1. 2019 global AML and financial crime techsprint (2019). https://www.fca.org.uk/events/techsprints/2019-global-aml-and-financial-crime-techsprint

2. Akoglu L, Tong H, Koutra D (2015) Graph based anomaly detection and description: a survey. Data Min Knowl Discov 29(3):626–688
3. Alexandre C (2018) A multi-agent system based approach to fight financial fraud: an application to money laundering. ArXiv
4. Chen Z, Van Khoa LD, Teoh EN, Nazir A, Karuppiah E, Lam KS (2018) Machine learning techniques for anti-money laundering (AML) solutions in suspicious transaction detection: a review. Knowl Inf Syst 57:245–285
5. Colladon AF, Remondi E (2017) Using social network analysis to prevent money laundering. Expert Syst Appl 67:49–58
6. Han J, Barman U, Hayes J, Du J, Burgin E, Wan D (2018) NextGen AML: distributed deep learning based language technologies to augment anti money laundering investigation. In: Proceedings of ACL 2018, system demonstrations. Association for Computational Linguistics, pp 37–42
7. Hanai M, Suzumura T, Tan WJ, Liu ES, Theodoropoulos G, Cai W (2019) Distributed edge partitioning for trillion-edge graphs. CoRR abs/1908.05855, http://arxiv.org/abs/1908.05855, 1908.05855
8. Jamshidi MB, Gorjiankhanzad M, Lalbakhsh A, Roshani S (2019) A novel multiobjective approach for detecting money laundering with a neuro-fuzzy technique. In: 2019 IEEE 16th international conference on networking, sensing and control (ICNSC), pp 454–458. https://doi.org/10.1109/ICNSC.2019.8743234
9. Liu W, Liu Z, Yu F, Chen P, Suzumura T, Hu G (2019) A scalable attribute-aware network embedding system. Neurocomputing 339:279–291
10. Ludwig H, Baracaldo N, Thomas G, Zhou Y, Anwar A, Rajamoni S, Ong Y, Radhakrishnan J, Verma A, Sinn M, Purcell M, Rawat A, Minh T, Holohan N, Chakraborty S, Whitherspoon S, Steuer D, Wynter L, Hassan H, Laguna S, Yurochkin M, Agarwal M, Chuba E, Abay A (2020) IBM federated learning: an enterprise framework white paper v0.1. 2007.10987
11. Molloy I, Chari S, Finkler U, Wiggerman M, Jonker C, Habeck T, Park Y, Jordens F, Schaik R (2016) Graph analytics for real-time scoring of cross-channel transactional fraud
12. Nayak K, Wang XS, Ioannidis S, Weinsberg U, Taft N, Shi E (2015) GraphSC: parallel secure computation made easy. In: 2015 IEEE symposium on security and privacy, pp 377–394. https://doi.org/10.1109/SP.2015.30
13. Page L, Brin S, Motwani R, Winograd T (1999) The PageRank citation ranking: bringing order to the web. Technical Report 1999-66, Stanford InfoLab. http://ilpubs.stanford.edu:8090/422/, previous number = SIDL-WP-1999-0120
14. Savage D, Wang Q, Chou PL, Zhang X, Yu X (2016) Detection of money laundering groups using supervised learning in networks. ArXiv abs/1608.00708
15. Truex S, Baracaldo N, Anwar A, Steinke T, Ludwig H, Zhang R (2018) A hybrid approach to privacy-preserving federated learning
16. Ueno K, Suzumura T, Maruyama N, Fujisawa K, Matsuoka S (2017) Efficient breadth-first search on massively parallel and distributed-memory machines. Data Sci Eng 2(1):22–35. https://doi.org/10.1007/s41019-016-0024-y
17. Weber M, Chen J, Suzumura T, Pareja A, Ma T, Kanezashi H, Kaler T, Leiserson CE, Schardl TB (2018) Scalable graph learning for anti-money laundering: a first look. CoRR abs/1812.00076, http://arxiv.org/abs/1812.00076, 1812.00076

Chapter 21
Federated Reinforcement Learning for Portfolio Management

Pengqian Yu, Laura Wynter, and Shiau Hong Lim

Abstract Financial portfolio management involves the constant redistribution of wealth over a set of financial assets and can, by its sequential nature, be modelled using reinforcement learning (RL). Federated learning allows traders to jointly train models without revealing their private data. We show on S&P500 market data how personalized, robust federated reinforcement learning using Fed+ produces trading policies that offer higher annual returns and Sharpe ratios than other methods.

21.1 Introduction

In reinforcement learning (RL), the goal is to learn a multi-step, or long-term, policy to control a system through trial and error. There are two key aspects to reinforcement learning that distinguish it from other areas in machine learning. Firstly, policies are multi-step decisions that generally will be used over a time horizon of interest. Secondly, the policy is trained not in a supervised manner but rather by trial and error, that is, in a semi-supervised manner. Specifically, RL algorithms have access to the system dynamics through sampling rather than through an analytical model of the dynamics. Information on the likely reward obtained when a policy leads an agent to a particular state is assumed to be given; hence, the reinforcement learning algorithm tries successive policies so as to maximize the long-term reward. Reinforcement learning solves the problem of correlating immediate actions with the delayed returns that they will eventually produce through sampling.

Deep reinforcement learning incorporates deep learning into the method of solving for optimal policies. In this case, the policy or other relevant functions are represented by a neural network. The use of neural networks to learn and represent the policy allows for a seamless use of federated learning in the

P. Yu · L. Wynter (✉) · S. H. Lim
IBM Research, Singapore, Singapore
e-mail: lwynter@sg.ibm.com; shonglim@sg.ibm.com

process: the federation never shares data on the agent experiences, or state–action pairs and their rewards, but it rather shares policy neural network parameters.

Federated reinforcement learning is of particular interest as reinforcement learning requires a tremendous amount of data to learn good policies via trial and error. Being able to leverage the data, or so-called "experiences" of the agents under different states and actions, without requiring the parties to explicitly share them, is a great advantage to training reinforcement learning policies in real-world settings, from robotics to navigation to financial trading.

Indeed, federated learning offers significant potential value for financial portfolio management for two main reasons: (i) Historical financial data is limited and so traders often augment public data using their own data models that they do not wish to share with other traders. Federated learning offers an avenue for them to jointly train policies on far more data, without revealing their private data. (ii) Financial markets are highly non-stationary. Federated learning on heterogeneous data offers the benefits of multi-task learning in a privacy-protected manner.

This chapter proceeds as follows. We begin by introducing the formulation and notation needed to describe federated reinforcement learning. Then, we introduce the model for RL-based financial portfolio management. After that we define the key methods used for data enrichment, an important aspect of the federated portfolio management application. Lastly, we provide extensive experimental results on federated portfolio management and conclude with the key take-home messages.

21.2   Deep Reinforcement Learning Formulation

In reinforcement learning (RL), the goal is to learn a policy to control a system modelled by a Markov decision process, which is defined as a 6-tuple $\langle S, \mathcal{A}, P, r, T, \gamma \rangle$. Here, $S = \bigcup_t S_t$ is the state space and $\mathcal{A} = \bigcup_t \mathcal{A}_t$ is the action space, both assumed to be finite dimensional and continuous; $P : S \times \mathcal{A} \times S \rightarrow \mathbb{R}$ is the transition kernel density and $r : S \times \mathcal{A} \rightarrow \mathbb{R}$ is the reward function; T is the (possibly infinite) decision horizon; and $\gamma \in (0, 1]$ is the discount factor. Deterministic policy gradient learns a deterministic target policy using deep neural networks with weights θ [16]. A policy parameterized by θ is a mapping $\mu_\theta : S \rightarrow \mathcal{A}$, specifying the action to choose in a particular state. At each time step $t \in \{1, \ldots, T\}$, the agent in state $s_t \in S_t$ takes a deterministic action $a_t = \mu_\theta(s_t) \in \mathcal{A}_t$, receives the reward $r(s_t, a_t)$, and transits to the next state s_{t+1} according to P

An RL agent's objective is to maximize its expected return given the starting distribution

$$J(\mu_\theta) \triangleq \mathbb{E}_{s_t \sim P} \left[\sum_{t=1}^{T} \gamma^{t-1} r(s_t, \mu_\theta(s_t)) \right].$$

Then, an optimal policy to achieve the objective can be found by applying the deterministic policy gradient theorem (see Theorem 1 in [16]), where the idea is to adjust the parameters θ of the policy in the direction of the performance gradients $\nabla_\theta J(\mu_\theta)$. Deterministic policy gradients can be estimated more efficiently than their stochastic counterparts (see [17]), thus avoiding solving a problematic integral over the action space.

21.3   Financial Portfolio Management

Portfolio management consists of sequentially allocating wealth to a collection of assets in consecutive trading periods [7, 12]. Due to the sequential decision-making nature of portfolio management, it is possible to apply reinforcement learning (RL) to model asset reallocation. Since rewards are delayed, and the process is time dependent, the RL formulation is furthermore a natural paradigm to use in developing trading policies. See, for example, recurrent RL [1], model-free off-policy RL [8], an optimal hedging framework [5], and state-augmented RL [18].

Federated learning offers significant potential value in this application for two main reasons, as mentioned above:

(i) Historical financial data on any particular asset of interest is limited. Consider the S&P500: the size of a training set for any S&P500 asset over the past 10 years is at most 2530 observations, as there are 253 trading days per year. Assets coming onto the market more recently will then have even less training data available. Traders thus augment public data using their own data models, which they do not wish to share with other traders. We describe several such data augmentation models in the next section. Federated learning offers an avenue for traders to jointly train policies on far more data, without revealing their private data.

(ii) Financial markets are highly non-stationary. As such the policies learned from the historical training data may not generalize well due to distribution shift over time. Federated learning, when involving independent parties jointly training models on heterogeneous data, offers the benefits of multi-task learning in a privacy-protected manner. Multi-task learning is known to improve the transferability of models as is needed in a non-stationary environment such as financial markets.

Assume from here on out that the financial market is sufficiently liquid such that any transactions can be executed immediately with minimal market impact.

Following [8], let t denote the index of asset trading days and $v_{i,t}$, $i = \{1, \ldots, \eta\}$, the closing price of the ith asset at time t, where η is the number of assets in a given asset universe. The price vector $\boldsymbol{v}_t$ consists of the closing prices of all n assets. An additional dimension (the first dimension indexed by 0) in $\boldsymbol{v}_t$, $v_{0,t}$, denotes the cash price at time t. We normalize all temporal variations in $\boldsymbol{v}_t$ with respect to cash so $v_{0,t}$ is constant for all t. Define the price relative vector at time t as $\boldsymbol{y}_t \triangleq \boldsymbol{v}_{t+1} \oslash \boldsymbol{v}_t = (1, v_{1,t+1}/v_{1,t}, \ldots, v_{\eta,t+1}/v_{\eta,t})^{\top}$, where $\oslash$ denotes element-wise division.

Define $\boldsymbol{w}_{t-1}$ as the portfolio weight vector at the beginning of time t, where its ith element $w_{i,t-1}$ represents the proportion of asset i in the portfolio after capital reallocation and $\sum_{i=0}^{\eta} w_{i,t} = 1$ for all t. The portfolio is initialized with $\boldsymbol{w}_0 = (1, 0, \ldots, 0)^{\top}$. At the end of time t, the weights evolve according to $\boldsymbol{w}_t' = (\boldsymbol{y}_t \odot \boldsymbol{w}_{t-1})/(\boldsymbol{y}_t \cdot \boldsymbol{w}_{t-1})$, where $\odot$ is element-wise. The reallocation from $\boldsymbol{w}_t'$ to $\boldsymbol{w}_t$ is gotten by selling and buying relevant assets. Paying all fees, this reallocation shrinks the portfolio value by $\beta_t \triangleq c\sum_{i=1}^{\eta} |w_{i,t}' - w_{i,t}|$, where $c = 0.2\%$ is the buy/sell fee; let ρ_{t-1} denote the portfolio value at the beginning of t and ρ_t' at the end, so $\rho_t = \beta_t \rho_t'$. The normalized close price matrix at t is $\boldsymbol{Y}_t \triangleq [\boldsymbol{v}_{t-l+1} \oslash \boldsymbol{v}_t | \boldsymbol{v}_{t-l+2} \oslash \boldsymbol{v}_t | \cdots | \boldsymbol{v}_{t-1} \oslash \boldsymbol{v}_t | \mathbf{1}]$, where $\mathbf{1} \triangleq (1, 1, \ldots, 1)^{\top}$ and l is the time embedding.

The financial portfolio management can then be formulated as a RL problem where, at time step t, the agent observes the *state* $s_t \triangleq (\boldsymbol{Y}_t, \boldsymbol{w}_{t-1})$, takes an *action* (portfolio weights) $a_t = \boldsymbol{w}_t$, and receives an immediate *reward* $r(s_t, a_t) \triangleq \ln(\rho_t/\rho_{t-1}) = \ln(\beta_t \rho_t'/\rho_{t-1}) = \ln(\beta_t \boldsymbol{y}_t \cdot \boldsymbol{w}_{t-1})$. Considering the policy μ_θ, the objective of RL agent is to maximize an objective function parameterized by θ: $\max_{\mu_\theta} J(\mu_\theta) = \max_{\mu_\theta} \sum_{t=1}^{T} \ln(\beta_t \boldsymbol{y}_t \cdot \boldsymbol{w}_{t-1})/T$. By deterministic policy gradient theorem [16], the optimal μ_θ can be found via the following update rule for parameters θ:

$$\theta \leftarrow \theta + \lambda \nabla_\theta J(\mu_\theta),$$

where λ is the learning rate.

Suppose now that there are N participants (parties), each with an RL agent performing portfolio management. Each party has its trading asset universe (data) $\mathcal{I}_n$ for $n \in \{1, 2, \ldots, N\}$, and the cardinality (number of assets) in every asset universe $\mathcal{I}_n$ is the same. The goal is to aggregate each party's model such that it can benefit from others. Since each party is trading on various asset universes, this multi-agent portfolio management problem is essentially a multi-task problem. In particular, each party's task is a Markov decision process $\langle \mathcal{S}_n, \mathcal{A}, P_n, r, T, \gamma \rangle$, where the action space $\mathcal{A}$ and the reward function r are common to all parties. The state space $\mathcal{S}_n$ depends on each party's trading asset universe $\mathcal{I}_n$ and the transition kernel P_n should be inferred from the underlying stock price dynamics of $\mathcal{I}_n$. Each party maximizes its objective and updates its RL agent using the deterministic policy gradient through the federation.

21.4 Data Augmentation Methods

Historical financial data is limited for any given asset. As noted above, the size of a training set for any S&P500 asset active for the past 10 years is only 2530 observations, as there are 253 trading days per year. One way that traders handle this relatively low volume of historical data is to augment it, by building generative models for each asset. Thus, such models are private to the trader, and the trader would seldom wish to share the data resulting from these models.

We present three approaches for training generative models to produce augmented financial price history of publicly traded assets. Specifically, we describe below Geometric Brownian Motion (GBM), Variable-Order Markov (VOM) model, and Generative Adversarial Network (GAN) approaches.

21.4.1 Geometric Brownian Motion (GBM)

Geometric Brownian Motion (GBM) is a continuous-time stochastic process in which the logarithm of the randomly varying quantity follows a Brownian motion with drift [9]. GBM is often used in mathematical finance to model stock prices in the Black–Scholes model [4] mainly because the expected returns of GBM are independent of the values of the stock price, which agrees with what we expect in reality [9]. In addition, a GBM process shows the same kind of "roughness" in its paths as we often observe in real stock prices. The close price of asset i follows a GBM if it satisfies the following stochastic differential equation: $dv_{i,t} = \mu_i v_{i,t} dt + \sigma_i v_{i,t} dW_t$. Here W_t is a Brownian motion, μ_i is the mean return of the stock prices given a historical date range, and σ_i is the standard deviation of returns of the stock prices in the same date range. The differential equation can be solved by an analytic solution: $v_{i,t} = v_{i,0} \exp((\mu_i - \sigma_i^2/2)t + \sigma_i W_t)$. We choose the initial price value $v_{i,0}$ to be the last day's close price of the asset in the training set, and we use the asset returns in the date range of RL training set to estimate μ_i and σ_i.

21.4.2 Variable-Order Markov (VOM)

Variable-Order Markov (VOM) models are an important class of models that extend the well-known Markov chain models [2], where each random variable in a sequence with a Markov property depends on a fixed number of random variables. In contrast, in VOM models this number of conditioning random variables may vary based on the specific observed realization. Given a sequence of returns $\{p_t\}_{t=0}^T$ where $p_{i,t} = (v_{i,t+1} - v_{i,t})/v_{i,t}$ for asset i in asset universe $\mathcal{I}$ at time t, the VOM model learns a model $\mathbb{P}$ that provides a probability assignment for each return in the sequence given its past observations. Specifically, the learner generates a conditional probability distribution $\mathbb{P}(\boldsymbol{p}_t|\boldsymbol{h}_{t'})$, where $h_{i,t'} = \{p_{i,t'}\}_{t'=t-k-1}^{t-1}$ represents a sequence of historical returns of length k up to time t. VOM models attempt to estimate conditional distributions of the form $\mathbb{P}(\boldsymbol{p}_t|\boldsymbol{h}_{t'})$ where the context length $|h_{i,t'}| = k$ varies depending on the available statistics. The changes in the logarithm of exchange rates, price indices, and stock market indices are usually assumed normal in the Black–Scholes model [4]. In this chapter, we make a similar assumption and let $\mathbb{P}$ be a multivariate log-normal distribution. That is, $\ln(\boldsymbol{p}_t|\boldsymbol{h}_{t'}) \sim \mathcal{N}(\boldsymbol{\mu}, \boldsymbol{\sigma}^2)$, where $\boldsymbol{\mu}$ and $\boldsymbol{\sigma}$ are the mean and covariance matrix of the assets' returns, respectively.

21.4.3 Generative Adversarial Network (GAN)

The Generative Adversarial Network (GAN) is a machine learning framework that uses two neural networks, pitting one against the other, in order to generate new, synthetic instances of data that can pass for real data [6]. One neural network, called the generator, is responsible for the generation of stock price paths, and the second one, the discriminator, has to judge whether the generated paths are synthetic or from the same underlying distribution as the data (i.e., the asset returns). We denote x as a collection of all assets' daily returns $p_{i,t}$ in asset universe $\mathcal{I}$ for $t = 0, 1, \ldots, T$. To learn the generator's distribution $\mathbb{P}_g$ over the asset returns x, we define a prior on input noise variables $\mathbb{P}_z(z)$ and then represent a mapping to data space as $G(z; \theta_g)$, where G is a differentiable function represented by dense layers with parameters θ_g. We use convolution layers [10] followed by dense layers for $D(x; \theta_d)$ that outputs a single scalar. $D(x)$ represents the probability that x came from the historical price returns rather than $\mathbb{P}_g$. We train D to maximize the probability of assigning the correct label to both training examples and samples from G. We simultaneously train G to minimize $\log(1 - D(G(z)))$. In other words, D and G play the following two-player minimax game: $\min_G \max_D \mathbb{E}_{x \sim \mathbb{P}(x)}[\log D(x)] + \mathbb{E}_{z \sim \mathbb{P}_z(z)}[\log(1 - D(G(z)))]$.

21.5 Experimental Results

In this section, we perform numerical experiments to illustrate the use of federated reinforcement learning to develop financial trading policies for portfolio management. The parties in the federation represent traders, or portfolio managers, each with its own approach for data enrichment using the data generation methods provided in the previous section. They each develop their own policies on their own portfolios but jointly train models by sharing parameters from the neural network policy mappings.

It is to be noted that each trader has a unique universe of assets, in this case taken from the S&P500. While there may be overlap of assets across the traders, this is not needed for the trained policies to transfer across traders and across asset universes. The neural network model proposed here allows for an invariance across the set of assets used, and a policy trained on one set of assets can be successfully used for trading on a different set, thanks to this invariance.

We show two main benefits in this section of federated reinforcement learning for portfolio management. The results demonstrate the benefit of reinforcement learning for this application as well as the benefit of federated learning; the profit accrued to parties using federated reinforcement learning is greater than that from policies learnt independently without federated learning, and also better than policies using standard financial trading baselines without reinforcement learning. Details of the experimental setup are provided below, followed by the results in terms of annualized return and Sharpe ratio.

21.5.1 Experimental Setup

Given N parties, each with its own trading asset universe and data I_n for $n \in \{1, 2, \ldots, N\}$. The number of assets in each universe I_n is the same. Each trader generates one year of additional, synthetic time-series data for each of their assets. The synthetic data is generated every 1000 RL local training iterations and appended to the last day's real closing prices of the assets. This combined synthetic–real data, which is strictly confidential to each party, is used by each party to train its RL agent.

The agent models follow the Ensemble of Identical Independent Evaluators (EIIE) topology with the online stochastic batch learning of [8], the latter of which samples mini-batches consecutively in time to train the EIIE networks. We choose the discount factor $\gamma = 0.99$ and time embedding $l = 30$ and use a batch size of 50 and a learning rate of 5×10^{-5} for all experiments. Other experimental details including the choice of time embedding and the RNN implementation of the EIIE for agent models are the same as those of [8]. We conduct our experiments on 10 virtual machines where each machine has 32 Intel® Xeon® Gold 6130 CPUs and 128 GM RAM and can support up to 5 party processes.

The experiments are performed on 50 assets from the S&P500 technology sector.[1] Each party n constructs its own asset universe I_n by randomly choosing 9 assets and pre-trains a private data augmentation method, i.e., GBM, VOM, or

---

[1] We use the following 50 assets from the S&P500 technology sector: AAPL, ADBE, ADI, ADP, ADS, AKAM, AMD, APH, ATVI, AVGO, CHTR, CMCSA, CRM, CSCO, CTL, CTSH, CTXS, DIS, DISH, DXC, FB, FFIV, FISV, GLW, GOOG, IBM, INTC, INTU, IPG, IT, JNPR, KLAC, LRCX, MA, MCHP, MSFT, MSI, NFLX, NTAP, OMC, PAYX, QCOM, SNPS, STX, T, TEL, VZ, WDC, WU, and XRX.

GAN, on their assets in $\mathcal{I}_n$. S&P price data from 2006 to 2018 is used for training data augmentation and the RL policies; 2019 price data is used for testing.

In each experiment, we assume all parties participate in the training. The number of global rounds $K = 800$ and local RL iterations $E = 50$. We use a fixed regularization parameter $\alpha = 0.01$ for each party's Local-Solve. In practice, we find that initializing the local model to a mixture model (i.e., using a small positive value of λ instead of the default $\lambda = 0$) at the beginning of every Local-Solve subroutine for each party yields good performance. Such a mixture model is computed using a convex combination of each party's local model and latest global model with weight $\lambda = 0.001$. In addition, we keep λ_k^t (also the diagonal entries for the diagonal matrix Λ_k^t) constant for all t. Results are based on training over a sufficiently wide grid of fixed λ_k^t (typically 10–13 values on a multiplicative grid of resolution 10^{-1} or 10^{-3}).

Results are for the best fixed λ_n^k selected individually for each experiment. Two metrics measure performance: the most intuitive is the geometric average return earned by an investment each year over a given period, i.e., annualized return. To take into account risk and volatility, we also report the Sharpe ratio [15], which in its simplest form is $\mathbb{E}[X]/\sqrt{\mathrm{var}[X]}$, where X is the (random) return of a portfolio. The Sharpe ratio is thus the additional return an investor receives per unit of increase in risk.

We make use of several federated learning algorithms for model fusion, but as we shall see the Fed+ family of methods is the best adapted to the financial portfolio management problem in that each trader wishes to obtain a policy that is optimized for the particular portfolio of assets. A single global model trained over all traders' portfolios need not perform best for any single trader's portfolio. In the remainder of this section we illustrate this by evaluating several federated learning fusion algorithms.

An additional benefit of the Fed+ family of algorithms for this application comes from the stability gained in the training process. In settings where data across parties may be very heterogeneous, it can happen that forcing convergence to a single model impacts negatively the training process, leading to a collapse of the training process. This can be a consequence of very large changes in the models from one training round to the next, as shown in Fig. 21.1.

Figure 21.2 takes a deeper dive into this phenomenon on the portfolio management application. The figure illustrates the behavior of the different algorithms as a function of how close the local party moves from a purely local model toward a common, central model. A local party update occurs in each subplot on the left side, at $\lambda = 0$. Observe that the local updates improve the performance from the previous aggregation indicated by the dashed lines. However, performance degrades after the subsequent aggregation, corresponding to the right-hand side of each subplot, where $\lambda = 1$. In fact, for FedAvg [13], RFA [14], and FedProx [11], performance of the subsequent aggregation is worse than the previous value (dashed line). Intermediate values of λ correspond to moving toward, but not reaching, the common, central model.

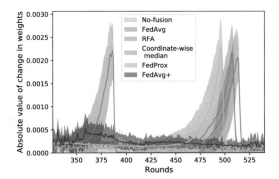

Fig. 21.1 Impact of federated learning aggregation on consecutive model changes. Absolute value of the model change (neural network parameters) before and after federated model aggregation on the financial portfolio optimization problem. Observe that FedAvg, RFA, coordinate-wise median, and FedProx cause large spikes in the parameter change that do not occur without federated learning or when using Fed+. The large spikes coincide precisely with training collapse, shown in Fig. 21.8 (bottom four figures)

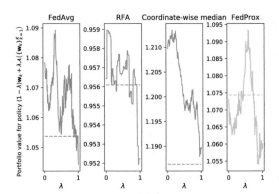

Fig. 21.2 Before and after aggregation, along the line given by varying $\lambda \in [0, 1]$ using a convex combination of local update and the common model. Performance on the financial portfolio optimization problem is shown as a function of locally shifting λ. Dashed lines represent the common model at the previous round. The right-hand side lower than the left-hand means that a full step toward averaging (or median) all parties, i.e., $\lambda = 1$, degrades local performance. This is the case with the standard FedAvg as well as with the robust methods

21.5.2 Numerical Results

We first consider the case where there are $N = 10$ parties, each with different asset universes. The learning curves obtained by training each party independently (no fusion) can be found in Fig. 21.3. Note that parties $2, 6, 7, 8, 9$, and 10 have unstable, diverging learning patterns as compared to parties $1, 3, 4$, and 5. As shown in Table 21.1, the averaged ℓ_2 and Frobenius norms of the covariance matrix for assets' returns in asset universes $\mathcal{I}_2, \mathcal{I}_6, \mathcal{I}_7, \mathcal{I}_8, \mathcal{I}_9$, and $\mathcal{I}_{10}$ are larger. In the

Fig. 21.3 Learning curve for each party in the financial portfolio management problem, without fusion. Notice that several parties experience some training issues, which combined, are compounded, and lead to training collapse in the federated training (see, for example, Fig. 21.5)

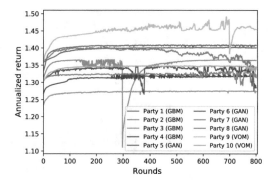

Table 21.1 Average norms of covariance matrix for assets' returns in different asset universes

Asset universes	I_1, I_3, I_4, I_5	$I_2, I_6, I_7, I_8, I_9, I_{10}$
Averaged ℓ_2-norm	1.35×10^{-3}	1.73×10^{-3}
Averaged Frobenius norm	1.51×10^{-3}	1.93×10^{-3}

Fig. 21.4 Average portfolio value in the 10-party financial portfolio management problem in the test period, across methods

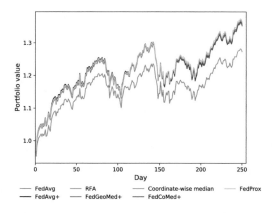

federated setting, the average portfolio value in the testing period for 10 parties is shown in Fig. 21.4. For baseline methods FedAvg [13], RFA [14], coordinate-wise median [19], and FedProx [11] and their Fed+ extensions, we average across the 10 different asset universes; the average performance is summarized in Table 21.2.

To investigate the learning behavior, we plot the average learning performance across 10 parties in Fig. 21.5. It is clear that the average performance for FedAvg training is significantly affected by participants with diverging learning behavior such as parties 2, 6, 7, 8, 9, and 10 as illustrated in Fig. 21.3. Results show that FedAvg degrades the performance of all parties due in a large part to a collapse in the training process (at training round 380). The robust federated learning methods, RFA, coordinate-wise median, and FedProx, achieve stable learning and yield better performance in 10-party federation. It is worthwhile to note that FedAvg+ stabilizes FedAvg as shown in Fig. 21.6. In addition, Fed+ algorithms improve the annual return by 1.91% and the Sharpe ratio by 0.07 on average as shown in Table 21.2.

Table 21.2 Average performances of different methods over 10 parties

Method	Annualized return	Sharpe ratio
FedAvg	27.38%	1.43
RFA	35.30%	1.77
Coordinate-wise median	36.03%	1.80
FedProx	36.56%	1.83
FedAvg+	35.62%	1.75
FedGeoMed+	36.01%	1.79
FedCoMed+	35.56%	1.78

Fig. 21.5 Average learning performance over 10 parties in the financial portfolio management problem, across methods. Note the training failure of the federated learning algorithms, which is not caused by adversarial parties or party-level failure, as evidenced by the single-party training curve in Fig. 21.3

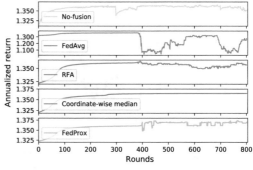

Fig. 21.6 Average learning performance of Fed+ algorithms on the financial portfolio optimization problem with 10 parties. Compare this with Fig. 21.5 using the baseline algorithms. No training collapse occurs with any of the Fed+ algorithm variants

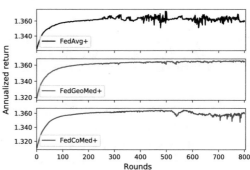

We next consider the case where there are $N = 50$ parties. The average performance with 50 parties is shown in Table 21.3 and the average portfolio value in the testing period is shown in Fig. 21.7. It is clear that Fed+ algorithms outperform FedAvg, RFA, coordinate-wise median, and FedProx baseline algorithms. The average learning curves can be found in Figs. 21.8 and 21.9. In this larger federation, the baseline (non-Fed+ algorithms) including both of the baseline robust federated learning methods, RFA and coordinate-wise median, experiences the same kind of collapse in training performance as seen previously. This suggests larger distribution mismatch across the 50 asset universes.

Table 21.3 Average performance of different methods over 50 parties

Method	Annualized return	Sharpe ratio
FedAvg	0.99%	0.11
RFA	4.90%	0.30
Coordinate-wise median	21.71%	1.17
FedProx	−1.46%	−0.02
FedAvg+	32.93%	1.67
FedGeoMed+	32.23%	1.64
FedCoMed+	32.47%	1.65

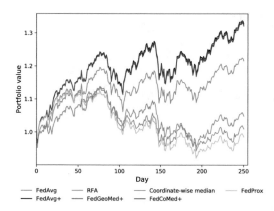

Fig. 21.7 Average portfolio value of the 50-party financial portfolio management problem during the test period, across methods. The Fed+ methods are superior to the baselines by a significant margin as regards the financial portfolio management application

As we have seen above, such sharp model change can lead to a collapse of the training process. Figure 21.8 (four bottom curves) shows the collapse of training, averaged over 50 parties, using FedAvg, RFA, coordinate-wise median, and FedProx. Note that this example *does not involve adversarial parties or party failure in any way*, as evidenced from the fact that the single-party training on the same dataset (top curve) does not suffer any failure. Rather, this is an example of federated learning on a real-world application where parties' data are not IID from a single dataset. As such, it is conceivable that federated model failure would be a relatively common occurrence in real-world applications using the vast majority of algorithms.

Data distribution similarity between each pair of universes (each asset universe is a multi-variate time series) was measured using Dynamic Time Warping (DTW; [3]). Unlike the Euclidean distance, DTW compares time series of variable size and is robust to shifts or dilatations across time. The averaged pairwise distance for the 50 asset universes is 154.51, which is 21.63% larger than that of the 10 asset

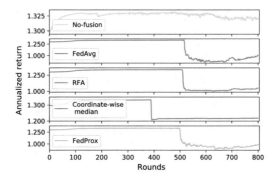

Fig. 21.8 Illustration of training collapse. Average training performance over 50 parties of the financial portfolio optimization problem. The top curve is the average of 50 single-party training processes, each party on its own data. The training processes collapse using all four federated learning algorithms. Note that there are no adversarial parties in the federation, nor are there any party-level failures across the 50 parties. This can be verified from the top curve that does not experience training collapse. Rather, the training collapse is due to the federated learning process itself and occurs when party-level training is forced to concur with a single, central model

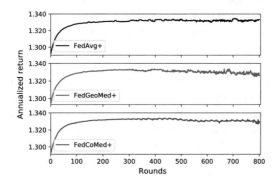

Fig. 21.9 Average learning performance of Fed+ algorithms on the financial portfolio optimization problem with 50 parties. Compare with Fig. 21.8 taken from the same problem using the baseline algorithms. No training collapse occurs with any of the Fed+ algorithm variants

universes. The DTW warping/alignment curves for asset universes I_1 and I_2 are illustrated in Fig. 21.10.

Performance across baseline methods is shown in Fig. 21.8 where the failure of the standard methods, FedAvg, RFA, coordinate-wise median, and FedProx, is clear. The parameter change before and after federated model aggregation is shown in Fig. 21.1. Observe that there are large spikes at a particular point in the training processes of the FedAvg, RFA, coordinate-wise median, and FedProx algorithms, implying that single-model methods may not work well in heterogeneous environments. Fed+, on the other hand, stabilizes each party's learning process and is robust, as shown in Fig. 21.9. Table 21.3 further shows that Fed+ and our variant algorithms outperform FedAvg, RFA, coordinate-wise median, and FedProx,

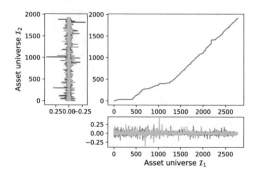

Fig. 21.10 Dynamic time warping/alignment curve for the data of the financial portfolio management problem, shown for returns of asset universes I_1 and I_2. Note that two multi-variate time series have different length. The graphic shows that the distance between data distributions given by the asset universes increases with the length of the time series, up to 2500 in this figure, leading to heterogeneity across parties

improving the annualized return by 26.01% and the Sharpe ratio by 1.26 on average. This improvement implies that each party benefits from sharing model parameters using Fed+, whereas the benefits were not seen with FedAvg, RFA, coordinate-wise median, and FedProx.

For the best performing variant of Fed+ on this application, FedAvg+, we further investigate how the risk-return performance improves as more parties share their local tasks. In particular, compared with single-party, no-federated-learning (no-fusion) training, the maximum improvements in annualized return are 4.06%, 12.57%, and 35.50%, and in Sharpe ratio are 0.19, 0.63, and 1.52 when $N = 5, 25$, and 50. This demonstrates that the benefit of using Fed+ increases as the number of parties increases.

21.6 Conclusion

This chapter presented an approach for federated reinforcement learning that makes use of deep learning models as a function estimator to represent the policy used by the RL agents. We demonstrated federated reinforcement learning on an application to financial portfolio management, whereby the trading strategies are policies to be trained using RL. The use of federated learning, and specifically the Fed+ algorithm family, allows portfolio managers to each improve their local models, i.e., their own policies, by benefiting from others, without sharing their private data. Private data in this setting comes from the use of private generative models used to augment public asset price data. We demonstrated the significant gains that can be achieved for portfolio managers through federated RL and Fed+, both in terms of annualized returns and Sharpe ratio. The gains can be seen with respect to independently training policies, i.e., without federated learning, as well as through the use of reinforcement learning to train the policies, as compared to standard (non RL-based) trading strategies.

Similar benefits from a federated approach can be expected in other application domains of reinforcement learning. Of particular interest are robotics and navigation. In robotics and navigation applications, simulation results and real-world experiences need not coincide, and hence federating multiple parties with their individual experiences can bring considerable improvements to each. As in the case of financial portfolio management, the Fed+ family of methods that allows for an emphasis on local models may prove to be critical to achieving the most dramatic benefits.

References

1. Almahdi S, Yang SY (2017) An adaptive portfolio trading system: a risk-return portfolio optimization using recurrent reinforcement learning with expected maximum drawdown. Expert Syst Appl 87:267–279
2. Begleiter R, El-Yaniv R, Yona G (2004) On prediction using variable order Markov models. J Artif Intell Res 22:385–421
3. Berndt DJ, Clifford J (1994) Using dynamic time warping to find patterns in time series. In: KDD workshop, Seattle, vol 10, pp 359–370
4. Black F, Scholes M (1973) The pricing of options and corporate liabilities. J Polit Econ 81(3):637–654
5. Buehler H, Gonon L, Teichmann J, Wood B (2019) Deep hedging. Quant Financ 19(8):1271–1291
6. Goodfellow I, Pouget-Abadie J, Mirza M, Xu B, Warde-Farley D, Ozair S, Courville A, Bengio Y (2014) Generative adversarial nets. In: Advances in neural information processing systems, pp 2672–2680
7. Haugen RA, Haugen RA (2001) Modern investment theory, vol 5. Prentice Hall, Upper Saddle River
8. Jiang Z, Xu D, Liang J (2017) A deep reinforcement learning framework for the financial portfolio management problem. arXiv preprint arXiv:170610059
9. Kariya T, Liu RY (2003) Options, futures and other derivatives. In: Asset pricing. Springer, Boston, MA, pp 9–26
10. Krizhevsky A, Sutskever I, Hinton GE (2012) Imagenet classification with deep convolutional neural networks. In: Advances in neural information processing systems, pp 1097–1105
11. Li T, Sahu AK, Zaheer M, Sanjabi M, Talwalkar A, Smith V (2020) Federated optimization in heterogeneous networks. Proc Mach Learn Syst 2:429–450
12. Markowitz H (1959) Portfolio selection: efficient diversification of investments, vol 16. Wiley, New York
13. McMahan B, Moore E, Ramage D, Hampson S, y Arcas BA (2017) Communication-efficient learning of deep networks from decentralized data. In: Artificial intelligence and statistics. PMLR, pp 1273–1282
14. Pillutla K, Kakade SM, Harchaoui Z (2019) Robust aggregation for federated learning. arXiv preprint arXiv:191213445
15. Sharpe WF (1966) Mutual fund performance. J Bus 39(1):119–138
16. Silver D, Lever G, Heess N, Degris T, Wierstra D, Riedmiller M (2014) Deterministic policy gradient algorithms. In: Proceedings of the 31st international conference on international conference on machine learning – JMLR.org, ICML'14, vol 32, pp I–387–I–395

17. Sutton RS, McAllester DA, Singh SP, Mansour Y (2000) Policy gradient methods for reinforcement learning with function approximation. In: Advances in neural information processing systems, pp 1057–1063
18. Ye Y, Pei H, Wang B, Chen PY, Zhu Y, Xiao J, Li B (2020) Reinforcement-learning based portfolio management with augmented asset movement prediction states. In: Proceedings of the AAAI conference on artificial intelligence, vol 34, pp 1112–1119
19. Yin D, Chen Y, Kannan R, Bartlett P (2018) Byzantine-robust distributed learning: towards optimal statistical rates. PMLR, Stockholmsmässan, Stockholm, vol 80. Proceedings of Machine Learning Research, pp 5650–5659. http://proceedings.mlr.press/v80/yin18a.html

Chapter 22
Application of Federated Learning in Medical Imaging

Ehsan Degan, Shafiq Abedin, David Beymer, Angshuman Deb, Nathaniel Braman, Benedikt Graf, and Vandana Mukherjee

Abstract Artificial intelligence and in particular deep learning have shown great potential in the field of medical imaging. The models can be used to analyze radiology/pathology images to assist the physicians with their tasks in the clinical workflow such as disease detection, medical intervention, treatment planning, and prognosis to name a few. Accurate and generalizable deep learning models are in high demand but require large and diverse sets of data. Diversity in medical images means images collected at various institutions, using several devices and parameter settings from diverse populations of patients. Thus, producing a diverse data set of medical images requires multiple institutions to share their data. Despite the universal acceptance of Digital Imaging and Communications in Medicine (DICOM) as a common image storage format, sharing large numbers of medical images between multiple institutions is still a challenge. One of the main reasons is strict regulations on storage and sharing of personally identifiable health data including medical images. Currently, large data sets are usually collected with participation of a handful of institutions after rigorous de-identification to remove personally identifiable data from medical images and patient health records. De-identification is time consuming, expensive, and error prone and in some cases can remove useful information. Federated Learning emerged as a practical solution for training of AI models using large multi-institute data sets without a need for sharing the data, thereby removing the need for de-identification while satisfying necessary regulations. In this chapter, we present several examples of federated learning for medical imaging using IBM Federated Learning.

E. Degan (✉) · S. Abedin · D. Beymer · V. Mukherjee
IBM Almaden Research Center, San Jose, CA, USA
e-mail: edehgha@us.ibm.com

A. Deb · B. Graf
IBM Watson Health Imaging, Cambridge, MA, USA

N. Braman
Tempus Labs, Chicago, IL, USA

22.1   Introduction

With the advent of deep learning, computer vision algorithms have taken a big leap forward. Similar progress is expected in the field of computer vision for medical images. Application of computer vision tasks such as detection, segmentation, and classification to medical images can be extremely helpful in assisting physicians in performing their tasks faster, more accurately, and more consistently. Many tasks such as disease detection, tumor localization, treatment planning, and prognosis, to name a few, can benefit from deep learning models (see [1] and references therein). However, to train accurate, reliable, and generalizable deep learning models, one requires a large data set of training samples from various sources. Although large and diverse sets of natural images are readily accessible in public domain [2], publicly available medical imaging data sets are relatively small and from a few sources [3, 4]. There are two main reasons for lack of such large and diverse training data sets: (1) requirement for labels and annotations, and (2) difficulty in sharing health data.

In order to conduct a supervised training, one requires labels or annotations for the images. Acquiring labels for natural images through crowd-sourcing is relatively easy and inexpensive. However, in the medical domain, labels and annotations should be produced by medical experts. Recently, Natural Language Processing (NLP) methods are used to analyze radiology reports and automatically produce labels at large scale [5, 6]. However, if detailed annotations, such as contouring around organs or tumors are required, the annotation task can become prohibitively expensive. Self-supervised and unsupervised learning methods are being developed to train models with less dependency on annotated data and overcome the first obstacle.

Sharing medical images across multiple institutions is still a challenge, despite worldwide acceptance of Digital Imaging and Communications in Medicine (DICOM) format [7]. The main reason for this challenge is that medical images, like other health records, may contain Protected Health Information (PHI) and are strictly guarded by laws and regulations such as Health Insurance Portability and Accountability Act (HIPAA) in the USA [8] or General Data Protection Regulation (GDPR) in Europe [9]. As a result, sharing medical images requires rigorous de-identification. De-identification is time consuming, expensive, and error prone and in some cases can remove useful information.

Federated learning is a machine learning technique that allows several parties to participate in model training without sharing their local sensitive data [10]. In a federated learning scenario, each training party trains a model using its local data and sends its model update, not the training data, to an aggregator that combines the updates coming from different parties into a single model (see Fig. 22.1). Federated learning allows us to train generalizable models using large and diverse data sets while satisfying security and privacy regulations by obviating the need to share sensitive data. Therefore, federated learning is a very attractive solution for medical imaging.

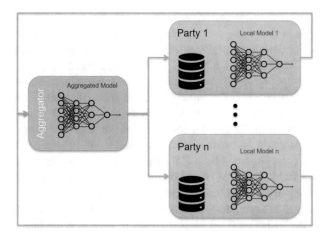

Fig. 22.1 Data flow diagram of Federated Learning

In this chapter, we demonstrate implementation of two of the most common computer vision applications in medical imaging: image classification and image segmentation. In the first task, an image is classified into positive or negative depending on the presence or absence of a collection of target image findings. In the second task, the model produces a binary mask delineating a target object.

We train our models using IBM Federated Learning [11], which provides infrastructure and coordination for federated learning. Although this framework applies to deep learning models as well as other machine learning methods, we strictly use it to train deep learning models.

In the following sections, we demonstrate the application of federated learning to the two above-mentioned tasks. We report our work on image segmentation to delineate pulmonary embolisms in volumetric CT images. We also carried out experiments on 2D and 3D image classification by training models to detect Pneumothorax in X-ray images and Emphysema in 3D CT images. For image segmentation and 3D image classification, we implemented a simulated federated learning scenario. In this scenario, the training data is recorded in a centralized repository, but it is partitioned into different groups and each group of data is exclusively used by a party to train the model. As data are kept in a centralized location, the trained model can be compared to a centrally trained one. For the 2D classification task, however, we used two sources of data kept in two geographically distant repositories to demonstrate a more realistic scenario.

22.2   Image Segmentation

Outlining an organ, abnormality, or other image findings is one of the main applications of computer vision in medical imaging. In order to demonstrate the capabilities of federated learning in such a task, we implemented a segmentation model to delineate pulmonary embolisms in contrast-enhanced chest CT images. Pulmonary Embolism (PE) is a blockage in the pulmonary arteries most likely caused by a blood clot. CT Pulmonary Angiogram (CTPA) is the imaging modality of choice to detect PE. Detecting clinically evident PE is important in quickly diagnosing patients with symptoms and signs of venous thromboembolism. Untreated clinically apparent PE has a nearly 30% mortality rate in contrast to an 8% mortality rate for those patients who receive treatment [12–16]. Although the mortality rate from PE alone is only 2.5% [17], timely detection and anticoagulation therapy improves the patient's outcome. Patients suspected of PE are recommended to take a D-Dimer test followed by a CT Pulmonary Angiography (CTPA) for high probability clinical assessment. A radiologist has to carefully inspect each branch of the pulmonary arteries for suspected PE. Consequently, diagnosis of PE hinges on the radiologist's experience, attention span, and eye fatigue, among others. Computer Aided Detection (CAD) software for PE detection has historically shown to help radiologists detect and diagnose PE [18–22]. In addition, detecting PE in CT angiography (CTA) images can be useful in a retrospective setup where the CAD software is used to detect missed findings.

PE usually has small-size irregular-shaped pathological patterns. Hence, the image region distinctive for PE classification may only account for a small portion of the imaging data even when image patches are used. Localizing the distinctive image region is critical for successful PE classification. Computer aided methods have been developed to automatically detect PE. These methods are often two-stage solutions where the first stage produces a group of PE candidates in the image and the second stage classifies the candidates into true PE and false positives [23–25]. The first stage can be implemented as a segmentation task in which a model analyzes the images and delineates candidate embolisms to be classified by the second stage.

In this section, we train a PE segmentation model as the first stage using 2 data sets in a federated setup. The first data set [26] consisted of 40 CTPA images, each from a different patient. The scans were acquired at Unidad Central de Radiodiagnóstico in Madrid, Spain, from several scanners using a local institutional CTPA protocol. Each CTPA volume was annotated by three board-certified radiologists with several years of experience, independently, and a reference standard was finally created by consolidating all three annotations. We used this data set at Party 1. The second data set [27] consisted of Computed Tomography Angiograms (CTA) images for pulmonary embolism of 35 different patients published by Ferdowsi University of Mashhad, Iran. Each CTA volume was annotated by two radiologists and consolidated to create a reference standard. We used this data set at Party 2.

For testing, we used a private data set that consisted of 334 volumes that include both CTA and CTPA volumes acquired from multiple scanners and hospitals. For annotating each PE positive volume, a team of 7 board-certified radiologists drew a contour around each embolism on slices spaced approximately 10 mm apart. The annotators had non-overlapping assignments, such that each CT volume is annotated by only one annotator.

For more efficient detection, we apply PE segmentation to identify embolism candidates. In order to provide more context for the annotated slices, we used a slab-

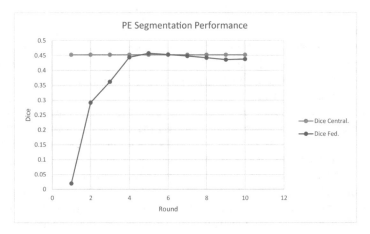

Fig. 22.2 Comparison of centralized and federated learning in terms of continuous Dice coefficient

based 2D segmentation method using U-Net [28]. Instead of using just 2D slices, a slab of nine slices is fed to the network with the corresponding binary mask as the ground truth. The U-Net model in our segmentation task consists of 70 layers with a contracting path with repeated 3×3 convolutions, each followed by a rectified linear unit (ReLU) and a 2×2 max pooling operation with a stride of 2 for downsampling. The expansive path consists of up-sampling of features followed by a 2×2 convolution, concatenation with the correspondingly cropped feature map from the contracting path and 3×3 convolutions, each followed by a ReLU. The probability map is computed by a pixel-wise softmax over the final feature map. For training, the continuous dice loss (DL) function [29] is used.

For the federated training, we trained the aforementioned network by both parties for 10 rounds of 50 epochs each. We used Federated Averaging [10] to aggregate the model updates from each party. For comparison, we also trained the segmentation model using the combined data set for 100 epochs, which reached a dice coefficient of 0.45 on our test set. After each round, we evaluated the aggregated model using the test data set. The results of the aggregated model after each round are compared to the results of the centrally trained model in Fig. 22.2. Remarkably, the aggregated model was able to achieve similar performance to the central model after just 5 rounds.

22.3   3D Image Classification

In this section, we used federated learning to train a classifier to detect emphysema in 3D chest CT scans. As opposed to segmentation, which estimates a dense binary mask indicating the localized presence of disease, classification requires just a single binary detection result per 3D volume. When training, this translates into a labeling requirement of a single disease detection per volume, which is much less burdensome than the dense 3D mask for segmentation. For labeling the training data, it also opens up possibilities for leveraging disease diagnosis codes in the patient medical records, and NLP on imaging radiology reports. Let us now explore disease classification of emphysema.

Emphysema is a disease of the lung where alveoli—the terminal chambers of the respiratory tract—collapse and merge, leaving regions of open airspace in the lungs. In CT scans, this appears as dark patches of low attenuation. The challenge for computer aided detection of emphysema is that the low attenuation regions may still have a variety of appearances, and emphysema lesions may only appear in a portion of the lung volume. Researchers have recently used machine learning approaches for emphysema detection and quantification, such as multiple instance learning [30–33], Convolutional Neural Networks (CNNs) [34–36], and Convolutional Long Short-Term Memory (ConvLSTM) [37].

The architecture for our classifier is based on ConvLSTM [38] taken from our previous work on detection of emphysema [37]. This architecture is a variant of the popular LSTM model that utilizes convolutional operations to distinguish variations in spatial patterns. For instance, ConvLSTM excels in detecting spatiotemporal patterns, such as video classification. Rather than applying ConvLSTM to time series image data, we instead propose its use to scan through a series of consecutive slices of an imaging volume to learn patterns of disease without manual annotations of its location. Our approach allows the detection of disease on and between slices, storing them through multiple bidirectional passes through a volume and output as a final set of features characterizing overall disease presence.

Our architecture, depicted in Fig. 22.3, consists of four units. Each unit features two 2D convolutional layers, which extract features from each slice separately, followed by max pooling, and then finally by a ConvLSTM layer to process the volume slice by slice. Each convolutional layer has a kernel size of 3×3 and

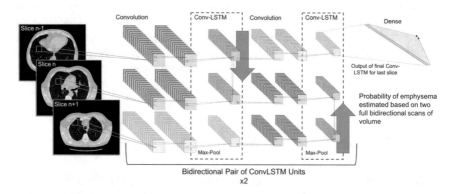

Fig. 22.3 Network architecture for detection of emphysema in 3D chest CT scans

Table 22.1 Emphysema
training and testing data

		Positive	Negative	Total
Training	Party 1	492	526	1018
	Party 2	477	540	1017
Testing		519	465	984

rectified linear unit (ReLU) activation, followed by batch normalization. The outputs from the convolutional layers for each slice are then processed sequentially by a ConvLSTM layer, with tanh activation and hard sigmoid recurrent activation. All layers within a unit share the same number of filters and process the volume in ascending or descending order. The four units have the following dimensionality and directionality: Ascending Unit 1: 32 filters, Descending Unit 1: 32 filters, Ascending Unit 2: 64 filters, Descending Unit 2: 64 filters. The final ConvLSTM layer outputs a single set of features, which summarizes the network's findings after processing through the imaging volume multiple times. A fully connected layer with sigmoid activation then computes probability of emphysema.

We use a simulated 2 party configuration for training our emphysema detection network in a federated manner. For training data, our images are from a large body of CT scans collected from our data providing partner. We divided a set of 2035 CT scans between the 2 parties—1018 for party 1 and 1017 for party 2, with a roughly even split between positive and negative examples for each party (see Table 22.1). For testing data, we used the same data source, forming a testing cohort of 984 CT scans with 465 negatives and 519 positives. For labeling the data either positive or negative for emphysema, we leveraged the associated imaging report. In order to detect emphysema in the reports, we searched for the word emphysema and a list of its synonyms. The results were manually verified.

The model is initially trained in a centralized manner, using a set of low dose CT scans from the National Lung Screening Trial (NLST) [39] data set, with 7100 scans for training and 1776 for validation. Low dose CT scans are more suitable for screening purposes and expose the subject to lower levels of radiation compared to regular dose CT. To help generalize the model to regular dose CT scans, we next applied transfer learning using a set of CT scans from the Lung Tissue Research Consortium (LTRC) [40] data set, with 858 training and 200 validation scans. We used the weights of this model as initial weights for federated learning.

To demonstrate federated learning, we perform a number of experiments with different configurations for parties 1 and 2, as well as a comparison with centralized training. In order to compare the models, we calculate the area under the receiver operating characteristic curve (AUC) using our test set. First, we train the model at each party, independently, for 5 epochs, starting from the initial model from the NLST/LTRC training. This essentially performs transfer learning, independently, for each of the parties, and the test set AUCs are shown in Fig. 22.4. Next, we perform federated learning with parties 1 and 2 with 20 rounds of 5 epochs each. Similar to the PE segmentation case, we used Federated Averaging [10] for model aggregation. For federated learning, the test set AUC is evaluated at the end of each

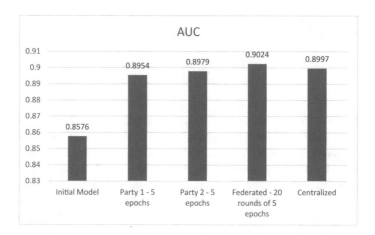

Fig. 22.4 Performance of emphysema detection model trained with different scenarios in terms of AUC

round, and the maximum AUC is reported in Fig. 22.4. Finally, for a comparison to centralized learning on the combined training sets of parties 1 and 2, we train the model in a standard centralized manner, yielding the test set AUC in Fig. 22.4, right. We also tested the initial model on the test set as shown in Fig. 22.4.

As it can be seen, the model trained in a federated setup outperforms each of the individually trained models and shows better generalizability. It also slightly outperforms the centrally trained model after only 5 rounds although the difference may be caused by the random nature of stochastic gradient descent and the difference between the number of epochs in the centralized training and the effective number of epochs in the federated learning scenario.

22.4   2D Image Classification

In this section, we implemented a more realistic federated learning scenario where the two parties were located at geographically distant locations and communicated over the internet with a secure connection. In this scenario, we trained a model to classify X-ray images for detection of Pneumothorax, a collapsed lung that occurs when air leaks between the lung and the chest wall. Severe Pneumothorax can be fatal and needs immediate medical intervention. Therefore, timely detection of Pneumothorax is vital. The most commonly utilized imaging modalities for the detection of Pneumothorax are chest X-ray, CT scan, and thoracic ultrasound.

We used two sources of data for our training. The first source was from a large body of X-ray images collected by our data providing partner. We use this data set as Party 1, located at IBM Rochester Data Center in Rochester, NY. Each imaging study was accompanied by its radiology report. We used an in-house NLP algorithm to analyze the reports and obtain disease labels for the images.

Table 22.2 Pneumothorax data composition

	Private		MIMIC	
	Positive	Negative	Positive	Negative
Training	260	743	206	747
Testing	2241	15,074	266	1476

For the second data set, we used MIMIC-CXR database [3, 4]. This is a large and publicly available database of X-ray images and their radiology reports collected at the Beth Israel Deaconess Medical Center in Boston, MA. This database contains 377,110 images corresponding to 227,835 radiographic studies, all anonymized to satisfy HIPPA requirements. We use this data set as Party 2, located at IBM Almaden Research Center, San Jose, CA. We labeled the studies in this data set based on the agreement of CheXpert [5] and NegBio [6] algorithms. We assumed a study was positive for Pneumothorax when both NLP algorithms classified it as positive and it was negative when both classified it as negative.

In order to provide a proof of concept quickly, we used only a fraction of the data available to us. Our training and testing data set is shown in Table 22.2. In total, we used 1003 randomly selected training samples from 1003 studies from our private data source and 953 randomly selected training samples from 953 patients from MIMIC data set. Positive and negative samples are randomly selected in a way to have a similar distribution between the two parties.

For testing samples from the MIMIC data set, images were selected from patients who did not contribute to the training samples. However, our private data set was anonymized in a way that it is impossible for us to group studies based on patients. Therefore, for testing samples from our private data set, we used images from studies that were not present in the training set. As we selected our test samples from different acquisition months than our training samples, the likelihood of leakage between training and testing is low, but not zero.

Each chest X-ray study may include frontal and lateral images. We only used frontal images for our training and testing. To this end, we used a deep learning model to classify X-ray images into frontal and lateral images. The frontal images were cropped to the area around the lungs and re-sampled to 1024×1024 pixels.

We started by using a DenseNet-121 [41] pre-trained on ImageNet [2]. Then, the model weights were updated using more than 100,000 images from our private data set in a centralized fashion. For this stage, we used CheXpert NLP [5] to create labels for 14 different categories of findings that included Pneumothorax. Lastly, we selected only the output corresponding to Pneumothorax and used this model as the initial model sent by the aggregator to the training parties to be trained by the data in Table 22.2 in a federated fashion.

For Party 1 we used a GPU server at IBM Rochester Data Center with 8 Nvidia GeForce GTX 1080 GPU devices. For Party 2, we used a GPU server at IBM Almaden Research Center with 8 Nvidia Tesla P100 GPU devices. We used a server without GPU as the aggregator. We used 4 GPU devices at Party 1 and 3 GPU devices at Party 2 to train with a batch size of 9. We used the Adam optimizer with a learning rate of 1e-4. All images were augmented on-the-fly with a random rotation of ±10°, height and width shift of ±10%, shear range of ±10%, and zoom between 1 and 1.1. Images were augmented with a probability of 80%. We used Federated Averaging [10] for model aggregation.

In the Federated Learning proof of concept implementation, the data sets that were used in training the algorithm by the remote training servers were treated as potentially containing PHI. These servers and the training data were operating in air-gapped networks with very restricted access. The aggregator and the participating training servers communicated via REST APIs. Even though the information exchanged via these APIs did not contain any sensitive information, because of the sensitive nature of the data in these training environments, only encrypted communication was allowed for any inbound or outbound network flows. This also allowed securing the information exchanged from any malicious on-path attacks. For the proof-of-concept implementation, all communications between the aggregator and remote training servers were encrypted by the Transport Layer Security (TLS) protocol and only were allowed over specific exposed ports. As an additional layer of security, only requests originating from certain white-listed server IP addresses were allowed and only a set of privileged users were allowed SSH access to these training servers via a strict access control mechanism. Figure 22.5 shows the different network flows that were involved in the communication between the aggregator and the remote training servers:

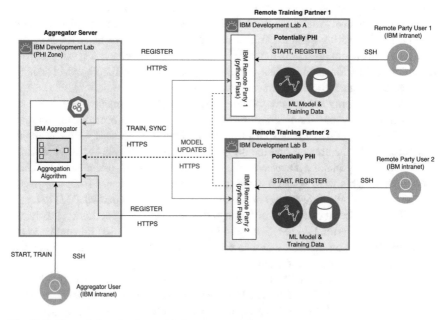

Fig. 22.5 Network flow for the proof of concept experiment

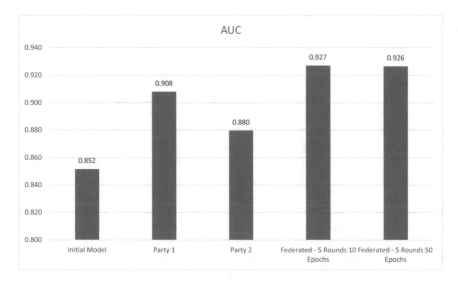

Fig. 22.6 Performance of models trained with different scenarios in terms of AUC

In order to demonstrate the performance of federated learning, we conducted the following experiments. For the first experiment, we trained the model at each party, independently, for 50 epochs. For the second experiment, we trained the model in a federated fashion for 5 rounds of 50 epochs each. Lastly, we trained the model for 5 rounds of 10 epochs at each party. In the latter two experiments, convergence was achieved before round 5. We used the test set in Table 22.2 and calculated the AUC for all the trained models as well as the initial model. Figure 22.6 shows the results. As it can be seen, the models trained in a federated fashion outperform both individually trained models. We did not train a central model as we could not share data between the two parties.

22.5   Discussion

In this chapter, we used IBM Federated Learning [11] for image classification and image segmentation tasks. These two tasks cover a wide range of medical imaging applications. IBM Federated Learning can easily integrate custom layers, loss functions, data loader and data augmentation methods as well as custom aggregation algorithms. Therefore, it can be easily used to train many other imaging tasks in a federated setup. As mentioned above, different number of GPU devices can be used at each party to accommodate the size of the data set as well as hardware availability.

For all the experiments in this chapter, we used Federated Averaging [10] as our aggregation method. However, IBM Federated Learning [11] supports several other aggregation methods, including user defined ones, suitable for different problems and applications.

As part of training or testing data for each of the experiments above, we used a private data set from our data providing partner. These images were collected over several months and from several hospitals. This private data set is anonymized at source. HIPAA is fully enforced and all data are handled according to the Declaration of Helsinki. As mentioned before, due to the specific data anonymization protocol used, we cannot correspond imaging studies to patients. As a result, we could only separate training and testing data sets at a study level, not a patient level.

For 3D classification and image segmentation tasks, we compared the federated model with a centrally trained one. Since we were conducting a simulated federated learning scenario, training a model using data from both parties was possible. For 2D classification task, however, we treated the problem as a real federated learning setup and did not move data between parties. Therefore, it was not possible to train a model using data from both parties. For this reason, we compared the model trained in a federated setup to two models trained at each of the parties, independently. As expected, the federated model outperforms both models on a test set that has data from both parties. Figure 22.6 shows that the model trained at Party 1 (using private data) performs better than the model trained at Party 2 on the test set. As it can be seen in Table 22.2, the test set contains more samples from Party 1 data distribution than that of Party 2. This can explain the superior performance of the model trained at Party 1 with private data on our test set.

In Sect. 22.1 we discussed that there are two main hurdles for training of medical imaging models on large and diverse collections of data—the difficulty in obtaining large-scale annotated data and the challenges in sharing medical images. As we demonstrated in the example experiments shown in this chapter, the latter obstacle can be overcome by using federated learning that obviates the need for sharing data. For our classification tasks, we used NLP algorithms to automatically analyze readily available radiology reports and provide labels for our training and testing samples. As NLP algorithms get better at their tasks, combination of NLP algorithms for data labeling and federated learning can provide a viable path toward scalable and generalizable computer vision for medical images. It should be mentioned, however, that the segmentation task still requires expert annotations. Unsupervised and semi-supervised methods can, hopefully, be combined with federated learning to provide a solution.

22.6 Conclusions and Future Work

In this chapter, we demonstrated the application of federated learning in two fundamental tasks of medical imaging: image segmentation and image classification. In the segmentation task, we showed that a federated model for segmenting PE can achieve a performance equivalent to a model trained on a centralized data set in as few as 5 communication rounds. In a realistic setup, we trained a model for classification of X-ray images in a federated setup between two parties at two different geographical locations. We used air-gapped servers and encrypted communication for security. The model trained in the federated setup outperforms each of the individual models, trained independently at each party. This demonstrates the importance of federated learning in training generalizable models when data sharing is not possible.

In the future, we plan on experimenting with different aggregation methods to investigate their effects on the convergence of the aggregated model. We will also train models at large scales with data contributions from various sources. In particular, we intend to use our NLP algorithm to label 2D and 3D images for various diseases and image findings using the available reports and use these automatically generated labels in a federated learning setup between several imaging centers for scalable and generalizable model training.

References

1. Kim M, Yun J, Cho Y, Shin K, Jang R, Bae HJ, Kim N (2019) Deep learning in medical imaging. Neurospine 16(4):657–668. https://doi.org/10.14245/ns.1938396.198
2. Russakovsky O, Deng J, Su H et al (2015) ImageNet large scale visual recognition challenge. Int J Comput Vis 115:211–252
3. Johnson A, Pollard T, Mark R, Berkowitz S, Horng S (2019) MIMIC-CXR Database (version 2.0.0). PhysioNet. https://doi.org/10.13026/C2JT1Q.
4. Johnson AEW, Pollard TJ, Berkowitz SJ et al (2019) MIMIC-CXR, a de-identified publicly available database of chest radiographs with free-text reports. Sci Data 6:317. https://doi.org/10.1038/s41597-019-0322-0
5. Irvin J, Rajpurkar P, Ko M, Yu Y, Ciurea-Ilcus S (2019) CheXpert: a large chest radiograph dataset with uncertainty labels and expert comparison. In: 33rd AAAI conference on artificial intelligence
6. Peng Y, Wang X, Lu L, Bagheri M, Summers RM, Lu Z (2018) NegBio: a high-performance tool for negation and uncertainty detection in radiology reports. In: AMIA 2018 informatics summit
7. (2001) DICOM reference guide. Health Dev 30:5–30
8. HIPAA (2020) US Department of Health and Human Services. https://www.hhs.gov/hipaa/index.html
9. GDPR (2016) Intersoft consulting. https://gdpr-info.eu
10. McMahan HB, Moore E, Ramage D, Hampson S, Arcas BAY (2017) Communication-efficient learning of deep networks from decentralized data. arXiv preprint arXiv:1602.05629
11. Ludwig H, Baracaldo N, Thomas G, Zhou Y, Anwar A, Rajamoni S, Ong Y, Radhakrishnan J, Verma A, Sinn M et al (2020) IBM federated learning: an enterprise framework white paper V0.1. arXiv preprint arXiv:2007.10987
12. Dalen JE (2002) Pulmonary embolism: what have we learned since Virchow? Natural history, pathophysiology, and diagnosis. Chest 122(4):1440–1456
13. Barritt DW, Jordan SC (1960) Anticoagulant drugs in the treatment of pulmonary embolism: a controlled trial. The Lancet 275(7138):1309–1312
14. Hermann RE, Davis JH, Holden WD (1961) Pulmonary embolism: a clinical and pathologic study with emphasis on the effect of prophylactic therapy with anticoagulants. Am J Surg 102(1):19–28

15. Morrell MT, Dunnill MS (1968) The post-mortem incidence of pulmonary embolism in a hospital population. Br J Surg 55(5):347–352
16. Coon WW, Willis PW 3rd, Symons MJ (1969) Assessment of anticoagulant treatment of venous thromboembolism. Ann Surg 170(4):559
17. Carson JL, Kelley MA, Duff A, Weg JG, Fulkerson WJ, Palevsky HI, Schwartz JS, Thompson BT, Popovich J Jr, Hobbins TE, Spera MA (1992) The clinical course of pulmonary embolism. N Engl J Med 326(19):1240–1245
18. Das M, Mühlenbruch G, Helm A, Bakai A, Salganicoff M, Stanzel S, Liang J, Wolf M, Günther RW, Wildberger JE (2008) Computer-aided detection of pulmonary embolism: influence on radiologists' detection performance with respect to vessel segments. Eur Radiol 18(7):1350–1355
19. Zhou C, Chan HP, Patel S, Cascade PN, Sahiner B, Hadjiiski LM, Kazerooni EA (2005) Preliminary investigation of computer-aided detection of pulmonary embolism in three-dimensional computed tomography pulmonary angiography images. Acad Radiol 12(6):782
20. Schoepf UJ, Schneider AC, Das M, Wood SA, Cheema JI, Costello P (2007) Pulmonary embolism: computer-aided detection at multidetector row spiral computed tomography. J Thorac Imaging 22(4):319–323
21. Buhmann S, Herzog P, Liang J, Wolf M, Salganicoff M, Kirchhoff C, Reiser M, Becker CH (2007) Clinical evaluation of a computer-aided diagnosis (CAD) prototype for the detection of pulmonary embolism. Acad Radiol 14(6):651–658
22. Engelke C, Schmidt S, Bakai A, Auer F, Marten K (2008) Computer-assisted detection of pulmonary embolism: performance evaluation in consensus with experienced and inexperienced chest radiologists. Eur Radiol 18(2):298–307
23. Liang J, Bi J (2007) Computer aided detection of pulmonary embolism with tobogganing and multiple instance classification in CT pulmonary angiography. In: Biennial international conference on information processing in medical imaging. Springer, Berlin/Heidelberg, pp 630–641
24. Tajbakhsh N, Gotway MB, Liang J (2015) Computer-aided pulmonary embolism detection using a novel vessel-aligned multi-planar image representation and convolutional neural networks. In: International conference on medical image computing and computer-assisted intervention. Springer, Cham, pp 62–69
25. Huang SC, Kothari T, Banerjee I, Chute C, Ball RL, Borus N et al (2020) PENet—a scalable deep-learning model for automated diagnosis of pulmonary embolism using volumetric CT imaging. NPJ Digit Med 3(1):1–9
26. Gonzalez G. CAD-PE challenge website. Available online: http://www.cad-pe.org
27. Masoudi M, Pourreza HR, Saadatmand-Tarzjan M, Eftekhari N, Zargar FS, Rad MP (2018) A new data set of computed-tomography angiography images for computer-aided detection of pulmonary embolism. Sci Data 5(1):1–9
28. Ronneberger O, Fischer P, Brox T (2015) U-Net: convolutional networks for biomedical image segmentation. In: International conference on medical image computing and computer-assisted intervention. Springer, Cham, pp 234–241
29. Milletari F, Navab N, Ahmadi SA (2016) V-Net: fully convolutional neural networks for volumetric medical image segmentation. In: 2016 fourth international conference on 3D vision (3DV). IEEE, pp 565–571
30. Ørting SN, Petersen J, Thomsen LH, Wille MMW, Bruijne de M (2018) Detecting emphysema with multiple instance learning. In: 2018 IEEE 15th international symposium biomedical imaging (ISBI), pp 510–513
31. Cheplygina V, Sørensen L, Tax DMJ, Pedersen JH, Loog M, de Bruijne M (2014) Classification of COPD with multiple instance learning. In: 2014 22nd international conference on pattern recognition, pp 1508–1513
32. Peña IP, Cheplygina V, Paschaloudi S et al (2018) Automatic emphysema detection using weakly labeled HRCT lung images. PLoS ONE 13(10). https://www.ncbi.nlm.nih.gov/pmc/articles/PMC6188751/

33. Bortsova G, Dubost F, Ørting S et al (2018) Deep learning from label proportions for emphysema quantification. In: Medical image computing and computer assisted intervention—MICCAI 2018, pp 768–776

34. Karabulut EM, Ibrikci T (2015) Emphysema discrimination from raw HRCT images by convolutional neural networks. In: 2015 9th international conference on electrical and electronics engineering, ELECO, pp 705–708. http://ieeexplore.ieee.org/document/7394441/

35. Negahdar M, Coy A, Beymer D (2019) An end-to-end deep learning pipeline for emphysema quantification using multi-label learning. In: 41st annual international conference on IEEE engineering in medicine biology society, EMBC 2019, pp 929–932

36. Humphries S, Notary A, Centeno JP, Strand M, Crapo J, Silverman E, Lynch D (2020) Deep learning enables automatic classification of emphysema pattern at CT. Radiology 294(2):434–444

37. Braman N, Beymer D, Degan E (2018) Disease detection in weakly annotated volumetric medical images using a convolutional LSTM network. arXiv preprint arXiv:1812.01087

38. Shi X, Chen Z, Wang H, Yeung DY, Wong WK, Woo WC (2015) Convolutional LSTM network: a machine learning approach for precipitation nowcasting. Adv Neural Inf Process Syst 2015:802–810

39. National Lung Screening Trial Research Team; Aberle DR, Berg CD, Black WC, Church TR, Fagerstrom RM, Galen B, Gareen IF, Gatsonis C, Goldin J, Gohagan JK, Hillman B, Jaffe C, Kramer BS, Lynch D, Marcus PM, Schnall M, Sullivan DC, Sullivan D, Zylak CJ (2011) The national lung screening trial: overview and study design. Radiology 258(1):243–253

40. Hohberger LA, Schroeder DR, Bartholmai BJ et al (2014) Correlation of regional emphysema and lung cancer: a lung tissue research consortium-based study. J Thorac Oncol Off Publ Int Assoc Study Lung Cancer 9(5):639–645

41. Huang G, Liu Z, Van Der Maaten L, Weinberger KQ (2017) Densely connected convolutional networks. In: 2017 IEEE conference on computer vision and pattern recognition (CVPR), pp 2261–2269. https://doi.org/10.1109/CVPR.2017.243

Chapter 23
Advancing Healthcare Solutions with Federated Learning

Amogh Kamat Tarcar

Abstract As the COVID19 pandemic began spreading, there were only pockets of information available with hospitals across geographies. Researchers attempting to analyze information scrambled to collaborate. These efforts were hampered due to regulations and privacy protection laws in various nations, which frequently confine access to clinical information. On the one hand machine learning and AI were helping doctors to make quicker diagnostic decisions using predictive models and on the other pharmaceutical companies could leverage AI for advancing drug discovery and vaccine research. COVID19 is one example among many research efforts for advancing treatment for ailments concerning prominent health issues such as cancer treatment and rare diseases. Yet, these AI systems were being developed in silos and their capabilities were hampered by the lack of a collaborative learning mechanism, thus limiting their potential. In this chapter we describe how multiple healthcare services can collaboratively build common global machine learning models using federated learning, without directly sharing data and not running afoul of regulatory constraints. This technique empowers healthcare organizations harness data from multiple diverse sources, much beyond the reach of a single organization. Furthermore, we will discuss some engineering aspects of FL project implementation such as data preparation, data quality management, challenges over governance of models developed with FL, and incentivizing the process. We also cover some challenges arising from data as well as model privacy concerns, which could be addressed with solutions such as differential privacy, Trusted Execution Environments, and homomorphic encryption.

A. K. Tarcar (✉)
Lead Data Scientist in AI Research, Persistent Systems Limited, Goa, India
e-mail: amogh_tarcar@persistent.com

23.1 Introduction

As the COVID19 pandemic began spreading far and wide, there were only pockets of information available with hospitals across geographies. Researchers attempting to analyze information scrambled to collaborate. These efforts were hampered due to regulations and privacy protection laws in various nations, which frequently confine access to clinical information. On the one hand machine learning and AI were helping doctors to make quicker diagnostic decisions using predictive models and on the other pharmaceutical companies could leverage AI for advancing drug discovery and vaccine research. COVID19 is one example among many research efforts for advancing treatment for ailments concerning prominent health issues such as cancer treatment and rare diseases. Yet, these AI systems were being developed in silos and their capabilities were hampered by the lack of a collaborative learning mechanism, thus limiting their potential.

Data scientists encountered numerous hurdles as they started exploring the idea of building image classifier with chest X-ray data for facilitating faster diagnosis as well as predict severity of COVID-19 infection. Even though researchers had narrowed down on X-ray medical imagery datasets from various healthcare data modalities such as pathological data, clinical data, and medical history data, it was hard to obtain this X-ray data as the healthcare industry is heavily regulated and this data is inherently sensitive. Furthermore there are multiple legal and ethical facets while working on aggregating this data for training machine learning models. Keeping pace with the diverse demographics affected by the pandemic was another challenge. For building competent machine learning models, the dataset needs to represent real-world data distribution. The dataset needs to be curated across various demographics covering varied age groups, ethnicities, and medical conditions. As the pandemic spread wide around the globe, data would need to be aggregated from geographies spreading multiple countries and continents for developing an impactful model.

Data anonymization and de-identification could perhaps help to work around the legalities of data aggregation, but the challenge of aggregating diverse demographic data is non-trivial. Data aggregation represents one component of the dataset preparation process. The other component includes curation and annotations by radiologists and medical professionals, which is equally challenging.

Another perspective on trying to understand why it is difficult to aggregate data from various healthcare businesses can be attributed to the growing awareness about the monetary value of private data. There is significant competitive advantage building healthcare solutions, which are enhanced with data-driven insights. Aggregating data from competing healthcare institutions is an important challenge when building AI solutions in healthcare. The business value perspective gets all the more emphasized as we explore AI solutions for advancing vaccine research. The pharmaceutical sector is extremely sensitive with respect to their proprietary data, yet challenging times such as fighting the COVID-19 pandemic has made it clear that collaboration is the only way for speeding up the process of drug discovery and vaccine research.

23.2 How Can Federated Learning Be Applied in Healthcare?

Federated Learning has the potential to flip these data aggregation challenges on its head. Instead of aggregating data to create a single ML model, Federated Learning aggregates ML models themselves. This ensures that data never leaves its source location, and it allows multiple parties to collaborate and build a common ML model without directly sharing sensitive data.

Academic researchers working in healthcare were quick in adopting Federated Learning techniques and applying it in the space of medical imaging tasks. Researchers from the University of Pennsylvania, USA, and Kings College, London, have demonstrated the feasibility of utilizing FL for brain tumor [1, 2] using decentralized MRI data. On similar lines, researchers improved a breast density classifier using mammograms [3].

Another interesting application of FL in the image domain includes analyzing pathology data from biomedical studies. The HealthChain consortium [4] in France is utilizing FL for helping oncologist devise better treatment for patients using histopathology and dermoscopy images. Considering the potential of FL technology, medical institutions have invested in programs such as the Trustworthy Federated Data Analytics (TFDA) project [5], the Federated Tumour Segmentation (FeTS) project [6], and the German Cancer Consortium's Joint Imaging Platform (JIP) [7].

Beyond imaging data, there have been interesting applications of FL in text mining and analysis as well. Researchers from Harvard Medical School demonstrated improved patient representational learning and phenotyping by utilizing clinical notes from different hospitals and clinics using FL [8]. In a similar vein, researchers worked with electronic health records to find clinically similar patients across hospitals [9].

On a parallel track, the pharmaceutical industry also has been embracing FL for advancing drug discovery and vaccine research by collaborating together with competitors. MELLODY (Machine learning ledger orchestration for drug discovery) is a project wherein 10 pharma companies have agreed to participate in collaborating together by providing access to insights on their proprietary data for various ML tasks using Federated Learning [10]. An example of application in drug discovery includes FL-based quantitative structure–activity relationship (QSAR) analysis [11].

23.3 Building a Healthcare FL Platform at Persistent with IBM FL

A collaborative platform powered by Federated Learning could help overcome hurdles of data access without running afoul of the regulatory constraints. For instance, if diagnostic models for detecting the severity of Covid infections were developed in the USA using traditional machine learning approaches, then their potential would be limited by the availability of data within the reach of this specific organization or geography. As the Covid infection spread in Italy before making headway in the USA, researchers could potentially use a collaboration platform to utilize vast amounts of data from Italy and the other regions of Europe and make faster progress while building diagnostic models. Thus, a FL platform could be leveraged in order to accelerate the development of performant machine learning models by enabling global collaboration. To put it in perspective, the progress made by researchers can be accelerated manifold with the help of this collaborative platform.

Cloud technologies such as containerization facilitate solution deployment by providing support for consistent application deployment environment. Platform components can be containerized and distributed using cloud infrastructure.

At Persistent, we developed a platform leveraging the IBM FL library, which could facilitate the development as well as the deployment of FL models. Our platform, following the solution architecture outlined in Fig. 23.1, consists of two core components: the first component is the aggregator container, which could be deployed on a hosted cloud environment. The second component is the client container, which is deployed either in customer premises or customer cloud infrastructure and is provided access to data. The aggregator container orchestrates federated training rounds over multiple client nodes. These global models are stored on the cloud. The client container node can request for latest global models using cloud API services and render predictions.

We have developed multiple prototype machine learning models using this platform, which healthcare organizations can use to predict recommendations for their on-duty medical staff. For instance, one model analyzes X-ray reports to predict

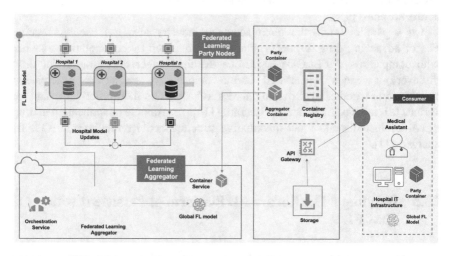

Fig. 23.1 Federated Learning Platform Reference Architecture

the severity of a COVID infection while another one analyzes gastrointestinal endoscopy images (GI tract) and predicts clinically significant labels.

This platform is a building block for the development of machine learning solutions alleviating problems concerning prominent health issues such as cancer treatment, heart ailments, etc. as well across various biomedical domains such as vaccine research and drug discovery. It empowers healthcare businesses to leverage data from multiple diverse sources and offers a key advantage of maintaining the data at its source while enabling the machine learning model to gain insights on data, from all participating nodes.

23.4   Guiding Principles for Building Platforms and Solutions for Enabling Application of FL in Healthcare

Based on our experience in building the prototype platform, we believe there is broad scope in developing engineering platforms and tools to facilitate building these FL solutions to compliment the core technical applicability of FL in healthcare. Especially considering cross silo collaboration between organizations, a platform should provide the necessary infrastructure for running FL frameworks and also integrate well with existing infrastructure and software components. The following are desirable features to consider while building a FL platform from a practitioner's point of view.

23.4.1   Infrastructure Design

The FL platform can enable the execution of projects in various party and aggregator topologies. In contrast to central machine learning a star FL topology with a central server and multiple collaborating parties would need a reinforced computing infrastructure and connectors with existing data components. As federated learning is often based on synchronous communication between collaborating nodes, in particular for scenarios not involving mobile devices, the network infrastructure is the first major module while building FL platform. While network connectivity during Core FL training rounds is critical, there could be various layers of connectivity desired between a central aggregator and parties to facilitate orchestration of FL tasks, which need to be built by building communication channels for exchanging meta data.

As the compute-intensive work of training models is now running on client nodes, the client infrastructure needs to be upgraded as required by the planned high level ML tasks. For instance, in medical imaging tasks, processing and training on high resolution images would need GPU equipped computing infrastructure. Another major component are data connectors that would integrate with existing data infrastructure to facilitate data flow into the client nodes.

23.4.2 Data Connectors Design

As the data residing in silos at each independent party is organized organically as required by the various applications generating and consuming this data, platforms need to be cognizant of loading data from diverse data sources. Also, depending on the maturity of the participating organizations, data might be arranged in on-premise data warehouses, in structured databases, or in unstructured cloud data lakes. Platforms need to have support for plugging into these party data sources as well as running various Extract, Transform, and Load (ETL) jobs. Depending on the federated machine learning task at hand, there needs to be support for running data preprocessing and standardization workloads as well. As the healthcare sector is heavily regulated, there are multiple data standardization regulations in effect. A recent regulation in the USA has suggested using FHIR [12] guidelines for the implementation of various APIs by care providers and payers. This would be a major advantage for data connectors with standardized protocol for interfacing with various healthcare organizations.

23.4.3 User Experience Design

When designing a user interface for FL platform, it is important to identify its using personas and analyze the interaction flow for each persona. Some common personas in healthcare include healthcare researchers, medical professionals, data owners, and data scientists. For instance, consider a case wherein a FL platform is deployed in a healthcare research laboratory, which analyzes histopathology slides and applied vision models for segmentation and classification of these slides [13]. The first persona could be of data scientist who would need an application view, which helps them focus on the detailed process of tweaking model parameters and fusion algorithms. A pathology subject matter expert persona would need to be provided with an application view, which abstracts the technical details of federated learning and underlying infrastructure and helps focus solely on the application end result of classification and explore facets such as visually inspect ML model performance using explainable AI tools. Likewise, data owners would prefer to have a clear application view of how data pipelines can be integrated with existing infrastructure and how data flow can be orchestrated in and out of the platform client nodes.

Beyond supporting the primary tasks of executing FL tasks, the platform should also support personas such as product managers and dev-ops personnel, who will be monitoring the model building process as well as maintaining numerous models and their versions across multiple FL projects.

23.4.4 Deployment Considerations

Careful considerations are required for crafting the production deployment of FL platform. The deployment should be easy to set up and facilitate adding more collaborating parties with minimal friction. Furthermore, for supporting multiple machine learning workloads it should be flexible to scale horizontally in party development environment. The FL platform should also have features for automation of various operational workloads and integrate with industry standard Continuous Integration and Continuous Delivery tools.

It is recommended to have platform components packaged as containers in a micro services-based architecture for ease of automation in operations and deployments. They offer the flexibility of deployment on-premises as well as on private or public cloud infrastructure.

23.5 Core Technical Considerations with FL in Healthcare

23.5.1 Data Heterogeneity

Machine learning models perform better when trained with Identical and Independently Distributed (IID) data. In the case where data is aggregated at a central location data scientists can work on analyzing the observed data distribution and craft preprocessing steps for converting data into a homogeneous distribution. Even in cases where the decentralized training is applied wherein multiple training nodes train in parallel for reducing the training time, data scientists attempt to distribute data across nodes to be IID.

Real-world data conditions are seldom identical or independent. They are often very diverse. Aggregating diverse data, which enables machine learning models to generalize well, is a challenging task. Depending on the geographies and medical protocols prevalent in a particular region, data will be inherently Non-Identical and Non-Independently distributed (Non-IID). There lies the advantage of applying federated learning for building models by federating over multiple diverse datasets.

The data will be Non-IID due to the diverse demographics exhibiting the healthcare problem as well as technical factors such as differences between medical equipment employed across various geographies. For instance, there could be image resolution differences depending on the medical equipment and calibration or bias based on the regions wherein the diagnostic labs are situated. This heterogeneity of data must be accounted for when designing machine learning models in Federated Learning setting. There are multiple solutions described in this book for attempting to address model convergence issues due to heterogeneity of the data. From an algorithmic point of view, Fed+ [13] and Siloed Federated Learning [14] could potentially help solving this challenge.

Another data concern is feature alignment across collaborating parties. In the case of horizontal federated learning, each party needs to present training data to the model with a set of predefined feature columns. The party data preparation process would involve setting up protocols for how to get feature alignment correct across all the parties as well make sure that the normalization needed for models is applied correctly. This could be facilitated by meta data exchange between collaborating parties.

23.5.2   Model Governance and Incentivization

From a machine learning model lifecycle perspective, it is desirable for federated learning to follow regular model dev-ops practices. A FL platform should organize the federated training process as a series of iterative runs, which produces model artifacts at the aggregator as well as party nodes. These iterative runs need to be parameterized for reproducibility and version control. This gains prominence when paired with regulations such as "GPDR right-to-be-forgotten" clause. At a high level, upon receiving a request to be forgotten, data owners are expected to comply by erasing the requester's data and updating systems that were built using it. For machine learning systems, this could trigger retraining models if necessary.

Federated learning enables building models collaboratively by learning shared insights across data silos. In a collaborative setup, some parties may contribute quantitatively, while others may enrich the global model by qualitatively diverse training data. Thus, there is a need for accountability and quantifiable measures of contributions made by each collaborating party. These quantifiable measures need to be implemented as a part of the technical protocol to be followed while orchestrating federated training rounds. For instance, a quantifiable measure could be the percentage increase in accuracy over a common mutually agreed test dataset, which will need to be calculated and entered in a ledger before aggregator fuses the models to update the global model. Optionally, this ledger could then be used for incentivizing collaborative parties by rewarding them or for revenue sharing in case the global model will be made available as a chargeable service in inference mode for other consumers. Depending on the level of trust in the consortium, this ledger could be implemented over Distributed Ledger Technology with smart contracts for processing contributions [15–17].

23.5.3   Trust and Privacy Considerations

The parties collaborating in federated learning could vary from being siloed offices belonging to the same company to institutions driven by social causes forming con-sortia for advancing state-of-the-art solutions to competitor companies collaborating for economic gains. Depending upon the nature of the consortium, there will be multiple levels of trust dynamics in effect when training models collaboratively. At a high level, we can bucket concerns into two broad categories: First, due to the distributed nature of the federated training process, there are concerns regarding the integrity of each collaborating party or the aggregator in executing the assigned task. Second, due to the sensitive nature of the data over which models are being trained there are concerns regarding the potential of information leakage.

As discussed in the earlier chapters of this book, the federated learning process can be fortified against these concerns driven by nefarious motives using various approaches such as differential privacy, secure multiparty Computation, homomorphic encryption [18], and Trusted Execution Environments (TEEs) [19]. Let us try to understand the nuances of trust and privacy with a couple of healthcare examples.

Consider having formed a consortium of healthcare research institutions with an aim of collaboratively building models with medical imaging data. The sensitive medical imaging data does not leave its source. However, the model parameter updates are being exchanged across the network and are fused together by the central aggregator to update the global model. In case there are no privacy preserving measures in place, the models built by this consortia could be vulnerable to adversarial attacks. A health insurance company with nefarious motives could attempt to run membership inference attacks on these models to determine whether a subset of their clientele was part of the training data. Obfuscating the model updates by implementing differential privacy will be considered as a good defense against such attempts for protecting information leakage. However, as discussed in the paper [18], adding differential privacy could lead to accuracy trade-offs in cross silo consortia with limited number of collaborating nodes. A solution suggested by this paper includes securing model updates by pairing homomorphic encryption with differential privacy for fortifying against information leakage.

In another scenario of a collaboration between pharmaceutical companies for building models, which helps in drug discovery, the trust dynamics in effect could raise concerns over integrity of execution by aggregator or collaborating party node. Typically, in this setting, parties allow models to be trained over proprietary data. The model updates carry insights, which are shared with the aggregator for performing model aggregation. The compromised execution of the aggregator protocol implementation could make the model updates shared by collaborating parties vulnerable to ulterior motives; this includes trying to retain and compare model updates for attempting to probe the model for leaking information about the private data. Another motive could be attempts to alter the aggregator protocol in an unethical way to perturb model updates or discard model updates to favor some collaborators over others or cause hindrance to model convergence. Implementing the fusion algorithms in trusted execution environments such as Intel SGX could help build trust in the process of federated aggregation of model updates [19]. Taking advantage of the attestation provided by hardware backed trusted execution environments, healthcare companies could adopt federated learning technology holistically as a platform for trustworthy win–win collaboration.

23.5.4 Conclusion

As discussed in this chapter, Federated Learning has a strong potential for building bridges between the silos that exist today across healthcare industry for developing impactful AI solutions. By solving the access to real-world data across demographics and private firewalls, FL unlocks the potential of ML models to enabling researchers to solve challenging problems across various disciplines ranging from improved clinical diagnosis to accelerating vaccine and drug discovery. However as prevalent in any emerging technology, federated learning, by virtue of its decentralized implementation, brings with it the complexity of distributed execution as well as security and privacy vulnerabilities, which will be addressed with time by robust engineering and rigorous research efforts. We are optimistic about the evolution of FL and its high impact in the field of advancing healthcare solutions.

References

1. Sheller MJ, Reina GA, Edwards B, Martin J, Bakas S (2018) Multi-institutional deep learning modeling without sharing patient data: a feasibility study on brain tumor segmentation. In: International MICCAI brain lesion workshop. Springer, pp 92–104
2. Li W et al (2019) Privacy-preserving federated brain tumor segmentation. In: International workshop on machine learning in medical imaging. Springer, pp 133–141
3. Medical Institutions Collaborate to Improve Mammogram Assessment AI (2020). https:// blogs.nvidia.com/blog/2020/04/15/federated-learning-mammogram-assessment
4. HealthChain consortium (2020). https://www.substra.ai/en/healthchain-project
5. Trustworthy federated data analytics (TFDA) (2020). https://tfda.hmsp.center/
6. The federated tumor segmentation (FETS) initiative (2020). https://www.fets.ai
7. Joint Imaging Platform (JIP) (2020). https://jip.dktk.dkfz.de/jiphomepage/
8. Liu D, Dligach D, Miller T (2019) Two-stage federated phenotyping and patient representation learning. In: Proceedings of the 18th BioNLP workshop and shared task. Association for Computational Linguistics, Florence, pp 283–291
9. Lee J, Sun J, Wang F, Wang S, Jun CH, Jiang X (2018) Privacy-preserving patient similarity learning in a federated environment: development and analysis. JMIR Med Inform 6:e20
10. Machine learning ledger orchestration for drug discovery (2022). https://www.melloddy.eu
11. Chen S et al (2020) FL-QSAR: a Federated Learning-based QSAR prototype for collaborative drug discovery. Bioinformatics 36:5492–5498
12. FHIR—Fast Healthcare Interoperability Resources (2022). https://www.hl7.org/fhir
13. Yu et al (2020) Fed+: a family of fusion algorithms for federated learning
14. Andreux M et al (2020) Siloed federated learning for multi-centric histopathology datasets. In: MICCAI 2020 DCL workshop
15. Drungilas V et al (2021) Towards blockchain-based federated machine learning: smart contract for model inference. Appl Sci 11:1010. https://doi.org/10.3390/app110310102019(2021)
16. Ma C et al (2020) When federated learning meets blockchain: a new distributed learning paradigm
17. Trustless federated learning (2020). https://www.scaleoutsystems.com/ai-blockchain
18. Baracaldo N et al (2019) A hybrid approach to privacy-preserving federated learning. In: AISec'19: proceedings of the 12th ACM workshop on artificial intelligence and security, pp 1–11
19. Ping An: Security Technology Reduces Data Silos (2020). https://www.intel.in/content/www/in/en/customer-spotlight/stories/ping-an-sgx-customer-story.html

Chapter 24
A Privacy-preserving Product Recommender System

Tuan M. Hoang Trong, Mudhakar Srivatsa, and Dinesh Verma

Abstract B2C companies with multiple business locations or retail stores make extensive use of recommender systems, which make suggestions to customers on what products to buy, provide coupons that most relevant to a customer, and offer discounts personalized to a user and similar functions. While traditional recommender systems are designed to mine customer transaction data from a central database, the high level of risk associated with data breach at a central site is motivating a need to create recommender systems using principles of federated learning. In distributed recommender systems, customer and transaction data is kept in many different partitions, e.g., one approach would be to keep all transaction data to be stored locally at each store for a retailer with many physical stores. However, such partitioning leads to unique challenges, e.g., data can be unbalanced and incomplete (lacking customer information or customer feedback) across partitions. In this chapter, we examine such a recommender system, which has been integrated into the IBM Federated Learning (FL) framework.

24.1 Introduction

A recommender system predicts the rating and preferences of a customer for a given item, where the items are products being sold to the customers. Example of items could include products sold at any retail outlet, music videos, and news articles. Recommender systems are used in both physical retail stores as exemplified by the discount coupons that get printed out at the cash register and online stores as exemplified by music recommendation, suggested news articles, or related product links.

Early work on recommender systems started with the development of Tapestry in the mid-1990s [1] which used content-based collaborative filtering techniques. In

T. M. Hoang Trong (✉) · M. Srivatsa · D. Verma
IBM Research, Yorktown Heights, NY, USA
e-mail: tmhoangt@us.ibm.com

this system, email users were able to provide a *filter* that enables them to select the right emails from various mailing lists. People "collaborated" to help each other create better filters by providing "annotations" to the email they read. These annotations could be accessed by the *filters*, which could map a user's annotations to those of users with similar annotations and use that to determine how the filter should operate.

With the growth of e-commerce, usage of recommender systems has become more widespread, with applications such as song recommendations in companies like Spotify, product recommendations in companies like Amazon, movies recommendations in companies like Netflix, and wine recommendations from FirstLeaf. Recommender systems are getting more complex and make use of a diverse set of inputs, including user-generated contents of various types (e.g., comments on products) [2], user-defined or service-defined constraints [3], and context-aware information such as time and location of input data [4, 5].

Typically, the development of a recommender system is a multidisciplinary effort that involves experts from various fields such as Artificial intelligence, Human-Computer Interaction, Information Technology, Data Mining, Statistics, Adaptive User Interfaces, Decision Support Systems, Marketing, or Consumer Behavior [6]. Most of the time, the target of a recommender system is an individual user; but it can also be a group of users [3]. Applications of group-based recommender systems can be music, movies, or travel destination recommendation [7].

With the increase in data collected about individual users, personalized-focused recommender systems are getting more complex using AI models such as deep learning [8, 9]. However, this also brings around issues related to user privacy [10]. It is customary in most industries to maintain the data at a central data warehouse to ease the task of mining. However, such centralized systems have a significant business risk associated with regulatory requirements and privacy loss if the central data warehouse is breached, as well as high maintenance cost for big-data systems such as Hadoop in processing centralized data [11, 12].

Federated Learning [13] provides an alternative approach where data is distributed among different parties and learns a shared model by aggregating locally computed updates via a central coordinating server called the *aggregator* [14]. For enhanced security, a differential policy can be used for adding noise to training data [15, 16]. Alternatively, encrypted communication such as homomorphic encryption (HE) can be used before sending the model update to the cloud or central server. Federated learning can be deployed in several data-constrained scenarios [17].

In this chapter, we share our work in building a product recommender system using federated learning as depicted in Fig. 24.1. This was based on the use case shared with us by our colleagues at a leading U.S. retailer. With the federated approach, customer transaction data is maintained at each store instead of aggregating all data to a central store. The federated learning framework enables the deployment of model training at each store. Each store trains the model on its local data, and the individual store can control what information should be used to create the model. After each training, the agent at the individual store sends the model parameters to the central location. The central location (aggregator) combines all

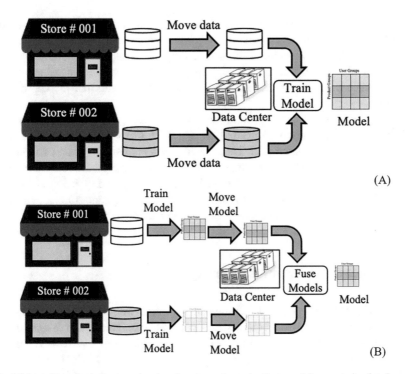

Fig. 24.1 A hypothetical scenario showing two stores in that models are trained using (**a**) centralized approach, and (**b**) federated approach

models and sends the result, in the form of a model update object, back to each store. The model update is used to construct a final AI model at the store. For every new transaction in the store, this final AI model suggests what products the customer is likely to buy next and generates a coupon for those products. Models are updated periodically, first at the store (using in-store new data), then aggregated across all the stores through the central aggregator.

24.2   Related Work

Federated learning is a natural requirement in the context of edge computing. Sheller et al. (2018) described their work in training neural net in the context of edge devices [18]. A hash ledger is used to track local data that are permitted in local training, with a trusted mechanism to allow global update only from trusted edge devices using techniques such as Byzantine Gradient Descent (BGD) [19]. Shiqiang et al. also discuss this aspect in the chapter on Efficient Federated Learning in Edge Computing Systems.

Some works have focused on reducing data transmission during the model update, in that the updated matrix has a predefined structure such as low-rank or sparse matrix [20] or the work discussed in a previous chapter on Communication-Efficient Federated Optimization Algorithms. While our work does not try to improve such aspects of federated learning, we also showed that the approach is also memory efficient.

Ammad-ud-din et al. (2019) introduced the first federated implementation of a collaborative filter using user profile [21]. Qu et al. (2020) introduced a federated news recommender system [22]. Here, local gradients from a group of randomly selected users are uploaded to the aggregator, from there the gradients are combined at the global news recommendation model. The same updated global model is then distributed to each user device for local model update. Random noise and local differential privacy are applied to the local models before sharing.

24.3 Federated Recommender System

Instead of a user profile-based recommender system, we propose a recommender system based on buying patterns. In our approach, we model the transaction data available as a buying matrix (B) of size TxP where each of the T rows represents a customer buy transaction and P is the number of products available at the store. For each of the transaction rows, context-aware information such as the buying time or specific customer attribute can also be added, but we will initially focus just on the raw transaction data without the context attributes. From a federated learning perspective, the ith store matrix B_i would have T_i rows.

We use a three-phase federated machine learning system: (1) to map individual's transactional data into buying pattern's behavior, upon which (2) the system then can generate buying pattern's product rating profile matrix and (3) uses this matrix to train the recommender system via the high memory efficiency and performance sparse linear memory algorithm (SLIM) [23, 24].

To deal with the large number of users, we map different users into groups identified by their buying patterns. Each store maps its transaction-item matrix B into a group rating matrix (A) of size GxP where G is the number of buying patterns and P is the number of products. The number of groups (G) is much smaller than the number of users and reduces the transaction matrix into each store into a matrix with common rows and columns across all of the stores. Each future transaction can be mapped into one of the existing buying patterns, and the SLIM algorithm can be used for making a product recommendation. This process is shown in Fig. 24.2.

By learning the relationship of products to a buying pattern, rather than to a user profile, we overcome the limitations of user profile-based approach. Specifically, we can eliminate personal data from the model, reduce data transmission cost, and handle unbalanced set of transactions across different stores better. When a customer with an unknown profile comes into a store and buys some products, this system is

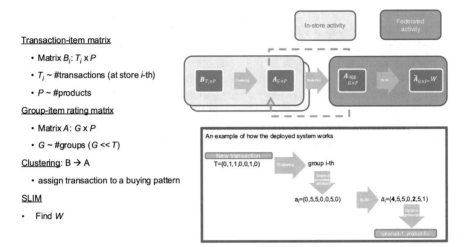

Fig. 24.2 The figure shows major components in the system: B_i is the transactional matrix at store ith, the buying pattern/product rating matrix A_i at each store; the buying pattern/product rating matrix A_{agg} after aggregated the data A_i from all the stores

able to determine the best matching buying pattern and make a recommendation despite having no associated profile.

An unsupervised clustering algorithm is used to determine the buying patterns. The initial stage of federated learning for recommender systems consists of a federated unsupervised clustering algorithm. Transactions at one store may cover some, but not all buying patterns. The federated learning mechanism runs the unsupervised clustering algorithms on local data, and then sends the model parameters over to the aggregator. The model parameters would capture details such as centroids of clusters and the radius of each cluster, the overall density of the clusters, as well as densities at different percentiles of the radius (e.g., 25% of the overall radius). By sending these parameters, the technique hides the details of the transactional data and any user information and reduces data amount.

The aggregator collects all of these parameters and identifies a global set of clusters using such parameters. It re-creates the samples using the parameters provided, after which any clustering mechanism can be applied onto the samples. These mechanisms include K-means or using a KD-tree method to identify the nearest centroids and merge them by finding the new centroids based on the sides of the two components clusters [25].

The result will be sent back to each store to calculate the "inertia" metric. The inertia is calculated as the mean-squared error of all the samples to the associated centroids. The local inertia is sent to the aggregator, which computes the global inertia. If the global inertia is improved, it will keep the new centroid, and move on to the next ones. This process repeats until no further improvement is observed. This global set of centroids are then sent back to individual stores to assess the buying pattern/product rating matrix using the local data.

Once the set of buying patterns have been calculated, the transaction data is converted to the group matrix which computes the frequency of products being bought in each buying pattern. Each row in the transaction matrix is binary with a 1 marking each product bought in the transaction and 0 indicating a product that is not bought. When buying patterns are identified, transactions belong to each buying pattern are used to calculate a rating matrix, where each entry in the row indicates the probability that the product will be bought according to that buying pattern. This rating matrix is then passed to the SLIM model to learn the aggregation matrix W which later will be sent back to each store for performing top-N product recommendation.

24.3.1 Algorithms

The pseudo-code of the three-phase federated machine learning system is shown in Fig. 24.3. In phase 1, the unsupervised clustering is used to map an individual's transactional data into a buying pattern. The number of buying patterns is the hyperparameter at this stage. The higher the number of buying patterns, the more accurate the relationship among products in a transaction is captured. If it is equal to the number of transactions, each transaction is a unique pattern. However, to reduce the size of the buying pattern matrix, a value much smaller than the number of products should be used.

Given the local data of transactions, we aim to generate a collection of buying patterns, Fig. 24.3—top box. The information from each buying patterns, as

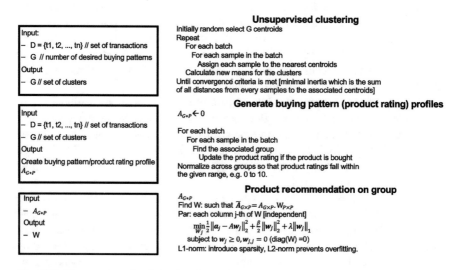

Fig. 24.3 The pseudo-code of the algorithm being used at each stage of the three-stage learning strategy

represented by the cloud of points in the P-dim space, is then aggregated based on the following statistics: centroid locations, the radius of the cluster, the density of the clusters, and the densities at different percentiles of the cluster's radius (e.g., 25% and 50% of the overall radius). Once a global buying pattern is achieved, through federated training using KD-tree [25], the rating profile matrix A is then generated in stage 2. The rating profile matrix captures the score of each product in each buying pattern. A product can appear in multiple buying patterns. This max-score is the hyperparameter in this stage. Essentially, it tells the range of values and how the normalization process is used to normalize the scores across buying patterns.

Through federated training, the global rating profile matrix is generated at the aggregator site, which is then fed into the high memory efficiency and performance sparse linear memory algorithm (SLIM). The SLIM model is then returned in the form of two sparse matrices, which captures the memory efficiency of sharing the model. There are two types of models that are trained and produced in this three-phase system. The first one is the grouping model and the second one is the rating model. At deployment, the grouping model is used to map a transaction from the transactional space to the buying pattern, along with its vector representation in the grouping space, and then the rating model is used to generate the recommended product for the vector representing the transaction in the grouping space.

24.3.2 *Implementation*

We have implemented the three-phase recommender system using the IBM FL framework, which is a Python-based library for implementing federated algorithm [14]. The framework provides for creating a fusion handler class at the centralized location, and a local training handler class to be used on individual sites, e.g., retail stores. The fusion handler is managed by the *aggregator* daemon *(ibmfl.aggregator.aggregator)*, running on the centralized server and the local training handler, which can link to an external AI model, is managed by the *party* daemon *(ibmfl.party.party)* running on each site. The *party* manages two more handlers: data handler and connection handler; while the *aggregator* manages one more handler: connection handler.

Each party runs on its dedicated machine at a site, connected via the Internet to an aggregator in a data center or on the Cloud. The daemons using this framework must select a networking mechanism which makes use of one of a few built-in mechanisms, such as Flask, TCP, or RabbitMQ, for coordinating activities between the individual *parties*, and the *aggregator*. The information is provided via a YAML config file which is passed to the commands to launch the daemons like below.

```
python -m ibmfl.aggregator.aggregator <agg_config>
python -m ibmfl.party.party <party_config>
```

The *aggregator's* possible actions are determined by its state which can be given from the console. A dedicated Aggregator class's method will be evoked for each command. At the console level, the state can be changed by going through a sequence of commands. At first, a server thread on the aggregator needs to be activated using the "START" command, waiting for connection from all parties.

The *party's* possible actions are also determined by its state, as well as the message type sent from the *aggregator*. At the console level, the state can be changed by going through a sequence of commands. At first, a dedicated thread on the party for listening to the message from the aggregator needs to be activated using the "START" command. The party then can register itself to the aggregator using the "REGISTER" command, which requires the aggregator to be in the "START" state first. The information needed for a party to connect to the aggregator is provided via a YAML file.

Once all parties are registered, the aggregator can initiate one of the following actions: SAVE_MODEL, SYNC_MODEL, EVAL_MODEL, and TRAIN; each triggers a corresponding action on the fusion handler. Among them, the "TRAIN" command which is linked to the *start_global_training()* method that needs to be implemented by the designated fusion handler class whose name is provided via the YAML file, e.g., *FedRecFusionHandler* class. Inside this method, the fusion handler class can setup payload, and use any of the following mechanism to evoke the action on the parties:

- query(func, payload, parties): ask all or a given list of parties to execute the given function "func" of the local training handler,
- query_parties(payload, parties): ask a group of parties to perform the train() method of the local training handler,
- query_all_parties(payload): ask all parties to perform the train() method of the local training handler.

Using query_all_parties() or query_parties() APIs triggers the *train()* in the local training handler that in turn evokes the *fit_model()* API of the AI model to be trained locally. In many federated learning scenarios where the single AI model is being trained, this pattern helps to simplify the model development as the user-defined model needs only to override the single *fit_model()* API of the FLModel-derived class.

In our use case, the party needs to execute a sequence of functions and coordinate the result with the aggregator. The sequence of commands is captured and explained in Fig. 24.4. This is achieved using a hook mechanism in that the fusion handler controls the sequence of execution using *query()* mechanism which can call an arbitrary function defined within the local training model, as shown in Fig. 24.5.

The details for setting the YAML config files are outlined in detail in [26]. Here, we briefly introduced the changes needed for our use case. The YAML file for the aggregator should contain four sections: *connection, fusion, hyperparams*, and *protocol_handler;* and two optional sections: *metrics* and *model*. In our example, we override two sections: *fusion* and *hyperparams*, as shown in Fig. 24.6.

The YAML file for each party should contain five sections: *aggregator, connection, local_training, data*, and *protocol_handler*. In our example, we override three sections: *data, local_training,* and *model*, as shown in Fig. 24.7. Unlike the

- *connect()* establish connection
- *learn1()* perform machine learning (ML) at individual stores (clustering) - [customer data remains at stores]
- *send1()* send cluster's statistics to federated agent [sale data is not exposed]
- Fed.learn1 () aggregate and revise clustering model
    - Find new centroids, send back to stores to assess the scores using local data
- Fed.send2 () send the suggested global centroids
- learn2() calculate the score using suggested new centroids, using local data; and send the score to federated agent.
- Fed.learn2() – stage 3 of learning

Fig. 24.4 The diagram showing the overall sequence of global training at each party sides and aggregator side: *connect()*—each party registers itself to the aggregator, *learn1()*—each party performs the first stage of learning, *send1()*—each party returns the ModelUpdate to the aggregator, *Fed.learn1()*—the aggregator performs its first stage of fusion, *Fed.send2()*—the aggregator sends its ModelUpdate to each party; *learn2()*—each party performs its second stage of learning; *Fed.learn2()*—the aggregator performs its second stage of fusion

```
class FedRecFusionHandler(FusionHandler):
  def start_global_training(self):
    max_score = self.params_global['max_score']
    num_groups = self.params_global['num_groups']
    results = self.query('update_params',
                {'max_score': max_score,
                 'num_groups': num_groups})
    data = self.query('learn1', {'test': True})
    self.learn1(data)
    self.query('update_centroids', {'centroids':self._centroids})
    data = self.query('learn2', {'test': True})
    self.learn2(data)
    self.send_global_model()
```

Fig. 24.5 The sequence of execution at the aggregator site to perform federated training

aggregator, which does not need to know the information of the parties, the parties need to know the *connection* information to the aggregator.

24.4   Results

In order to evaluate the performance of the federated learning mechanism, we used synthetic data that was generated and vetted for reasonableness with our colleagues from a leading U.S. retailer. Actual transaction data was not used to protect against user privacy.

```
fusion:
  name: FedRecFusionHandler
  path: ibmfl.aggregator.fusion.fed_rec_fusion_handler
hyperparams:
  global:
    max_timeout: 60 # max waiting time aggregator to receive parties' replies
    parties: 2 # number of register parties (#max_connection)
    num_groups: 10 # num buying patterns to learn
    max_score: 10
```

Fig. 24.6 A portion of the YAML config file for the aggregator. Here, the FedRecFusionHandler class is the user-defined class, derived from FusionHandler base class, with dedicated APIs to be implemented for the use case

```
data:
  info: # provides the data files' location
    store_info: { store_id: 134 }
    location: "./examples/retailproduct/data"
    name: RetailProductDataHandler # a class that can load the local data file
    path: ibmfl.util.data_handlers.retailproduct_data_handler # path to the class
  local_training:
    name: FedRecLocalTrainingHandler
    path: ibmfl.party.training.fed_rec_local_training_handler
  model:
    name: RecSystemFLModel
    path: ibmfl.model.rec_system_model
```

Fig. 24.7 A portion of the YAML config file for a party. Here, the "store_id" needs to be unique for each store, which is used to retrieve the data for the demo

Synthetic data was generated for two stores: one representing an urban area (store-1) and the other representing a suburban area (store-2). The urban store had 3 times as many transactions as the suburban store. As shown in Fig. 24.8, both stores had different buying patterns. There were 700 users in store-1 and 200 users in store-2. We perform the model evaluation by generating synthetic transaction data representing 10 buying patterns of consumers (numbered 0–9) across 100 products at both stores. Each buying pattern is a propensity for a user to preferentially buy a specific group of products. The set of products which had a non-zero fraction of being bought in each group (buying pattern) are as shown below:

- group 0 [1 2 5 7 8 11 13 16 23 41 47 48 49 52 54 63 66 78 80 87 91 94 97],
- group 1 [6 10 12 19 22 24 30 59 79 85],
- group 2 [29 31 37 43 73 81 83 93 98 99],
- group 3 [28 30 34 44 79 85 86 88],
- group 4 [4 14 15 20 44 70 84 86 88],

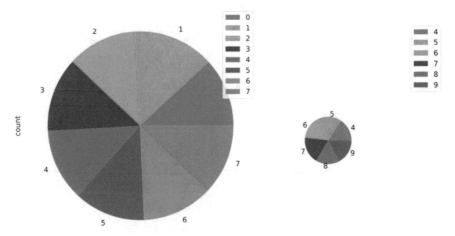

Fig. 24.8 The fraction of the number of transactions in each buying pattern. The size of the circle reflects the amount of data available in each store, and the size of the pie reflects the amount of data for each buying pattern in a given store

- group 5 [0 10 21 24 25 30 31 46 50 57 67 69 73 76 96 98],
- group 6 [3 10 26 31 35 40 55 56 57 58 72 75 89 90 92 95],
- group 7 [9 12 18 29 32 39 42 60 68 71 77 79 82 85 93 98],
- group 8 [17 27 33 36 38 45 51 61 62 64 74],
- group 9 [5 6 22 30 33 49 51 53 62 65 97].

Each transaction was generated following the steps by randomly drawing the group index, then randomly drawing the number of products purchased from that group. Each user transaction had a probability of 80% of buying products from one buying pattern, and 20% from a second randomly chosen pattern. Each user had about 3000 transactions.

Different mixes of buying patterns at two different stores were simulated. A mix represents how the ten buying patterns were distributed across stores, Table 24.1. One store had thrice as many transactions as the other. Both approaches (federated and centralized recommendation) were implemented, and their approaches in identifying and grouping users into buying patterns were compared. By shuffling the buying patterns at each store, we estimate how often the buying pattern group identified by two approaches match. Using the generated data, by averaging the five different mixes, the result shows overall matching of 94% in Fig. 24.9.

In the second experiment, we examined the amount of data that would be sent to/from the central data server in both approaches. Each store is assumed to have a million transactions, and there is no data compression during the data transmission. Here, we estimated that federated approach for 1000 stores and we estimated that the federated way uses 1/250th of inbound bandwidth and 1/6th of outbound bandwidth compared to the centralized approach (Fig. 24.10). Several other approaches potentially also aimed to provide the communication-

Table 24.1 Mixing of different buying patterns

Mix	Buying patterns in store 1	Buying patterns in store 2
1	[0–4], 5, 6, 7	5, 7, 8, 9, 10
2	[0–4], 5, 6, 8	5, 7, 8, 9, 10
3	[0–4], 5, 8, 9	5, 7, 8, 9, 10
4	[0–4], 5, 7, 9	5, 7, 8, 9, 10
5	[0–4], 6, 7, 9	5, 7, 8, 9, 10

Mix 5   100%
Mix 4   100%
Mix 3   90%
Mix 2   100%
Mix 1   90%

Fig. 24.9 The matching between results from centralized approach and federated approach at each mix

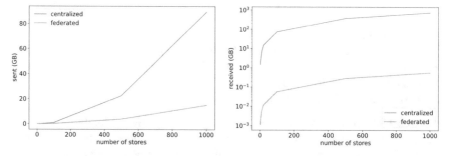

Fig. 24.10 The data sent and received by the aggregator is measured, by varying the number of parties

efficient result as reviewed in chapter titled Communication-Efficient Federated Optimization Algorithms.

24.5 Conclusion

We described in this chapter a multi-stage, federated, non-deep-learning-based recommender system that can be quickly deployed on conventional hardware and runs on available compute infrastructure at a retail store. The method is based on the well-tested algorithm in a recommender system called SLIM. The result shows that a federated product recommender system is a viable approach for retail product recommendation in that the results are comparable to the traditional centralized approach. Importantly, data privacy is maintained by design in the federated recommender system. A federated recommender system is more bandwidth-efficient than a centralized recommender system. In general, a federated AI is not limited to product recommendation but has many more applications that are mentioned in other chapters in the book.

References

1. Goldberg D, Nichols D, Oki BM, Terry D (1992) Using collaborative filtering to weave an information tapestry. Commun ACM 35(12):61–70
2. Xu Y, Yin J (2015) Collaborative recommendation with user generated content. Eng Appl Artif Intell 45:281–294
3. Felfernig A, Boratto L, Stettinger M, Tkalčič M (2018) Group recommender systems: an introduction. Springer
4. Villegas NM, Sánchez C, Díaz-Cely J, Tamura G (2018) Characterizing context-aware recommender systems: a systematic literature review. Knowl-Based Syst 140:173–200
5. Adomavicius G, Tuzhilin A (2011) Context-aware recommender systems. In: Recommender systems handbook. Springer, Boston, MA, pp 217–253
6. Ricci F, Rokach L, Shapira B, Kantor PB (2011) Recommender systems handbook. Springer
7. Sriharsha Dara C (2020) Ravindranath Chowdary, "a survey on group recommender systems". J Intell Inf Syst 54:271–295
8. Zhang S, Yao L, Sun A, Tay Y (2019) Deep learning based recommender system: a survey and new perspectives. ACM Comput Surv (CSUR) 52(1):1–38
9. Khan ZY, Niu Z, Sandiwarno S, Prince R (2020) Deep learning techniques for rating prediction: a survey of the state-of-the-art. Artif Intell Rev:1–41
10. https://ec.europa.eu/info/law/law-topic/data-protection/data-protection-eu_en, Retrieved Feb, 19, 2021
11. Zhao Z-D, Shang M-S (2010) User-based collaborative-filtering recommendation algorithms on Hadoop. In: 2010 third international conference on knowledge discovery and data mining. IEEE, pp 478–481
12. Dahdouh K, Dakkak A, Oughdir L, Ibriz A (2019) Large-scale e-learning recommender system based on spark and Hadoop. J Big Data 6(1):1–23
13. Jakub K, Brendan McMahan H, Ramage D, Richtárik P (2016) Federated optimization: distributed machine learning for on-device intelligence. arXiv preprint arXiv:1610.02527
14. Ludwig H, Baracaldo N, Thomas G, Zhou Y, Anwar A, Rajamoni S, Ong Y et al (2020) IBM federated learning: an enterprise framework white paper v0.1. arXiv preprint arXiv:2007.10987
15. Rodríguez-Barroso N, Stipcich G, Jiménez-López D, Ruiz-Millán JA, Martínez-Cámara E, González-Seco G, Victoria Luzón M, Veganzones MA, Herrera F (2020) Federated Learning and Differential Privacy: software tools analysis, the Sherpa. ai FL framework and methodological guidelines for preserving data privacy. Inform Fusion 64:270–292
16. https://github.com/IBM/differential-privacy-library
17. Yang Q, Liu Y, Chen T, Tong Y (2019) Federated machine learning: concept and applications. ACM Trans Intell Syst Technol (TIST) 10(2):1–19
18. Sheller M, Cornelius C, Martin J, Huang Y, Wang S-H Methods and apparatus for federated training of a neural network using trusted edge devices. Intel Corp. https://patents.google.com/patent/US20190042937A1
19. Alistarh D, Allen-Zhu Z, Li J (2018) Byzantine stochastic gradient descent. arXiv preprint arXiv:1803.08917
20. Hugh BM, David MB, Jakub K, Xinna Y (2020) Efficient communication among agents. Google, LLC. https://patents.google.com/patent/GB2556981A/en

21. Ammad-Ud-Din M, Ivannikova E, Khan SA, Oyomno W, Fu Q, Tan KE, Flanagan A (2019) Federated collaborative filtering for privacy-preserving personalized recommendation system. arXiv preprint arXiv:1901.09888
22. Qi T, Wu F, Wu C, Huang Y, Xie X (2020) Privacy-preserving news recommendation model learning. In: Proceedings of the 2020 conference on empirical methods in natural language processing: findings, pp 1423–1432
23. Sculley D (2010) Web scale K-means clustering. In: Proceedings of the 19th international conference on world wide web
24. Ning X, Karypis G (2011) Slim: sparse linear methods for top-n recommender systems. In: 2011 IEEE 11th international conference on data mining. IEEE, pp 497–506
25. Bentley JL (1975) Multidimensional binary search trees used for associative searching (KD-tree). Commun ACM 18(9)
26. https://github.com/IBM/federated-learning-lib

Chapter 25
Application of Federated Learning in Telecommunications and Edge Computing

Utpal Mangla

Abstract Federated Learning is gaining significant prominence in the Telecommunication Industry as Communication Service Providers (CSPs) look at harnessing their data assets, while maintaining data privacy requirements and building new use cases to monetize opportunities made possible by this data.

One of the biggest assets that CSPs have is data. The top 50 carriers globally contain data of over five billion consumers worldwide. As telecommunication companies use Artificial Intelligence and Machine Learning (AI/ML) technologies to extract analytical and predictive capabilities, federated learning is becoming an important imperative for building centralized models with distributed training data. 5G and Edge computing enable significantly improved network capacity, lower latency, higher speeds, and increased efficiency.

25.1 Overview

Federated Learning is gaining significant prominence in the Telecommunication Industry as Communication Service Providers (CSPs) look at harnessing their data assets, while maintaining data privacy requirements and building new use cases to monetize opportunities made possible by this data.

One of the biggest assets that CSPs have is data. The top 50 carriers globally contain data of over five billion consumers worldwide. As telecommunication companies use Artificial Intelligence and Machine Learning (AI/ML) technologies to extract analytical and predictive capabilities, federated learning is becoming an important imperative for building centralized models with distributed training data. 5G and Edge computing enable significantly improved network capacity, lower latency, higher speeds, and increased efficiency [1].

U. Mangla (✉)
IBM, Toronto, ON, Canada
e-mail: utpal.mangla@ca.ibm.com

© The Author(s), under exclusive license to Springer Nature Switzerland AG 2022
H. Ludwig, N. Baracaldo (eds.), *Federated Learning*,
https://doi.org/10.1007/978-3-030-96896-0_25

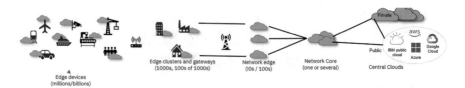

Fig. 25.1 5G and edge landscape

Data and AI models will be distributed across multiple nodes in 5G Edge computing but sharing the information can be complex for security, bandwidth, storage, and other constraints so Federated Learning is ideal for such an environment.

The illustration above (Fig. 25.1) depicts a typical edge environment integrated into the CSP's 5G infrastructure. The far edge consists of many edge devices which capture the essential data and transmit them to edge clusters located close to the far edge. The edge cluster will consist of Multi-Edge Compute Nodes (MEC) [2]. The far edge has limited compute power and will usually only run the trained models. Trained models will also exist at the MEC where further processing of the data can occur as the edge clusters will have greater compute power than the edge devices. The core network will have greater compute power permitting even more powerful models processing larger data sets. The CSPs will generally own and manage the network in the edge environment while the applications will be customized for different industries. Managing the AI models in this distributed environment can be greatly enhanced by Federated Learning where an aggregator, when placed at the appropriate node, can distill a model which can be shared and then distribute it to the relevant nodes.

We will now elaborate on a few Federated Learning use cases applicable to the telecommunications industry and then provide a unified use case which brings them together. We will conclude by discussing some of the challenges the industry will face when implementing Federated Learning to monetize the massive amounts of data they have.

25.2   Use Cases

There are many Federated Learning use cases and the following are some examples. We will start with some core Telcom use cases which will benefit from federated learning and then provide an integrated use case which builds on the core use case to create an end-to-end solution for CSPs.

25.2.1 Vehicular Networks

Vehicular networks are growing in importance in intelligent transportation systems as it permits vehicles to operate more effectively through data sharing. IEEE predicts that autonomous cars will comprise 75% of total traffic on the road by the year 2040 [3]. These vehicles have to deal with large amounts of data from multiple data sources in various formats. Providing the essential data in a timely, secure fashion is critical and Federated Learning will play an important role. Normally, Federated training in vehicular networks includes the following processes: Initialization, Local Training, and Global Aggregation [4]. Initialization sets up the environment and determines the training needed. During local training the models are trained using local data. The parameters are then loaded to the aggregator by the participants where the global model is created. The updated parameters are then sent back to the participants. In synchronous mode all vehicles upload their training parameters to the server periodically at the end of predefined interval. In asynchronous mode, each vehicle will complete its training and upload the parameters when sufficient information is collected locally and the FL server will update the global model upon receipt of a set of parameter uploads. Vehicular networks provide many benefits including early warning signals for motorist, better provisioning in transit, inter-vehicle and road-vehicle communication as well as enabling autonomous vehicles to operate effectively.

25.2.2 Cross-Border Payment

The telecom industry is impacted by government regulations such as Federal Communications Commission in the USA and regulatory bodies in the EU. In addition, telecom companies have to follow standards set up by various bodies such as European Telecommunications Standards Institute. To enable greater transparency, visibility, and to prevent fraud, it is imperative that various telecom companies globally work in close coordination with each other. As consumers travel globally and start using greater roaming services and perform financial transactions supported by their respective telecom companies, protection of data privacy and data security becomes extremely critical.

New use cases are being created by telecom providers to generate monetization opportunities by tapping into customer data that they possess. Cross-border payment system (CBPS) is an opportunity that carriers are interested in tapping into [5]. Figure 25.2 illustrates some of the key players that participate in cross-border contactless payments and the interactions among them.

Federated learning will become extremely critical to solving this problem and ensuring that personalized models are built between banks, carriers, and networks. Federated learning will become the foundation to building a cross-enterprise, cross-data, and cross-domain platform in which an ecosystem is created by ensuring the

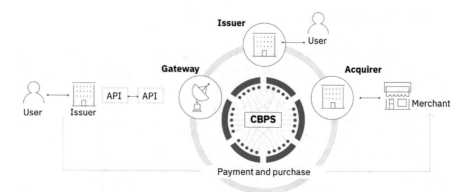

Fig. 25.2 Roles within a cross-border payment system. Source: IBM Institute for Business Value. [5]

appropriate models can be created and enhanced across the different players while ensuring data integrity and security is maintained.

25.2.3 Edge Computing

Edge computing is built on computing resources where the workload is placed closer to where the data is created permitting actions which can then be taken in response to an analysis of the data. By harnessing and managing the compute power that is available on remote premises, such as factories, stadiums, or vehicles, developers can create applications that reduce latencies, lower demands on network bandwidth, increase privacy of sensitive information, and enable operations even when networks are disrupted. Federated learning can be used for AI/ML models to learn from data residing across multiple edges, without sharing the raw data, thereby offering higher levels of data privacy. For example, models can be trained at different edge nodes for a video surveillance system. Models running on cameras (edge device) or the MEC (edge cluster) at certain locations can be trained to identify certain objects due to clarity of images at that location. The learning for a set of devices using localized data can occur at the edge cluster and an aggregator can be present at the nodes further up-stream including the core network or in the cloud. The aggregator can distill a model that recognizes well common objects occurring, which can be shared and then distribute it to relevant new or existing nodes so these nodes do not have to go through the retraining. It is important that the aggregator be placed at the right nodes, and this is one of the challenges that need to be addressed especially as the number of nodes grows.

25.2.4 Cyberattack

Cyberattacks can seriously impact mobile edge networks so detecting and correcting the attacks is critical. Nguyen et al. [6] describe how Deep Learning techniques can accurately detect a wide range of attacks. This detection, however, needs the right dataset in sufficient amounts in order to train the models but this data can be sensitive given the nature of security data. Abeshu and Chilamkurti [7] describes how Federated Learning can create effective models to detect cyberattack models. Each edge node owns a set of data for intrusion detection. Each model is trained at the different edge nodes and then sent to the Federated Learning system. The system will aggregate all parameters from the participants and send the updated global model back to all the edge nodes. Each edge node can therefore learn from other edge nodes without a need of sharing its real data. A key benefit of using Federated Learning to detect cyberattacks is improved accuracy in detecting attacks while maintaining privacy at the edge nodes.

Bagdasaryan et al. [8] have identified Federated Learning being vulnerable to backdoor attacks. For example, malicious training samples can be introduced to the edge nodes resulting in the final model being tainted. Secure protocols have been developed to guard against malicious attacks (e.g., defensive distillation and adversarial training regularization). These protocols are enhanced by integrating them with Blockchain; with its immutability and traceability, it can be an effective tool to prevent malicious attacks in federated learning [9]. Changes made by each node to its local model can be chained together on the distributed ledger offered by a blockchain such that those model updates are audited. This provides traceability to participants using Blockchains inherent characteristics which help the detection of tamper attempts and malicious model substitutes.

25.2.5 6G

6G is still being developed but there are many areas where Federated Learning can be applied to improve the rollout of 6G [10]. Some key enhancements that 6G will provide over 5G and impact Federated Learning include the following:

- Better Performance and Availability: 6G will deliver 1 Tbps data rate per user, ultra-low end-to-end delay, and high energy efficiency networking. It will also support more extensive networking including less dense areas such as the underwater environment. In addition, 6G communications will support highly diversified data, thus extending the types of applications that can be built. 6G is therefore being designed to support more devices which will be able to handle the new data types and applications which will be available.
- Greater Security: Data security and privacy will be an important part of 6G communication. As more and more data will be managed at edge nodes, it is essential that this data is protected during transmission and when stored at

different locations. Wang et al. [11] discuss key aspects of 6G networks including edge computing with the relevant security and privacy issues including that vehicle networks should consider not only the network environment but also the physical environment in a 6G environment.

- Intelligent Services: 6G will be much more complex to handle as the complexity of the network, applications, and services provided will be greater. The 6G network and applications running on the network will therefore need to automatically identify and correct themselves in a close loop automated fashion with minimal human intervention using AI [10].

The above will permit implementing new use cases which are either not possible now or can only currently be implemented in a limited fashion due to network shortcomings. Examples of such use cases include sophisticated augmented/virtual reality interactions, real-time eHealth permitting advanced medical procedures, complex industrial implementations that extend beyond Industry 4.0, and very extensive unmanned mobility integrating mobile sensors, autonomous land vehicles, and aerial devices such as drones.

To achieve the above, it is essential that the large number of far edge devices using different services can collaboratively train a shared global model using their own local datasets. When 6G is rolled out, the devices will be much more sophisticated than what is currently available so will be able to run much more sophisticated models than current systems handling more complex data. Examples of such models could include root cause network analysis, predictive network models, and video analytics thus offloading some of the activities which may run on the MEC to the far edge device. Model updates of all selected devices need to be managed for aggregation to obtain new global models which can then be used to further enhance the far edge models. The aggregation can take place in a peer-to-peer way or be sent to a central location in the core network or the cloud.

25.2.6 *"Emergency Services" Use Case to Demonstrate the Power of Federated Learning*

We will now look at a use case that integrates the key points mentioned above to illustrate how federated learning will be used in an actual implementation.

Today, a 911 emergency response time can take anywhere from 37.5 to 43 min according to a study conducted by the University of California, San Francisco, based on 63,000 cardiac arrest cases—a life-threatening condition—occurring outside of the hospital premises [12]. With the advent of 5G, an end-to-end 5G network slice can be established between the 911 caller, Emergency Medical Technician (EMT), and the hospital alongside an optimal driving route that supports rich multimedia communication between the EMT and the attending physician; it is possible to make critical real-time decisions on behalf of the patient.

With the confluence of 5G, MEC, and AI/ML, such scenarios are closer to reality. To achieve that, the network should support a common architecture from Core to Edge, for both Network and IT workloads, with the flexibility to move workloads across the network, as per the ebbs and flows of network and customer traffic supported by dynamic network slicing [13]. It should also support the flexibility of a "build once and deploy anywhere" on the Core to Edge spectrum to provide the best experience even with bandwidth intensive use cases. This should be complemented with DevSecOps [14] methodology and robust security framework to ensure 5G and Edge network security, cloud and container security, devices, application, and data security.

There are various types of data collected for analysis, such as the vital statistics of the patient, the status and location of the ambulance, the video feed from the ambulance to the hospital, etc. This generates a large quantity of data to help train the models to better determine and predict the health of the patient and changes to be made to the ambulance operation. Changes could include determining if the ambulance is on the right route to be associated with the optimal network signal or network slice, avoiding disruptions such as a tunnel that can interrupt the signal and traffic jams that can delay the patient reaching the hospital. A key observation is that diverse representative data that can truly enhance machine learning models are encountered rarely [15]; the rest of the data is fairly predictable and redundant, and hence a summary of the most important data, accounting for data privacy, can suffice to train the models. Sending the summary data will also help reduce network traffic and prevent overloading the systems receiving the data, as only relevant data is sent to the receiving systems.

The following set of activities are occurring (Fig. 25.3).

1. Data is created as the ambulance moves across different locations.
2. The data is transferred through the 5G network to the network edge.
3. Summary data is sent from the network edge to the central cloud.

There are various types of data in large volumes collected for analysis, such as the vital statistics of the patient, the status and location of the ambulance, and the video feed from the ambulance to the hospital. To get the best results from the models, it is best to use diverse geographical data that spans multiple edges:

- Not only from city roads, but also highways.

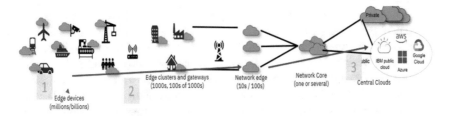

Fig. 25.3 Sample flow across 5G and edge landscape

- Various weather conditions such as rain, snow, and sunny days; day and nighttime, etc.
- Patient data of the illness for various demographics of age, gender, race, physical conditions, pre-existing medical conditions, etc.
- Network data as the ambulance moves through different locations, and thus span multiple vRANs (Radio Access Networks), IP/transport layers, and multiple MEC servers (Mobile Edge Computing).

Fusion of data across multiple geographical edges and patients helps us train the models to improve the overall treatment provided to the patient in transit. However, there are some key points to be considered here:

- How do we ensure the security and privacy of the edge data? This data can be used to train the models at the edge, but private data should not be shared with others in its raw form.
- How can the models running at a given edge be enhanced by the learnings at other edges and while ensuring the data is not shared? There has to be some level of collaboration between these edge nodes to make this possible.
- How to scale model training in a cloud and multi-edge environment? With some of the training occurring at the edge while others at the central cloud, how does one scale this mode of training while keeping the data secure?
- How to dynamically orchestrate resources across cloud and multi-edges to facilitate model training and scoring in a telco network cloud? Such solutions should leverage the inherent strengths of 5G networks such as dynamic network slicing, improving the bandwidth or QoS of a slice, or migrating the ambulance traffic to another suitable slice.

Federated learning will enable us to address many of the points above. For example, the server is designed to have no visibility into an agents' local data and training process because of privacy concerns. The aggregation algorithm used is usually weighted averaging of the agent updates to generate the global model. In the ambulance use case, there is both data/resource similarity and diversity. These characteristics make these data points ideal candidates for federated learning models.

To make the most efficient use of the limited computation and communication resources in edge computing systems, researchers have explored the design of efficient federated learning algorithms [16]. These algorithms track the resource variation and data diversity at the edge, based on which optimal parameter aggregation frequency, gradient sparsity, and model size are determined. The key idea is that model parameters only need to be aggregated by the server when necessary. In cases where different agents have similar data, the amount of exchanged information can be significantly less than cases where agents have diverse data. The optimal degree of information exchange also depends on the trade-off between computation and communication. When there is abundant computation resource (CPU cycles, memory, etc.) but low communication bandwidth, it is beneficial to compute more and communicate less, and vice versa. Various experiments show that our algorithms can significantly reduce the model training time, compared to the standard (non-optimized) federated averaging (FedAvg) algorithm [17]. When combined with model pruning [18], we can also largely improve the inference time, making it possible to deploy trained models to resource-limited edge devices.

While federated learning has many benefits, it adds a new vulnerability in the form of misbehaving agents. Studies have explored the possibility of an adversary controlling a small number of agents, with the goal of corrupting the trained model [19]. The adversary's objective is to cause the jointly trained global model to misclassify a set of chosen inputs with high confidence, i.e., it seeks to poison the global model in a targeted manner. Since the attack is targeted, the adversary also attempts to ensure that the global model converges to a point with good performance on the test or validation data. Such attacks emphasize the need for admitting trusted agents in federated learning and develop model training algorithms that are robust to adversarial attacks [20]. These defenses are being increasingly adapted into the federated learning setup (e.g., in the design of robust model fusion algorithms) to detect and secure the learning process against such attacks.

25.3 Challenges and Future Directions

Federated learning is clearly applicable to many areas of telecommunications but there are also many challenges. Some of the challenges and future research directions are discussed below.

25.3.1 Security and Privacy Challenges and Considerations

The above use cases illustrate that telecom systems require different machine learning models to run on numerous devices distributed across the far edge, near edge, core data centers, and public cloud. This also makes the system vulnerable to malicious attacks. Telcos have a lot of data about their subscribers including call details, mobility information, and network of contacts. In addition, critical and sensitive financial, health, and first responder data is transmitted by the telecom network. Federated Learning is also vulnerable to communication security issues such as Distributed Denial-of-Service (DoS) and jamming attacks [4]. Security and privacy is therefore paramount to telcos and will be critically examined before implementing a solution using Federated Learning. It is possible for models to not reveal their data using a secure aggregation algorithm aggregating the encrypted local models without the need for decrypting them in the aggregator [21]. However, the adoption of these approaches sacrifices the performance and also requires significant computation on participating mobile devices. There currently is a trade-off between privacy guarantee and system performance when implementing

Federated Learning, which needs to be addressed to implement robust solutions using Federated Learning for Telecom.

25.3.2   Environment Considerations

For most current studies on FL applications, the focus mainly lies in the federated training of the learning model, with the implementation challenges neglected. Telecom systems will often have limited communication and computation resources. There may not be enough local learners at an edge node to participate in the global update, systems will have to trade-off between model performance and resource preservation, and the model updates might still be large in size for low-powered devices. In the existing approaches, mobile devices need to communicate with the server directly and this may increase the energy consumption. In summary, the telecom environment including wireless applications, communication cost, and quality of the energy costs should also be considered in the model creation.

25.3.3   Data Considerations

Telecom data can vary a lot and can be inconsistent. Many Federated Learning approaches assume the wireless connections of participants are always available but this may not be the case due to network, device, or other issues. A large number of dropped data from the training participation can significantly degrade the performance [22] so the algorithms need to be robust to accommodate for the varying data quality during training. Labeled data is often needed for supervised learning but the data generated in the network may be unlabeled or mislabeled [23] so this poses a challenge to identify participants with appropriate data for model training. The above are just examples of data considerations that need to be examined further to successfully implement Federated Learning in Telecom.

25.3.4   Regulatory Consideration

There are many regulations that telecom companies need to adhere to. The United States has the Federal Communications Commission which governs regulations in the US. It is therefore essential that any implementation adheres to the regulations. One example is network failure for disaster communications. Most federated models are, like regular deep neural networks and some forms of ensemble models, black-box models and the unexplainable predictions or decisions output by the black-box model may cause huge losses to users. 5G networks are beginning to rely on AI models to operate. A failure can have regulatory implications, and it is important

to explain why the systems responded the way it did but current tools are limited in providing the explanations. In addition, these models will introduce biases over time so need to be explained and corrected. Given the complex network system supported by telecoms and the huge impact if the systems fail, the development of an interpretable federated models will become important.

25.4 Concluding Remarks

In summary, federated learning provides a secure model training strategy for many telecom use cases which can impact multiple industries. The training of the models is distributed between multiple agents on the edge which allows learning across multiple edges, without having to actually share the raw data between them, often resulting in better models for deployment at the edge. Doing this enables telecoms to manage and monetize their data at the edge, thus enabling them to significantly improve their rollout of next-generation networks.

References

1. https://developer.ibm.com/articles/edge-computing-architecture-and-use-cases/
2. Multi-access edge computing. https://www.etsi.org/technologies/multi-access-edge-computing
3. Look Ma, no hands. http://www.ieee.org/about/news/2012/5september_2_2012.html
4. Tan K, Bremner D, Krenec JL, Imran M (2020) Federated machine learning in vehicular networks: a summary of recent applications. In: 2020 international conference on UK-China emerging technologies (UCET)
5. https://www.ibm.com/thought-leadership/institute-business-value/report/contactless-payment-tele
6. Nguyen KK, Hoang DT, Niyato D, Wang P, Nguyen D, Dutkiewicz E (2018) Cyberattack detection in mobile cloud computing: a deep learning approach. In: Proceedings of the IEEE WCNC, pp 1–6
7. Abeshu A, Chilamkurti N (2018) Deep learning: the frontier for distributed attack detection in fog-to-things computing. IEEE Commun Mag 56(2):169–175
8. Bagdasaryan E, Veit A, Hua Y, et al (2019) How to backdoor federated learning. ArXiv Preprint ArXiv:1807.00459. https://arxiv.org/abs/1807.00459
9. Preuveneers D, Rimmer V, Tsingenopoulos I et al (2018) Chained anomaly detection models for federated learning: an intrusion detection case study. Appl Sci. https://doi.org/10.3390/app8122663
10. Liu Y, Yuan X, Xiong Z, Kang J, Wang X, Niyato D (2020) Federated learning for 6G communications: challenges, methods, and future directions. China Commun 17(9):105–118. https://doi.org/10.23919/JCC.2020.09.009
11. Wanga M, Zhua T, Zhang T, Zhangb J, Yua S, Zhoua W (2020) Security and privacy in 6G networks: new areas and new challenges. Digital Commun Netw 6(3):281–291
12. wendychong-ibm-2020.medium.com/ai-edge-saving-grace-with-advanced-ai-ml-accelerators-for-5g-edge-computing-use-cases

13. Dynamic network slicing: challenges and opportunities. https://link.springer.com/chapter/10.1007/978-3-030-49190-1_5
14. DevSecOps. https://www.ibm.com/cloud/architecture/adoption/devops. Aug 2020
15. ibm.medium.com/ai-edge-coreset-intelligent-data-sampling-for-ai-ml-at-the-5g-edge
16. S. Wang, T. Tuor, T. Salonidis, K. K. Leung, C. Makaya, T. He, and K. Chan, "Adaptive federated learning in resource constrained edge computing systems," IEEE J Select Areas Commun, vol. 37, no. 6, pp. 1205–1221, Jun. 2019
17. Konečný J, McMahan HB, Yu FX, Richtárik P, Suresh AT, Bacon D (2016) Federated learning: strategies for improving communication efficiency. arXivpreprint arXiv:161005492
18. Wenyuan X, Fang W, Ding Y, Zou M, Xiong N (2021) Accelerating federated learning for IoT in big data analytics with pruning, quantization and selective updating. IEEE Access
19. Bhagoji A, Chakraborty S, Mittal P, Calo S (2019) Analyzing federated learning through an adversarial lens. In: International conference on machine learning (ICML)
20. ibm.medium.com/ai-edge-federated-learning-learn-on-private-and-partitioned-dataset
21. Bonawitz K et al (2016) Practical secure aggregation for federated learning on user-held data. In: NIPS workshop on private multi-party machine learning
22. McMahan HB et al (2016) Communication-efficient learning of deep networks from decentralized data. [Online]. Available: arXiv:1602.05629
23. Gu Z et al (2019) Reaching data confidentiality and model accountability on the Caltrain. In: Proceedings International Conference on Dependable Systems and Networks, pp 336–348

Printed in the United States
by Baker & Taylor Publisher Services